CISM COURSES AND LECTURES

The series presents lecture notes, monographs, edited works and proceedings in the field of Mechanics, Engineering, Computer Science and Applied Mathematics.
Purpose of the series is to make known in the international scientific and technical community results obtained in some of the activities organized by CISM, the International Centre for Mechanical Sciences.

INTERNATIONAL CENTRE FOR MECHANICAL SCIENCES

COURSES AND LECTURES - No. 399

CREEP AND DAMAGE IN MATERIALS AND STRUCTURES

EDITED BY

HOLM ALTENBACH
MARTIN LUTHER UNIVERSITY

JACEK J. SKRZYPEK
CRACOW UNIVERSITY OF TECHNOLOGY

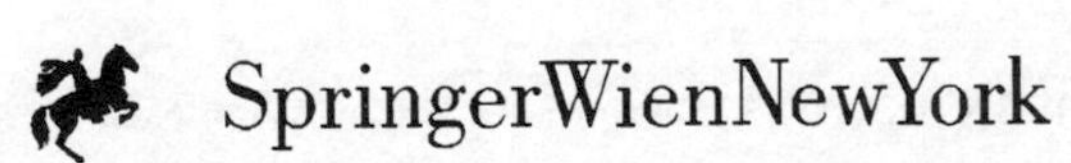

This volume contains 128 illustrations

Printed in Italy
SPIN 10745131

In order to make this volume available as economically and as rapidly as possible the authors' typescripts have been reproduced in their original forms. This method unfortunately has its typographical limitations but it is hoped that they in no way distract the reader.

ISBN 3-211-83321-8 Springer-Verlag Wien New York

PREFACE

The theory of time-dependent engineering materials, considering material deterioration and changes of thermo-mechanical properties in time, is a quickly developing branch of solid mechanics. Several monographs and textbooks have been published during last years and a great number of scientific papers have dealt both with a constitutive modelling of materials and its experimental calibration, as well as a computerized FEM and FDM application to the analysis and design of structural elements for creep-damage, elastic-plastic-damage, creep-fatigue and other conditions. State and evolution equations capable of analysing of simple structures and lifetime predicting for complex thermo-mechanical loading conditions are usually either a mechanisms-based when an intuitive phenomenological approach is used, or the thermo-dynamic-based when the rigorous thermodynamics of irreversible processes is employed. Creating a bridge between the theoretical material modelling on one hand, and the experimental validation and computer simulation on the other, is the main objective of this course.

The idea to organize the CISM Advanced School on "Modelling of Creep and Damage Processes in Materials and Structures", held in Udine (Italy) from 7 to 11 September, 1998, was born in 1996 when the coordinators were invited to read lectures on CEEPUS Summer School on "Analysis of Elastomers and Creep and Flow of Glass and Metals", organized in Zilina (Slovakia) from 19 to 30 August, 1996, by prof. Vlado Kompis. The CISM Advanced School eventually collected six lecturers, professors: Holm Altenbach (Germany), Jean-Louis Chaboche (France), Peter Gummert (Germany), David Hayhurst (United Kingdom), Erhard Krempl (USA) and Jacek Skrzypek (Poland), who presented lectures to circa 50 participants from 16 countries. The course has provided a comprehensive survey of one- and three-dimensional constitutive material models based on the continuum mechanics, creep-damage mechanics and thermodynamics. On the other hand, a possibility of numerical applications of various time-dependent material behaviour problems, including creep, creep-damage, elastic-brittle-damage, elastic-plastic-damage, creep-fatigue, thermo-creep-damage, etc., has been reviewed and discussed. Experimental observations of the primary creep, the steady-state creep, the tertiary creep, the cyclic creep, and other behaviours, have been the starting point of the one-dimensional modelling and its calibration and discussion. In addition, for practical use of the time-dependent material constitutive equations we need their three-dimensional generalization, which requires a tensorial representation for the anisotropic, in general, constitutive and evolution equations of damaged metallic or non-metallic materials.

In these Lecture Notes six contributions of the authors are included in such a way that various approaches to constitutive and structural modelling for creep, damage and other conditions are developed and demonstrated, starting from the most general material behaviour, principles and state variables (P. Gummert, H. Altenbach, J. Skrzypek) through experimentally-based as well as advanced thermodynamically-based constitutive modelling of the crystalline metallic and ceramic or composite materials for high-temperature creep, damage, fatigue and other conditions (J.-L. Chaboche, D. Hayhurst, E. Krempl). Each of the six capable of describing metallic, non-metallic or composite materials. When applied to the metallic materials, ductile plastic damage and creep-fatigue damage are presented in a detailed form. Concise discussion of brittle-damage models applied to metallic matrix composites or ceramic/ceramic composites is presented. Application of CDM methods to the inelastic damage structural analysis, the lifetime prediction, and the local approach to fracture, are also included.

In Chapter 5: "Material Data Bases and Mechanisms-Based Constitutive Equations for Use in Design" by David R. Hayhurst (UMIST, Manchester), single- or multi-state variables mechanism-based CDM creep-damage constitutive equations for the super-computer simulation of high-temperature design of engineering components, are reviewed. Particular emphasis is placed upon the aluminium alloys and nickel superalloys applications. Experimental techniques and procedures are discussed for selection of the dominant mechanism, and calibration of these equations for the accurate predictions and extrapolations. Capability of CDM approach to the high temperature creep crack growth analysis and lifetime prediction is also discussed and recommended. The example of a ridged test piece is used to demonstrate the power of the method and to highlight the importance of carrying out a reliable laboratory testing and creating good material data.

Chapter 6: "Cyclic Creep and Creep-Fatigue Interaction" by Erhard Krempl (Rensselaer Polytechnic Institute, Troy, N.Y.) provides a survey of the experimental results from low-cycle fatigue testing of various metallic materials (copper, 304 stainless steel, different engineering alloys, etc.) in low or high homologous temperature conditions. The unified approaches of the viscoplasticity theory with state variables and the viscoplasticity theory based on overstress (VBO) are systematically developed and applied to modelling of primary, secondary and tertiary creep at very high temperature. The results give confidence that VBO model is capable of predicting the long-term cyclic behaviour and lifetime of structural components at high temperature. Three appendices are attached to give details for the standard linear solids, modelling of rate independence and negative rate sensitivity and excerpts from the WRC bulletin.

The book is addressed to young researches and scientists working in the field of mechanics of inelastic materials and structures as well as to Ph.D. students in computational mechanics, mechanical, environmental and civil engineering, and material science. The book interlinks the material science foundations, the constitutive modelling and computer simulation applied to analysis and design of simple structural components for high temperature creep, damage, fatigue and other conditions. This publication may be recommended as an interesting textbook which shows the creep and continuum damage mechanics as a rapidly developing discipline, although the wish list in this field is long and open - to mention only material anisotropy, creep damage of composites, unilateral damage response, damage-fracture interaction, application of CDM approach to crack growth prediction, local and non-local approaches, probabilistic approaches, etc.

Holm Altenbach
Jacek Skrzypek

ACKNOWLEDGEMENTS

The coordinators and editors of the book wish to express their thanks to all invited professors whose excellent contributions brought essential impact to the success of the course and have enabled to present this interesting reference text to the reader. Special appreciations are offered to Dr. Adam Wróblewski the technical editor of the book, for great amount of efforts and thorough work when unifying the manuscripts for the LaTeX standard and final formatting. Additionally, the editors acknowledge the assistance and help of Dr. Artur Ganczarski when creating instructions for the authors and editing some parts of the text, as well as for Dr. Jan Bielski who proofread English language. Finally, editors express their gratitude to Mikolaj Skrzypek for useful graphics edition when preparing the camera-ready manuscript and integrating all figures to the text.

CONTENTS

Page

GENERAL CONSTITUTIVE EQUATIONS FOR SIMPLE AND NON-SIMPLE MATERIALS

P.R. Gummert
Technical University Berlin, Berlin, Germany

ABSTRACT

Stress and deformation are related by constitutive equations. Using the symmetric stress tensor field $\mathbf{S}(\mathbf{X}, t)$ at the place of $\mathbf{X}$ and at the present time t and a (symmetric) deformation field, e.g. $\mathbf{B}(\mathbf{X}, \tau)$ at the same place of $\mathbf{X}$ and the past time τ, it is necessary and possible to postulate six equations $\mathbf{S}(\mathbf{B})$. Together with three equations of NEWTON's law (resp. with three equilibrium conditions) and six equations of displacement–deformation conditions between $\mathbf{B}(\mathbf{X}, \tau)$ and the displacement field $\mathbf{u}(\mathbf{X}, \tau)$ an array of fifteen equations is generated to solve the fifteen unknowns as scalar components of the two tensor fields $\mathbf{S}(\mathbf{X}, t)$, $\mathbf{B}(\mathbf{X}, \tau)$ and the vector field $\mathbf{u}(\mathbf{X}, \tau)$. Embedded in a consistent mathematical and physical frame theory, this paper is an attempt to derive constitutive laws in a general way and to classify the materials in a systematical way. The knowledge about a material is complete, if in addition to the constitutive equations a procedure of determination of the appropriate material functions and/or parameters is provided (material identification).

1. KINEMATICS

A body $\mathcal{B}$ is a three–dimensional differentiable manifold. The elements of $\mathcal{B}$ are called elements Z and one of them representing all Z is called X. The body $\mathcal{B}$ is mapped to the EUKLIDean space $\Re_3$. There we find $\mathcal{B}$ (see Fig. 1.1)

(i) in a timeless and arbitrary reference placement $\chi(\mathrm{X}, t_0) = \boldsymbol{\kappa}$

(ii) in a past placement $\chi(\mathrm{X}, \tau)$ at the time τ

(iii) in an instantaneous placement $\chi(\mathrm{X}, t)$

with $-\infty < \tau \leq t$. Let $s = t - \tau$ the past time, starting at the present time $\tau = t$; $s = 0$ and running to the infinite past $\tau \to -\infty$, $s \to \infty$, thus $\infty > s \geq 0$.

In every placement we find all elements Z of the body at an unique position $\chi(\mathrm{Z}, t)$. One element never can be at two different places at the same time s, and two different elements never can be at the same spot within the same time s. The manifold of the placements, mapped to $\Re_3$, are called *configurations* $\chi(\mathrm{Z}, t)$ and are described by EUKLIDean vectors $\mathbf{x}(\mathrm{X}, \tau) = \mathbf{x}(\mathrm{X}, t - s)$. The special placement of X in the reference configuration is denoted by $\mathbf{X}(\mathrm{X}, t_0)$.

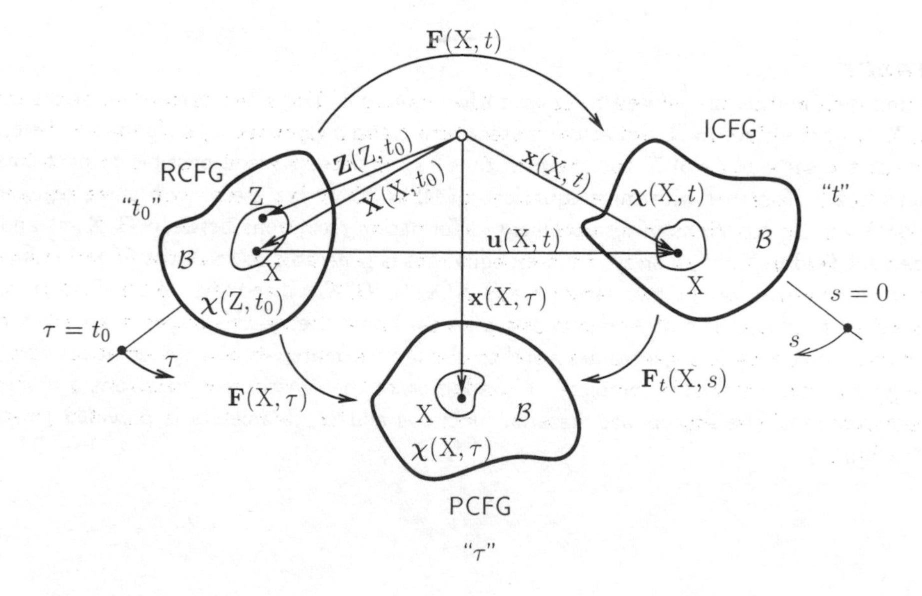

Figure 1.1: Configuration

If a field $\Phi(\mathrm{X}, \tau)$ is referred to the placement of X at the PCFG (ICFG is included, as $\tau = t$ is included), then the field is described by $\Phi = \Phi(\mathbf{x}, \tau)$ with $\mathbf{x} = x_i \mathbf{e}_i$, and the co–ordinates x_i are called *spatial*.

If a field $\Phi(\mathrm{X},\tau)$ is referred to the placement of X at the RCFG, then the field is described by $\Phi = \Phi(\mathrm{X},\tau)$ with $\mathbf{X} = X_j \mathbf{e}_j$, and the co–ordinates X_j are called *material*. This way the field $\Phi(\mathbf{X},\tau)$ at the time τ is described by the position $\mathbf{X}$ the element X has been placed in the RCFG.

In particular a *motion* of a body is a sequence of configurations $\chi(\mathrm{X},\tau)$ for all X denoted either by $\chi(\mathrm{X},\tau) = \chi(\mathrm{x},\tau)$ in spatial representation or by $\chi(\mathrm{X},\tau) = \chi(\mathbf{X},\tau)$ in material representation, whereas $\chi(\mathbf{X},\tau) = \mathrm{x}(\tau)$ is the place of X at τ described by the co–ordinates X_j referring to the RCFG.

Additionally it is presumed that the derivatives of x_i and X_j exist and are continuous in a sufficient number. For the derivatives with respect to x_i, that means for the *spatial gradients* follows

$$\begin{aligned} &\mathbf{Grad}\,\mathbf{X} = \mathbf{X}\nabla = \frac{\partial \mathbf{X}}{\partial \mathbf{x}} = (X_j \mathbf{e}_j)\left(\mathbf{e}_i \frac{\partial}{\partial x_i}\right) = \frac{\partial X_j}{\partial x_i}\mathbf{e}_j\mathbf{e}_i = \widehat{\mathbf{F}} \\ &\mathbf{Grad}\,\mathbf{x} = \mathbf{I} \end{aligned} \tag{1.1}$$

and analogue with respect to X_j resp. for the material gradients follows

$$\begin{aligned} &\mathbf{Grad}\,\mathbf{x} = \mathbf{x}\nabla = \frac{\partial \mathbf{x}}{\partial \mathbf{X}} = (x_i \mathbf{e}_i)\left(\mathbf{e}_j \frac{\partial}{\partial X_j}\right) = \frac{\partial x_i}{\partial X_j}\mathbf{e}_i\mathbf{e}_j = \mathbf{F} = \widehat{\mathbf{F}}^{-1} \\ &\mathbf{Grad}\,\mathbf{X} = \mathbf{I} \end{aligned} \tag{1.2}$$

where $\mathbf{F}$ is called the "deformation gradient"[1]. It maps all geometrical and physical states from RCFG to PCFG (ICFG) and allows to convert material co–ordinates to spatial co–ordinates and vice versa.

The second–order tensor $\mathbf{F}$ is regular with the consequence of $\det \mathbf{F} \neq 0$. So we get (Fig. 1.2)

$$\mathrm{d}\mathbf{x} = \frac{\partial \mathbf{x}}{\partial \mathbf{X}}\mathrm{d}\mathbf{X} = \mathbf{Grad}\,\mathbf{x}\,\mathrm{d}\mathbf{X} = \mathbf{F}\,\mathrm{d}\mathbf{X} \tag{1.3}$$

$$\mathrm{d}a\,\mathbf{n} = (\det \mathbf{F})\,\mathbf{F}^{-\mathrm{T}}\,\mathrm{d}A\,\mathbf{n}_0 \tag{1.4}$$

$$\mathrm{d}v = (\det \mathbf{F})\,\mathrm{d}V \tag{1.5}$$

Moreover for the scalar–product of two vectors $\mathrm{d}\mathbf{x}$ and $\mathrm{d}\mathbf{y}$ results

$$\begin{aligned} \mathrm{d}\mathbf{x}\cdot\mathrm{d}\mathbf{y} &= (\mathbf{F}\mathrm{d}\mathbf{X})\cdot(\mathbf{F}\mathrm{d}\mathbf{Y}) = (\mathbf{F}\mathrm{d}\mathbf{X})^{\mathrm{T}}\cdot(\mathbf{F}\mathrm{d}\mathbf{Y}) = \left(\mathrm{d}\mathbf{X}\mathbf{F}^{\mathrm{T}}\right)\cdot(\mathbf{F}\mathrm{d}\mathbf{Y}) \\ &= \mathrm{d}\mathbf{X}\cdot\mathbf{F}^{\mathrm{T}}\mathbf{F}\cdot\mathrm{d}\mathbf{Y} = \mathrm{d}\mathbf{X}\cdot\mathbf{C}\cdot\mathrm{d}\mathbf{Y} \end{aligned} \tag{1.6}$$

It is shown by comparison of two motions of the same body differing by a (pure) time depending vector $\mathbf{c}(t)$ that $\mathbf{F}$ is invariant to (pure, rigid) translation. Let $\mathbf{x}_2(\mathrm{X},t) = \mathbf{x}_1(\mathrm{X},t) + \mathbf{c}(t)$, then

$$\mathbf{F}_2 = \mathbf{Grad}\,\mathbf{x}_2 = \mathbf{Grad}\,[\mathbf{x}_1 + \mathbf{c}(t)] = \mathbf{Grad}\,\mathbf{x}_1 = \mathbf{F}_1 = \mathbf{F} \tag{1.7}$$

[1] Since $\mathbf{F}$ contains more than the deformation of the body it should be better called "configuration gradient"

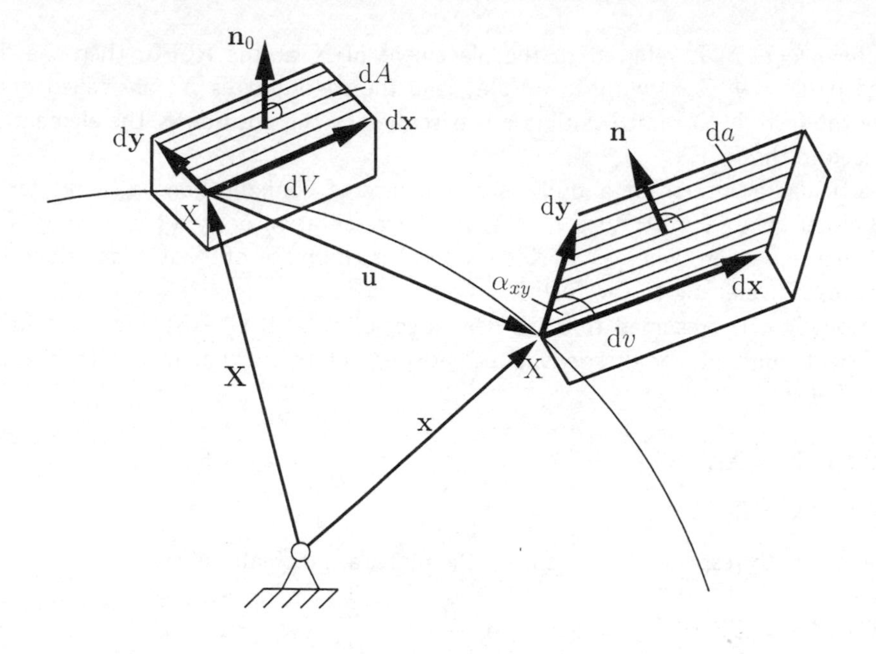

Figure 1.2: Element mappings

Then $\mathbf{F}$ must still contain the (rigid) rotation and the (proper) deformation. These two properties can be separated now by a unique multiplicative decomposition:

$$\mathbf{F}(\mathrm{X}, \tau) = \mathbf{R}(\tau)\, \mathbf{U}(\mathrm{X}, \tau), \tag{1.8}$$

in which is $\mathbf{R}$ (proper) orthogonal with $\mathbf{R}\mathbf{R}^{\mathrm{T}} = \mathbf{I}$; $\det \mathbf{R} = +1$, describing the rigid body rotation, and

$$\begin{aligned} \mathbf{U} &= \mathbf{U}^{\mathrm{T}} = \mathbf{F}^{\mathrm{T}}\mathbf{R} = \mathbf{R}^{\mathrm{T}}\mathbf{F} \\ \mathbf{U}^2 &= \mathbf{U}^{\mathrm{T}}\mathbf{U} = \left(\mathbf{F}^{\mathrm{T}}\mathbf{R}\right)\left(\mathbf{R}^{\mathrm{T}}\mathbf{F}\right) = \mathbf{F}^{\mathrm{T}}\mathbf{F} =: \mathbf{C} \end{aligned} \tag{1.9}$$

($\mathbf{U}$ — right stretching tensor; $\mathbf{C}$ — right CAUCHY tensor), describing the deformation. It is caused by (1.6) that $\mathbf{C} = \mathbf{F}^{\mathrm{T}}\mathbf{F}$ is a basic quality to determine all deformation processes (stretching and shearing angles) of an element with

$$\begin{aligned} \mathrm{d}\mathbf{x} \cdot \mathrm{d}\mathbf{y} &= (\mathrm{d}x_i \mathbf{e}_i) \cdot (\mathrm{d}x_j \mathbf{e}_j) = \mathrm{d}x_i \mathrm{d}x_j \cos\alpha_{ij} = \mathrm{d}X_i\, \mathbf{C}\, \mathrm{d}X_j = C_{ij} \mathrm{d}X_i \mathrm{d}X_j \\ C_{ij} &= \begin{cases} \dfrac{\mathrm{d}^2 x_i}{\mathrm{d}X_j^2}; & i = j \\ \dfrac{\mathrm{d}x_i\, \mathrm{d}x_j \cos\alpha_{ij}}{\mathrm{d}X_i\, \mathrm{d}X_j}; & i \neq j \end{cases} \end{aligned} \tag{1.10}$$

Using the material formulation with the "detour"

$$\mathbf{x} = \mathbf{X} + \mathbf{u} \tag{1.11}$$

whereas $\mathbf{u}$ is the displacement–vector (see Figs. 1.1, 1.2) and

$$\mathbf{x}\nabla = \mathbf{F} = (\mathbf{X} + \mathbf{u})\,\nabla = \mathbf{X}\nabla + \mathbf{u}\nabla = \mathbf{I} + \mathbf{H} \tag{1.12}$$

defining the *displacement gradient*:

$$\mathbf{H} := \mathbf{u}\nabla \neq \mathbf{H}^{\mathrm{T}} = \nabla\mathbf{u} \tag{1.13}$$

we get for arbitrary deformations expressed by the right CAUCHY tensor (1.9)

$$\begin{aligned} \mathbf{C} &= \mathbf{F}^{\mathrm{T}}\mathbf{F} = (\mathbf{I} + \mathbf{H})^{\mathrm{T}}\,(\mathbf{I} + \mathbf{H}) = \mathbf{I} + \mathbf{H} + \mathbf{H}^{\mathrm{T}} + \mathbf{H}^{\mathrm{T}}\mathbf{H} = \mathbf{C}^{\mathrm{T}} \\ &= \mathbf{I} + \mathbf{u}\nabla + \nabla\mathbf{u} + (\nabla\mathbf{u})\,(\mathbf{u}\nabla) \end{aligned} \tag{1.14}$$

The advantage of this form is that the influence of the geometrical non–linearity is evidently shown by the last (non–linear) term $\mathbf{H}^{\mathrm{T}}\mathbf{H}$ and that the non–linearity is considered by this squared term completely and only.

The physical meaning of (1.8) $\mathbf{F} = \mathbf{RU}$ is an sequential mapping, separating the configuration–gradient in a (right) stretching history $\mathbf{U}\,(\mathrm{X}, \tau)$ first and in an arbitrary rigid body rotation $\mathbf{R}$ later. The uniqueness of the mapping demands that the same result would be obtained by a rotation first and a stretching second:

$$\mathbf{F}\,(\mathrm{X}, \tau) = \mathbf{V}\,(\mathrm{X}, \tau)\,\mathbf{R}(\tau) \tag{1.15}$$

Here $\mathbf{V}\,(\mathrm{X}, \tau)$ is the left stretching tensor with the relations

$$\mathbf{V} = \mathbf{V}^{\mathrm{T}} = \mathbf{F}\mathbf{R}^{\mathrm{T}} = \mathbf{R}\mathbf{F}^{\mathrm{T}} = \mathbf{R}\mathbf{U}\mathbf{R}^{\mathrm{T}} \tag{1.16}$$

$$\begin{aligned} \mathbf{V}^2 &= (\mathbf{R}\mathbf{U}\mathbf{R}^{\mathrm{T}})\,(\mathbf{R}\mathbf{U}\mathbf{R}^{\mathrm{T}}) = \mathbf{R}\mathbf{U}\mathbf{U}\mathbf{R}^{\mathrm{T}} = \mathbf{R}\mathbf{U}^2\,\mathbf{R}^{\mathrm{T}} \\ &= \mathbf{R}\mathbf{C}\mathbf{R}^{\mathrm{T}} \\ &= \mathbf{R}\mathbf{U}\mathbf{R}^{\mathrm{T}}\,\mathbf{R}\mathbf{U}\mathbf{R}^{\mathrm{T}} = \mathbf{F}\mathbf{I}\mathbf{F}^{\mathrm{T}} = \mathbf{F}^{\mathrm{T}} =: \mathbf{B} \end{aligned} \tag{1.17}$$

where $\mathbf{B}$ is consistently the left CAUCHY deformation tensor

$$\begin{aligned} \mathbf{B} &= \mathbf{F}\mathbf{F}^{\mathrm{T}} = (\mathbf{I} + \mathbf{H})\,(\mathbf{I} + \mathbf{H})^{\mathrm{T}} \\ &= \mathbf{I} + \mathbf{H} + \mathbf{H}^{\mathrm{T}} + \mathbf{H}\mathbf{H}^{\mathrm{T}} = \mathbf{B}^{\mathrm{T}} \\ &= \mathbf{I} + \mathbf{u}\nabla + \nabla\mathbf{u} + (\mathbf{u}\nabla)\,(\nabla\mathbf{u}) \end{aligned} \tag{1.18}$$

$\mathbf{U}$ and $\mathbf{V}$ have the same eigen–values k and the eigen–directions $\mathbf{k}$ are rotated with $\mathbf{k}_V = \mathbf{R}\mathbf{k}_U$.

As (1.14) includes the identity–mapping between RCFG and ICFG, what could be interpreted as the rigid body motion up to t, only the terms $\mathbf{H}$, $\mathbf{H}^{\mathrm{T}}$ and $\mathbf{H}^{\mathrm{T}}\mathbf{H}$ contain the proper deformation. So it is obvious to define a measurement of deformation without the identity, called the right and left GREEN–tensors:

$$\begin{aligned} \mathbf{G}^{\mathrm{r}} &= \tfrac{1}{2}\left[\mathbf{F}^{\mathrm{T}}\mathbf{F} - \mathbf{I}\right] = \tfrac{1}{2}\left[\mathbf{C} - \mathbf{I}\right] = \tfrac{1}{2}\left[\mathbf{H} + \mathbf{H}^{\mathrm{T}} + \mathbf{H}^{\mathrm{T}}\mathbf{H}\right] \\ &= \tfrac{1}{2}\left[\mathbf{u}\nabla + \nabla\mathbf{u} + (\nabla\mathbf{u})\,(\mathbf{u}\nabla)\right] \end{aligned} \tag{1.19}$$

$$\begin{aligned} \mathbf{G}^{\mathrm{l}} &= \tfrac{1}{2}\left[\mathbf{F}\mathbf{F}^{\mathrm{T}} - \mathbf{I}\right] = \tfrac{1}{2}\left[\mathbf{B} - \mathbf{I}\right] = \tfrac{1}{2}\left[\mathbf{H} + \mathbf{H}^{\mathrm{T}} + \mathbf{H}\mathbf{H}^{\mathrm{T}}\right] \\ &= \tfrac{1}{2}\left[\mathbf{u}\nabla + \nabla\mathbf{u} + (\mathbf{u}\nabla)(\nabla\mathbf{u})\right] \end{aligned} \tag{1.20}$$

For reasons of linearization a magnitude of $\mathbf{H}$ is introduced by

$$\varepsilon = \sup \|\mathbf{H}\| = \sup \left[\mathrm{tr}\left(\mathbf{H}\mathbf{H}^{\mathrm{T}}\right)\right]^{1/2} \tag{1.21}$$

Then $\mathbf{H}$ is of order $O(\varepsilon)$; $\mathbf{H}^{\mathrm{T}}$ is of order $O(\varepsilon)$; but $\mathbf{H}^{\mathrm{T}}\mathbf{H}$ and $\mathbf{H}\mathbf{H}^{\mathrm{T}}$ are of order $O(\varepsilon^2)$. From (1.20) e.g. we get

$$\mathbf{G}^{\mathrm{l}} = \tfrac{1}{2}\left[\mathbf{u}\nabla + \nabla\mathbf{u} + (\mathbf{u}\nabla)(\nabla\mathbf{u})\right] = O(\varepsilon) + O(\varepsilon) + O\left(\varepsilon^2\right) \tag{1.22}$$

If a problem is allowed to be geometrically linearized, then the terms of $O(\varepsilon^2)$ are neglected in comparison with $O(\varepsilon)$. It results in

$$\mathrm{lin}\,\mathbf{G}^{\mathrm{l}} = \tfrac{1}{2}\left[\mathbf{u}\nabla + \nabla\mathbf{u}\right] = \tfrac{1}{2}\left(\mathbf{H} + \mathbf{H}^{\mathrm{T}}\right) = \mathbf{E} \tag{1.23}$$

$$\mathrm{lin}\,\mathbf{G}^{\mathrm{r}} = \tfrac{1}{2}\left[\mathbf{u}\nabla + \nabla\mathbf{u}\right] = \tfrac{1}{2}\left(\mathbf{H} + \mathbf{H}^{\mathrm{T}}\right) = \mathrm{lin}\,\mathbf{G}^{\mathrm{l}} = \mathbf{E} \tag{1.24}$$

$$\mathbf{E} = \tfrac{1}{2}\left[\mathbf{u}\nabla + \nabla\mathbf{u}\right] = \tfrac{1}{2}\left(\mathbf{H} + \mathbf{H}^{\mathrm{T}}\right) \tag{1.25}$$

as the *infinitesimal strain* tensor (engineering strains).

Because $\mathbf{R}\mathbf{R}^{\mathrm{T}} = \mathbf{I}$ and $\det\mathbf{R} = +1$, we get furthermore $\det\mathbf{U} = \det\mathbf{V} = \det\mathbf{F}$; $\det\mathbf{C} = \det\mathbf{B} = (\det\mathbf{F})^2$

For the relative configuration (see Fig. 1.1) at the past time τ resp. s we get by chain rule and with (1.2)

$$\nabla\mathbf{x}(\tau) = \mathbf{F}(\tau) = \frac{\partial\mathbf{x}(\tau)}{\partial\mathbf{x}(t)}\frac{\partial\mathbf{x}(t)}{\partial\mathbf{X}(t)} = \mathbf{F}_t(s)\,\mathbf{F}(t) \tag{1.26}$$

Here $\mathbf{F}_t(s)$ is the relative *configuration gradient*, which maps the instantaneous configuration to the past configuration. The polar decomposition theorem takes the form

$$\mathbf{F}_t(s) = \mathbf{R}_t(s)\,\mathbf{U}_t(s) = \mathbf{V}_t(s)\,\mathbf{R}_t(s) \tag{1.27}$$

The deformation tensors $\mathbf{C}$ resp. $\mathbf{B}$ are now related to this relative gradient by

$$\mathbf{C}(\tau) = \mathbf{F}^{\mathrm{T}}(t)\;\mathbf{C}_t(s)\,\mathbf{F}(t) \qquad \text{and} \qquad \mathbf{B}(\tau) = \mathbf{F}_t(s)\,\mathbf{B}(t)\,\mathbf{F}_t^{\mathrm{T}}(s) \tag{1.28}$$

As we need the deformation rate additionally, it comes together with (1.26) for the derivative with respect to τ

$$\dot{\mathbf{F}}(\tau) = \frac{\partial}{\partial\tau}\mathbf{F}(\tau) = \dot{\mathbf{F}}_t(s)\;\mathbf{F}(t)\,; \qquad \dot{\mathbf{F}}(t) = \left|\frac{\partial}{\partial\tau}\mathbf{F}(\tau)\right|_{\tau=t} = \dot{\mathbf{F}}_t(0)\;\mathbf{F}(t) \tag{1.29}$$

or by eliminating $\dot{\mathbf{F}}_t(0)$

$$\dot{\mathbf{F}}_t(0) = \dot{\mathbf{F}}_t(t)\,\mathbf{F}^{-1}(t) =: \mathbf{L} = \mathbf{Grad}\,\dot{\mathbf{x}}(t) = \mathbf{Grad}\,\mathbf{v} = \mathbf{Grad}\,\mathbf{v}\;\mathbf{F}(t) \tag{1.30}$$

Using decomposition $\mathbf{L} = \mathbf{R}\mathbf{U}$, $\mathbf{L}$ can be written as

$$\mathbf{L} = \dot{\mathbf{F}}\mathbf{F}^{-1} = (\mathbf{RU})^{\cdot}(\mathbf{RU})^{-1} = \left(\dot{\mathbf{R}}\mathbf{U} + \mathbf{U}\dot{\mathbf{R}}\right)\left(\mathbf{U}^{-1}\mathbf{R}^{-1}\right) = \dot{\mathbf{R}}\mathbf{R}^{-1} + \mathbf{R}\dot{\mathbf{U}}\mathbf{U}^{-1}\mathbf{R}^{\mathrm{T}}$$

The symmetric part of $\mathbf{L}$ is the *deformation–rate tensor*

$$\mathbf{D} = \tfrac{1}{2}\left(\mathbf{L} + \mathbf{L}^{\mathrm{T}}\right) = \operatorname{sym}\mathbf{L} = \tfrac{1}{2}\left[\mathbf{R}\left(\dot{\mathbf{U}}\mathbf{U}^{-1} + \mathbf{U}^{-1}\dot{\mathbf{U}}\right)\mathbf{R}^{\mathrm{T}}\right] \tag{1.31}$$

and the skew part of $\mathbf{L}$ is called the *spin tensor* $\mathbf{W}$

$$\mathbf{W} = \tfrac{1}{2}\left(\mathbf{L} - \mathbf{L}^{\mathrm{T}}\right) = \operatorname{skew}\mathbf{L} = \dot{\mathbf{R}}\mathbf{R}^{\mathrm{T}} + \tfrac{1}{2}\left[\mathbf{R}\left(\dot{\mathbf{U}}\mathbf{U}^{-1} + \mathbf{U}^{-1}\dot{\mathbf{U}}\right)\mathbf{R}\right] \tag{1.32}$$

Higher time–derivatives could be denoted in a similar way by $\mathbf{L}_n = \mathbf{Grad}\ \overset{(n)}{\mathbf{x}}$.

2. GENERAL CONSTITUTIVE EQUATION

2.1. Fundamentals

A *mechanical process*[2] is described completely, if the stress field $\mathbf{S}(\mathrm{X}, t)$ and the kinematical field represented by the configuration $\chi(\mathrm{X}, \tau)$ or by the displacement vector $\mathbf{u}(\mathrm{X}, \tau)$ is known. This process might be called *admissible*, if it is possible to determine one of these fields if the other is given. In general the order of priority is arbitary, but here we will submit the order to be determined by the *kinematical* history first and the *kinetic* body state (stress) is to be subsequent.

With the balance of the stress field (CAUCHY I), we have three equations, and with the displacement–deformation conditions (e.g. (1.14), (1.19)) we have additional six equations to solve the mechanical process with the symmetric stress tensor $\mathbf{S} = \mathbf{S}^{\mathrm{T}}$ (six unknowns), the displacement vector (three unknowns) and the appropriate symmetric deformation resp. deformation–rate tensors $\mathbf{C}$, $\mathbf{B}$, $\mathbf{L}$, or others (six unknowns).

Hence we have nine equations for fifteen unknowns — therefore six equations are missing. Those are expected from the condition of admissibility of a mechanical process resp. by the relation of stress and configuration in a continuous media. They are called the material laws or the *constitutive equations.* To generate these equations we postulate *principles* which have to be satisfied necessarily if the mechanical process is an admissible one.

2.2. Principle of determinism (PDET)

The stress $\mathbf{S}$ of an element X as a part of a body $\mathcal{B}$ at the time t is determined by the entire kinematical motion–history of all elements Z of $\mathcal{B}$, thus $\chi(\mathrm{Z}, \tau)$

$$\mathbf{S}(\mathrm{X}, t) = \underset{\tau=-\infty}{\overset{t}{\mathcal{F}}}\left[\chi(\mathrm{Z}, \tau), \mathrm{X}, t\right] \tag{2.1}$$

If a material is homogenous and non–aging there is no influence of an explicite X and of the present time t to the material properties:

$$\mathbf{S}(\mathrm{X}, t) = \underset{\tau=-\infty}{\overset{t}{\mathcal{F}}}\left[\chi(\mathrm{Z}, \tau), \mathrm{X}, t\right] = \underset{s=0}{\overset{\infty}{\mathcal{F}}}\left[\chi(\mathrm{Z}, t - s)\right] \tag{2.2}$$

[2] In thermodynamics the number of fields and the number of variables is higher, since at least the internal energy, the entropy, the temperature, and the heat flux vector must be taken into account.

One consequence of (2.2) is: if any two motions of two elements X1 and X2 are the same for all times $s > 0$, then the value of the functional $\mathcal{F}$, and therefore, the stress $\mathbf{S}$ is the same.

The functional is a tensor–valued functional of rank two, but it is not necessarily a second–order tensor function, as the entire history $s > 0$ has its influence to the present value $(s = 0,\ t)$ of $\mathbf{S}$. It is the *response functional* to arbitrary, but purely mechanical processes, that means to arbitrary motions of a continuous, deformable body in respect to the element X at the time t.

Since (2.2) is the general constitutive law, it describes the material behaviour. This behaviour and within the material properties may not depend on the observer resp. on the base these properties are related to. Therefore the next step will be the postulation of this invariance.

2.3. Principle of frame–indifference (PFIN)

Let OB1 and OB2 are two observers of the same kinematical history χ with the assumption that they describe their observations related to the bases [1] resp. [2]. Let us assume that OB1 is fixed resp. that he describes the process in respect to a fixed base. OB2 may be moved arbitrarily, that means, he or his base differs from OB1 by an arbitrary translation $\mathbf{c}(t)$ and a (rigid) rotation $\mathbf{Q}(t)$ (Fig. 2.1). If the process is objective (same physical process related to different bases), the process–variable must satisfy the following conditions:

$$\begin{array}{lll} \text{scalar } s, & \text{objective if } s^* = s - s_0 & \text{(shifted base)} \\ \text{vector–field } \mathbf{v}, & \text{objective if } \mathbf{v}^* = \mathbf{Q}\mathbf{v} & \text{(rotated base)} \\ \text{tensor–field } \mathbf{T}, & \text{objective if } \mathbf{T}^* = \mathbf{Q}\mathbf{T}\mathbf{Q}^{\mathrm{T}} & \text{(rotated base)} \end{array} \tag{2.3}$$

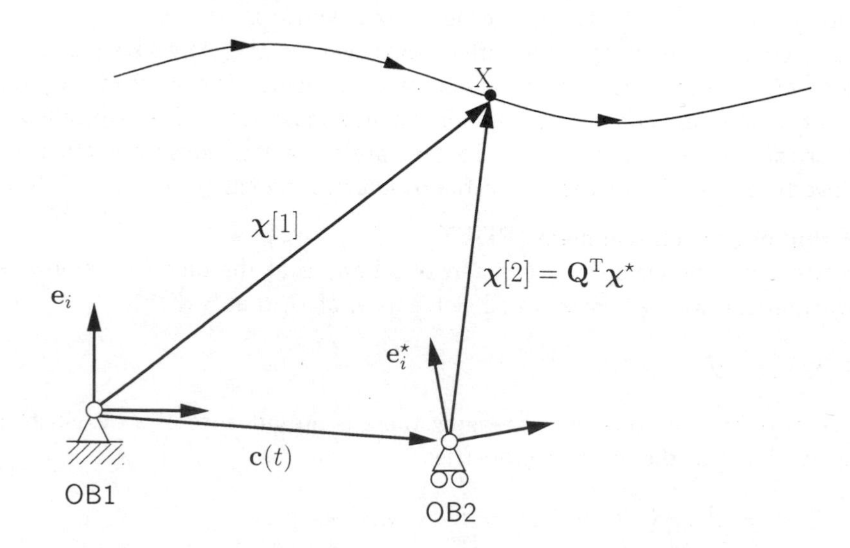

Figure 2.1: Observer transformation

For the configurations yield with $\chi^* = \mathbf{Q}\chi$

$$\chi[1] = \mathbf{c}(t) + \chi[2] = \mathbf{c}(t) + \mathbf{Q}^{\mathrm{T}}\chi^* \tag{2.4}$$

$$\chi^* = \mathbf{Q}\chi[2] = \mathbf{Q}\chi[1] - \mathbf{c}(t) \tag{2.5}$$

If now observer OB2 takes PDET (2.2) introducing his observation χ^*, he gets "his" stress "S". This stress must be objective in the sense of (2.3), hence "S" $= \mathbf{S}^*$

$$\text{“S”}\,(\mathrm{X},t) = \overset{\infty}{\underset{s=0}{\mathcal{F}}}\left[\chi\left(\mathrm{Z},t-s\right)\right] = \mathbf{S}^* = \mathbf{Q}\mathbf{S}\mathbf{Q}^{\mathrm{T}} \tag{2.6}$$

t may be replaced by t^* (see (2.3)), if t is shifted objectively.

Since the observation of OB1 is χ, he gets the stress $\mathbf{S} = \mathbf{Q}^{\mathrm{T}}\mathbf{S}^*\mathbf{Q} = \mathcal{F}[\chi]$. Together with (2.2) the two observations can be merged either by

$$\overset{\infty}{\underset{s=0}{\mathcal{F}}}\left[\chi^*\left(\mathrm{Z},t-s\right)\right] = \mathbf{S}^* = \mathbf{Q}\mathbf{S}\mathbf{Q}^{\mathrm{T}} = \mathbf{Q}\mathcal{F}\left[\chi\left(\mathrm{Z},t-s\right)\right]\mathbf{Q}^{\mathrm{T}} \tag{2.7}$$

or by

$$\overset{\infty}{\underset{s=0}{\mathcal{F}}}\left[\chi\left(\mathrm{Z},t-s\right)\right] = \mathbf{Q}^{\mathrm{T}}\overset{\infty}{\underset{s=0}{\mathcal{F}}}\left[\chi^*\left(\mathrm{Z},t-s\right)\right]\mathbf{Q} = \mathbf{Q}^{\mathrm{T}}\overset{\infty}{\underset{s=0}{\mathcal{F}}}\left\{\mathbf{Q}\left[\chi\left(\mathrm{Z},t-s\right) - \mathbf{c}\right]\right\}\mathbf{Q} \tag{2.8}$$

Not any functional is a constitutive equation, but those and *only those which satisfy the functional–restrictions (2.7) or (2.8) are matter–functionals,* as the stress caused by the motion must be objective (PFIN). This must be valid for all elements $\mathbf{Q}$ of the rotation group and for all only time–depending vectors $\mathbf{c}\,(t-s)$. Then it must be right for $\mathbf{Q} = \mathbf{I}$ and $\mathbf{c}(t) = \chi(\mathrm{X},t)$ either. By this we deduce from (2.8) the constitutive equation

$$\mathbf{S}\,(\mathrm{X},t) = \overset{\infty}{\underset{s=0}{\mathcal{F}}}\left[\chi\left(\mathrm{Z},t-s\right)\right] = \overset{\infty}{\underset{s=0}{\mathcal{F}}}\left[\chi\left(\mathrm{Z},t-s\right) - \chi\left(\mathrm{X},t-s\right)\right] \tag{2.9}$$

and the stress becomes a functional of the motion–difference–history of all Z in respect to X of the body.

If the body is *rigid,* then must be $\chi(\mathrm{Z},t-s) = \chi(\mathrm{X},t-s)$ (identical motion of all elements) with the consequence

$$\mathbf{S}\,(\mathrm{X},t) = \overset{\infty}{\underset{s=0}{\mathcal{F}}}\left[\chi\left(\mathrm{Z},t-s\right)\right] = \overset{\infty}{\underset{s=0}{\mathcal{F}}}\left[\chi\left(\mathrm{X},t-s\right) - \chi\left(\mathrm{X},t-s\right)\right] = \mathcal{F}\left[\mathbf{0}\right] := \mathbf{0} \tag{2.10}$$

That includes the statement that an arbitrary motion history does not produce any stress in a body of rigid material, resp. the stress in a rigid body is not determined by an arbitrary motion of this body.

2.4. Principle of neighbourhood (PNBH)

From (2.9) we see that all particles Z of $\mathcal{B}$ have an influence to the matter–functional and herewith to the stress.

Let $N(\mathrm{X})$ a neighbourhood of X (see Fig. 1.1). The timeless reference configuration may be denoted by $\boldsymbol{\kappa}$, then $N(\mathbf{x})$ is representable by the difference position vector of Z as the *localization* in $N(\mathrm{X})$ of the configuration $\boldsymbol{\kappa}$

$$\mathbf{Y}(\mathrm{Z}) = \mathbf{Z} - \mathbf{X} = \boldsymbol{\kappa}(\mathrm{Z}) - \boldsymbol{\kappa}(\mathrm{X}) \tag{2.11}$$

If the body is moved with time, the neighbourhood is changed to

$$\mathbf{y}(Z, \tau) = \mathbf{z} - \mathbf{x} = \boldsymbol{\chi}(Z, \tau) - \boldsymbol{\chi}(X, \tau) \tag{2.12}$$

Now $\mathbf{y}(Z, \tau)$ describes the motion of a neighbourhood of X as it appears to an observer moving with the element X, since $\mathbf{y}(X, \tau) = \mathbf{0}$. Hence (2.12) can be converted into material representation, considering that $\mathbf{y}$ differs from $\boldsymbol{\chi}$ only by translation

$$\mathbf{y}(\mathbf{Z}, \tau) = \boldsymbol{\chi}(\mathbf{Z} - \mathbf{X}, \tau) = \boldsymbol{\chi}(\mathbf{Y}, \tau) \tag{2.13}$$

and that the gradient of $\boldsymbol{\chi}$ coincides with the gradient of $\mathbf{y}$:

$$\begin{aligned} \mathbf{F}(\mathbf{Y}, \tau) &= \mathbf{Grad}\, \boldsymbol{\chi}(\mathbf{Y}, \tau) = \mathbf{F}(\mathbf{Z}, \tau) - \mathbf{F}(\mathbf{X}, \tau) \\ \mathbf{F}(\mathbf{X}, \tau) &= \mathbf{Grad}\, \boldsymbol{\chi}(\mathbf{X}, \tau) = \mathbf{Grad}\, \mathbf{y}(\mathbf{0}, \tau) \end{aligned} \tag{2.14}$$

An equivalent representation with the relative configuration and the relative configuration–gradient $\mathbf{F}_t(\mathbf{X}, s)$ is always possible (cp. (1.27)).

The localization is helpful in order to get informations about the influence of the more or less larger neighbourhood $\mathbf{y}$ of X to X itself. If the influence of the motion $\boldsymbol{\chi}$ of the neighbourhood $N(X)$ is small, or if the neighbourhood $\mathbf{y}(Z, \tau)$ considered as decisive for this influence is small, then the material is called a *local* material, otherwise it is called *non–local*.

Analogue to a TAYLOR–expansion for scalar functions

$$f(\mathbf{z}) = f(\mathbf{x} + \Delta\mathbf{x}) = f(\mathbf{x}) + f'(\mathbf{x})\Delta\mathbf{x} + \tfrac{1}{2}f''(\mathbf{x})\Delta\mathbf{x}\Delta\mathbf{x} + \ldots + \mathrm{Res}(\Delta\mathbf{x}) \tag{2.15}$$

we expand the motion $\boldsymbol{\chi}(\mathbf{Z}, .)$ at any time in the neighbourhood $\mathbf{y}$ of X:

$$\boldsymbol{\chi}(\mathbf{Z}, .) = \boldsymbol{\chi}(\mathbf{X} + \mathbf{Y}, .) = \boldsymbol{\chi}(\mathbf{X}, .) + \frac{\partial \boldsymbol{\chi}}{\partial \mathbf{X}}\mathbf{Y} + \frac{1}{2}\frac{\partial^2 \boldsymbol{\chi}}{\partial \mathbf{X}^2}\mathbf{Y}\mathbf{Y} + \ldots + \mathrm{Res}(\mathbf{Y}) \tag{2.16}$$

The terms $\dfrac{\partial^n \boldsymbol{\chi}}{\partial \mathbf{X}^n}$ are the derivatives of $\boldsymbol{\chi}(\mathbf{X}, \tau)$ with respect to $\mathbf{X}$ of the order $n = 1, 2, 3, \ldots, N$.

For $n = 1$ we get again (see (1.2))

$$\frac{\partial \boldsymbol{\chi}\,(\mathbf{X}, .)}{\partial \mathbf{X}} = \boldsymbol{\chi}\nabla = \mathbf{Grad}\, \boldsymbol{\chi} = \mathbf{F}\,(\mathbf{X}, .) \tag{2.17}$$

For higher $n > 1$ the derivatives are

$$n = 2: \quad \frac{\partial^2 \boldsymbol{\chi}}{\partial \mathbf{X}^2} = \frac{\partial}{\partial \mathbf{X}}\left(\frac{\partial \boldsymbol{\chi}}{\partial \mathbf{X}}\right) = \boldsymbol{\chi}\nabla\nabla = \mathbf{Grad}\,\mathbf{Grad}\, \boldsymbol{\chi} = \overset{2}{\mathbf{F}}\,(\mathbf{X}, .) \tag{2.18}$$

$$n = 3: \quad \frac{\partial^3 \boldsymbol{\chi}}{\partial \mathbf{X}^3} = \frac{\partial^2}{\partial \mathbf{X}^2}\left(\frac{\partial \boldsymbol{\chi}}{\partial \mathbf{X}}\right) = \boldsymbol{\chi}\nabla\nabla\nabla = \mathbf{Grad}\,\mathbf{Grad}\,\mathbf{Grad}\, \boldsymbol{\chi} = \overset{3}{\mathbf{F}}\,(\mathbf{X}, .) \tag{2.19}$$

$$n: \quad \frac{\partial^n \boldsymbol{\chi}}{\partial \mathbf{X}^n} = \boldsymbol{\chi}\nabla \ldots (n) \ldots \nabla = \mathbf{Grad} \ldots (n) \ldots \mathbf{Grad}\, \boldsymbol{\chi} = \overset{n}{\mathbf{F}}\,(\mathbf{X}, .) \tag{2.20}$$

the material gradients $\overset{n}{\mathbf{F}}\,(\mathbf{X}, .)$ of the configuration.

If the order of differentiation is n, the appertaining gradients are tensors of rank $(n+1)$. They are not functions of $\mathbf{Z}$ anymore, but of $\mathbf{X}$ now[3].

The more terms n are taken, the more we know about the kinematical environment $\mathbf{Y}$ and its influence to the local element X at $\mathbf{X}$, the more precise the influence of the neighbourhood is described, and the error of the expansion represented by the neglected remainder "Res" becomes smaller.

If it is assumed or ascertained that the elements Z with far distance $|\mathbf{Y}|$ have less or no influence to the state at X, or if this influence is restricted to a small(er) region at all, the higher gradients $N > n$ loose their importance and they can be neglected. This case is postulated by a special *principle of local action* (PLOA).

It may be mentioned that the expansion is only analogue to a TAYLOR–expansion, as we want to know only which and of what kind the independent variables expected in the matter functional are. The investigation of convergency of the TAYLOR–expansion or the representation by other convergent series, e.g. by multiplicating the higher gradients $\mathbf{F}$ with fitting parameters or factors, is not necessarily considered, if we demand informations about the motion–difference–history in (2.9) as the argument of the functional. For the motion–difference–history as the argument in (2.9), we obtain with (2.15)

$$\begin{aligned}&\chi(\mathbf{Z},t-s)-\chi(\mathbf{X},t-s)=\chi(\mathbf{X}+\mathbf{Y},t-s)-\chi(\mathbf{X},t-s)\\&=\alpha_1(\chi\nabla)\mathbf{Y}+\alpha_2(\chi\nabla\nabla)\mathbf{YY}+\ldots+\alpha_n(\chi\nabla\ldots(n)\ldots\nabla)\mathbf{Y}\ldots(n)\ldots\mathbf{Y}\\&+\mathrm{Res}\,(n,\mathbf{Y})\end{aligned} \tag{2.21}$$

For the matter–functional in the updated form (2.9) we obtain with (2.15)...(2.21)

$$\begin{aligned}\mathbf{S}(\mathbf{X},t) &= \mathcal{F}[\chi(\mathbf{Z},t-s)-\chi(\mathbf{X},t-s)]\\ &= \mathcal{F}[\chi\nabla,\chi\nabla\nabla,\chi\nabla\nabla\nabla,\ldots,\chi\nabla\ldots(n)\ldots\nabla,\mathbf{Y}]\end{aligned} \tag{2.22}$$

with $\chi=\chi(\mathbf{X},t-s)$. The gradients can be eliminated by the configuration–tensors $\mathbf{F}(\mathbf{X},t)$ according to (2.21):

$$\begin{aligned}\mathbf{S}(\mathbf{X},t) &= \mathcal{F}[\chi\nabla,\chi\nabla\nabla,\ldots,\chi\nabla\ldots(n)\ldots\nabla,\mathbf{Y}]\\ &= \mathcal{F}\left[\mathbf{F}(),\overset{2}{\mathbf{F}}(),\overset{3}{\mathbf{F}}(),\ldots,\overset{n}{\mathbf{F}}(),\mathbf{Y}\right]\end{aligned} \tag{2.23}$$

A material of this type (2.23) which is described by the first n gradients up to $\overset{n}{\mathbf{F}}(\mathbf{X},t-s)$ is defined as a *material of grade* n.

Of course, (2.23) is a material equation, if and only if the principle of frame–indifference (2.7) is satisfied additionally.

2.5. Materials of grade $n(>1)$

It is to expect that, if the funcional contains higher gradients n as tensors of rank $(n+1)$, the stress–tensors as the consequence of Eq. (2.23) are of higher rank than two

[3]For some authors this is the argument to aver that the theory looses its character of non–locality hereby. If non–locality, however, is a more or less influence of a more or less kinematical neighbourhood, then this influence is considered completely by the higher gradients in an appropriate theory.

also. But, for higher stress tensors are no equations of balance of momentum available, unless for the "classical" EULER–CAUCHY stress tensor of rank two.

For this yields

$$\nabla \cdot \mathbf{S} + \mathbf{k} = \rho\ddot{\mathbf{x}} = \rho\ddot{\mathbf{u}} \tag{2.24}$$

Together with (2.23) it would come

$$\nabla \cdot \mathcal{F}[\mathbf{F}(\mathbf{X}, t-s), \overset{2}{\mathbf{F}}(), \overset{3}{\mathbf{F}}(), \ldots, \overset{n}{\mathbf{F}}(), \mathbf{Y}] + \mathbf{k} = \rho\,\ddot{\mathbf{u}} \tag{2.25}$$

Relation (2.25) leads to a differential equation of order $(n+1)$ in respect to the spot. As the number of boundary conditions does not increase in the same way as necessary to determine all integration–constants and to solve the appropriated boundary problem, the formulation of a non–local constitutive theory by (2.25) would cause additional problems, unless an analogue equation of momentum–balance as in (2.24) would exist and found for each rank $n > 1$, for instance in the form:

$$\begin{aligned} n = 1 \quad &: \quad \nabla \cdot \mathbf{S} + \mathbf{k} = \rho\ddot{\mathbf{x}} = \rho\ddot{\mathbf{u}} \\ n = 2 \quad &: \quad \nabla \cdot \overset{2}{\mathbf{S}} + \overset{2}{\mathbf{K}} = \rho_2\,(\mathbf{u}\nabla)^{\cdot\cdot} = \rho_2 \mathbf{H}^{\cdot\cdot} \\ n \quad &: \quad \nabla \cdot \overset{n}{\mathbf{S}} + \overset{n}{\mathbf{K}} = \rho_n\,(\mathbf{u}\nabla \ldots (n) \ldots \nabla)^{\cdot\cdot} = \rho_n\,\overset{(n-1)}{\mathbf{H}}{}^{\cdot\cdot} \end{aligned} \tag{2.26}$$

Those balance equations are neither defined nor known as fundamental theorems. That is the reason why we will not follow this way.

Another possibility of balancing, proposed by [8], is to stay at the classical field equation (2.24) (CAUCHY I)

$$\nabla \cdot \mathbf{T} + \mathbf{k} = \rho\,\ddot{\mathbf{x}} \tag{2.27}$$

but to summarize all stress tensors $\mathbf{S}$ in a suitable way (see later) with the result of a *total stress tensor* $\mathbf{T}$ of rank two. Each of the added stress tensors is declared as a member of a

kinetic set $\quad \mathbf{S} : \left\{ \mathbf{S}, \overset{2}{\mathbf{S}}, \overset{3}{\mathbf{S}}, \ldots, \overset{n}{\mathbf{S}} \right\}$, and is generated by a

kinematical set $\quad \mathbf{U} : \left\{ \mathbf{u}, \overset{2}{\mathbf{U}}, \overset{3}{\mathbf{U}}, \ldots, \overset{m}{\mathbf{U}} \right\}$, where $\overset{m}{\mathbf{U}}$ are m–independent

kinematical tensors of rank m, which are co–ordinated to X at every time. If the elements of the kinematical set are separated into a displacement part $\mathbf{u}(\mathrm{X}, t)$ and an element–rotated part $\omega(\mathrm{X}, t)$, which is the axial vector of the skew part of $\overset{m}{\mathbf{U}}$, even COSSERAT–media resp. polar media of grade m are describable. The special case $m = 1$ for a *non–polar* media is included. In the ongoing treatment only non–polar materials will be investigated further. Therefore is $m = 1$ and the kinematical set reduces to $\mathbf{U} = \mathbf{u}$ and the gradients:

$$\mathbf{u}\nabla = \mathbf{H} = \mathbf{F} - \mathbf{I}; \qquad \mathbf{u}\nabla\nabla = (\mathbf{u}\nabla) = \mathbf{H}\nabla = \overset{2}{\mathbf{H}}; \qquad \mathbf{u}\nabla ... (n) ... \nabla = \overset{n}{\mathbf{H}} \tag{2.28}$$

The members of the kinetic set $\mathbf{S}$ are now computed by an energy–balance instead of the missing momentum–balance for higher stress tensors. Herewith the theory bases on a energy–equivalence and guarantees this way the satisfaction of the energy–resp. energy–flux criteria (first theorem of thermodynamics reduced to mechanical processes) [8, 9, 14, 15].

With the definition of the kinetic energy $E = \int e \, \mathrm{d}m$, the deformation energy $W = \int w \, \mathrm{d}m$, and the work $P = P_V + P_A$, where the work is divided into the work of the volume forces P_V and the work of the external surface forces P_A, the energy balance is

$$P = E + W = \int [e + w] \, \mathrm{d}m \tag{2.29}$$

resp. written as energy–flux relation

$$\dot{P} = \dot{E} + \dot{W} = \int [\dot{e} + \dot{w}] \, \mathrm{d}m \tag{2.30}$$

If the kinematical set $\mathbf{U}$ has one member only, that means $m = 1$, hence

$$P_V = \int_V \mathbf{k} \cdot \mathbf{u} \, \mathrm{d}V \tag{2.31}$$

For polar materials ($m > 1$) the equivalent expression would be

$$P_V = \int_V \left[\mathbf{k} \cdot \mathbf{u} + \overset{2}{\mathbf{K}} \cdot\cdot (\mathbf{u}\nabla) + \overset{3}{\mathbf{K}} \cdot\cdot\cdot (\mathbf{u}\nabla\nabla) + \ldots \right] \mathrm{d}V \tag{2.32}$$

If the kinetic set $\mathbf{S}$ has one member only, that means $n = 1$, hence

$$P_A = \int_{A(V)} \boldsymbol{\sigma}_\mathrm{n} \cdot \mathbf{u} \, \mathrm{d}A = \int_{A(V)} (\mathbf{n} \cdot \mathbf{S}) \cdot \mathbf{u} \, \mathrm{d}A = \int_{A(V)} \mathbf{n} \cdot (\mathbf{S} \cdot \mathbf{u}) \, \mathrm{d}A \tag{2.33}$$

we would have to expand (2.33) for non–local materials $n > 1$ by an analogue expression,

$$P_A = \int_{A(V)} \mathbf{n} \cdot \left[\mathbf{T} \cdot \mathbf{u} + \overset{3}{\mathbf{T}} \cdot\cdot (\mathbf{u}\nabla) + \overset{4}{\mathbf{T}} \cdot\cdot\cdot (\mathbf{u}\nabla\nabla) + \ldots \right] \mathrm{d}V \tag{2.34}$$

wheras $\overset{n}{\mathbf{T}}$ are the total stress tensors of rank n according to (2.27). For the time derivative $\dot{P}$ the vector $\mathbf{u}$ in (2.31) is to substitute by $\dot{\mathbf{u}} = \mathbf{v}$, hence for $m = 1$ and for all n

$$\dot{P}_V = \int_V \mathbf{k} \cdot \dot{\mathbf{u}} \, \mathrm{d}V \tag{2.35}$$

$$\dot{P}_A = \int_{A(V)} \mathbf{n} \cdot \left[\sum_{k=0}^{n-1} \left(\overset{k+2}{\mathbf{T}} \cdot (k+1) \cdot (\dot{\mathbf{u}}\nabla \ldots (k) \ldots \nabla) \right) \right] \mathrm{d}A \tag{2.36}$$

The integral over the surface in (2.36) is replaced by an integral over the volume for every multiplication "$\otimes$" by theorem of GAUSS

$$\int_{A(V)} \mathbf{n} \otimes \Phi \, \mathrm{d}V = \int_V \nabla \otimes \Phi \, \mathrm{d}V \tag{2.37}$$

$$\dot{P}_A = \int_V \nabla \cdot \left[\sum_{k=0}^{n-1} \left(\overset{k+2}{\mathbf{T}} \cdot (k+1) \cdot (\dot{\mathbf{u}}\nabla \ldots (k) \ldots \nabla) \right) \right] \mathrm{d}V \tag{2.38}$$

Between the higher gradients and their multiple scalar products with an arbitrary tensor $\mathbf{A}$ the following identity is proved by product–rule:

$$\overset{2}{\mathbf{A}} \cdot\cdot (\dot{\mathbf{u}}\nabla) = \nabla \cdot \left(\overset{2}{\mathbf{A}} \cdot \dot{\mathbf{u}}\right) - \left(\nabla \cdot \overset{2}{\mathbf{A}}\right) \cdot \dot{\mathbf{u}} \tag{2.39}$$

$$\overset{3}{\mathbf{A}} \cdot\cdot\cdot (\dot{\mathbf{u}}\nabla\nabla) = \nabla \cdot \left[\overset{3}{\mathbf{A}} \cdot\cdot (\dot{\mathbf{u}} \cdot \nabla)\right] - \nabla \cdot \left[\left(\nabla \cdot \overset{3}{\mathbf{A}}\right) \cdot \dot{\mathbf{u}}\right] + \left[\nabla \cdot \left(\nabla \cdot \overset{3}{\mathbf{A}}\right)\right] \cdot \nabla \tag{2.40}$$

Introducing (2.39) and (2.40) to (2.38) and comparing the result with the equivalent form of the classical field theory ($n = 1$), thus

$$\dot{P}_A = \int_V \nabla \cdot (\mathbf{S} \cdot \dot{\mathbf{u}})\, \mathrm{d}V$$

the "total stress tensors" $\mathbf{T}$ must be related with the set–members $\overset{n}{\mathbf{S}}$ under energy–equivalence conditions by

$$\overset{k+2}{\mathbf{T}} = \overset{k+2}{\mathbf{S}} - \nabla \cdot \overset{k+3}{\mathbf{S}} + \nabla \cdot \left(\nabla \cdot \overset{k+4}{\mathbf{S}}\right) - + \ldots = \overset{k+2}{\mathbf{S}} - \nabla \cdot \overset{k+3}{\mathbf{T}} \tag{2.41}$$

in particular, for $k = 0$

$$\mathbf{T} = \overset{2}{\mathbf{S}} - \nabla \cdot \overset{3}{\mathbf{S}} + \nabla \cdot \left(\nabla \cdot \overset{4}{\mathbf{S}}\right) - + \ldots \tag{2.42}$$

For $\mathbf{T}$ equation (2.27) holds, hence

$$\nabla \cdot \left[\overset{2}{\mathbf{S}} - \nabla \cdot \overset{3}{\mathbf{S}} + \nabla \cdot \left(\nabla \cdot \overset{4}{\mathbf{S}}\right) - + \ldots\right] + \mathbf{k} = \rho \ddot{\mathbf{u}} \tag{2.43}$$

The set–members $\overset{m}{\mathbf{S}} = \overset{2}{\mathbf{S}}, \overset{3}{\mathbf{S}}, \ldots$ follow from the constitutive equation (2.23).

Equation (2.43) may be interpreted as an extended local balance (2.24) for the volume, where the local stress $\mathbf{S}$ is replaced by the total stress $\mathbf{T}$ (comp. (2.27)):

$$\nabla \cdot \mathbf{T} + \mathbf{k} = \rho \ddot{\mathbf{u}}$$

and by an equivalent surface work expression (cp. (2.36)):

$$\begin{aligned} \dot{P}_A &= \int_{A(V)} \mathbf{n} \cdot \left[\mathbf{T} \cdot \dot{\mathbf{u}} + \overset{3}{\mathbf{T}} \cdot\cdot (\dot{\mathbf{u}}\nabla) + \overset{4}{\mathbf{T}} \cdot\cdot\cdot (\dot{\mathbf{u}}\nabla\nabla) + \ldots\right] \mathrm{d}A \\ &= \int_{A(V)} \mathbf{n} \cdot \left[\sum_{k=0} \overset{k+2}{\mathbf{T}} \cdot (k+1) \cdot (\dot{\mathbf{u}}\nabla \ldots (k) \ldots \nabla)\right] \mathrm{d}A \end{aligned} \tag{2.44}$$

In case of geometrical linear problems and of *hyper–elastic* materials, the stress tensors are delivered by a potential function as a strain energy function

$$w(\mathrm{X}, t) = w(\mathbf{E}, \nabla\mathbf{E}, \nabla\nabla\mathbf{E}, \nabla \ldots (n-1) \ldots \nabla\mathbf{E}) \tag{2.45}$$

whereas with (1.22)

$$\begin{aligned} &\mathbf{E} = \tfrac{1}{2}(\mathbf{u}\nabla + \nabla\mathbf{u}) = \tfrac{1}{2}\left(\mathbf{H} + \mathbf{H}^{\mathrm{T}}\right) \\ &\nabla \ldots (n-1) \ldots \nabla\mathbf{E} = \left[\nabla \ldots (n-1) \ldots \nabla(\mathbf{u}\nabla + \nabla\mathbf{u})\right] \end{aligned} \tag{2.46}$$

In particular we get for

$$\overset{k+2}{\mathbf{S}} = \frac{\partial w}{\partial(\nabla \ldots (k) \ldots \nabla\mathbf{E})} \tag{2.47}$$

and again with (2.41), thus

$$\overset{k+2}{\mathbf{T}} = \overset{k+2}{\mathbf{S}} - \nabla \cdot \overset{k+3}{\mathbf{S}} + \nabla \cdot \left(\nabla \cdot \overset{k+4}{\mathbf{S}}\right) - + \ldots = \overset{k+2}{\mathbf{S}} - \nabla \cdot \overset{k+3}{\mathbf{T}} \tag{2.48}$$

for $\dot{W} = \int \rho\dot{w}\,\mathrm{d}V$ follows together with (2.39) and (2.41)

$$\dot{W} = \int\limits_V \rho\left[\frac{\partial w}{\partial\mathbf{E}} \cdot\cdot (\dot{\mathbf{u}}\nabla) + \frac{\partial w}{\partial(\nabla\mathbf{E})} \cdot\cdot\cdot (\dot{\mathbf{u}}\nabla\nabla) + \ldots\right] \mathrm{d}V \tag{2.49}$$

$$\begin{aligned} \dot{W} &= \int\limits_V \left\{\nabla \cdot \left(\rho\frac{\partial w}{\partial\mathbf{E}} \cdot \dot{\mathbf{u}}\right) - \nabla \cdot \left[\nabla \cdot \rho\frac{\partial w}{\partial(\nabla\mathbf{E})} \cdot \dot{\mathbf{u}}\right] + \nabla \cdot \left[\rho\frac{\partial w}{\partial(\nabla\mathbf{E})} \cdot\cdot \dot{\mathbf{u}}\nabla - +\right]\right\} \mathrm{d}V \\ &- \int\limits_V \left\{\nabla \cdot \left(\rho\frac{\partial w}{\partial\mathbf{E}}\right) \cdot \dot{\mathbf{u}} - \nabla \cdot \left[\nabla \cdot \rho\frac{\partial w}{\partial(\nabla\mathbf{E})}\right] \cdot \dot{\mathbf{u}} + \nabla \cdot \left[\rho\frac{\partial w}{\partial(\nabla\mathbf{E})}\right] \cdot\cdot \dot{\mathbf{u}}\nabla - +\right\} \mathrm{d}V \end{aligned}$$

The first integral is, after transferring with GAUSS–theorem (2.36), the surface work $\dot{P}_A$, and the second integral becomes the momentum–balance (see (2.43)). After introducing (2.47) and by comparing with (2.48) we prove the result (2.42):

$$\mathbf{T} = \mathbf{S} - \nabla \cdot \overset{3}{\mathbf{S}} + \nabla \cdot \left(\nabla \cdot \overset{4}{\mathbf{T}}\right) - + \ldots \tag{rep. (2.42)}$$

2.6. Applications

In an elastic shear beam structure the shear force is proportional to the stress and the stress is proportional to the third derivative of the displacement

$$\mathbf{S} \sim \mathbf{Q} = \left(EI\frac{\partial^3\mathbf{u}}{\partial x^3}\right)\mathbf{e}_z \sim \mathbf{u}\nabla\nabla\nabla \cdot\cdot\cdot \mathbf{e}_z = \overset{3}{\mathbf{F}} \cdot\cdot\cdot \mathbf{e}_z \tag{2.50}$$

hence the structure can be comprehended as a material–structure of grade $n = 3$.

Investigations of this non–local theory considering higher gradients up to the order $n = 2$ and $n = 3$ have been made about COUETTE–flow of REINER–RIVLIN fluids [14, 15], of blood–stream and special visco–elastic materials (animal and human bones). The results show that special properties and phenomena, e.g. the velocity–profile of a shear flow can be described correctly with higher gradients better, resp. at all. While a solution by a local theory (REINER–RIVLIN) with ($\mathbf{D}$, see (1.31))

$$\mathbf{S} = (p\mathbf{I}) + y_1(D_{\mathrm{I}}, D_{\mathrm{II}}, , D_{\mathrm{III}})\mathbf{D} + y_2(D_{\mathrm{I}}, D_{\mathrm{II}}, D_{\mathrm{III}})\mathbf{D}^2 \tag{2.51}$$

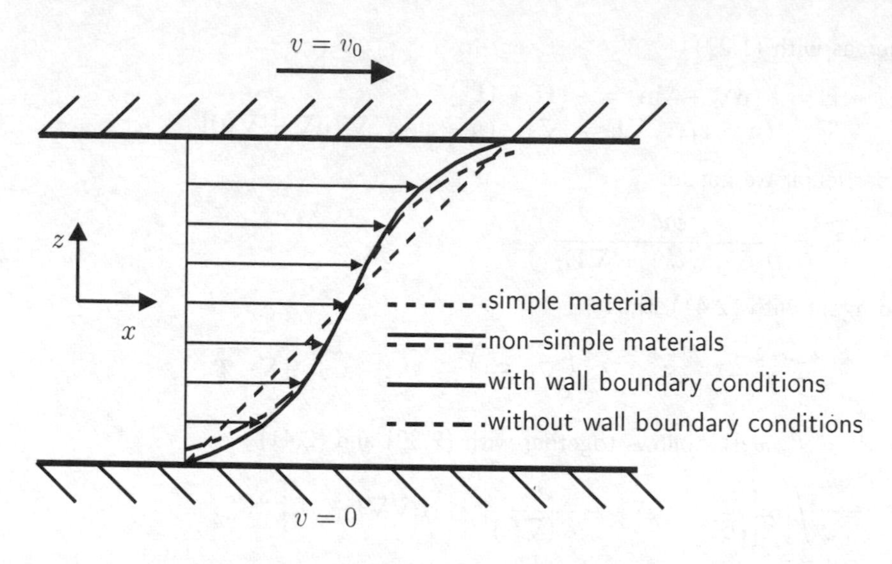

Figure 2.2: Velocity profile in COUETTE–flow

leads to a linear profile (see Fig. 2.2):

$$\mathbf{v}(z) = v_0 z \mathbf{e}_x \tag{2.52}$$

The real profile observed in experiments is non–linear and is related to a non–local material at least of grade $n = 3$, better $n = 5$, with the constitutive equation:

$$\mathbf{S} = (p\mathbf{I}) + \alpha_1 \mathbf{D} + \alpha_2 \nabla\nabla\mathbf{D} + \alpha_3 \nabla\nabla\nabla\nabla\mathbf{D} \tag{2.53}$$

resulting in a profile (see Fig. 2.2)

$$\mathbf{v}(z) = [v_0 z + C \sinh(\alpha z)]\, \mathbf{e}_x \tag{2.54}$$

2.7. Materials of grade $n = 1$, simple materials

Starting with (2.23), a local material is influenced by an infinitesimal neighbourhood only. The higher gradients $n > 1$ are not considered anymore. The only argument in the functional becomes with (2.21)

$$\boldsymbol{\chi}(\mathbf{Z}, t-s) - \boldsymbol{\chi}(\mathbf{X}, t-s) = \boldsymbol{\chi}(\mathbf{X} + \mathrm{d}\mathbf{X}, t-s) - \boldsymbol{\chi}(\mathbf{X}, t-s) = (\boldsymbol{\chi}\nabla)\,\mathrm{d}\mathbf{X} = \mathbf{F}\,\mathrm{d}\mathbf{X} \tag{2.55}$$

(*principle of local action*). The constitutive equation (2.23) is simplified to

$$\mathbf{S}(\mathbf{X}, t) = \underset{s=0}{\overset{\infty}{\mathcal{F}}} [\boldsymbol{\chi}\nabla, \mathrm{d}\mathbf{X}] = \underset{s=0}{\overset{\infty}{\mathcal{F}}} [\mathbf{F}(\mathbf{X}, t-s)] \tag{2.56}$$

The material is of grade $n = 1$ and is called a *simple material*. Compared to PFIN (2.7), the variable has changed from $\boldsymbol{\chi}^*$ to $\boldsymbol{\chi}^*\nabla$. This is the configuration gradient of an arbitrary moved observer (OB2). Because it is not yet proved, whether this gradient is the objective gradient $\mathbf{F}^*$, let's call at first

$$\boldsymbol{\chi}^*\nabla = \text{“}\mathbf{F}\text{”} \tag{2.57}$$

Since PFIN (2.6) is a necessary condition for the matter functional, it follows with (2.3)

$$\begin{aligned} \text{“}\mathbf{F}\text{”} &= \boldsymbol{\chi}^{*}\nabla = \{\mathbf{Q}\left[\boldsymbol{\chi}\left(\mathbf{X}, t-s\right)\right] - \mathbf{c}\left(t\right)\}\nabla \\ &= \left(\mathbf{Q}\nabla\right)\left[\boldsymbol{\chi}\left(\mathbf{X}, t-s\right) - \mathbf{c}\left(t\right)\right] + \mathbf{Q}\left\{\left[\boldsymbol{\chi}\left(\mathbf{X}, t-s\right) - \mathbf{c}\left(t\right)\right]\nabla\right\} \qquad (2.58) \\ \text{“}\mathbf{F}\text{”} &= \mathbf{Q}\left[\boldsymbol{\chi}\left(\mathbf{X}, t-s\right)\nabla\right] = \mathbf{Q}\mathbf{F}\left(\mathbf{X}, t-s\right) \end{aligned}$$

We obtain “$\mathbf{F}$” $= \mathbf{QF} \neq \mathbf{F}^{*} = \mathbf{QFQ}^{\mathrm{T}}$ with the result that “$\mathbf{F}$” is not objective. Therefore the constitutive equation for simple materials has to fulfil (2.7) in the form

$$\mathcal{F}\left[\mathbf{Q}\left(t-s\right)\mathbf{F}\left(\mathbf{X}, t-s\right)\right] = \mathbf{Q}\left(t\right)\mathcal{F}\left[\mathbf{F}\left(\mathbf{X}, t-s\right)\right]\mathbf{Q}^{\mathrm{T}}\left(t\right) \qquad (2.59)$$

or

$$\mathcal{F}\left[\mathbf{F}\left(\mathbf{X}, t-s\right)\right] = \mathbf{Q}^{\mathrm{T}}\left(t\right)\mathcal{F}\left[\mathbf{Q}\left(t-s\right)\mathbf{F}\left(\mathbf{X}, t-s\right)\right]\mathbf{Q}\left(t\right) \qquad (2.60)$$

Inserting the polar decompostion (1.8) resp. (1.15) the consequence is

$$\mathbf{S} = \mathcal{F}\left[\mathbf{F}\left(\mathbf{X}, t-s\right)\right] = \mathbf{Q}^{\mathrm{T}}\left(t\right)\mathcal{F}\left[\mathbf{Q}\left(t-s\right)\mathbf{R}\left(t-s\right)\mathbf{U}\left(\mathbf{X}, t-s\right)\right]\mathbf{Q}\left(t\right) \qquad (2.61)$$

and because $\mathbf{Q}$ as well as $\mathbf{R}$ are orthogonal tensors with $\mathbf{QQ}^{\mathrm{T}} = \mathbf{R}^{\mathrm{T}}\mathbf{R} = \mathbf{I}$, adapted to the rigid body rotation at every time, it is allowed to substitute $\mathbf{Q}$ by $\mathbf{R}^{\mathrm{T}}$ resp. $\mathbf{Q}^{\mathrm{T}}$ by $\mathbf{R}$ at every time.

As the objective material equation we obtain

$$\mathbf{S} = \mathcal{F}\left[\mathbf{F}\left(\mathbf{X}, t-s\right)\right] = \mathbf{R}\left(t\right)\mathcal{F}\left[\mathbf{U}\left(\mathbf{X}, t-s\right)\right]\mathbf{R}^{\mathrm{T}}\left(t\right) \qquad (2.62)$$

This is postulated as a principle of *material objectivity* (PMOB) for simple materials. The independent variable of the functional is now the right stretching tensor $\mathbf{U}$ resp. ($\mathbf{U}^2 = \mathbf{C}$) the right CAUCHY deformation tensor $\mathbf{C}$:

$$\mathbf{S} = \mathcal{F}\left[\mathbf{F}\left(\mathbf{X}, t-s\right)\right] = \mathbf{R}\left(t\right)\mathcal{F}\left[\mathbf{C}\left(\mathbf{X}, t-s\right)\right]\mathbf{R}^{\mathrm{T}}\left(t\right) \qquad (2.63)$$

Sometimes the *convected stress tensor* $\mathbf{S}^{\mathrm{C}}$ and the *rotated stress tensor* $\mathbf{S}^{\mathrm{R}}$ are introduced by

$$\mathbf{S}^{\mathrm{C}} = \mathbf{F}^{\mathrm{T}}\mathbf{S}\mathbf{F} = (\mathbf{R}\mathbf{U}^{\mathrm{T}})\mathbf{S}(\mathbf{R}\mathbf{U}) = \mathbf{U}^{\mathrm{T}}\mathbf{R}^{\mathrm{T}}\mathbf{S}\mathbf{R}\mathbf{U} = \mathbf{U}(\mathbf{R}^{\mathrm{T}}\mathbf{S}\mathbf{R})\mathbf{U} \qquad (2.64)$$

and

$$\mathbf{S}^{\mathrm{R}} = \mathbf{R}^{\mathrm{T}}\mathbf{S}\mathbf{R} \qquad (2.65)$$

With those stress tensors equation (2.62) gets the form

$$\mathbf{S}^{\mathrm{C}} = \mathbf{U}(t)\mathcal{F}\left[\mathbf{U}\left(\mathbf{X}, t-s\right)\right]\mathbf{U}(t) \qquad (2.66)$$

and respectively using (2.65)

$$\mathbf{S}^{\mathrm{R}} = \mathbf{R}^{\mathrm{T}}\left[\mathbf{R}\mathcal{F}\left[\mathbf{U}\left(\mathbf{X}, t-s\right)\right]\mathbf{R}^{\mathrm{T}}\right]\mathbf{R} = \mathcal{F}\left[\mathbf{U}\left(\mathbf{X}, t-s\right)\right] \qquad (2.67)$$

Evidently a material is determined objectively by the right deformation tensors $\mathbf{U}$ or $\mathbf{C}$, although neither $\mathbf{U}$ or $\mathbf{C}$ are objective, how we see from this relations:

$$\begin{aligned} \text{“}\mathbf{F}\text{”} &= \text{“}\mathbf{R}\text{”}\,\text{“}\mathbf{U}\text{”} = \mathbf{QF} = \mathbf{QRU} = \mathbf{QVR} = \mathbf{QVQ}^{\mathrm{T}} = \mathbf{V}^{*} \qquad (2.68) \\ \text{“}\mathbf{C}\text{”} &= \text{“}\mathbf{F}^{\mathrm{T}}\text{”}\,\text{“}\mathbf{F}\text{”} = (\mathbf{QRU})^{\mathrm{T}}(\mathbf{QRU}) \\ &= \mathbf{U}^{\mathrm{T}}\mathbf{R}^{\mathrm{T}}\mathbf{R}\mathbf{R}^{\mathrm{T}}\mathbf{R}\mathbf{U} = \mathbf{U}^{\mathrm{T}}\mathbf{U} = \mathbf{U}^2 = \mathbf{C} \qquad (2.69) \\ \text{“}\mathbf{U}\text{”} &= \mathbf{U} \quad \text{and} \quad \text{“}\mathbf{R}\text{”} = \mathbf{QR} \qquad (2.70) \end{aligned}$$

The left deformation tensors are not arguments in the material functional, although they are objective

$$\begin{aligned} \text{“}\mathbf{B}\text{”} &= \text{“}\mathbf{F}\text{”}\text{“}\mathbf{F}^{\mathrm{T}}\text{”} = (\mathbf{QRU})(\mathbf{QRU})^{\mathrm{T}} \\ &= \mathbf{R}^{\mathrm{T}}\mathbf{RUU}^{\mathrm{T}}\mathbf{R}^{\mathrm{T}}\mathbf{R} = \mathbf{U}^2 = \mathbf{C} \end{aligned} \tag{2.71}$$

$$\begin{aligned} \text{“}\mathbf{B}\text{”} &= (\mathbf{QF})(\mathbf{QF})^{\mathrm{T}} = \mathbf{QFF}^{\mathrm{T}}\mathbf{Q}^{\mathrm{T}} = \mathbf{QBQ}^{\mathrm{T}} = \mathbf{B}^* \\ &= \mathbf{R}^{\mathrm{T}}\mathbf{V}^2\mathbf{R} = \mathbf{QV}^2\mathbf{Q}^{\mathrm{T}} = (\mathbf{V}^*)^2 = (\text{“}\mathbf{V}\text{”})^2 \end{aligned} \tag{2.72}$$

$$\text{“}\mathbf{V}\text{”} = \mathbf{V}^* = \mathbf{U} = \text{“}\mathbf{U}\text{”} = \mathbf{RU}^*\mathbf{R}^{\mathrm{T}} \neq \mathbf{V} = \mathbf{U}^* \tag{2.73}$$

Considering the right deformation history $\mathbf{C}(t-s)$ we will express the functional relation by the relative configuration PCFG, using (1.26), (1.27), et. al.

$$\begin{aligned} \mathbf{F}(t) &= \mathbf{R}(t)\mathbf{U}(t) \\ \mathbf{F}(\tau) &= \mathbf{F}(t-s) = \mathbf{F}_t(s)\mathbf{F}(t) = \mathbf{R}_t(s)\mathbf{U}_t(s)\mathbf{F}(t) = \mathbf{R}_t(s)\mathbf{U}_t(s)\mathbf{R}(t)\mathbf{U}(t) \end{aligned} \tag{2.74}$$

With $\mathbf{U}_t^*(s) = \mathbf{Q}\mathbf{U}_t(s)\mathbf{Q}^{\mathrm{T}}$ follows

$$\mathbf{F}(t-s) = \mathbf{R}_t(s)\mathbf{R}(t)\mathbf{U}_t^*(s)\mathbf{R}^{\mathrm{T}}(t)\mathbf{R}(t)\mathbf{U}(t) = \mathbf{R}(t-s)\mathbf{U}_t^*(s)\mathbf{U}(t) \tag{2.75}$$

The material objectivity demands (see (2.59))

$$\mathcal{F}\left[\mathbf{Q}\,(t-s)\,\mathbf{F}\,(\mathrm{X}, t-s)\right] = \mathbf{Q}\,(t)\;\mathcal{F}\left[\mathbf{F}\,(\mathbf{X}, t-s)\right]\,\mathbf{Q}^{\mathrm{T}}\,(t) \tag{2.76}$$

hence for the left side with (2.75) and $\mathbf{Q}^{\mathrm{T}}(t) = \mathbf{R}$

$$\mathcal{F}\left[\mathbf{Q}\,(t-s)\,\mathbf{F}\,(\mathrm{X}, t-s)\right] = \mathcal{F}\left[\mathbf{Q}\,(t-s)\,\mathbf{R}\,(t-s)\,\mathbf{U}_t^*(s)\mathbf{U}(t)\right] \tag{2.77}$$

and the material law can be writen as

$$\mathbf{S} = \mathbf{R}\mathcal{F}\left[\mathbf{U}_t^*(s)\mathbf{U}(t)\right]\mathbf{R}^{\mathrm{T}} \quad \text{or} \quad \mathbf{S}^{\mathrm{R}} = \mathcal{F}\left[\mathbf{U}_t^*(s)\mathbf{U}(t)\right] \tag{2.78}$$

or in an equivalent representation, expressed by $\mathbf{C} = \mathbf{U}^2$

$$\mathbf{S} = \mathbf{R}\mathcal{F}\left[\mathbf{C}_t^*(s)\mathbf{C}(t)\right]\mathbf{R}^{\mathrm{T}} \quad \text{or} \quad \mathbf{S}^{\mathrm{R}} = \mathcal{F}\left[\mathbf{C}_t^*(s)\mathbf{C}(t)\right] \tag{2.79}$$

If the relative right GREEN tensor is used analogue to (1.19)

$$\mathbf{G}^*(s) = \mathbf{C}_t^*(s) - \mathbf{I} \tag{2.80}$$

we can finally generate the constitutive equation for simple (local) materials under satisfying all foregoing principles

$$\mathbf{S}^{\mathrm{R}} = \mathcal{F}\left[\mathbf{I} + \mathbf{G}^*\,(s)\,\mathbf{C}\,(t)\right] = \mathcal{F}\left[\mathbf{C}\,(t) + \mathbf{G}^*\,(s)\,\mathbf{C}\,(t)\right] \tag{2.81}$$

As the objective material equation we obtain

$$\begin{aligned} \mathbf{S}\,(\mathbf{X}, t) &= \mathcal{F}\left[\mathbf{C}\,(t)\right] + \mathcal{F}\left[\mathbf{G}^*\,(s)\,\mathbf{C}\,(t)\right] \\ &= \mathbf{f}\left[\mathbf{C}\,(\mathbf{X}, t)\right] + \mathcal{F}\left[\mathbf{G}^*\,(s)\,\mathbf{C}\,(t)\right] \end{aligned} \tag{2.82}$$

Here $\mathbf{f}[\mathbf{C}(\mathbf{X}, t)]$ is a tensor–valued function of the (history–less) instantaneous right CAUCHY tensor and $\mathcal{F}\left[\mathbf{G}^*\,(s)\,\mathbf{C}\,(t)\right]$ is the functional containing the relative right GREEN deformation history and the right CAUCHY deformation at the presence.

If the material has *no memory* to events of the past, the tensor function $\mathbf{f}$ is the only relation between the stress and the deformation — the material reacts spontaneously to the instantaneous deformation state (*super–elasticity*).

If the material has a memory, then the actual state of stress depends on all events ever happened in the past with a more or less influence to the presence. This way the memory can be a *permanent* one, as nothing is forgotten, or it can be a *fading* memory with the consequence that: the "older" (larger s) the event $\mathbf{G}^*(\mathbf{X}, s)$ is, the less is its influence to the stress state at the present time t as a response to all events (and vice versa).

It may be mentioned that the matter–functionals contain special deformation tensors which are the results of satisfying the different principles. This answers the question: what are the "right" or "best" deformation tensors to describe the material and which of the numerous proposals of deformation qualities are consistently related to the stress in a constitutive equation.

Finally and last but not least, this relations succeed in a possibility of classification of materials in a general and systematical way, embedded in a consistent mathematical and physical frame theory.

2.8. Material symmetry

The properties of a material can be influenced or determined by orientation of the material. In order to describe this influence of the grade of anisotropy, we change the reference configuration from the origin one to another by an initial transformation (mapping). If the functional, containing the material properties is changed too, the material has *no symmetry*. If the functional under a modified reference by a special transformation is not changed, the material is symmetric related to this transformation.

Let RCFG1 and 2 two reference configurations and let $\mathcal{B}$ one body which is related to those RCFG by two different vectors $\mathbf{X}$ and $\overline{\mathbf{X}}$ (cp. Fig. 2.3).

Assuming the same motion history of the same material referred to both reference configurations R1 and R2, we get:

$$\begin{aligned} &\text{RCFG1:}\quad \chi(\mathbf{X}, t) \rightarrow \chi(\mathbf{X}, t)\nabla = \mathbf{F}(\mathbf{X}, t) \\ &\text{RCFG2:}\quad \chi(\overline{\mathbf{X}}, t) \rightarrow \chi(\overline{\mathbf{X}}, t)\nabla = \overline{\mathbf{F}}(\overline{\mathbf{X}}, t) \end{aligned}$$

By chain–rule it is shown, that

$$\overline{\mathbf{F}} = \frac{\partial \chi}{\partial \overline{\mathbf{X}}} = \frac{\partial \chi}{\partial \mathbf{X}} \frac{\partial \mathbf{X}}{\partial \overline{\mathbf{X}}} = \mathbf{F}\mathbf{K} \tag{2.83}$$

with $K = \dfrac{\partial \mathbf{X}}{\partial \overline{\mathbf{X}}}$, as the tensor of the 'reference–configuration–transition'. For arbitrary $\mathbf{K}$ we get always

$$\mathcal{F}[\mathbf{F}] = \overline{\mathcal{F}}\left[\overline{\mathbf{F}}\right] = \overline{\mathcal{F}}[\mathbf{F}\mathbf{K}] \tag{2.84}$$

This can be considered as a proof, that a simple material remains a simple material even under the modification of the reference, because the response of the material depends on the history of motion only through the history of the deformation gradient $\mathbf{F}$ resp. $\overline{\mathbf{F}}$. Only the form of the functional is changed from $\mathcal{F}$ to $\overline{\mathcal{F}}$ by a history–independent configuration–change tensor $\mathbf{K}$.

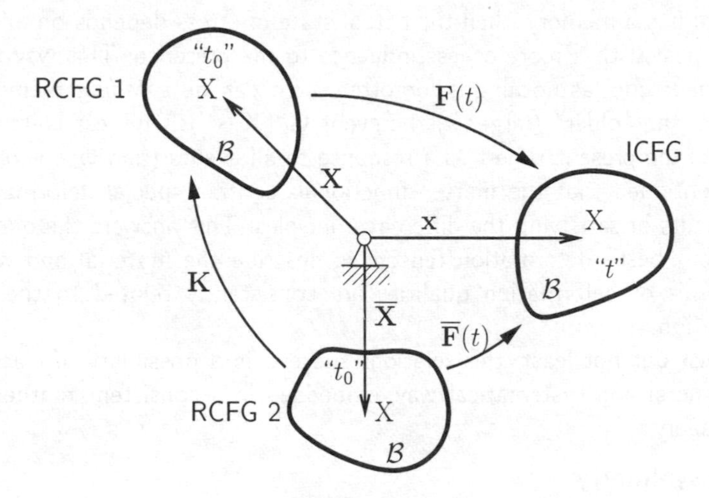

Figure 2.3: Changing of reference configuration

If now the transition of the RCFG even leads to the same functional $\mathcal{F} = \overline{\mathcal{F}}$, so that the material has the same properties when we choose to refer its motion to different RCFGs, then we call this material *symmetric*, and we denote the specific tensors $\mathbf{K} = \mathbf{M}$ (members of the *material–symmetry group*). The same property of the material demands the identical constitutive equation as a stress–deformation relation. Hence for simple materials we postulate a *principle of material symmetry* (PMAS):

$$\overline{\mathbf{S}} = \overline{\mathcal{F}}\left[\,\overline{\mathbf{F}}\,\right] = \overline{\mathcal{F}}\left[\mathbf{F}\mathbf{M}\right] = \mathcal{F}\left[\,\overline{\mathbf{F}}\,\right] = \mathcal{F}\left[\mathbf{F}\mathbf{M}\right] = \mathcal{F}\left[\mathbf{M}\right] = \mathbf{S} \tag{2.85}$$

Herewith the tensors $\mathbf{M}$ are those for which the material equation is invariant under change of reference configuration expressed by $\mathbf{M}$. The preceded 'motion' $\mathbf{M}$ is without any influence to the stress–strain relation.

If one member $\mathbf{M}$ is found, another member under transition of the reference can be generated by

$$\overline{\mathcal{F}}\left[\,\overline{\mathbf{F}}\,\right] = \mathcal{F}\left[\mathbf{F}\mathbf{K}\right] = \mathcal{F}\left[(\mathbf{F}\mathbf{M})\,\mathbf{K}\right] = \mathcal{F}\left[(\overline{\mathbf{F}}\mathbf{K}^{-1})\,(\mathbf{M}\mathbf{K})\right] = \overline{\mathcal{F}}\left[\,\overline{\mathbf{F}}\overline{\mathbf{M}}\,\right] \tag{2.86}$$

thus

$$\overline{\mathbf{M}} = \mathbf{K}^{-1}\mathbf{M}\mathbf{K} \tag{2.87}$$

We can arrange the members of $\mathbf{M}$ into groups of different material symmetry with different properties of the members. It is possible that one particular property belongs to different groups simultaneously or that one group is completely included into another group. Every group should contain at least one member (Fig. 2.4)

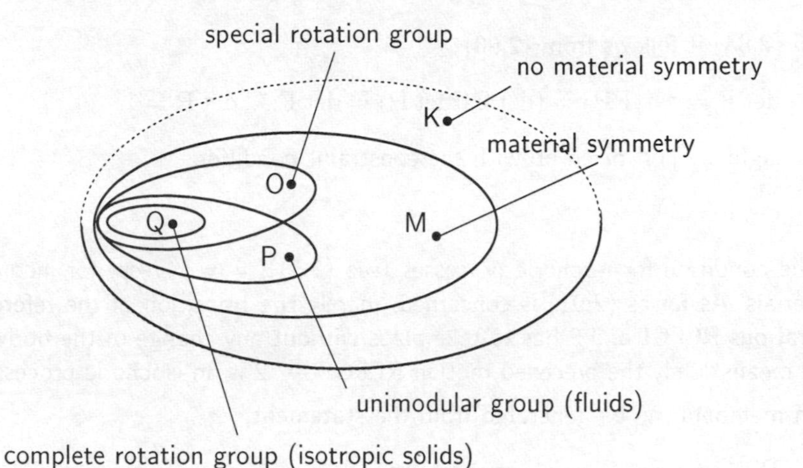

Figure 2.4: Material symmetry groups

1. If there is one element only, and this element is the identity

$$\mathbf{M} = \mathbf{I}, \qquad \mathcal{F}[\mathbf{FM}] = \mathcal{F}[\mathbf{F}] \tag{2.88}$$

then the M–group defines a complete *anisotropic* material or the *triclinic* crystal class. In any other 'direction' by changing the RCFG we obtain other material properties.

2. If there is a special group of proper (at least one) rotation

$$\mathbf{M} = \mathbf{O} = \mathbf{Q}_k^\alpha \quad (k = 1, 2, 3; \quad \alpha = \lambda\pi) \neq \mathbf{I} \tag{2.89}$$

$$\mathcal{F}[\mathbf{F}] = \mathcal{F}[\mathbf{FQ}_k^\alpha]$$

the reference configurations RCFG1 and 2 are called undistored, the material is a *simple solid*, and its symmetry is called *aelotropic*.

It denotes a neither completely anisotropic, nor a completely isotropic solid.

(a) I.e. for $\alpha = \pi$ and $k = 1, 2$: $\mathbf{Q}_1^\pi \mathbf{Q}_2^\pi$
an *orthotropic* solid and the *rhombic* crystal class

(b) I.e. for $\alpha = \pi$ or $\pi/2$ and $k = 1, 3$: $\mathbf{Q}_3^{\pi/2} \mathbf{Q}_1^\pi$
an *orthotropic* solid and the *tetragonal* crystal class

(c) I.e. for $\alpha = \pi/2$ and $k = 1, 2, 3$: $\mathbf{Q}_1^{\pi/2} \mathbf{Q}_2^{\pi/2} \mathbf{Q}_3^{\pi/2}$
an *orthotropic* solid and the *hextetrahedral* crystal class

3. If the group has the property

$$\det \overline{\mathbf{F}} = \det \mathbf{F}; \qquad \mathcal{F}[\mathbf{F}] = \mathcal{F}[\mathbf{FP}] \tag{2.90}$$

this group is called the *unimodular* group and it defines a *simple fluid*.

With (2.84) it follows from (2.90)

$$\det \overline{\mathbf{F}} = \det(\mathbf{FP}) = (\det \mathbf{F})(\det \mathbf{P}) = \det \mathbf{F}; \quad \det \mathbf{P} = 1 \tag{2.91}$$

This might be not mixed up with the constraint condition

$$\det \mathbf{F} = 1 \tag{2.92}$$

as the condition for isochoric processes (see (1.5)); $v = V$ resp. for incompressible materials. As far as (2.91) is concerned, merely the transition of the reference configurations RCFG1 and 2 has to take place without any change of the body–volume, that means: Only the preceded motion RCFG1 $\rightarrow$ 2 is an isochoric process.
Such members can be generated from the statement:

$$\mathbf{P} = (\det \mathbf{F})^m \mathbf{F}^n; \qquad m, n \text{ arbitrary} \tag{2.93}$$

Eq. (2.93) together with (2.90) yield

$$\mathbf{FP} = \mathbf{F}[(\det \mathbf{F})^m \mathbf{F}^n] = (\det \mathbf{F})^m \mathbf{F}\mathbf{F}^n \tag{2.94}$$

and, for instance, with $n = -1$

$$\mathbf{FP} = (\det \mathbf{F})^m \mathbf{I}; \qquad \mathbf{P} = (\det \mathbf{F})^m \mathbf{F}^{-1} \tag{2.95}$$

and finally together with (2.91) we get

$$\det \mathbf{P} = 1 = \det\left[(\det \mathbf{F})^m \mathbf{F}^{-1}\right] = (\det \mathbf{F})^{3m} (\det \mathbf{F})^{-1} = (\det \mathbf{F})^{3m-1} \tag{2.96}$$

thus

$$3m - 1 = 0 \qquad \text{or} \qquad m = 1/3. \tag{2.97}$$

Introducing the results (2.97) in (2.93) the symmetry group is

$$\mathbf{P} = (\det \mathbf{F})^{1/3} \mathbf{F}^{-1} = (F_{\mathrm{III}})^{1/3} \mathbf{F}^{-1} \tag{2.98}$$

It follows from (1.5) that $\mathbf{F}_{\mathrm{III}} = \det \mathbf{F} = v(t)/V = \rho/\rho(t)$, thus Eq. (2.98) can be written as

$$\mathbf{P} = [\rho/\rho(t)] \mathbf{F}^{-1} \tag{2.99}$$

Roughly, the response functional (2.90) changes to

$$\mathbf{S} = \mathcal{F}[\mathbf{F}] = \mathcal{F}[\mathbf{FP}] = \mathcal{F}[\mathbf{F}[\rho/\rho(t)]\mathbf{F}^{-1} = \mathcal{F}[\rho(t), \rho] \tag{2.100}$$

In a simple fluid the stress depends on the density only, if the RCFG is identified with the instantaneous configuration (ICFG) at (every) present time t (EULER–representation)[4].

[4]if not, see later (2.125) at al.

4. If the group $\mathbf{M}$
$$\mathbf{M}=\mathbf{Q} \quad \text{or} \quad \mathbf{M}=\mathbf{R}=\mathbf{Q}^{\mathrm{T}} \qquad \mathcal{F}[\mathbf{F}]=\mathcal{F}[\mathbf{F}\mathbf{Q}^{\mathrm{T}}] \tag{2.101}$$
contains the entire proper rotation group $\mathbf{Q}\mathbf{Q}^{\mathrm{T}}=\mathbf{I}$; $\det\mathbf{Q}=\det\mathbf{Q}^{\mathrm{T}}=+1$, we receive the special aelotropic property of *isotropy*. The material has for all directions (every rotation is allowed) the same behaviour — it is *isotropic*.

Because all elements of $\mathbf{Q}$ satisfy the condition $\det\mathbf{Q}=+1$, the isotropic group is a sub–group of the $\mathbf{P}$–group (2.90) with the consequence that *every simple fluid is isotropic*, but only those materials satisfying (2.101) *are simple isotropic solids.*

2.9. Conclusions

Whenever the transition $\mathbf{M}$ is executed and whatever the symmetry group $\mathbf{M}$ in (2.85) is, if there is any symmetry at all, the constitutive relation for simple materials (2.59) does not depend on this RCFG–change. This leads from (2.63) to
$$\mathbf{S}(\mathbf{X},t)=\mathcal{F}[\mathbf{F}]=\mathcal{F}[\mathbf{F}\mathbf{M}]=\mathbf{R}\mathcal{F}[\mathbf{Q}\mathbf{F}\mathbf{M}]\mathbf{R}^{\mathrm{T}} \tag{2.102}$$
resp.
$$\mathbf{S}^{\mathrm{R}}(\mathbf{X},t)=\mathcal{F}[\mathbf{Q}\mathbf{F}]=\mathcal{F}[\mathbf{Q}\mathbf{F}\mathbf{M}] \tag{2.103}$$
In application to *isotropic solids*, we get from (2.103) with $\mathbf{M}=\mathbf{R}=\mathbf{Q}^{\mathrm{T}}$
$$\mathcal{F}\Bigg[\underbrace{\mathbf{Q}\mathbf{F}}_{\text{“}\mathbf{F}\text{”}}\Bigg]=\mathcal{F}\Bigg[\underbrace{\mathbf{Q}\mathbf{F}\mathbf{Q}^{\mathrm{T}}}_{\mathbf{F}^*}\Bigg]=\underbrace{\mathbf{Q}\mathcal{F}[\mathbf{F}]\mathbf{Q}^{\mathrm{T}}}_{\mathcal{F}^*} \tag{2.104}$$
$$\mathcal{F}\left[\mathbf{Q}\mathbf{F}\mathbf{Q}^{\mathrm{T}}\right]=\mathcal{F}[\mathbf{F}^*]=\mathcal{F}^*[\mathbf{F}]=\mathbf{Q}\mathcal{F}[\mathbf{F}]\mathbf{Q}^{\mathrm{T}} \tag{2.105}$$
(2.105) stands for the decisive statement: For simple isotropic solids the functional of the objective argument $\mathbf{F}^*$ must be the objective functional $\mathcal{F}^*$ of the (non–transformed) argument. Functionals of this quality are called *isotropic functionals.*

Under satisfaction of all principles PDET, PFIN, PLOA, PMOB, and PMAS the following constitutive equation for simple materials is accomplished:
$$\mathbf{S}^{\mathrm{R}}(\mathbf{X},t)=\mathcal{F}\left[\mathbf{Q}\mathbf{F}(\mathbf{X},t-s)\mathbf{Q}^{\mathrm{T}}\right]=\mathcal{F}\left[\mathbf{F}^*(\mathbf{X},t-s)\right]=\mathcal{F}^*\left[\mathbf{F}(\mathbf{X},t-s)\right] \tag{2.106}$$
If we would now use (2.67), where $\mathbf{U}(\mathbf{X},t-s)$ has been the appropriate deformation tensor, we would obtain from (2.103):
$$\begin{aligned}\mathcal{F}[\mathbf{Q}\mathbf{F}] &= \mathcal{F}[\mathbf{Q}\mathbf{R}\mathbf{U}]=\mathbf{Q}\mathcal{F}[\mathbf{R}\mathbf{U}]\mathbf{Q}^{\mathrm{T}}\\ &= \mathcal{F}[\mathbf{R}^{\mathrm{T}}\mathbf{R}\mathbf{U}]=\mathcal{F}[\mathbf{U}]=\mathbf{Q}\mathcal{F}[\mathbf{R}\mathbf{U}]\mathbf{Q}^{\mathrm{T}}=\mathcal{F}^*[\mathbf{R}\mathbf{U}]\end{aligned} \tag{2.107}$$
If this relation would fulfil (2.105), then $\mathbf{U}$ must be $\mathbf{U}=(\mathbf{R}\mathbf{U})^*$. But, due to $(\mathbf{R}\mathbf{U})^*=\mathbf{Q}(\mathbf{R}\mathbf{U})\mathbf{Q}^{\mathrm{T}}=\mathbf{R}^{\mathrm{T}}\mathbf{R}\mathbf{U}\mathbf{R}=\mathbf{U}\mathbf{R}\neq\mathbf{U}$, the calculation shows that $\mathcal{F}[\mathbf{U}]$ is not an isotropic matter functional.

Using, however, $\mathbf{F}=\mathbf{V}\mathbf{R}=\mathbf{V}\mathbf{Q}^{\mathrm{T}}$, and starting again from (2.103) we would obtain
$$\mathcal{F}[\mathbf{Q}\mathbf{F}]=\mathcal{F}[\mathbf{Q}\mathbf{V}\mathbf{R}]=\mathcal{F}[\mathbf{Q}\mathbf{V}\mathbf{Q}^{\mathrm{T}}]=\mathbf{Q}\mathcal{F}[\mathbf{V}\mathbf{R}]\mathbf{Q}^{\mathrm{T}} \tag{2.108}$$

Since $\mathbf{QVQ}^{\mathrm{T}} = \mathbf{V}^*$ and $\mathcal{F}[\mathbf{VR}] = \mathcal{F}[\mathbf{VM}] = \mathcal{F}[\mathbf{V}]$ with $\mathbf{M} = \mathbf{R}$ in case of isotropic material symmetry, it is deduced that

$$\mathcal{F}[\mathbf{QVQ}^{\mathrm{T}}] = \mathcal{F}[\mathbf{V}^*] = \mathbf{Q}\mathcal{F}[\mathbf{V}]\mathbf{Q}^{\mathrm{T}} = \mathcal{F}^*[\mathbf{V}] \tag{2.109}$$

is an isotropic functional indeed. The stress–strain relation becomes

$$\mathbf{S}^{\mathrm{R}} = \mathcal{F}[\mathbf{V}(\mathbf{X}, t-s)] \qquad \text{resp.} \qquad \mathbf{S} = \mathbf{Q}\mathcal{F}[\mathbf{V}(\mathbf{X}, t-s)]\mathbf{Q}^{\mathrm{T}} \tag{2.110}$$

and we ascertain: the rotated stress tensor for those materials is a functional of the left stretching history $\mathbf{V}(\mathbf{X}, t-s)$.

Equivalent again are the forms with $\mathbf{B} = \mathbf{V}^2$

$$\mathbf{S}^{\mathrm{R}} = \mathcal{F}[\mathbf{B}(\mathbf{X}, t-s)] \qquad \text{resp.} \qquad \mathbf{S} = \mathbf{Q}\mathcal{F}[\mathbf{B}(\mathbf{X}, t-s)]\mathbf{Q}^{\mathrm{T}} \tag{2.111}$$

and with $\mathbf{G}^{\mathrm{l}} = \frac{1}{2}[\mathbf{B} - \mathbf{I}]$

$$\mathbf{S}^{\mathrm{R}} = \mathcal{F}\left[\mathbf{G}^{\mathrm{l}} + \mathbf{I}\right] = \mathcal{F}[\mathbf{G}^{\mathrm{l}}(\mathbf{X}, t-s)] + \mathcal{F}[\mathbf{I}] = \mathbf{f}_0 + \mathcal{F}[\mathbf{G}^{\mathrm{l}}(\mathbf{X}, t-s)] \tag{2.112}$$

Apparently an isotropic functional must be a functional of an objective deformation–history.

By equation (2.68) and (2.71) it has been proved that the right deformation tensors "C" = $\mathbf{C}$ and "U" = $\mathbf{U}$, that means: they are not objective. On reverse, the left deformation tensors have been "B" = $\mathbf{B}^*$ and "V" = $\mathbf{V}^*$, they are objective and therefore, they are the arguments of the matter–functional.

According to the intermediate (relative) configuration at the past time $\tau = t - s$ we receive with the relations (1.26), (1.27)

$$\begin{gathered}\mathbf{F}(t-s) = \mathbf{F}_t(s)\mathbf{F}(t); \quad \mathbf{F}(t) = \mathbf{RU} = \mathbf{VR}; \quad \mathbf{F}_t(s) = \mathbf{R}_t(s)\mathbf{U}_t(s) \\ \mathcal{F}[\mathbf{QF}(\mathbf{X}, t-s)\mathbf{M}] = \mathcal{F}[\mathbf{Q}(t-s)\mathbf{R}_t(s)\mathbf{U}_t(s)\mathbf{V}(t)\mathbf{R}(t)\mathbf{Q}]\end{gathered} \tag{2.113}$$

If this holds for every rotation, it must be valid for $\mathbf{Q}(t-s) = \mathbf{R}_t(s)$ too, and (2.78) becomes

$$\mathcal{F}[\mathbf{R}_t^{\mathrm{T}}(s)\mathbf{R}_t(s)\mathbf{U}_t(s)\mathbf{V}(t)\mathbf{R}(t)\mathbf{R}^{\mathrm{T}}] = \mathcal{F}[\mathbf{U}_t(s)\mathbf{V}(t)] \tag{2.114}$$

Due to (2.106), it is

$$\mathcal{F}[\mathbf{F}^*] = \mathbf{Q}(t)\mathcal{F}[\mathbf{F}]\mathbf{Q}^{\mathrm{T}}(t) = \mathbf{Q}(t)\mathbf{S}\mathbf{Q}^{\mathrm{T}}(t) = \mathcal{F}^*[\mathbf{F}] \tag{2.115}$$

Since $\mathbf{Q}(t-s) = \mathbf{R}_t^{\mathrm{T}}(s)$, consistently follows $\mathbf{Q}(t) = \mathbf{R}_t^{\mathrm{T}}(0) = \mathbf{I}$, and the constitutive equation is formulated as

$$\mathbf{S}(\mathbf{X}, t) = \mathbf{R}_t^{\mathrm{T}}(0)\mathcal{F}[\mathbf{U}_t(s)\mathbf{V}(t)]\mathbf{R}_t^{\mathrm{T}}(0) = \mathcal{F}[\mathbf{U}_t(s)\mathbf{V}(t)] \tag{2.116}$$

respectively

$$\mathbf{S}(\mathbf{X}, t) = \mathcal{F}[\mathbf{C}_t(s)\mathbf{B}(t)] \tag{2.117}$$

or with the relative GREEN–history (2.80)

$$\begin{aligned}\mathbf{S}(\mathbf{X}, t) &= \mathcal{F}[\mathbf{G}(s) + \mathbf{I})\mathbf{B}(t)] = \mathcal{F}[\mathbf{B}(t)] + \mathcal{F}[\mathbf{G}(s)\mathbf{B}(t)] \\ &= \mathbf{f}[\mathbf{B}(t)] + \mathcal{F}[\mathbf{G}(s)\mathbf{B}(t)]\end{aligned} \tag{2.118}$$

The stress becomes a tensor–valued tensor functions of the actual left CAUCHY deformation plus a functional (memory part) of left relative GREEN deformation–history and the actual left CAUCHY deformation.

The necessity to introduce the relative right deformation history and the absolute left actual deformation in i.e. (2.117) is substantiated by the properties of $\mathbf{C}_t(s)$ and of $\mathbf{B}(t)$ in connection with the material symmetry elements of $\mathbf{M}$. Thus, for

$$\mathbf{C}_t(s) = \mathbf{F}_t^{\mathrm{T}}(s)\mathbf{F}_t(s) = \left[\mathbf{F}(\tau)\mathbf{F}^{-1}(t)\right]^{\mathrm{T}}\left[\mathbf{F}(\tau)\mathbf{F}^{-1}(t)\right]$$

we get

$$\mathbf{C}_t(s) = \mathbf{F}^{-\mathrm{T}}(t)\mathbf{F}^{\mathrm{T}}(\tau)\mathbf{F}(\tau)\mathbf{F}^{-1}(t) \tag{2.119}$$

On the other hand results for

$$\begin{aligned}
\overline{\mathbf{C}}_t(s) &= \overline{\mathbf{F}}_t^{\mathrm{T}}(s)\overline{\mathbf{F}}_t(s) = \left[\overline{\mathbf{F}}(\tau)\,\overline{\mathbf{F}}^{-1}(t)\right]^{\mathrm{T}}\left[\overline{\mathbf{F}}(\tau)\,\overline{\mathbf{F}}^{-1}(t)\right] \\
&= \left[(\mathbf{F}(\tau)\mathbf{M})\,(\mathbf{F}(t)\mathbf{M})^{-1}\right]^{\mathrm{T}}\left[(\mathbf{F}(\tau)\mathbf{M})\,(\mathbf{F}(t)\mathbf{M})^{-1}\right] \\
&= \left[\mathbf{F}(\tau)\mathbf{M}\mathbf{M}^{-1}\mathbf{F}^{-1}(t)\right]^{\mathrm{T}}\left[\mathbf{F}(\tau)\mathbf{M}\mathbf{M}^{-1}\mathbf{F}^{-1}(t)\right] \qquad (2.120) \\
&= \mathbf{F}^{-\mathrm{T}}(t)\left(\mathbf{M}\mathbf{M}^{-1}\right)^{\mathrm{T}}\mathbf{F}^{\mathrm{T}}(\tau)\mathbf{F}(\tau)(\mathbf{M}\mathbf{M}^{-1})\mathbf{F}^{-1}(t) \\
\overline{\mathbf{C}}_t(s) &= \mathbf{F}^{-\mathrm{T}}(t)\mathbf{F}^{\mathrm{T}}(\tau)\mathbf{F}(\tau)\mathbf{F}^{-1}(t)
\end{aligned}$$

Comparison of (2.120) with (2.119) shows that $\overline{\mathbf{C}}_t(s) = \mathbf{C}_t(s)$ is invariant to the RCFG, and this holds for every *material symmetry* $\mathbf{M}$, and for simple solids as well as for simple fluids. For the actual left deformation, however, we get $\mathbf{B}(t) = \mathbf{F}(t)\mathbf{F}^{\mathrm{T}}(t)$, but

$$\overline{\mathbf{B}}(t) = \overline{\mathbf{F}}(t)\,\overline{\mathbf{F}}^{\mathrm{T}}(t) = (\mathbf{F}(t)\,\mathbf{M})\,(\mathbf{F}(t)\,\mathbf{M})^{\mathrm{T}} = \mathbf{F}(t)\,\mathbf{M}\mathbf{M}^{\mathrm{T}}\mathbf{F}^{\mathrm{T}}(t) \tag{2.121}$$

In general, $\overline{\mathbf{B}}(t)$ is not $\mathbf{B}(t)$, unless $\mathbf{M}\mathbf{M}^{\mathrm{T}}$ could be indentfied as the unit tensor $\mathbf{I}$. This is valid for $\mathbf{M} = \mathbf{Q}$, thus for isotropic solids. For those we get especially

$$\overline{\mathbf{B}}(t) = \mathbf{F}(t)\mathbf{M}\mathbf{M}^{\mathrm{T}}\mathbf{F}^{\mathrm{T}}(t) = \mathbf{F}(t)\mathbf{Q}\mathbf{Q}^{\mathrm{T}}\mathbf{F}^{\mathrm{T}}(t) = \mathbf{F}(t)\mathbf{F}^{\mathrm{T}}(t) = \mathbf{B}(t) \tag{2.122}$$

This can be considered as a proof that $\mathbf{C}_t(s)$ and $\mathbf{B}(t)$ are the appertaining deformation qualities for simple isotropic solids as in (2.117):

$$\mathbf{S}(\mathbf{X}, t) = \mathcal{F}[\mathbf{C}_t(s)\mathbf{B}(t)] \tag{2.123}$$

For simple fluids, which are described by the intermediate configuration (PCFG), $\mathbf{C}_t(s)$ remains the needed deformation tensor. But introducing the fluid group $\mathbf{M} = \mathbf{P}$ into (2.121) the deformation tensor $\mathbf{B}(t)$ has to be replaced by

$$\begin{aligned}
\overline{\mathbf{B}}(t) &= \mathbf{F}(t)\mathbf{P}\mathbf{P}^{\mathrm{T}}\mathbf{F}^{\mathrm{T}}(t) = \mathbf{F}(t)\left[(\det\mathbf{F})^m\,\mathbf{F}^{-1}\left(\det\mathbf{F}^{\mathrm{T}}\right)^m\mathbf{F}^{-\mathrm{T}}\right]\mathbf{F}^{\mathrm{T}}(t) \\
&= (\det\mathbf{F})^{2m}\left[\mathbf{F}\mathbf{F}^{-1}\mathbf{F}^{-\mathrm{T}}\mathbf{F}^{\mathrm{T}}\right] = (\det\mathbf{F})^{2m}\,\mathbf{I} \qquad (2.124) \\
&= \text{i.e. } (\det\mathbf{F})^{2/3}\,\mathbf{I} = (\rho/\rho(t))^{2/3}\,\mathbf{I}
\end{aligned}$$

and the material relation (2.117) or (2.123) for isotropic fluids turns into

$$\mathbf{S}(\mathbf{X}, t) = \mathcal{F}[\mathbf{C}_t(s); \rho(t)] \tag{2.125}$$

$\mathbf{C}_t(s)$ does not change by the transition of reference, but e.g. the density $\rho(t)$ does.

If the fluid does not even change the density (simple, *incompressible* fluid, thus $\det \mathbf{F} = 1$), it permits the choice of any arbitrary and changing RCFG and the actual $\mathbf{B}$ is to substitute by a multiple of the identity $\mathbf{I}$, following (2.124): $\mathbf{B} = p\mathbf{I}$, and for the constitutive law results

$$\mathbf{S}(\mathbf{X}, t) = \mathbf{f}[p\mathbf{I}] = \text{e.g.} = -p\mathbf{I} \tag{2.126}$$

The example $\mathbf{S} = -p\mathbf{I}$ is the constitutive law for a PASCAL–fluid without shear stresses and with an isotropic pressure distribution p in all directions.

3. SPECIAL CLASSES OF ISOTROPIC SIMPLE MATERIALS

With (2.123) for simple, isotropic solids and with (2.125) for simple fluids, we have deduced the general constitutive equations under satisfaction of all postulated principles.

The representation and the special form of the response functional to the motion–history remains open, as long as we make no restrictions and assumptions about the special kind of the memory the material has. Now we demand systematically memory properties of the material, and we will obtain *different classes of materials* and their valid matter equations.

3.1. Geometrically and physically non–linear, compressible materials without memory (super–elasticity)

If the material has *no memory* at all, the functional of 's' becomes a tensor–valued tensor–function of the appropriate actual deformation tensor. The material is not influenced by the past but only by the presence — it forgets all and remembers nothing — it reacts spontaneously.

The material law (2.117) resp. (2.118) is reduced that way to

$$\mathbf{S}(\mathbf{X}, t) = \mathcal{F}[\mathbf{B}(t)] + \mathcal{F}[\mathbf{C}_t(s)\,\mathbf{B}(t)] = \mathbf{f}[\mathbf{B}(\mathbf{X}, t)] \tag{3.1}$$

This tensor function could be e.g. a complete polynomial series of arbitrary order m:

$$\mathbf{f}[\mathbf{B}(\mathbf{X}, t)] = f_0\mathbf{B}^0 + f_1\mathbf{B}^1 + f_2\mathbf{B}^2 + f_3\mathbf{B}^3 + \ldots + f_m\mathbf{B}^m = \sum_{k=0}^{m} f_k\mathbf{B}^k \tag{3.2}$$

where f_k is a scalar–valued tensor function of $\mathbf{B}$, but only, if it is provable that (3.2) is an isotropic tensor function. Then it must hold (2.107):

$$\begin{gathered}\mathbf{f}(\mathbf{B}^*) = \mathbf{f}(\mathbf{Q}\mathbf{B}\mathbf{Q}^{\mathrm{T}}) = \mathbf{Q}\mathbf{f}\,(\mathbf{B})\,\mathbf{Q}^{\mathrm{T}} = \mathbf{f}^*\,(\mathbf{B}) \\ \sum\left[\mathbf{Q}(f_k\mathbf{B}^k)\mathbf{Q}^{\mathrm{T}}\right] = \mathbf{Q}\left[\sum f_k\mathbf{B}^k\right]\mathbf{Q}^{\mathrm{T}}\end{gathered} \tag{3.3}$$

The objectivity of the scalar function demands $f_k = f_k^*$, or in other words: the function f_k depends on the only qualities of $\mathbf{B}$, which are invariant to objective tensor transformations, that means to orthognal transformations resp. rigid rotations $\mathbf{Q}$. These are the invariants of $\mathbf{B}$. Hence, the material functions must be

$$f_k = f_k\,(B_{\mathrm{I}}, B_{\mathrm{II}}, B_{\mathrm{III}}) \qquad \text{or} \qquad f_k\,(\mathrm{I}_B, \mathrm{II}_B, \mathrm{III}_B) \tag{3.4}$$

introducing the principle invariants

$$\mathrm{I}_B = \mathrm{tr}(\mathbf{B}); \qquad \mathrm{II}_B\, \mathrm{tr}(\mathbf{B}^2); \qquad \mathrm{III}_B = \mathrm{tr}(\mathbf{B}^3) \tag{3.5}$$

From (3.3) we receive

$$f_k^* \sum \left[\mathbf{Q}(\mathbf{B}^k)\mathbf{Q}^{\mathrm{T}}\right] = f_k(B_{\mathrm{I}}, B_{\mathrm{II}}, B_{\mathrm{III}},)\mathbf{Q}\left[\sum \mathbf{B}^k\right]\mathbf{Q}^{\mathrm{T}}$$

Since

$$\begin{aligned} \mathbf{QB}^k\mathbf{Q}^{\mathrm{T}} &= \mathbf{QBB}\ldots(k)\ldots\mathbf{BQ}^{\mathrm{T}} \\ &= \mathbf{QB}\left(\mathbf{Q}^{\mathrm{T}}\mathbf{Q}\right)\mathbf{B}\left(\mathbf{Q}^{\mathrm{T}}\mathbf{Q}\right)\ldots(k)\ldots\left(\mathbf{Q}^{\mathrm{T}}\mathbf{Q}\right)\mathbf{BQ}^{\mathrm{T}} \\ &= \mathbf{QBQ}^{\mathrm{T}}\,\mathbf{QBQ}^{\mathrm{T}}\ldots(k)\ldots\mathbf{QBQ}^{\mathrm{T}} = \left(\mathbf{QBQ}^{\mathrm{T}}\right)^k, \end{aligned} \tag{3.6}$$

it is proved by (3.6) that (3.3) is fulfilled, and that a tensor polynom in $\mathbf{B}$ is an isotropic tensor function. The stress–strain relation becomes:

$$\mathbf{S}(\mathbf{X},t) = \mathbf{f}[\mathbf{B}(\mathbf{X},t)] = \sum_{k=0}^{m} f_k\left(B_{\mathrm{I}}, B_{\mathrm{II}}, B_{\mathrm{III}}\right)\mathbf{B}^k \tag{3.7}$$

With the reiteration condition (no expansion, no approximation) of CAYLEY–HAMILTON

$$\mathbf{B}^k = B_{\mathrm{I}}\mathbf{B}^{k-1} - B_{\mathrm{II}}\mathbf{B}^{k-2} + B_{\mathrm{III}}\mathbf{B}^{k-3} \tag{3.8}$$

we can express every order k of $\mathbf{B}$ by the next three lower powers of $\mathbf{B}$ and the invariants B_{I}, B_{II}, B_{III}. We start inserting (3.8) with $k = m$ and end up with $k = 3$. The functions f_k are modified to other functions Φ_k, but they remain functions of the three invariants. We conclude finally, instead of (3.7) with

$$\begin{aligned} \mathbf{S}(\mathbf{X},t) &= \Phi_0\left(B_{\mathrm{I}}, B_{\mathrm{II}}, B_{\mathrm{III}}\right)\mathbf{B}^0 + \Phi_1\left(B_{\mathrm{I}}, B_{\mathrm{II}}, B_{\mathrm{III}}\right)\mathbf{B}^1 + \Phi_2\left(B_{\mathrm{I}}, B_{\mathrm{II}}, B_{\mathrm{III}}\right)\mathbf{B}^2 \\ &= \Phi_0\left(B_{\mathrm{I}}, B_{\mathrm{II}}, B_{\mathrm{III}}\right)\mathbf{I} + \Phi_1\left(B_{\mathrm{I}}, B_{\mathrm{II}}, B_{\mathrm{III}}\right)\mathbf{B}^1 + \Phi_2\left(B_{\mathrm{I}}, B_{\mathrm{II}}, B_{\mathrm{III}}\right)\mathbf{B}^2 \end{aligned} \tag{3.9}$$

If the inversion of $\mathbf{B}^{-1}$ is prefered instead of the squared form $\mathbf{B}^2$, the reiteration can be continued to the next lower power with the result

$$\mathbf{S}(\mathbf{X},t) = \Psi_{-1}\left(B_{\mathrm{I}}, B_{\mathrm{II}}, B_{\mathrm{III}}\right)\mathbf{B}^{-1} + \Psi_0\left(B_{\mathrm{I}}, B_{\mathrm{II}}, B_{\mathrm{III}}\right)\mathbf{I} + \Psi_1\left(B_{\mathrm{I}}, B_{\mathrm{II}}, B_{\mathrm{III}}\right)\mathbf{B} \tag{3.10}$$

If finally the left GREEN–tensor (1.22) is introduced again, we get from (3.9) the alternative form

$$\begin{aligned} \mathbf{S}(\mathbf{X},t) &= \Phi_0\mathbf{I} + \Phi_1\left(2\mathbf{G}^{\mathrm{l}} + \mathbf{I}\right) + \Phi_2\left(2\mathbf{G}^{\mathrm{l}} + \mathbf{I}\right)^2 \\ &= \left(\Phi_0 + \Phi_1 + \Phi_2\right)\mathbf{I} + 2\left(\Phi_1 + \Phi_2\right)\mathbf{G}^{\mathrm{l}} + 4\Phi_2\left(\mathbf{G}^{\mathrm{l}}\right)^2 \\ &= g_0\left(G_{\mathrm{I}}, G_{\mathrm{II}}, G_{\mathrm{III}}\right)\mathbf{I} + g_1\left(G_{\mathrm{I}}, G_{\mathrm{II}}, G_{\mathrm{III}}\right)\mathbf{G}^{\mathrm{l}} + g_2\left(G_{\mathrm{I}}, G_{\mathrm{II}}, G_{\mathrm{III}}\right)\left(\mathbf{G}^{\mathrm{l}}\right)^2 \end{aligned} \tag{3.11}$$

The functions Φ_k, Ψ_k, or g_k are the materials functions (e.g. no parameters), which have to be determined by fitting multi–axial experiments and by consequences and conclusions from energy balances and criterias.

Often (3.9) is called the MOONEY–RIVLIN–material equation. It contains the complete geometrically and physically non–linear behaviour of a non–memorizing (elastic, spontaneous) simple solid material.

3.2. Geometrically and physically non–linear, incompressible materials without memory

The special case (2.92) according to (1.5) of isochoric processes $\det \mathbf{F} = 1$ with

$$B_{\mathrm{III}} = \det \mathbf{B} = \det \mathbf{F}\mathbf{F}^{\mathrm{T}} = (\det \mathbf{F})^2 = 1 = \mathrm{const}$$

describes an *incompressible* material. The hydrostatic pressure $p\mathbf{I}$ remains undeterminable from the motion and the constitutive equation (3.9) reduces to

$$\begin{aligned} \mathbf{S}(\mathbf{X},t) &= (p+\phi_0)\mathbf{I} + \phi_1(B_{\mathrm{I}},B_{\mathrm{II}})\mathbf{B} + \phi_2(B_{\mathrm{I}},B_{\mathrm{II}})\mathbf{B}^2 \\ \mathbf{S}(\mathbf{X},t) &= A\mathbf{I} + \phi_1(B_{\mathrm{I}},B_{\mathrm{II}})\mathbf{B} + \phi_2(B_{\mathrm{I}},B_{\mathrm{II}})\mathbf{B}^2 \end{aligned} \tag{3.12}$$

or (3.11) is specialized as

$$\mathbf{S}(\mathbf{X},t) = A\mathbf{I} + \gamma_1(G_{\mathrm{I}},G_{\mathrm{II}})\mathbf{G} + \gamma_2(G_{\mathrm{I}},G_{\mathrm{II}})\mathbf{G}^2 \tag{3.13}$$

The numbers of the material functions is diminished from 3 to 2. They are functions of the two first invariants only. (3.12) is sometimes used to describe rubber–like materials, but often not very sucessful, because compressibility and memory effects in reality contradict to the presumptions of this class.

3.3. Geometrically–linear, compressible material without memory

The case of geometrical linearization is realized with (1.22): the GREEN–tensors are substituted by the infinitesimal deformation tensor (engineering strains). Under these conditions, it would result from (3.11):

$$g_0\,(G_{\mathrm{I}},G_{\mathrm{II}},G_{\mathrm{III}})\,\mathbf{I} + g_1(G_{\mathrm{I}},G_{\mathrm{II}},G_{\mathrm{III}})\mathbf{E} + g_2(G_{\mathrm{I}},G_{\mathrm{II}},G_{\mathrm{III}})\mathbf{E}^2 \tag{3.14}$$

But this would be an inconsistent linearization, as some terms (i.e. $\mathbf{H}\mathbf{H}^{\mathrm{T}}$) of order $O(\varepsilon^2)$ are already neglected, however, the squared term $\mathbf{E}^2$ is of order $O(\varepsilon^2)$, too. It also must vanish in order to achieve a consistent geometrical linearized constitutive condition. Additionally, the invariants of the GREEN tensor are transferred to the invariants of the classical strain tensor E_{I}, E_{II} and E_{III}. So we finally get:

$$\mathbf{S}(\mathbf{X},t) = h_0\,(E_{\mathrm{I}},E_{\mathrm{II}},E_{\mathrm{III}})\,\mathbf{I} + h_1(E_{\mathrm{I}},E_{\mathrm{II}},E_{\mathrm{III}})\mathbf{E} \tag{3.15}$$

3.4. Geometrically–linear, incompressible material without memory

Again the hydrostatic pressure is not determined from the motion, since an arbitrary pressure could be superposed changing the stress but not the strain. On the other hand, now the first invariant $E_{\mathrm{I}} = \varepsilon_1+\varepsilon_2+\varepsilon_3 = \operatorname{div}\mathbf{u} = \nabla\cdot\mathbf{u} = 0$ describes the volume dilatation $v = V = \mathrm{const}$, with the consequence that $E_{\mathrm{I}} = 0$, and that the material relation becomes here:

$$\mathbf{S}(\mathbf{X},t) = A\mathbf{I} + h_1(E_{\mathrm{II}},E_{\mathrm{III}})\mathbf{E} \tag{3.16}$$

3.5. Physically–linear, compressible material without memory

If there is a linear connection between the stress and the deformation, we call those behaviour *physical–linear*. As the invariants I, II, III are of power one, two, and three in the deformations, the first invariant $\Phi_0(B_{\mathrm{I}})$ i.e. in (3.9) connected with the identity is the only one which can occur in the material relations. In case of physically–linear problems the material function Φ_1 connected with the deformation tensor itself must be a material constant Φ_{10} and the material function Φ_2 must vanish, since it is linked with the squared deformation tensor. The constitutive equation (3.9) becomes

$$\mathbf{S}(\mathbf{X},t) = \Phi_0(B_\mathrm{I})\mathbf{I} + \Phi_{10}\mathbf{B} \tag{3.17}$$

A pure stress–strain linearity, however, is achieved only if $\Phi_0(B_\mathrm{I})$ is linear in B_I, thus

$$\Phi_0(B_\mathrm{I}) = \Phi_{00}B_\mathrm{I} = \Phi_{00}\,\mathrm{tr}\,\mathbf{B} \tag{3.18}$$

The last two equations together lead to

$$\mathbf{S}(\mathbf{X},t) = \Phi_{00}\,(\mathrm{tr}\,\mathbf{B})\,\mathbf{I} + \Phi_{10}\mathbf{B} \tag{3.19}$$

(3.19) is not geometrically linearized, although the term $\mathbf{B}^2$ is not contained anymore, but $\mathbf{B}$ still represents the geometrical nonlinearity with respect to large displacement derivatives ($\mathbf{H}$, $\mathbf{H}^\mathrm{T}$, and $\mathbf{H}\mathbf{H}^\mathrm{T}$).

3.6. Physically–linear, incompressible material without memory

As before, the factor of the unit tensor remains undetermined and the hydrostatic pressure and Φ_{00} are merged to a common, motion–independent, and initial state–depending quality A. This case is related by

$$\mathbf{S}(\mathbf{X},t) = A\mathbf{I} + \Phi_{10}\mathbf{B} \tag{3.20}$$

3.7. Geometrically and physically–linear, compressible material without memory

If we start with (3.14) and add the constraints according to physical linearity (3.17), we immediately formulate the completely linearized stress–strain–relation as

$$\mathbf{S}\,(\mathbf{X},t) = h_{00}\,(\mathrm{tr}\,\mathbf{E})\,\mathbf{I} + h_{10}\mathbf{E} \tag{3.21}$$

With $h_{00} = \lambda$ and $h_{10} = 2\mu$, we have derived HOOKE's law in the well–known LAMÉ form:

$$\mathbf{S}\,(\mathbf{X},t) = \lambda\,(\mathrm{tr}\,\mathbf{E})\,\mathbf{I} + 2\mu\mathbf{E} \tag{3.22}$$

or with $h_{00} = 2G\nu/\,(1-\nu)$ and $h_{10} = 2G$ (shear modulus), we obtain the elastic material equation in the form:

$$\mathbf{S}\,(\mathbf{X},t) = 2G\left[\mathbf{E} + \frac{\nu}{1-\nu}\,(\mathrm{tr}\,\mathbf{E})\,\mathbf{I}\right] \tag{3.23}$$

3.8. Geometrically and physically–linear, incompressible material without memory

With the constraint condition $(\det\mathbf{F} = 1)$ and/or $(\mathrm{tr}\,\mathbf{E} = 0)$ again, it is deduced at once that the material law gets the form:

$$\mathbf{S}\,(\mathbf{X},t) = A\mathbf{I} + 2G\mathbf{E} \tag{3.24}$$

To avoid the terms $A\ \mathbf{I}$ in every incompressible material equation, it is senseful to modify these equations with the stress deviator

$$\widehat{\mathbf{S}} = \mathbf{S} - \tfrac{1}{3}\,(\mathrm{tr}\,\mathbf{S})\,\mathbf{I} \tag{3.25}$$

with the consequence that the deformation terms remain in the stress–strain relations only.

3.9. Non–linear, compressible materials with short memory (differential–type materials)

With the assumption of a (very) short memory, the relative deformation history in (2.116) for a simple material

$$\mathbf{S}(\mathbf{X},t) = \mathcal{F}\left[\mathbf{U}_t(s)\,\mathbf{V}(t)\right] \tag{3.26}$$

has its influence to the stress, but the value of $\mathcal{F}$ depends on the values of $\mathbf{U}_t(s)$ for s very near to zero only. We use a TAYLOR expansion for $\mathbf{U}_t(s)$ around $s = 0$ up to the order r

$$\mathbf{U}_t(s) = \mathbf{U}_t(0) + a_1 \frac{\partial \mathbf{U}_t}{\partial s} s + a_2 \frac{\partial^2 \mathbf{U}_t}{\partial s^2} s^2 + \ldots + a_r \frac{\partial^r \mathbf{U}_t}{\partial s^r} s^r \tag{3.27}$$

With (1.27), (1.30), and (1.31) it was shown that the following relations hold:

$$\dot{\mathbf{F}}_t(s)\Big|_{s=0} = \dot{\mathbf{F}}_t(0) = \dot{\mathbf{F}}(t)\,\mathbf{F}^{-1}(t) = \mathbf{L} = \mathbf{D}(t) + \mathbf{W}(t) \tag{3.28}$$

On the other hand, we get from (2.76)

$$\mathbf{F}_t(s) = \mathbf{R}_t(s)\,\mathbf{U}_t(s); \qquad \dot{\mathbf{F}}_t(s) = \dot{\mathbf{R}}_t(s)\,\mathbf{U}_t(s) + \mathbf{R}_t(s)\,\dot{\mathbf{U}}_t(s) \tag{3.29}$$

with the first derivative with respect to s and with $\mathbf{R}_t(0) = \mathbf{U}_t(0) = \mathbf{I}$

$$\dot{\mathbf{F}}_t(0) = \dot{\mathbf{R}}_t(0)\,\mathbf{U}_t(0) + \mathbf{R}_t(0)\,\dot{\mathbf{U}}_t(0) = \dot{\mathbf{U}}_t(0) + \dot{\mathbf{R}}_t(0) \tag{3.30}$$

Equations (3.28) and (3.30) are two expressions for the same tensor $\mathbf{L} = \dot{\mathbf{F}}_t(0)$. The comparison of both leads to

$$\mathbf{D}(t) = \dot{\mathbf{U}}_t(0) \qquad \text{and} \qquad \mathbf{W}(t) = \dot{\mathbf{R}}_t(0) \tag{3.31}$$

We define the higher derivative of $\mathbf{U}_t$ (relative stretching velocity) as

$$\overset{r}{\mathbf{D}}(t) = \frac{\partial^r \mathbf{U}_t(s)}{\partial s^r}\Big|_{s=0} = \overset{(r)}{\mathbf{U}}_t(0) \tag{3.32}$$

then the TAYLOR–series (3.27) has the formulation

$$\mathbf{U}_t(s) = \mathbf{I} + a_1 \mathbf{D}(t)\,s + a_2\,\overset{2}{\mathbf{D}}(t)\,s^2 + \ldots + a_r\,\overset{r}{\mathbf{D}}(t)\,s^r \tag{3.33}$$

The functional $\mathcal{F}$ becomes a functional of the deformation–rate tensors $\overset{r}{\mathbf{D}}(t)$ up to the time–derivative r. The number r is called the *complexity of the material.*

Since the time–variable of $\mathbf{D}$ is the present time t now, the functional of $\mathbf{U}_t(s)$ is a tensor–valued tensor function $\mathbf{f}$ of $\mathbf{D}(t)$ again:

$$\begin{aligned} \mathbf{S}(\mathbf{X},t) &= \mathcal{F}\left[\mathbf{U}_t(s)\,\mathbf{V}(t)\right] = \mathcal{F}\left[\mathbf{I}, \mathbf{D}(t), \overset{2}{\mathbf{D}}(t), ..., \overset{r}{\mathbf{D}}(t), \mathbf{V}(t)\right] \\ \mathbf{S}(\mathbf{X},t) &= \mathbf{f}\left[\mathbf{D}(\mathbf{X},t), \overset{2}{\mathbf{D}}(\mathbf{X},t), ..., \overset{r}{\mathbf{D}}(\mathbf{X},t), \mathbf{V}(\mathbf{X},t)\right] \end{aligned} \tag{3.34}$$

For simple isotropic *solids* of the differential–type class the actual left deformation tensor $\mathbf{V}(\mathbf{X},t)$ resp. $\mathbf{B}(\mathbf{X},t)$ are together with the actual deformation rate tensors the arguments of the tensor function determining the stress.

For simple *fluids* of the differential–type class with short–time memory the actual left deformation, i.e. $\mathbf{B}(\mathbf{X},t)$ is to substitute by the density $\rho(t)$ (see (2.124)) again:

$$\mathbf{S}(\mathbf{X},t) = \mathbf{f}\left[\mathbf{D}(\mathbf{X},t), \overset{2}{\mathbf{D}}(\mathbf{X},t), ..., \overset{r}{\mathbf{D}}(\mathbf{X},t), \rho(t)\right] \tag{3.35}$$

For example, we take an isotropic solid with the complexity $r = 1$. Using isotropic tensor polynoms, like in (3.2), for the two objective tensors $\mathbf{D}$ and $\mathbf{V}$, we would obtain in analogy to (3.7) – (3.9) and after reiteration with CAYLEY–HAMILTON a reduced form up to the order 2, i.e.:

$$\begin{aligned} \mathbf{S}(\mathbf{X},t) &= \mathbf{f}\left[\mathbf{D}(\mathbf{X},t), \mathbf{V}(\mathbf{X},t)\right] = \mathbf{f}_1\left[\mathbf{D}(\mathbf{X},t)\right]\mathbf{f}_2\left[\mathbf{V}(\mathbf{X},t)\right] \\ &= \left[a_0(D_K)\,\mathbf{I} + a_1(D_K)\,\mathbf{D} + a_2(D_K)\,\mathbf{D}^2\right] \\ &\quad \left[b_0(V_K)\,\mathbf{I} + b_1(V_K)\,\mathbf{V} + b_2(V_K)\,\mathbf{V}^2\right] \\ &= f_{00}(D_K, V_K)\,\mathbf{I} + f_{10}()\,\mathbf{D} + f_{20}()\,\mathbf{D}^2 + f_{01}()\,\mathbf{V} + f_{02}()\,\mathbf{V}^2 \\ &\quad + f_{11}()\,(\mathbf{DV} + \mathbf{VD}) + f_{21}()\left(\mathbf{D}^2\mathbf{V} + \mathbf{VD}^2\right) \\ &\quad + f_{12}()\left(\mathbf{DV}^2 + \mathbf{V}^2\mathbf{D}\right) + f_{22}()\left(\mathbf{D}^2\mathbf{V}^2 + \mathbf{V}^2\mathbf{D}^2\right) \end{aligned} \tag{3.36}$$

The constitutive relation contains 9 material functions of the 3 invariants D_K and the 3 invariants V_K.

A special case would follow, if for $r = 1$ a simple fluid is considered. The deformation $\mathbf{V}$ is replaced by the density ρ. Thus, the stress (3.35) would be expressed by

$$\mathbf{S}(\mathbf{X},t) = \mathbf{g}\left[\mathbf{D}(\mathbf{X},t), \rho(t)\right] = g_{00}(D_K)\,\mathbf{I} + g_{10}(D_K)\,\mathbf{D} + g_{20}(D_K)\,\mathbf{D}^2 \tag{3.37}$$

The result is well–known as a REINER–RIVLIN fluid. The material law is analogue to the MOONEY–RIVLIN material for an isotropic solid without memory (compare (3.9)). If the second order term $\mathbf{D}^2$ is neglected resp. if the material function g_{20} is determined by zero, the special case of a generalized compressible NEWTON fluid is obtained, where the stress is proportional to the deformation rate $\mathbf{D}$

$$\mathbf{S}(\mathbf{X},t) = h_{00}(D_K)\,\mathbf{I} + h_{10}(D_K)\,\mathbf{D} \tag{3.38}$$

3.10. Physically–linear materials with memory (rate–type materials)

The general constitutive relation for a simple isotropic material according to (2.110) resp. (2.116) had the form

$$\mathbf{S}^{\mathrm{R}}(\mathbf{X},t) = \mathcal{F}\left[\mathbf{V}(\mathbf{X},t-s)\right] = \mathcal{F}\left[\mathbf{U}_t^*(s)\,\mathbf{V}(t)\right] \tag{3.39}$$

whereas

$$\mathbf{U}_t^* (s) = \mathbf{Q} (t) \mathbf{U}_t (s) \mathbf{Q}^{\mathrm{T}} (t) = \mathbf{R}^{\mathrm{T}} (t) \mathbf{U}_t (s) \mathbf{R} (t) \tag{3.40}$$

The definition (3.32)

$$\overset{r}{\mathbf{D}} (t) = \left. \frac{\partial^r \mathbf{U}_t (s)}{\partial s^r} \right|_{s=0} = \overset{(r)}{\mathbf{U}}_t (0) \tag{3.41}$$

leads together with (3.40) to

$$\overset{(r)}{\mathbf{U}^*}_t (0) = \mathbf{R}^{\mathrm{T}} (t) \overset{(r)}{\mathbf{U}}_t (s) \mathbf{R} (t) \tag{3.42}$$

The solution of the functional equation (3.39), $\mathbf{S} (t)$ aş dependend on $\mathbf{U}_t (s)$, may be restricted by a differential equation, where $p = 0, 1, 2, ..., q$ is the order of differentiation of the stress tensor with respect to time t, and $k = 0, 1, 2, ..., r$ (as above) is the r–th time–derivative of the relative stretching (3.41). This restriction for (3.39) as an admissible mechanical process should be

$$\mathbf{d} \left\{ \mathbf{S}, \dot{\mathbf{S}}, \ddot{\mathbf{S}}, \ldots, \overset{(q)}{\mathbf{S}}; \mathbf{U}_t (0) = \mathbf{I}, \dot{\mathbf{U}}_t (0), \ddot{\mathbf{U}}_t (0), ..., \overset{(r)}{\mathbf{U}}_t (0) \right\} = \mathbf{0} \tag{3.43}$$

or

$$\overset{(q)}{\mathbf{S}} = \mathbf{d} \left\{ \mathbf{S}, \dot{\mathbf{S}}, \ldots, \overset{(q-1)}{\mathbf{S}}; \dot{\mathbf{U}}_t (0), \ddot{\mathbf{U}}_t (0), ..., \overset{(r)}{\mathbf{U}}_t (0) \right\} \tag{3.44}$$

The class of materials with this restriction are called *materials of the rate type*. The derivatives should exist in a sufficient number, and the differential equation should have a solution that is determined except for the initial conditions

$$\mathbf{S} (0), \dot{\mathbf{S}} (0), ..., \overset{(q-1)}{\mathbf{S}} (0).$$

Because the general constitutive equation has to satisfy the PMOB, we have to demand that the derivative of the stresses are objective and that the restriction (3.44) is an isotropic function, too. This way we postulate the co–rotational stress rate, analogue to (2.6), by

$$^1\overset{\circ}{\mathbf{S}} = \frac{\partial}{\partial s} \left[\mathbf{R}_t^{\mathrm{T}} (s) \mathbf{S} (t) \mathbf{R}_t (s) \right] \Big|_{s=0} \tag{3.45}$$

and in an appropriate way for the higher derivatives

$$^q\overset{\circ}{\mathbf{S}} = \frac{\partial^q}{\partial s^q} \left[\mathbf{R}_t^{\mathrm{T}} (s) \mathbf{S} (t) \mathbf{R}_t (s) \right] \Big|_{s=0} \tag{3.46}$$

If (3.45) is performed, by product rule we get together with (3.31) and by the property of a skew tensor

$$\begin{aligned} \mathbf{W}(t) &= \dot{\mathbf{R}}_t(0) \qquad \text{and} \qquad \mathbf{W} = -\mathbf{W}^{\mathrm{T}} \\ {}^1\overset{\circ}{\mathbf{S}} &= \frac{\partial}{\partial s}\left[\mathbf{R}_t^{\mathrm{T}}(s)\,\mathbf{S}(t)\,\mathbf{R}_t(s)\right]\Big|_{s=0} = \left[\dot{\mathbf{R}}_t^{\mathrm{T}}\mathbf{S}\mathbf{R}_t + \mathbf{R}_t^{\mathrm{T}}\dot{\mathbf{S}}\mathbf{R}_t + \mathbf{R}_t^{\mathrm{T}}\mathbf{S}\dot{\mathbf{R}}_t\right]\Big|_{s=0} \\ &= \dot{\mathbf{S}} + \mathbf{S}\mathbf{W} - \mathbf{W}\mathbf{S} \end{aligned} \tag{3.47}$$

As $\mathbf{S}^{\mathrm{R}}(\mathbf{X},t) = \mathbf{R}^{\mathrm{T}}(t)\,\mathbf{S}\mathbf{R}(t)$ and $\mathbf{U}_t^*(s) = \mathbf{R}^{\mathrm{T}}(t)\,\mathbf{U}_t(s)\,\mathbf{R}(t)$ the constitutive law must be satisfied if $\mathbf{S}^{\mathrm{R}}$ is replaced by $\mathbf{S}$ and $\mathbf{U}_t^*(s)$ by $\mathbf{U}_t(s)$. In other words: if the material satisfies PMOB and PMAS, then the argument can be replaced by its passive transformation for any arbitrary orthogonal $\mathbf{R}(t)$ (cp. (2.105)). The same is valid for the higher derivatives. Hence $\mathbf{d}$ in (3.44) is an isotropic function describing an isotropic material of the rate type, if

$$\begin{aligned} {}^q\overset{\circ}{\mathbf{S}} &= \mathbf{d}\left\{\mathbf{S}^1, \overset{\circ}{\mathbf{S}}=, \ldots, {}^{q-1}\overset{\circ}{\mathbf{S}}; \dot{\mathbf{U}}_t(0), \ddot{\mathbf{U}}_t(0), \ldots \overset{(r)}{\mathbf{U}}_t(0); \mathbf{V}(t)\right\} \\ &= \mathbf{d}\left\{\mathbf{S}, {}^1\overset{\circ}{\mathbf{S}}=, \ldots, {}^{q-1}\overset{\circ}{\mathbf{S}}; \mathbf{I}, \mathbf{D}, \overset{2}{\mathbf{D}}, \ldots, \overset{r}{\mathbf{D}}; \mathbf{V}\right\} \end{aligned} \tag{3.48}$$

at every time t. Fluids follow an equation that is formulated analogously as (3.48), but with $\rho(t)$ instead of $\mathbf{V}(t)$.

Investigating some special cases, it is obvious that for solids with $q = 0$ (cp. (3.34))

$$\mathbf{S} = \mathbf{d}\left\{\mathbf{D}, \overset{2}{\mathbf{D}}, \ldots, \overset{r}{\mathbf{D}}; \mathbf{V}\right\} \tag{3.49}$$

the rate–type material is identical with the differential–type material. For solids with $q = 1$ we obtain for a rate–type meterial of the complexity r

$${}^1\overset{\circ}{\mathbf{S}} = \dot{\mathbf{S}} + \mathbf{S}\mathbf{W} - \mathbf{W}\mathbf{S} = \mathbf{d}\left\{\mathbf{D}, \overset{2}{\mathbf{D}}, \ldots, \overset{r}{\mathbf{D}}; \mathbf{V}\right\} \tag{3.50}$$

and in particular for the complexity $r = 1$

$$\begin{aligned} {}^1\overset{\circ}{\mathbf{S}} &= \dot{\mathbf{S}} + \mathbf{S}\mathbf{W} - \mathbf{W}\mathbf{S} &&= \mathbf{d}\{\mathbf{S}, \mathbf{D}, \mathbf{V}\} \\ \dot{\mathbf{S}} &= \mathbf{d}\{\mathbf{S}, \mathbf{D}, \mathbf{W}, \mathbf{V}\} &&= \mathbf{d}\{\mathbf{S}, \mathbf{L}, \mathbf{V}\} \end{aligned} \tag{3.51}$$

As herewith stress $\mathbf{S}$ and stress rate $\overset{\circ}{\mathbf{S}}$ are related to deformation $\mathbf{V}$ and deformation rate $\mathbf{D}$ by a differential equation of first order, (3.51) describes basically *linear visco-elastic* materials of MAXWELL–like and KELVIN– like bodies.

The combinations of these material models result in higher parametric bodies like THOMPSON, LETHERSICH, and for $q = 2$ and $r = 2$

$$^{2}\overset{\circ}{\mathbf{S}} = \mathbf{d}\left\{\mathbf{S}, {}^{1}\overset{\circ}{\mathbf{S}}, \mathbf{D}, \overset{2}{\dot{\mathbf{D}}}\right\} \tag{3.52}$$

in a BURGER material, which is a suitable model to describe linear visco–elastic materials with a sufficient number (eight) of material parameters. These are adapted in a process of material–identification to the behaviour of the matter in reality. Initial (spontaneous) effects are reproducable as well as creep–, retardation– and relaxations– phenomenas.

The memory of this material class is a long–time memory with the special character of exponential functions. Those are monotonous functions that fulfil the restriction: the less time has passed since the motion event (deformation) has happened, the more influence the event to the presence has, and vice versa. The memory kernels in the solution of the differential equation (i.e. (3.48)) are LAPLACE–kernels with negative exponents in the exp–functions

$$h(s) = \exp\left[-\alpha(t-\tau)\right] = \exp(-\alpha s) \tag{3.53}$$

If (3.51) is linear in $\mathbf{D}$ and the isotropic function $\mathbf{d}$ is independent on $\mathbf{V}$ (i.e. simple fluid), the constitutive equation (3.51) is reducable to

$$^{1}\overset{\circ}{\mathbf{S}} = \mathbf{d}\{\mathbf{S}, \mathbf{D}\} \qquad \text{resp.} \qquad \dot{\mathbf{S}} = \mathbf{d}\{\mathbf{S}, \mathbf{L}\} \tag{3.54}$$

Those materials are called *hypo–elastic* materials.

The linearity in $\mathbf{D}$ and the application of CAYLEY–HAMILTON relation allows a form of (3.54) by a general representation–theorem (RIVLIN–ERICKSEN):

$$\begin{aligned} \mathbf{d}(\mathbf{S}, \mathbf{D}) \;=\; & \left[d_{00}\,\mathrm{tr}\,\mathbf{D} + d_{10}\,\mathrm{tr}(\mathbf{SD}) + d_{20}\,\mathrm{tr}\left(\mathbf{S}^2\mathbf{D}\right)\right]\mathbf{I} \\ & + \left[d_{01}\,\mathrm{tr}\,\mathbf{D} + d_{11}\,\mathrm{tr}(\mathbf{SD}) + d_{21}\,\mathrm{tr}\left(\mathbf{S}^2\mathbf{D}\right)\right]\mathbf{S} \\ & + \left[d_{02}\,\mathrm{tr}\,\mathbf{D} + d_{12}\,\mathrm{tr}(\mathbf{SD}) + d_{22}\,\mathrm{tr}\left(\mathbf{S}^2\mathbf{D}\right)\right]\mathbf{S}^2 \\ & + \left[d_{03}\mathbf{D}\right] + \left[d_{13}(\mathbf{SD} + \mathbf{DS})\right] + \left[d_{23}\left(\mathbf{DS}^2 + \mathbf{S}^2\mathbf{D}\right)\right] \end{aligned} \tag{3.55}$$

(3.54) and (3.55) are an access to plasticity (material with permanent memory) or by shifting the time scale $s \to f(s)$ to endochronical inelasticity [12].

3.11. Non–linear materials with memory (integral–type materials)

Let us resume the general constitutive equation for simple materials in the form (2.81)

$$\mathbf{S}^{\mathrm{R}}(\mathbf{X}, t) = \mathbf{f}\left[\mathbf{C}(\mathbf{X}, t)\right] + \mathcal{F}\left[\mathbf{G}^*(\mathbf{X}, s)\,\mathbf{C}(\mathbf{X}, t)\right] \tag{3.56}$$

If the functional in (3.56) is at least one integral over the time $0 < s < \infty$, separating the argument of the functional in a polynom of the tensor function $\mathbf{g}(s; \mathbf{C})$ and a polynom of the deformation history $\mathbf{G}^*(\mathbf{X}, s)$, thus

$$\mathcal{F}_{s=0}^{\infty}\left[\mathbf{G}^*(\mathbf{X}, s)\,\mathbf{C}(\mathbf{X}, t)\right] = \int_0^{\infty}\left[\mathbf{g}(s; \mathbf{C})\right]\left[\mathbf{G}^*(\mathbf{X}, s)\right]\mathrm{d}s \tag{3.57}$$

the simple material described by (3.57) is of the *integral–type*.

In application to simple *isotropic* media we restrict (3.57) together with the constitutive law (2.116) or (2.117) to

$$\mathbf{S}(\mathbf{X},t) = \mathcal{F}\left[\mathbf{C}_t(s)\,\mathbf{B}(t)\right] = \mathbf{f}\left[\mathbf{B}(t)\right] + \mathcal{F}\left[\left(\mathbf{G}(s)\,\mathbf{B}(t)\right)\right] \tag{3.58}$$

$$\mathcal{F}\left[\mathbf{G}(\mathbf{X},s)\,\mathbf{B}(\mathbf{X},t)\right] = \int_0^\infty \left[\mathbf{g}(s;\mathbf{B}(t))\right]\left[\mathbf{G}(\mathbf{X},s)\right]\mathrm{d}s \tag{3.59}$$

This time it is necessary that the integrand of (3.59) is an isotropic tensor function, hence

$$\mathbf{Q}\left[\mathbf{g}(s;\mathbf{B})\,\mathbf{G}\right]\mathbf{Q}^{\mathrm{T}} = \mathbf{g}\left(s;\mathbf{Q}\mathbf{B}\mathbf{Q}^{\mathrm{T}}\right)\left(\mathbf{Q}\mathbf{G}(\mathbf{X},s)\,\mathbf{Q}^{\mathrm{T}}\right) \tag{3.60}$$

For a simple fluid (tensor $\mathbf{B}$ replaced by scalar p) it comes from (3.60)

$$\mathbf{Q}\left[\mathbf{g}(s)\,\mathbf{G}\right]\mathbf{Q}^{\mathrm{T}} = \mathbf{g}(s)\left(\mathbf{Q}\mathbf{G}\mathbf{Q}^{\mathrm{T}}\right) \tag{3.61}$$

With the assumption of a first order polynom in $\mathbf{G}$, (3.57) is rewritten as

$$\begin{aligned}\mathbf{S}(\mathbf{X},s) &= \left\{\mathbf{f}\left[\mathbf{B}(t)\right] + \mathcal{F}\left[\mathbf{G}(s)\,\mathbf{B}(t)\right]\right\} = -p\mathbf{I} + \mathcal{F}\left[\mathbf{G}(s)\right]\\ &= -p\mathbf{I} + \int_0^\infty \mathbf{g}(s)\,\mathbf{G}(\mathbf{X},s)\,\mathrm{d}s\end{aligned} \tag{3.62}$$

For a simple, isotropic solid (3.58) is reduced by (3.59)

$$\begin{aligned}\mathbf{S}(\mathbf{X},s) &= \mathbf{f}\left[\mathbf{B}(t)\right] + \mathcal{F}\left[\mathbf{G}(s)\,\mathbf{B}(t)\right]\\ &= \mathbf{f}\left[\mathbf{B}(t)\right] + \int_0^\infty \left[\mathbf{g}(s;\mathbf{B})\right]\left[\mathbf{G}(\mathbf{X},s)\right]\mathrm{d}s\end{aligned} \tag{3.63}$$

If the solid is of first order (linear in $\mathbf{G}$), we get instead of (3.63)

$$\mathbf{S}(\mathbf{X},s) = \mathbf{f}\left[\mathbf{B}(t)\right] + \int_0^\infty \left[\mathbf{h}(s;\mathbf{B})\right]\left[\mathbf{G}(\mathbf{X},s)\right]\mathrm{d}s \tag{3.64}$$

(3.64) is interpretable as the response of an event $\mathbf{G}(\mathbf{X},s)$ at the past time $\tau = t - s$, during the time $\mathrm{d}s$, weighted (remembered, memorized) with the memory function $\mathbf{h}(s)$, taken at the time s that has passed between the event and the presence.

The function $\mathbf{h}(s)$ is the memory function as the kernel of the integral equation. It can be every monotonous function, i.e. like exp–function (LAPLACE–kernels, compare chapter above), LIOUVILLE–kernels, or ABEL–kernels as an expression of the appertaining *fading memory*. The extreme would be a permanent memory, that means $\mathbf{h}(s) = \mathrm{const}$ for all s. In Fig. 3.1 the memory function or the influence function $h(s) = \|\mathbf{h}(s)\|$ under the conditions

1. $h(s) > 0$, positive and real
2. $h(0) = 1$, normalized
3. $\lim (s \to \infty) [s^p h(s)] = 0$, monotonous

is sketched for different material classes and different kind of memories.

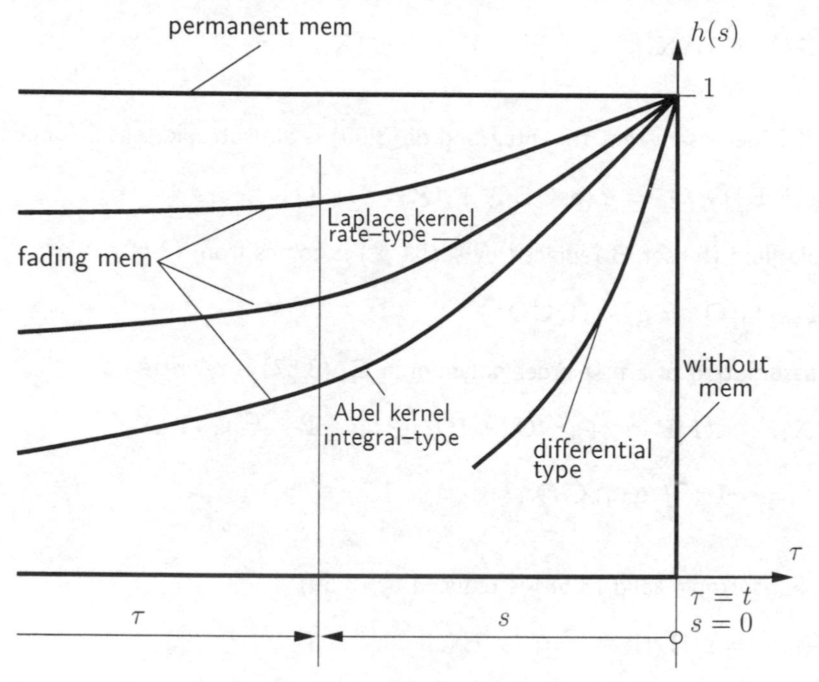

Figure 3.1: Memory (kernel) functions

As mentioned above, special cases would be

LAPLACE kernels : $h(s) = \exp(-\alpha s)$
ABEL kernels : $h(s) = 1/(\alpha s + 1)^\beta$

The linearity in $\mathbf{G}$ (3.64) could be regarded as a result of a *weak* (GATEAU–) derivative of the functional. That is a variation of the functional with respect to a scalar parameter in the argument. The result is always linear in the increment $\mathbf{G}$. In opposite, if the functional is differentiated to the argument $\mathbf{G}$ itself, it is a *strong* (FRECHET–) derivative. Using left GREEN tensors $\mathbf{G}^l = \mathbf{G}$ instead of left CAUCHY tensors $\mathbf{B} = 2\mathbf{G} + \mathbf{I}$ (see (1.20)), the functional in (3.63) becomes

$$\mathcal{F}[\mathbf{G}(s)\,\mathbf{B}(t)] = \mathcal{F}[\mathbf{G}(s)\,(2\mathbf{G}(t) + \mathbf{I})] = \mathcal{F}[\mathbf{G}(t)\,;\mathbf{G}(s)] \quad (3.65)$$

Defining the relative–difference GREEN deformation by

$$\mathbf{G}_\mathrm{d}(s) = \mathbf{G}(\tau) - \mathbf{G}(t) = \mathbf{G}(t - s) - \mathbf{G}(t) \quad (3.66)$$

with the particular value

$$\mathbf{G}_{\mathrm{d}}(0) = \mathbf{G}(t-0) - \mathbf{G}(t) = 0 \tag{3.67}$$

we rewrite (3.59) resp. (3.65)

$$\mathcal{F}[\mathbf{G}(s)\mathbf{B}(t)] = \mathcal{F}[\mathbf{G}(t);\mathbf{G}_{\mathrm{d}}(s)] \tag{3.68}$$

and we force the linearity in $\mathbf{G}_{\mathrm{d}}(s)$ of the argument in the functional by a weak derivative

$$\begin{aligned} \mathcal{F}[\mathbf{G}(t);\mathbf{G}_{\mathrm{d}}(s)] &= \mathcal{F}[\mathbf{G}_{\mathrm{d}}(s);0] + \frac{\partial}{\partial\lambda}[\mathcal{F}[\mathbf{G}(t);\mathbf{G}_{\mathrm{d}} + \lambda\mathbf{G}_{\mathrm{d}}(s)]]\big|_{\lambda=0} \\ &= \mathbf{g}[\mathbf{G}(t)] + \mathfrak{G}[\mathbf{G}(t);\mathbf{G}_{\mathrm{d}}(s)] \end{aligned} \tag{3.69}$$

with the essential difference that in (3.69) the functional is linear in $\mathbf{G}_{\mathrm{d}}(s)$. This is an equivalent form to (3.64) and the constitutive equation for simple, isotropic solids with memory of the integral–type becomes:

$$\mathbf{S}(\mathbf{X},t) = \mathbf{g}[\mathbf{G}(t)] + \int_0^\infty \underbrace{[\mathbf{g}(s,\mathbf{G}(t))][\mathbf{G}_{\mathrm{d}}(\mathbf{X},s)]}_{\mathbf{J}}\,\mathrm{d}s \tag{3.70}$$

As a consequence of its linearity in $\mathbf{G}_{\mathrm{d}}(s)$ the integrand $\mathbf{J}$ of (3.70) can be described by the RIVLIN–ERICKSEN representation theorem (cp. (3.55)).

$$\begin{aligned} \mathbf{J} &= \left(\alpha_0\mathbf{I} + \alpha_1\mathbf{G} + \alpha_2\mathbf{G}^2\right)\left(\beta_0\mathbf{I} + \beta_1\mathbf{G}_{\mathrm{d}}(s)\right) \\ &= \phi_0\mathbf{I} + \phi_1\mathbf{G} + \phi_2\mathbf{G}^2 + \phi_3\mathbf{G}_{\mathrm{d}}(s) + \phi_4[\mathbf{G}_{\mathrm{d}}(s)\mathbf{G} + \mathbf{G}\mathbf{G}_{\mathrm{d}}(s)] \\ &\quad + \phi_5\left[\mathbf{G}_{\mathrm{d}}(s)\mathbf{G}^2 + \mathbf{G}^2\mathbf{G}_{\mathrm{d}}(s)\right] \end{aligned} \tag{3.71}$$

$$\begin{aligned} \phi_0 &= \phi_{01}\,\mathrm{tr}[\mathbf{G}_{\mathrm{d}}(s)] + \phi_{02}\,\mathrm{tr}[\mathbf{G}_{\mathrm{d}}(s)\mathbf{G}(t)] + \phi_{03}\,\mathrm{tr}\left[\mathbf{G}_{\mathrm{d}}(s)\mathbf{G}^2(t)\right] \\ \phi_1 &= \phi_{11}\,\mathrm{tr}[\mathbf{G}_{\mathrm{d}}(s)] + \phi_{12}\,\mathrm{tr}[\mathbf{G}_{\mathrm{d}}(s)\mathbf{G}(t)] + \phi_{13}\,\mathrm{tr}\left[\mathbf{G}_{\mathrm{d}}(s)\mathbf{G}^2(t)\right] \\ \phi_2 &= \phi_{21}\,\mathrm{tr}[\mathbf{G}_{\mathrm{d}}(s)] + \phi_{22}\,\mathrm{tr}[\mathbf{G}_{\mathrm{d}}(s)\mathbf{G}(t)] + \phi_{23}\,\mathrm{tr}\left[\mathbf{G}_{\mathrm{d}}(s)\mathbf{G}^2(t)\right] \end{aligned} \tag{3.72}$$

With a tensor polynom reduced to second order for $\mathrm{g}[\mathbf{G}(t)]$ (see(3.11)), we get finally [19]

$$\begin{aligned} \mathbf{S} &= \psi_0\mathbf{I} + \psi_1\mathbf{G} + \psi_2\mathbf{G}^2 \\ &\quad + \int_{s=0}^{\infty} \{[\phi_{01}\,\mathrm{tr}(\mathbf{G}_{\mathrm{d}}(s)) + \phi_{02}\,\mathrm{tr}(\mathbf{G}_{\mathrm{d}}(s)\mathbf{G}(t)) + \phi_{03}\,\mathrm{tr}\left(\mathbf{G}_{\mathrm{d}}(s)\mathbf{G}^2(t)\right)]\,\mathbf{I} \\ &\quad + \left[\phi_{11}\,\mathrm{tr}(\mathbf{G}_{\mathrm{d}}(s)) + \phi_{12}\,\mathrm{tr}(\mathbf{G}_{\mathrm{d}}(s)\mathbf{G}(t)) + \phi_{13}\,\mathrm{tr}\left(\mathbf{G}_{\mathrm{d}}(s)\mathbf{G}^2(t)\right)\right]\mathbf{G} \\ &\quad + \left[\phi_{21}\,\mathrm{tr}(\mathbf{G}_{\mathrm{d}}(s)) + \phi_{22}\,\mathrm{tr}(\mathbf{G}_{\mathrm{d}}(s)\mathbf{G}(t)) + \phi_{23}\,\mathrm{tr}\left(\mathbf{G}_{\mathrm{d}}(s)\mathbf{G}^2(t)\right)\right]\mathbf{G}^2 \\ &\quad + \left[\phi_{31}\mathbf{G}_{\mathrm{d}}(s) + \phi_{32}(\mathbf{G}_{\mathrm{d}}(s)\mathbf{G} + \mathbf{G}\mathbf{G}_{\mathrm{d}}(t)) + \phi_{33}\left(\mathbf{G}_{\mathrm{d}}(s)\mathbf{G}^2 + \mathbf{G}^2\mathbf{G}_{\mathrm{d}}(s)\right)\right]\}\,\mathrm{d}s \end{aligned} \tag{3.73}$$

whereas $\phi_{01}, ..., \phi_{33}$ are 12 material functions for the memory part with $\phi_{ij} = \phi_{ij}(G_{\mathrm{I}}, G_{\mathrm{II}}, G_{\mathrm{III}}, s)$ and ψ_1, ψ_2, ψ_3 are 3 material functions for the spontaneous part with $\psi_i = \psi_i(G_{\mathrm{I}}, G_{\mathrm{II}}, G_{\mathrm{III}})$.

Using the order of magnitude ε (cp. (1.21)) again, we can classify all terms of (3.73) this way, and we obtain

$$\begin{array}{ll} \mathbf{G}_{\mathrm{d}}(s) = O(\varepsilon) & \mathbf{G}^2\mathbf{G}_{\mathrm{d}}(s) = O(\varepsilon^3) \\ \mathbf{G}_{\mathrm{d}}(s)\,\mathbf{G} = O(\varepsilon^2) & \mathrm{tr}\,\mathbf{G}, \mathrm{tr}\,\mathbf{G}_{\mathrm{d}} = O(\varepsilon) \\ \mathbf{G}^2 = O(\varepsilon^2) & \mathrm{tr}\,\mathbf{G}(\mathbf{G}\mathbf{G}_{\mathrm{d}}) = \mathrm{tr}(\mathbf{G}_{\mathrm{d}}\mathbf{G}) = O(\varepsilon^2) \\ \mathrm{tr}(\mathbf{G}_{\mathrm{d}}\mathbf{G}^2) = O(\varepsilon^3) & \end{array}$$

recognizing that the highest order $\mathbf{G}_{\mathrm{d}}(s)\,\mathbf{G}^2$ is of order $O(\varepsilon^3)$. Herewith reductions of lower order become possible. For instance, neglecting this highest order, we accomplish a *consistent theory of 2nd order*:

$$\begin{aligned} \mathbf{S}(\mathbf{X}, t) &= \psi_0\mathbf{I} + \psi_1\mathbf{G} + \psi_2\mathbf{G}^2 \\ &+ \int_0^\infty \{[\phi_{01}\mathrm{tr}(\mathbf{G}_{\mathrm{d}}) + \phi_{02}\mathrm{tr}(\mathbf{G}_{\mathrm{d}}\mathbf{G})]\,\mathbf{I} \\ &+ [\phi_{11}\mathrm{tr}(\mathbf{G}_{\mathrm{d}})]\,\mathbf{G} + [\phi_{31}\mathbf{G}_{\mathrm{d}} + \phi_{32}(\mathbf{G}_{\mathrm{d}}\mathbf{G} + \mathbf{G}\mathbf{G}_{\mathrm{d}})]\}\,\mathrm{d}s + O(\varepsilon^3) \end{aligned} \tag{3.74}$$

with 8 material functions $\psi_i = \psi_i(G_{\mathrm{I}}, G_{\mathrm{II}}, G_{\mathrm{III}})$; $\phi_{ij} = \phi_{ij}(G_{\mathrm{I}}, G_{\mathrm{II}}, G_{\mathrm{III}})$.

If even the terms of order $O(\varepsilon^2)$ are neglected, a reduced *consistent theory of 1st order* would be achieved. This could be i.e. physically linearized (see above) with the result

$$\mathbf{S}(\mathbf{X}, t) = \psi_0(G_{\mathrm{I}})\,\mathbf{I} + \psi_{10}\mathbf{G} + \int_0^\infty [\phi_{01}\mathrm{tr}\mathbf{G}_{\mathrm{d}}(s)\,\mathbf{I} + \phi_{31}\mathbf{G}_{\mathrm{d}}(s)]\,\mathrm{d}s + O(\varepsilon^2) \tag{3.75}$$

with 4 materials values $\psi_{10}, \phi_{01}, \phi_{31} = \mathrm{const}$, $\psi_0 = \psi_{00}\mathrm{tr}(\mathbf{G})$.

All 'classical' viscous and visco–elastic materials (MAXWELL, KELVIN) and even higher parametric visco–elastic materials, with a generalized memory behaviour and not only with LAPLACE–kernels as material functions ϕ are included.

If the memory part (the integral) is not considered at all, we come back to elastic materials and the result of (3.75) would be HOOKE's law again.

In order to reduce the number of material functions in (3.74) and/but to avoid the assumption of incompressible materials ($\det\mathbf{F} = 1$), which is difficult to prove or to measure in real bodies especially for large deformations, we formulate a constraint condition for *weak–compressible* matters:

$$\det\mathbf{F} = F_{\mathrm{III}} = 1 + \Delta \tag{3.76}$$

Since the relation holds

$$\det\mathbf{F} = F_{\mathrm{III}} = 1 + \mathrm{I_H} + \mathrm{II_H} + \mathrm{III_H}$$

the increment Δ in (3.76) can be identified as the first principal invariant $\mathrm{I_H}$, if the higher invariants are neglected in a theory of 2nd order. But as $\mathrm{I_H}$ is of order $O(\varepsilon)$ itself, it becomes clear that only the terms $\mathbf{G}, \mathbf{G}_{\mathrm{d}}(s), \mathrm{tr}\mathbf{G}, \mathrm{tr}\mathbf{G}_{\mathrm{d}}(s), \mathbf{G}^2$, and $\mathbf{G}\mathbf{G}_{\mathrm{d}}(s)$ accordingly a theory of magnitude $O(\varepsilon^2)$ are taken into account. From (3.74) follows

$$\begin{aligned} \mathbf{S}(\mathbf{X},t) &= \psi_0 \mathbf{I} + \psi_1 \mathbf{G}(t) + \psi_2 \mathbf{G}^2(t) \\ &+ \int_0^\infty \{\phi_{01}\,\mathrm{tr}(\mathbf{G}_{\mathrm{d}}(s))\,\mathbf{I} + \phi_{31}\mathbf{G}_{\mathrm{d}}(s) + \phi_{32}[\mathbf{G}_{\mathrm{d}}(s)\mathbf{G}(t) + \mathbf{G}(t)\mathbf{G}_{\mathrm{d}}(s)]\}\,\mathrm{d}s \end{aligned} \tag{3.77}$$

with only 6 material functions

$$\begin{aligned} &\phi_{01} = \phi_0; \quad \phi_{31} = \phi_1; \quad \phi_{32} = \phi_2 \\ &\psi_i = \psi_i(G_{\mathrm{I}}, G_{\mathrm{II}}, G_{\mathrm{III}}); \quad \phi_i = \phi_i(G_{\mathrm{I}}, G_{\mathrm{II}}, G_{\mathrm{III}}, s) \end{aligned}$$

If the integral–type material is *incompressible*, the ψ_0 and ϕ_0 are no material functions and within a consistent theory of order two (3.74) is simplified to [13]

$$\begin{aligned} \mathbf{S}(\mathbf{X},t) &= \psi_0 \mathbf{I} + \psi_1 \mathbf{G}(t) + \psi_2 \mathbf{G}^2(t) \\ &+ \int_0^\infty \{\phi_0\,\mathrm{tr}(\mathbf{G}_{\mathrm{d}}(s))\,\mathbf{I} + \phi_{31}\mathbf{G}_{\mathrm{d}}(s) + \phi_{32}[\mathbf{G}_{\mathrm{d}}(s)\mathbf{G}(t) + \mathbf{G}(t)\mathbf{G}_{\mathrm{d}}(s)]\}\,\mathrm{d}s \end{aligned} \tag{3.78}$$

where we find 4 material functions $\psi_i = \psi_i(G_{\mathrm{I}}, G_{\mathrm{II}})$; $\phi_i = \phi_i(G_{\mathrm{I}}, G_{\mathrm{II}}, s)$.

To separate the influence of time s and motion (deformation) $\mathbf{G}$, the material functions can be expanded like the functional in the neighbourhood of the non–motion (rest), thus

$$\begin{aligned} \phi_{ij}(G_{\mathrm{I}}, G_{\mathrm{II}}, G_{\mathrm{III}}, s) &= \phi_{ij}(0,0,0,s) + \left.\frac{\partial \phi_{ij}}{\partial G_{\mathrm{I}}}\right|_{0,s}(G_{\mathrm{I}} - 0) + \left.\frac{\partial \phi_{ij}}{\partial G_{\mathrm{II}}}\right|_{0,s}(G_{\mathrm{II}} - 0) \\ &+ \left.\frac{\partial \phi_{ij}}{\partial G_{\mathrm{III}}}\right|_{0,s}(G_{\mathrm{III}} - 0) + \frac{1}{2}\left\{\left.\frac{\partial^2 \phi_{ij}}{\partial G_{\mathrm{I}}^2}\right|_{0,s} G_{\mathrm{I}}^2 + \left.\frac{\partial^2 \phi_{ij}}{\partial G_{\mathrm{II}}^2}\right|_{0,s} G_{\mathrm{II}}^2 + \left.\frac{\partial^2 \phi_{ij}}{\partial G_{\mathrm{I}}\partial G_{\mathrm{II}}}\right|_{0,s} G_{\mathrm{I}}G_{\mathrm{II}} + \ldots\right\} \end{aligned} \tag{3.79}$$

As all ϕ are linked with at least $\mathrm{tr}\mathbf{G}_{\mathrm{d}}(s)$, in the sense of a *2nd order theory* the expansion is considered up to the first derivative only

$$\begin{aligned} \phi_{ij}(G_{\mathrm{I}}, G_{\mathrm{II}}, G_{\mathrm{III}}, s) &= \phi_{ij}(0,0,0,s) + \phi_{i1}G_{\mathrm{I}} + O(\varepsilon^2) \\ &= \phi_{ij}(0,0,0,s) + \phi_{i1}(s)(\mathrm{tr}\mathbf{G}) \end{aligned}$$

Consequently, the material functions become functions of s only. The number of them can be reduced to the number three, since the spontaneous part is always the initial value of the entire historical part (partial integration). These three functions are to determine from experiments on multi–axial testing machines (tension and/or torsion and internal pressure). So finally we obtain for *weak–compressible, integral–type materials of order 2*:

$$\begin{aligned} \mathbf{S}(\mathbf{X},t) &= \phi_0(0)\,\mathbf{I} + \phi_1(0)\,\mathbf{G} + \phi_2(0)\,\mathbf{G}^2 \\ &+ \int_{s=0}^\infty \{\phi_0(s)\,\mathrm{tr}\mathbf{G}_{\mathrm{d}}(s)\,\mathbf{I} + \phi_1(s)\,\mathbf{G}_{\mathrm{d}}(s) + \phi_2(s)[\mathbf{G}_{\mathrm{d}}(s)\mathbf{G} + \mathbf{G}\mathbf{G}_{\mathrm{d}}(s)]\}\,\mathrm{d}s \end{aligned} \tag{3.80}$$

with 3 material functions that depend on s only.

The constitutive equations can be applied in case of *slow motion* (slow deformation processes) with the replacement of $\mathbf{D}$ instead of $\mathbf{G}$ either. For geometrical linearized deformation $\mathbf{G}_{\mathrm{d}}(s) = \mathbf{G}(t-s) - \mathbf{G}(t)$ is to substitute by $\mathbf{E}_{\mathrm{d}}(s) = \mathbf{E}(t-s) - \mathbf{E}(t)$ and the squared term $\mathbf{E}_{\mathrm{d}}^2$ is consistently scratchable.

Retardation, relaxation and creep as well as all time–depending stress–strain effects are successfully described with these material equations e.g. in application to rubber and plastomers.

Often the problem is to find a sufficient number of experimental tests and data to determine the material functions in a process of multiaxial material identification. Testing machines which are invariant–controlled are not (yet) available. But by tension and/or torsion experiments at least the first and/or the second invariant can be controlled separately. The interaction of tension and torsion is an additional experiment, as the problem is non–linear and this is why the interaction is more than the sum of both tests separated.

4. NON–LINEAR DAMAGED MATERIALS

Analogue to continuum mechanics methods in order to postulate, generate and determine material equations for homogeneous and undamaged materials, it is possible to deduce those even for materials with voids and cracks in the sense of damaged matters.

We can find an access to this *damage mechanics* by introducing an additional, virtual or *fictitious* configuration (FCFG). This should have the property that it is the mapping of the real damaged body (ICFG) to a fictitious undamaged body in FCFG (Fig. 4.1), [16] under equivalence of geometry and deformation. The stresses are co–ordinated the same

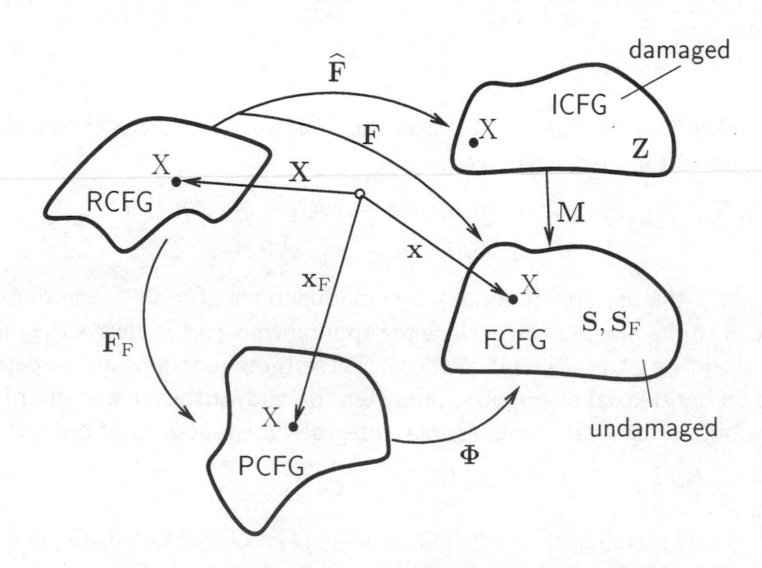

Figure 4.1: Configuration and their mappings

way, strictly speaking: the real stress in the damaged material $\mathbf{Z}(\mathrm{X},t)$ is adapted to the stress tensor $\mathbf{S}(\mathrm{X},t)$ in the virtual configuration, which is the stress tensor used above, satisfying the theorems CAUCHY I and CAUCHY II, thus for the *equilibrium stress*

$$\nabla \cdot \mathbf{S} + \mathbf{k} = \rho \ddot{\mathbf{u}} \qquad \text{and} \qquad \mathbf{S} = \mathbf{S}^{\mathrm{T}}$$

and where ρ is the mass density and $\mathbf{k}$ the volume force of the damaged body. The stress tensors $\mathbf{S}$ in FCFG and $\mathbf{Z}$ in ICFG are linked by a mean tensor $\mathbf{M} = \mathbf{M}^{\mathrm{T}}$ in a unique and bijective way. This tensor represents the homogenisation effect of the stress with the transition to FCFG.

$$\mathbf{Z} = \mathbf{M}^{-1}\mathbf{S} \qquad \text{resp.} \qquad \mathbf{S} = \mathbf{M}\mathbf{Z} \tag{4.1}$$

$\mathbf{Z}$ should exist at every X, unless X is damaged. Herewith follows as damage condition

$$\det \mathbf{M} = 0 \tag{4.2}$$

In order to formulate the material equation we define a symmetric stress tensor for the fictitious configuration (*effective stress*)

$$\mathbf{S}_F = \mathbf{Z}\mathbf{M}^{-\mathrm{T}} \tag{4.3}$$

Together with (4.1) it comes

$$\mathbf{S}_F = \mathbf{M}^{-1}\mathbf{S}\mathbf{M}^{-\mathrm{T}} \tag{4.4}$$

Since $\mathbf{Z}$, $\mathbf{S}$, and $\mathbf{S}_F$ are related by (4.3) and (4.4), one of them can be eliminated, e.g. the damage stress tensor $\mathbf{Z}$. The constitutive equation is writable as a relation of $\mathbf{S}_F$ and $\mathbf{M}$ now. Depending on the behaviour of the material, it could be elastic, plastic, or visco–elastic, we use one of the equation of the material classes above. I.e. for a rate–type material of $q = 1$ and $r = 1$, we get from (3.50) for the undamaged material [16]

$$\overset{\circ}{\mathbf{M}} = \mathbf{d}(\mathbf{S}_F, \mathbf{M}) \tag{4.5}$$

The motion–history from RCFG $\mathbf{X} = \mathbf{x}(\mathbf{X}, t_0)$ to FCFG $\mathbf{x}(\mathbf{X}, t)$ is now considered as a two–step history

1. from RCFG to the PCFG: $\mathbf{x}(\mathbf{X}, t_0) \to \mathbf{x}_F = \mathbf{x}_F(\mathbf{X}, t)$
2. from PCFG to the FCFG: $\mathbf{x}_F(\mathbf{X}, t) \to \mathbf{x} = \mathbf{x}(\mathbf{x}_F, t)$,

and we get as configuration gradients:

$$\mathbf{F}_F = \mathbf{x}_F\nabla \quad \text{and as before} \quad \mathbf{F} = \mathbf{x}\nabla \tag{4.6}$$

The total configuration gradient is the one–by–one mapping (chain–rule)

$$\mathbf{F} = \mathbf{x}\nabla = \partial\mathbf{x}/\partial\mathbf{X} = (\partial\mathbf{x}/\partial\mathbf{x}_F)(\partial\mathbf{x}_F/\partial\mathbf{X}) = \Phi\mathbf{F}^F \tag{4.7}$$

The damage process should occur during the second step, thus from PCFG to FCFG(ICFG), with the consequence that constitutive law for the *damaged* material depends on the configuration gradient between PCFG and FCFG, that is Φ. Hence, the objective left CAUCHY–deformation is here

$$\mathbf{B}_D = \Phi\Phi^{\mathrm{T}} = (\partial\mathbf{x}/\partial\mathbf{x}_F)\,(\partial\mathbf{x}/\partial\mathbf{x}_F)^{\mathrm{T}} \tag{4.8}$$

and the constitutive equation for the damaged part of the material follows analogue to (3.50) as

$$\overset{\circ}{\mathbf{B}}_D = \mathbf{d}\,(\mathbf{S}_F, \mathbf{B}_D) \tag{4.9}$$

As (4.9) is linear in $\mathbf{B}_D$, the matter law can be written with the representation theorem (3.55) for isotropic tensor functions

$$\begin{aligned} \mathbf{d}\,(\mathbf{S}_F, \mathbf{B}_D) = \; & [d_{00}\mathrm{tr}\mathbf{B}_D + d_{10}\,\mathrm{tr}\,(\mathbf{S}_F\mathbf{B}_D) + d_{20}\,\mathrm{tr}\,(\mathbf{S}_F^2\mathbf{B}_D)]\,\mathbf{I} \\ & + [d_{01}\mathrm{tr}\mathbf{B}_D + d_{11}\mathrm{tr}\,(\mathbf{S}_F\mathbf{B}_D) + d_{21}\mathrm{tr}\,(\mathbf{S}_F^2\mathbf{B}_D)]\,\mathbf{S}_F \\ & + [d_{02}\mathrm{tr}\mathbf{B}_D + d_{12}\mathrm{tr}\,(\mathbf{S}_F\mathbf{B}_D) + d_{22}\mathrm{tr}\,(\mathbf{S}_F^2\mathbf{B}_D)]\,\mathbf{S}_F^2 \\ & + [d_{03}\mathrm{tr}\mathbf{B}_D + d_{13}\mathrm{tr}\,(\mathbf{S}_F\mathbf{B}_D + \mathbf{B}_D\mathbf{S}_F) + d_{23}\mathrm{tr}\,(\mathbf{B}_D\mathbf{S}_F^2 + \mathbf{B}_D)] \end{aligned} \tag{4.10}$$

With the assumption that (4.10) is linear in $\mathbf{S}_F$ either, the equation is simplified again like in Chap. 3.11:

$$\begin{aligned} \mathbf{d}\,(\mathbf{S}_F, \mathbf{B}_D) = \; & [d_{00}\mathrm{tr}\mathbf{B}_D + d_{10}\mathrm{tr}\,(\mathbf{S}_F\mathbf{B}_D)]\,\mathbf{I} + [d_{01}\mathrm{tr}\mathbf{B}_D + d_{11}\mathrm{tr}\,(\mathbf{S}_F\mathbf{B}_D)]\,\mathbf{S}_F \\ & + [d_{03}\mathrm{tr}\mathbf{B}_D] + [d_{13}\mathrm{tr}\,(\mathbf{S}_F\mathbf{B}_D + \mathbf{B}_D\mathbf{S}_F)] \end{aligned} \tag{4.11}$$

With the stress tensors $\mathbf{S}$ and $\mathbf{S}_F$, with the mean tensor M, and the unknowns x and xF as the placements of X in the different configurations and the deformation tensor $\mathbf{B}_D$ we obtain an array of 42 unknowns. On the other hand we have 3 equations by CAUCHY I, 3 non–trivial equations from CAUCHY II, 9 material (4.5) for the undamaged, 9 material (4.9) for the damaged material, 9 displacement–deformation equations with (4.8), and 9 relations (4.4) between the two stress tensors, thus we have 42 equations. If the symmetry of all four tensors is considered at once, the array would have 30 unknowns with 30 relations only. By a simultaneous solution, i.e. in the order $\mathbf{S}$, $\mathbf{x}$, $\mathbf{x}_F$, $\mathbf{S}_F$, $\mathbf{B}_D$, $\overset{\circ}{\mathbf{B}}_D$, and $\mathbf{M}$ the damage process in materials with elastic or inelastic properties can be described in a similar way and by the same methods as in continuous media.

REFERENCES

1. Truesdell C. and Toupin R. : Classical Field Theories, in: Encyclopedia of Physics (Edited by W. Flügge), volume III/1, Springer Verlag, Berlin et al., 1960 .
2. Truesdell C. and Noll W. : Non–Linear Field Theories, in: Encyclopedia of Physics (Edited by W. Flügge), volume III/3, Springer Verlag, Berlin et al., 1965 .
3. Eringen C.A. : Mechanics of Continua, J. Wiley & Sons, Inc., New York et al., 1965.

4. Eringen C.A. : Nonlinear Theory of Continuous Media, McGraw-Hill Book Co., New York et al., 1962.
5. Leigh D.C. : Nonlinear Continuum Mechanics, McGraw-Hill Book Co., New York, 1968.
6. Truesdell C. : Rational Continuum Mechanics, volume 1, Academic Press, New York et al., 1977.
7. Truesdell C. and Bell J.F. : Mechanics of Solids, in: Encyclopedia of Physics (Edited by W. Flügge), volume VIa/1, Springer Verlag, Berlin et al., 1973 .
8. Trostel R. : Gedanken zur Konstruktion mechanischer Theorien I, II, Institut für Mechanik Techni߀che Universität Berlin, Forschungsberichte 1985/1988.
9. Trostel R. : Mathematische Grundlagen der Technischen Mechanik I, Vektor– und Tensor–Algebra, Vieweg Verlag, Braunschweig, 1993.
10. de Boer R. : Vektor– und Tensorrechnung für Ingenieure, Springer Verlag, Berlin et al., 1988.
11. Bertram A. : Axiomatische Einführung in die Kontinuumsmechanik, B.I. Wissenschaftsverlag, Mannheim et al., 1989.
12. Haupt P. : Viskoelastizität und Plastizität, Springer Verlag, Berlin et al., 1977.
13. Haupt P. : Viskoelastizität inkompressibler isotroper Stoffe, Diss., TU Berlin D 83, 1972.
14. Silber G. : Eine Systematik nichtlokaler kelvinhafter Fluide vom Grade drei auf der Basis eines klassischen Kontinuummodelles, volume Fortschrittberichte VDI, Reihe 18 / Nr. 26 VDI Verlag, Düsseldorf, TU Berlin, 1986.
15. Alexandru C. : Systematik nichtlokaler kelvinhafter Fluide vom Grade 2 auf der Basis eines COSSERAT Kontinuumsmodells, Fortschrittberichte VDI, Reihe 18 / Nr. 61, VDI Verlag, Düsseldorf, 1989.
16. Imiela C. : Ein Beitrag zur materialtheoretischen Interpretation der Schüdigungsmechanik, Diss., Institut für Mechanik TU Berlin, 1998.
17. Gummert P. : Reckling, K.-A. Mechanik VIEWEG Verlag, Braunschweig 1986/1989/1993.
18. Gummert P. : Materialgesetze des Kriechens und der Relaxation Zur Problematik des zeitabhüngigen Stoffverhaltens, Fortschrittberichte VDI, Reihe 38 / Nr. 7 VDI Verlag, Düsseldorf, 1987.
19. Gummert P. : A Constitutive Equation for Non–Linear Visco-Elastic Materials with Fading Memory, in: Proceedings of IUTAM–conference: 'Creep in Structures', Cracow 1990, Springer Verlag, New York et al., 1990 .
20. Altenbach J. and Altenbach H. : Einführung in die Kontinuumsmechanik, TEUBNER Studienbücher B.G. TEUBNER, Stuttgart, 1994.

CLASSICAL AND NON-CLASSICAL CREEP MODELS

H. Altenbach
Martin Luther University, Halle-Wittenberg, Germany

ABSTRACT

The following paper gives a short introduction into classical and non–classical creep models and their application to structural mechanics calculations. The analysis of creep processes is becoming more and more important in engineering practice. This development is connected with extended exploitation conditions and increasing safety standards. The quality of the predictions is influenced by the reliability of the material and structural models. In other words, it depends strongly on the possibilities to describe the creep problem, which should be analysed with the help of an adequate mathematical model.

In solid mechanics two types of equations are generally used - material independent and material dependent equations. The latter one should contain relations, which are able to reflect the individual response of the materials to external loadings. One goal in mechanics is to formulate suitable constitutive and, if necessary, evolution equations describing the material behaviour phenomenologically. Such an approach should be adopted for the material behaviour by an identification procedure which allows to find relations between the parameters in the equations and the experimentally determined characteristics of the material. The following considerations are related to these questions.

Classical creep equations are insensitive to the kind of loading. They describe, for example, identical behaviour under tension and compression (only the sign of the creep deformations is opposite). In addition, constitutive equations, which are able to reflect differences in the material behaviour with respect to the kind of loading, are discussed.

1. MOTIVATION

In engineering practice, an increasing number of different materials is used. Their real behaviour, especially their deformation and strength characteristics, can be described only approximately by classical material models. The extended application of light alloys, polymers, composites, ceramics, etc., requires an extension of the classical phenomenological constitutive equations and strength criteria. Certain materials show some effects in tests, which cannot be described by classical models. For example, different behaviour in tension and compression or the Poynting-Swift effect [1] (shear stresses result in axial strains) are obtained. In most cases, classical constitutive equations are based on potential formulations and equivalent stresses, which may be quadratic forms of the von Mises equivalent stress type and therefore insensitive to the sign of loading. So we can conclude that the von Mises-type theories cannot describe different behaviour in tension and compression. In addition, some effects cannot be modelled by tensorial linear equations, hence tensorial nonlinear constitutive equations must be formulated even in the case of small strains.

Until now no unique terminology is in use and such effects are called second order effects [2] or non–classical effects [3] in the literature. In all cases, a dependence of the material behaviour on the kind of loading can be established. It seems that it is necessary to connect the classification of the additional effects with experimental observations. Other definitions, partly given only by the type of the mathematical expressions (e.g., tensorial linear or nonlinear equations) describing the material behaviour, are not satisfying. For this purpose a proposal for a definition of the kind of material behaviour is presented in section 3. From the methodological point of view similar definitions are formulated for isotropic and anisotropic material behaviour. Additionally, the creep-damage coupling with respect to non–classical models is discussed.

The non–classical effects are connected with different types of material behaviour (elastic, plastic, creep) or limit states (strength, damage, fatigue). Here, we direct our attention to the creep behaviour and the main effects obtained in tests, which are the different behaviour in tension and compression, different equivalent stress-equivalent strain curves in tension and torsion, the dependency on the hydrostatic stress state, the compressibility and the Poynting-Swift effect. These effects are not only considered in the case of primary and secondary creep, but also in the case of creep-damage coupling, assuming isotropic damage with one scalar-valued damage variable. In the literature the creep-damage coupling behaviour is also named tertiary creep. By analogy we can discuss other models (anisotropic damage, more than one damage parameter or inner variables).

Starting with classical creep models a systematic extension of the classical models is presented. Here we introduce only models which correspond to the phenomenological approach. In the case of isotropic behaviour the classical creep law, the Norton's law, is based on the creep exponent and one additional parameter. Both are identified for the given material, loading and temperature conditions. The proposed generalised creep law is based on the creep exponent and six additional parameters. Special cases with a reduced number of parameters are deduced. All models are related to the assumptions of isothermal processes, quasi–static loadings and monotonic loading path. It is necessary that in the next years new theories for non-isothermal processes and cyclic loadings should be worked out, because there are many applications, e.g., in mechanical engineering.

After a short introduction into material modelling and the discussion of some experimental observations the main part is directed to classical and non–classical creep models without or with damage. All models are proposed for isotropic and anisotropic behaviour. Examples of applications are reported briefly. For a better understanding, the following section contains a short introduction into tensor calculus with Cartesian tensors.

2. SOME IMPORTANT FORMULAE IN TENSOR ALGEBRA AND ANALYSIS

The contents of this section is related to some rules of calculations with tensors in the three-dimensional Euclidean space. Restricting the derivations to the case of orthogonal Cartesian coordinates the direct (symbolic) and the component notation of tensor quantities are used. The direct notation is independent from the choice of the coordinate system. More details about the tensor calculus can be taken from [4, 5].

2.1. Scalars, vectors and tensors

In mechanics several tensorial variables of different rank are used. Examples are:

- the density ρ, the temperature T, the energy W, ...- scalars (zeroth rank tensors),
- the radius vector $\boldsymbol{r}$, the displacement vector $\boldsymbol{u}$, ...- vectors (first rank tensors),
- the stress tensor $\boldsymbol{\sigma}$, the deformation tensor $\boldsymbol{\varepsilon}$, ...- dyads (second rank tensors),
- the Hookean tensor ${}^{(4)}\boldsymbol{E}$, the material tensor ${}^{(4)}\boldsymbol{b}$, ...- fourth rank tensors and
- the material tensor ${}^{(6)}\boldsymbol{c}$ - sixth rank tensor.

Scalars are variables, which are fully independent on the choice of coordinate system (invariant variables) because they have no orientation. Vectors can be written as

$$\boldsymbol{a} = a_1\boldsymbol{e}_1 + a_2\boldsymbol{e}_2 + a_3\boldsymbol{e}_3, \qquad \boldsymbol{a} = (a_1, a_2, a_3), \qquad \boldsymbol{a} = (a_i),\ i = 1, 2, 3. \tag{2.1}$$

The a_i are the coordinates of the vector, which are related to the vector basis $\boldsymbol{e}_i$ with respect to the given coordinate system. This vector basis is assumed to be an orthonormal basis

$$|\boldsymbol{e}_i| = 1, \quad \boldsymbol{e}_i \cdot \boldsymbol{e}_j = \begin{cases} 1 & i = j \\ 0 & i \neq j \end{cases}. \tag{2.2}$$

For shorter writing we introduce the Einstein's summation convention

$$\boldsymbol{a} = a_1\boldsymbol{e}_1 + a_2\boldsymbol{e}_2 + a_3\boldsymbol{e}_3 = \sum_{i=1}^{3} a_i\boldsymbol{e}_i = a_i\boldsymbol{e}_i, \tag{2.3}$$

i is a dummy index.

The scalar product (inner product, dot product) of two vectors $\boldsymbol{a}$ and $\boldsymbol{b}$ is defined as

$$\boldsymbol{a}\cdot\boldsymbol{b} = a_i\boldsymbol{e}_i\cdot b_j\boldsymbol{e}_j = a_ib_j\boldsymbol{e}_i\cdot\boldsymbol{e}_j = a_ib_j\delta_{ij} = a_ib_i = \alpha; \quad \delta_{ij} = \begin{cases} 1 & i=j \\ 0 & i\neq j \end{cases}. \tag{2.4}$$

The second product is the dyadic product of two vectors

$$\boldsymbol{ab} = a_i\boldsymbol{e}_ib_j\boldsymbol{e}_j = a_ib_j\boldsymbol{e}_i\boldsymbol{e}_j = T_{ij}\boldsymbol{e}_i\boldsymbol{e}_j = \boldsymbol{T}. \tag{2.5}$$

In some textbooks for this product the following designation is used

$$\boldsymbol{a}\otimes\boldsymbol{b} \equiv \boldsymbol{ab}. \tag{2.6}$$

With the help of the dyadic product the second rank tensor $\boldsymbol{T}$ can be introduced

$$\boldsymbol{T} = \boldsymbol{ab} = a_ib_j\boldsymbol{e}_i\boldsymbol{e}_j = T_{ij}\boldsymbol{e}_i\boldsymbol{e}_j. \tag{2.7}$$

In a similar way we represent tensors of higher ranks:

- the 4th rank tensor

$$^{(4)}\boldsymbol{A} = \boldsymbol{abcd} = a_ib_jc_kd_l\boldsymbol{e}_i\boldsymbol{e}_j\boldsymbol{e}_k\boldsymbol{e}_l = \boldsymbol{TS} = T_{ij}S_{kl}\boldsymbol{e}_i\boldsymbol{e}_j\boldsymbol{e}_k\boldsymbol{e}_l = A_{ijkl}\boldsymbol{e}_i\boldsymbol{e}_j\boldsymbol{e}_k\boldsymbol{e}_l \tag{2.8}$$

- and the 6th rank tensor

$$^{(6)}\boldsymbol{B} = \boldsymbol{abcdgf} = \boldsymbol{TSP} = T_{ij}S_{kl}P_{mn}\boldsymbol{e}_i\boldsymbol{e}_j\boldsymbol{e}_k\boldsymbol{e}_l\boldsymbol{e}_m\boldsymbol{e}_n = B_{ijklmn}\boldsymbol{e}_i\boldsymbol{e}_j\boldsymbol{e}_k\boldsymbol{e}_l\boldsymbol{e}_m\boldsymbol{e}_n. \tag{2.9}$$

For the second rank tensors $\boldsymbol{T}$ and $\boldsymbol{S}$ we define the following products:

- the contraction or tensor product (scalar product)

$$\boldsymbol{T}\cdot\boldsymbol{S} = T_{ij}\boldsymbol{e}_i\boldsymbol{e}_j\cdot S_{kl}\boldsymbol{e}_k\boldsymbol{e}_l = T_{ij}S_{kl}\boldsymbol{e}_i\delta_{jk}\boldsymbol{e}_l = T_{ij}S_{jl}\boldsymbol{e}_i\boldsymbol{e}_l = M_{il}\boldsymbol{e}_i\boldsymbol{e}_l, \tag{2.10}$$

  which leads to a second rank tensor,

- the double contraction (double scalar product)

$$\boldsymbol{T}\cdot\cdot\,\boldsymbol{S} = T_{ij}\boldsymbol{e}_i\boldsymbol{e}_j\cdot\cdot\, S_{kl}\boldsymbol{e}_k\boldsymbol{e}_l = T_{ij}S_{kl}\delta_{jk}\delta_{il} = T_{ij}S_{ji} = \alpha, \tag{2.11}$$

  resulting in a scalar, and

- the dyadic product

$$\boldsymbol{TS} = T_{ij}\boldsymbol{e}_i\boldsymbol{e}_jS_{kl}\boldsymbol{e}_k\boldsymbol{e}_l = T_{ij}S_{kl}\boldsymbol{e}_i\boldsymbol{e}_j\boldsymbol{e}_k\boldsymbol{e}_l = A_{ijkl}\boldsymbol{e}_i\boldsymbol{e}_j\boldsymbol{e}_k\boldsymbol{e}_l, \tag{2.12}$$

  which leads to a fourth rank tensor.

2.2. Special second rank tensors

The following special tensors can be introduced:

- the unit tensor (identity tensor) $\boldsymbol{I}$

$$\boldsymbol{I} = \delta_{ij}\boldsymbol{e}_i\boldsymbol{e}_j = \boldsymbol{e}_1\boldsymbol{e}_1 + \boldsymbol{e}_2\boldsymbol{e}_2 + \boldsymbol{e}_3\boldsymbol{e}_3, \ \boldsymbol{e}_i\cdot\boldsymbol{I}\cdot\boldsymbol{e}_j = \delta_{ij}, \tag{2.13}$$

- the transposed tensor $\boldsymbol{T}^{\mathsf{T}}$

$$\boldsymbol{T} = \boldsymbol{ab} \Longrightarrow \boldsymbol{T}^{\mathsf{T}} = \boldsymbol{ba}, \quad \boldsymbol{T} = T_{ij}\boldsymbol{e}_i\boldsymbol{e}_j \Longrightarrow \boldsymbol{T}^{\mathsf{T}} = T_{ij}\boldsymbol{e}_j\boldsymbol{e}_i = T_{ji}\boldsymbol{e}_i\boldsymbol{e}_j, \tag{2.14}$$

- the symmetric and the antisymmetric (skew) tensors
  - the symmetric tensor $\boldsymbol{T}^{\mathsf{S}}$
$$\boldsymbol{T} = \boldsymbol{T}^{\mathsf{T}}, \quad T_{ij} = T_{ji} \Longrightarrow \boldsymbol{T} = \boldsymbol{T}^{\mathsf{S}} \tag{2.15}$$
  - and the antisymmetric tensor $\boldsymbol{T}^{\mathsf{A}}$
$$\boldsymbol{T} = -\boldsymbol{T}^{\mathsf{T}}, \quad T_{ij} = -T_{ji} \Longrightarrow \boldsymbol{T} = \boldsymbol{T}^{\mathsf{A}}, \tag{2.16}$$
- the trace of a tensor
$$\mathrm{tr}\boldsymbol{T} = \boldsymbol{I} \cdot\cdot \boldsymbol{T} = \delta_{kl}\boldsymbol{e}_k\boldsymbol{e}_l \cdot\cdot T_{ij}\boldsymbol{e}_i\boldsymbol{e}_j = T_{ii} = T_{11} + T_{22} + T_{33}. \tag{2.17}$$
- the spherical part $\boldsymbol{T}^{\mathsf{K}}$ of a tensor $\boldsymbol{T}$ and the deviator $\boldsymbol{T}^{\mathsf{D}}$
A second rank tensor can be represented by a spherical part
$$\boldsymbol{T}^{\mathsf{K}} = \frac{1}{3}(\boldsymbol{I} \cdot\cdot \boldsymbol{T})\boldsymbol{I} = \frac{1}{3}T_{kk}\delta_{ij}\boldsymbol{e}_i\boldsymbol{e}_j \tag{2.18}$$
and a deviator
$$\boldsymbol{T}^{\mathsf{D}} = \boldsymbol{T} - \boldsymbol{T}^{\mathsf{K}} = \boldsymbol{T} - \frac{1}{3}(\boldsymbol{I} \cdot\cdot \boldsymbol{T})\boldsymbol{I} = (T_{ij} - \frac{1}{3}T_{kk}\delta_{ij})\boldsymbol{e}_i\boldsymbol{e}_j \tag{2.19}$$
in the following unique way
$$\boldsymbol{T} = \boldsymbol{T}^{\mathsf{K}} + \boldsymbol{T}^{\mathsf{D}}, \tag{2.20}$$
or
$$T_{ij}\boldsymbol{e}_i\boldsymbol{e}_j = \frac{1}{3}T_{kk}\delta_{ij}\boldsymbol{e}_i\boldsymbol{e}_j + (T_{ij} - \frac{1}{3}T_{kk}\delta_{ij})\boldsymbol{e}_i\boldsymbol{e}_j, \tag{2.21}$$

2.3. Invariants of a second rank tensor

Invariant terms are independent on the choice of the coordinate system. Such a system of invariants can be related to the coefficients of the characteristic equation
$$\det(\boldsymbol{T} - \lambda\boldsymbol{I}) = \lambda^3 - J_1(\boldsymbol{T})\lambda^2 + J_2(\boldsymbol{T})\lambda - J_3(\boldsymbol{T}) = 0. \tag{2.22}$$
The J_i are called principal invariants. So we introduce

- the linear principal invariant
$$J_1(\boldsymbol{T}) = \mathrm{tr}\boldsymbol{T} \equiv \boldsymbol{T} \cdot\cdot \boldsymbol{I} \equiv T_{ii}, \tag{2.23}$$
- the quadratic principal invariant
$$J_2(\boldsymbol{T}) = \frac{1}{2}\left[J_1^2(\boldsymbol{T}) - J_1(\boldsymbol{T}^2)\right] = \frac{1}{2}(T_{ii}T_{jj} - T_{ij}T_{ji}), \tag{2.24}$$
- and the cubic principal invariant
$$\begin{aligned} J_3(\boldsymbol{T}) &= \frac{1}{3}\left[J_1(\boldsymbol{T}^3) + 3J_1(\boldsymbol{T})J_2(\boldsymbol{T}) - J_1^3(\boldsymbol{T})\right] \\ &= \frac{1}{3}J_1(\boldsymbol{T}^3) - \frac{1}{2}J_1(\boldsymbol{T}^2)J_1(\boldsymbol{T}) + \frac{1}{6}J_1^3(\boldsymbol{T}) \\ &= \det(T_{ij}). \end{aligned} \tag{2.25}$$

Other systems of invariants can be developed, e.g.:

- the basic invariants

$$\begin{array}{llcll} \text{linear basic invariant} & I_1(\boldsymbol{T}) & = & T_{ii}, & I_1(\boldsymbol{T}) = \boldsymbol{T} \cdot\cdot \boldsymbol{I}, \\ \text{quadratic basic invariant} & I_2(\boldsymbol{T}) & = & T_{ij}T_{ji}, & I_2(\boldsymbol{T}) = \boldsymbol{T} \cdot\cdot \boldsymbol{T}, \\ \text{cubic basic invariant} & I_3(\boldsymbol{T}) & = & T_{ij}T_{jk}T_{ki}, & I_3(\boldsymbol{T}) = (\boldsymbol{T} \cdot \boldsymbol{T}) \cdot\cdot \boldsymbol{T}, \end{array} \tag{2.26}$$

- or the modified basic invariants

$$\tilde{I}_1(\boldsymbol{T}) = \mathrm{tr}(\boldsymbol{T}),\ \tilde{I}_2(\boldsymbol{T}) = \frac{1}{2}\mathrm{tr}(\boldsymbol{T}^2),\ \tilde{I}_3(\boldsymbol{T}) = \frac{1}{3}\mathrm{tr}(\boldsymbol{T}^3). \tag{2.27}$$

The following relations between the principal and the basic invariants are existing

$$\begin{array}{lcllcl} J_1 & = & I_1, & I_1 & = & J_1, \\ J_2 & = & \dfrac{1}{2}(I_1^2 - I_2), & I_2 & = & J_1^2 - 2J_2, \\ J_3 & = & \dfrac{1}{3}I_3 - \dfrac{1}{2}I_1I_2 + \dfrac{1}{6}I_1^3, & I_3 & = & 3J_3 - 3J_1J_2 + J_1^3. \end{array} \tag{2.28}$$

For the deviator we can derive the principal and the basic invariants by analogy:

- principal invariants of the deviator

$$J_1(\boldsymbol{T}^{\mathsf{D}}) = I_1(\boldsymbol{T}^{\mathsf{D}}) = 0,\quad J_2(\boldsymbol{T}^{\mathsf{D}}) = -\frac{1}{2}I_2(\boldsymbol{T}^{\mathsf{D}}),\quad J_3(\boldsymbol{T}^{\mathsf{D}}) = \frac{1}{3}I_3(\boldsymbol{T}^{\mathsf{D}}) = \det\boldsymbol{T}^{\mathsf{D}}, \tag{2.29}$$

- and basic invariants

$$\begin{array}{lclcl} I_1(\boldsymbol{T}^{\mathsf{D}}) & = & T_{ii}^{\mathsf{D}} & = & \boldsymbol{T}^{\mathsf{D}} \cdot\cdot \boldsymbol{I} = 0, \\ I_2(\boldsymbol{T}^{\mathsf{D}}) & = & T_{ij}^{\mathsf{D}}T_{ji}^{\mathsf{D}} & = & \boldsymbol{T}^{\mathsf{D}} \cdot\cdot \boldsymbol{T}^{\mathsf{D}}, \\ I_3(\boldsymbol{T}^{\mathsf{D}}) & = & T_{ij}^{\mathsf{D}}T_{jk}^{\mathsf{D}}T_{ki}^{\mathsf{D}} & = & (\boldsymbol{T}^{\mathsf{D}} \cdot \boldsymbol{T}^{\mathsf{D}}) \cdot\cdot \boldsymbol{T}^{\mathsf{D}}. \end{array} \tag{2.30}$$

The following relations between the invariants can be deduced

$$\begin{array}{lclcl} I_1(\boldsymbol{T}^{\mathsf{K}}) & = & J_1(\boldsymbol{T}^{\mathsf{K}}) & \equiv & J_1(\boldsymbol{T}), \\ I_2(\boldsymbol{T}^{\mathsf{D}}) & = & -2J_2(\boldsymbol{T}^{\mathsf{D}}) & = & -2J_2(\boldsymbol{T}) + \dfrac{2}{3}J_1^2(\boldsymbol{T}), \\ I_3(\boldsymbol{T}^{\mathsf{D}}) & = & 3J_3(\boldsymbol{T}^{\mathsf{D}}) & = & 3J_3(\boldsymbol{T}) - J_1(\boldsymbol{T})J_2(\boldsymbol{T}) + \dfrac{2}{9}J_1^3(\boldsymbol{T}). \end{array} \tag{2.31}$$

2.4. Eigenvalue problem for a second rank tensor

The eigenvalues λ and the eigendirections $\boldsymbol{n}$ for a second rank tensor $\boldsymbol{T}$ can be obtained from the solution of the following equations

$$(\boldsymbol{T} - \lambda\boldsymbol{I}) \cdot \boldsymbol{n} = \boldsymbol{0},\ \boldsymbol{n} \cdot \boldsymbol{n} = 1,\quad (T_{ij} - \lambda\delta_{ij})n_j = 0,\ n_jn_j = 1. \tag{2.32}$$

The eigenvalues follow from the condition that nontrivial solutions are existing, which leads to the characteristic equation

$$\det(\boldsymbol{T} - \lambda\boldsymbol{I}) = 0,\ \det(T_{ij} - \lambda\delta_{ij}) = 0. \tag{2.33}$$

The roots of this equation $\lambda_{(\alpha)}, \alpha = I, II, III$ are called principal values. It can be shown that in the case of symmetric second rank tensors all principle values are real [5]. For each root we get the eigendirections (principal directions) $n_j^{(\alpha)}, \alpha = I, II, III$ from the system

$$\begin{array}{rcrcrcl} (T_{11} - \lambda)n_1 & + & T_{12}n_2 & + & T_{13}n_3 & = & 0, \\ T_{21}n_1 & + & (T_{22} - \lambda)n_2 & + & T_{23}n_3 & = & 0, \\ T_{31}n_1 & + & T_{32}n_2 & + & (T_{33} - \lambda)n_3 & = & 0, \\ n_1^2 & + & n_2^2 & + & n_3^2 & = & 1. \end{array} \qquad (2.34)$$

The second rank tensor satisfies his characteristic equation (Cayley-Hamilton theorem)

$$\boldsymbol{T}^3 - J_1(\boldsymbol{T})\boldsymbol{T}^2 + J_2(\boldsymbol{T})\boldsymbol{T} - J_3(\boldsymbol{T})\boldsymbol{I} = 0, \qquad (2.35)$$

which enables the representation of $\boldsymbol{T}^n$ $(n \geq 3)$ as a linear function of $\boldsymbol{T}^2, \boldsymbol{T}, \boldsymbol{T}^0 = \boldsymbol{I}$, e.g.,

$$\boldsymbol{T}^3 = J_1(\boldsymbol{T})\boldsymbol{T}^2 - J_2(\boldsymbol{T})\boldsymbol{T} + J_3(\boldsymbol{T})\boldsymbol{I}. \qquad (2.36)$$

2.5. Transformation rules for tensors

The rules of transformation from one coordinate system to a rotated system for tensors of the rank 2, 4 or 6 are (all indices range from 1 to 3)

$$\begin{array}{rcl} a'_{ij} & = & \alpha_{mi}\alpha_{nj}a_{mn}, \\ b'_{ijkl} & = & \alpha_{mi}\alpha_{nj}\alpha_{sk}\alpha_{tl}b_{mnst}, \\ c'_{ijklop} & = & \alpha_{mi}\alpha_{nj}\alpha_{sk}\alpha_{tl}\alpha_{uo}\alpha_{vp}c_{mnstuv}. \end{array} \qquad (2.37)$$

The α_{ij} are the elements of the transformation matrix:

$$\alpha_{ij} = \cos(\boldsymbol{e}_i{}', \boldsymbol{e}_j). \qquad (2.38)$$

2.6. Functions of a tensor argument

We can introduce the following linear functions

$$\begin{array}{rcll} \psi & = & \boldsymbol{B} \cdot\cdot \boldsymbol{D} & \text{linear scalar function}, \\ \boldsymbol{c} & = & {}^{(3)}\boldsymbol{B} \cdot\cdot \boldsymbol{D} & \text{linear vector function}, \\ \boldsymbol{P} & = & {}^{(4)}\boldsymbol{B} \cdot\cdot \boldsymbol{D} & \text{linear tensorial function}. \end{array} \qquad (2.39)$$

A linear scalar function can also be represented by a quadratic form of a second rank tensor

$$\psi[\boldsymbol{P}(\boldsymbol{D})] = \psi({}^{(4)}\boldsymbol{B} \cdot\cdot \boldsymbol{D}) = ({}^{(4)}\boldsymbol{B} \cdot\cdot \boldsymbol{D}) \cdot\cdot \boldsymbol{D} = B_{klmn}D_{nm}D_{lk}. \qquad (2.40)$$

If $B_{klmn} = B_{mnkl}$ and the symmetry conditions

$$\boldsymbol{D} = \boldsymbol{D}^{\mathsf{T}} \Longrightarrow P_{st} = B_{stmn}D_{nm}, \quad B_{klmn} = B_{klnm} \qquad (2.41)$$

and

$$\boldsymbol{P} = \boldsymbol{P}^{\mathsf{T}} \Longrightarrow P_{st} = P_{ts} = B_{stmn}D_{nm}, \quad B_{stmn} = B_{tsmn} \qquad (2.42)$$

hold true, the number of linear independent coordinates of a fourth rank tensor can be reduced from 81 to 21 coordinates.

2.7. Derivatives of the invariants of a second rank tensor

A scalar-valued function of a second rank tensor can be represented by

$$\psi = \psi(\boldsymbol{D}) = \psi(D_{11}, D_{22}, \dots, D_{31}). \tag{2.43}$$

Then we can calculate the derivative by the following equation

$$\psi_{,\boldsymbol{D}} = \frac{\partial \psi}{\partial \boldsymbol{D}} = \frac{\partial \psi}{\partial D_{kl}} \boldsymbol{e}_k \boldsymbol{e}_l. \tag{2.44}$$

On the other hand the derivatives of the invariants are

$$\begin{aligned} J_1(\boldsymbol{D})_{,\boldsymbol{D}} &= \boldsymbol{I}, \quad J_1(\boldsymbol{D}^2)_{,\boldsymbol{D}} = 2\boldsymbol{D}^{\mathsf{T}}, \quad J_1(\boldsymbol{D}^3)_{,\boldsymbol{D}} = 3\boldsymbol{D}^{2^{\mathsf{T}}}, \\ J_2(\boldsymbol{D})_{,\boldsymbol{D}} &= J_1(\boldsymbol{D})\boldsymbol{I} - \boldsymbol{D}^{\mathsf{T}}, \\ J_3(\boldsymbol{D})_{,\boldsymbol{D}} &= \boldsymbol{D}^{2^{\mathsf{T}}} - J_1(\boldsymbol{D})\boldsymbol{D}^{\mathsf{T}} + J_2(\boldsymbol{D})\boldsymbol{I} = J_3(\boldsymbol{D})(\boldsymbol{D}^{\mathsf{T}})^{-1}. \end{aligned} \tag{2.45}$$

So, we finally get

$$\psi[J_1, J_2, J_3]_{,\boldsymbol{D}} = \left(\frac{\partial \psi}{\partial J_1} + J_1 \frac{\partial \psi}{\partial J_2} + J_2 \frac{\partial \psi}{\partial J_3} \right) \boldsymbol{I} - \left(\frac{\partial \psi}{\partial J_2} + J_1 \frac{\partial \psi}{\partial J_3} \right) \boldsymbol{D}^{\mathsf{T}} + \frac{\partial \psi}{\partial J_3} \boldsymbol{D}^{2^{\mathsf{T}}}. \tag{2.46}$$

These calculations can be helpful for the use of the representation theorem of an isotropic function [6]

$$\boldsymbol{P} = \boldsymbol{F}(\boldsymbol{A}) = \nu_0 \boldsymbol{I} + \nu_1 \boldsymbol{A} + \nu_2 \boldsymbol{A}^2. \tag{2.47}$$

The coefficients ν_i itself are functions of the invariants

$$\nu_i = \nu_i[J_1(\boldsymbol{A}), J_2(\boldsymbol{A}), J_3(\boldsymbol{A})]. \tag{2.48}$$

3. INTRODUCTION INTO MATERIAL BEHAVIOUR MODELLING

Experimental observations are important for understanding the behaviour of different materials. Restricting our discussion to the case of phenomenological models, we only need information from macroscopic tests (standard tests in material testing). During these tests, loads and changes of the geometry are measured. On the other hand, the constitutive equations are formulated in terms of stresses and strains. Therefore, these stresses and strains have to be calculated from the acting forces or momentums and the elongations etc. of the specimen. This calculation is correct, presuming the existence of a homogeneous stress-strain state in the specimen. It is difficult to realise such states, but in many standard tests this assumption is approximately fulfilled.

Approximations in material testing are connected with the classification of the observed effects depending on their significance. At first, only the main effects, neglecting secondary influences, should be investigated. Thus we can introduce a first approximation of the material behaviour, which leads to simple constitutive equations with only a few parameters. For the identification of these parameters different standard tests in mechanical material testing can be used. Such tests are the uniaxial tension test, from which the axial strain and the transverse strain (transverse contraction) can be derived, or the torsion test, from which we get the shear strains. Both tests can be performed without difficulties and we can find a lot of test data in the literature. More difficulties are connected with other standard tests (e.g., uniaxial compression). Their realization demands high accuracy of the measurements and/or special equipment.

In addition, for many materials we obtain the same stress-strain curves in tension and compression tests (only the signs are opposite), but for some materials the behaviour is different. This leads to the conclusion that test data should be analysed and treated carefully.

3.1. Experimental motivation

Every engineering analysis and calculation is based on models, which can, in general, be described by some mathematical expressions (equations). With respect to the mechanics we can distinguish between two main groups:

- the material independent equations and
- the material dependent equations.

The first group reflects the geometrical relations (the strain-displacement equations) and the equilibrium conditions (static or dynamic). If we count the number of equations and the number of independent mechanical variables, we can see that the system of the material independent equations is undetermined (we have more variables than equations). It is thus necessary to introduce additional information to complete the system of governing equations. From our experience we know, for example, that a beam made of steel or made of rubber shows different deflections under the same transverse load, if the geometrical properties are identical: we obtain individual responses depending on the material used. To find an adequate mathematical description for the individual response (material behaviour) is the main topic in material modelling. These models must satisfy mathematical restrictions and reflect physics and materials science knowledge, but they have to be simple enough for engineering applications.

The development of material behaviour models is connected with different experimental observations. The first models start from macroscopic observations and are formulated basing on pure empiricism. After some developments in mathematics and mechanics, mathematically and mechanically based models on the phenomenological level were derived. In the last years we also got new proposals for models of the material behaviour from the atomic modelling (or physical modelling) using microscopic observations [7]. The different approaches give several answers to the question, in which way the model of the material behaviour should be formulated and what kind of experimental observations should be taken into account. On the other hand the different approaches lead to different possibilities to describe several effects of the material behaviour. The physical approach is connected with the investigation of deformation, damage and fracture mechanisms on the microscopic (or submicroscopic) level. The materials science approach, which is related to the mesoscopic level, describes the correlation between the structure of the material and their properties. The continuum mechanics approach uses the macroscopic level for the formulation of phenomenological models based on observations in standard material tests [8]. Until now the last approach is most convenient for the use in engineering applications (analysis of the stress-strain-state in constructions or structural elements). The different approaches are summarised in Table 3.1.

Table 3.1: Classification of material behaviour models

LEVEL	Microscopic	Mesoscopic	Macroscopic
CHARACTERISTIC LENGTH, SCALES	$10^{-10} - 10^{-7}$ m Atoms, Molecules	$10^{-7} - 10^{-4}$ m Assemblage of Grains	$10^{-4} - \ldots$ m Structure, Specimens
TYPICAL DEFECTS	Vacancies, Dislocations	Microcracks, Cavities, Voids	Macrocracks
SCIENCE	Solid physics	Materials science	Solid mechanics
AIM OF RESEARCH	Mechanisms of deformation and fracture	Relations between structure of the materials and their properties	Phenomenological constitutive equations

Every discussion in material modelling starts with material testing. The first question is - what kind of tests do we have to take into account. There is no unique answer, because it depends, for example, on the given experimental equipment and the facilities in the laboratories, etc. In general, the recommendation can be given to start with the so-called basic tests in mechanical testing (uniaxial tension, uniaxial compression, torsion).

In this section the explanations are focused on the standard tensile test. In Fig. 3.1 stress-strain curves for different materials are schematically shown. All curves are related to

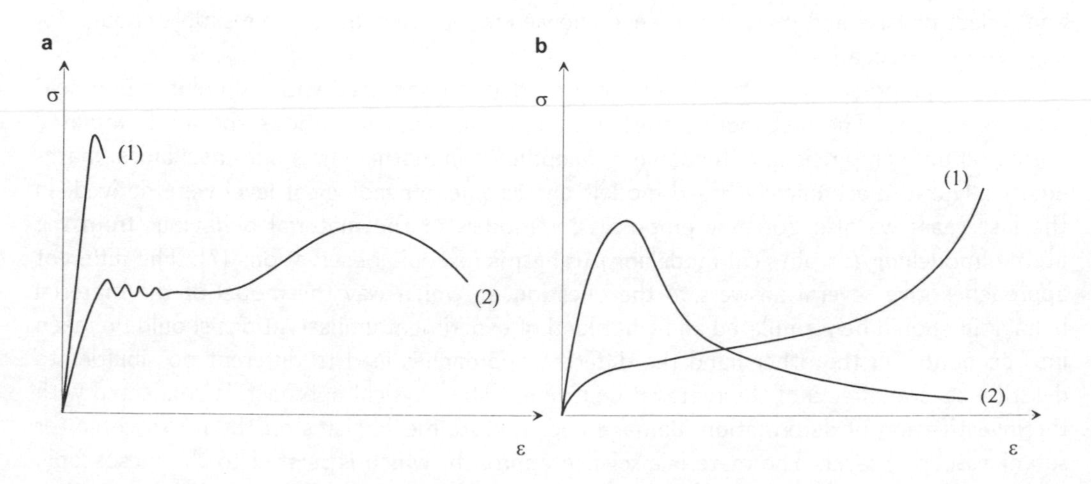

Figure 3.1. Stress-strain curves for various materials. a – brittle (1) and ductile (2) material, b – rubber-like (1) and viscoelastic (2) material

the nominal stress and the Cauchy's strain (both are computed for the initial geometrical values). These curves lead to the conclusion that the mathematical description of material behaviour must result in different mathematical expressions useable for engineering applications. The problem is to find equations suitable and convenient for application, which are detailed enough to reflect the main effects and in the same time simple enough for an easy handling. The solution of this problem in the engineering sense is the introduction of some simplification. The real experimental curves can be divided into sections, each of them can be presented approximately by a simple model. Examples of such models are the linear or the nonlinear elastic model, the perfectly plastic model, the linear elastic-perfectly plastic model, the plastic model with linear or non-linear hardening, etc. All these simple phenomenological material models can be formulated by engineering or continuum mechanics based methods. In fundamental research models of solid physics or materials science are used, too.

3.2. Classification of material models

Several kinds of classification for the material behaviour can be given. For example, in engineering applications material behaviour can be classified as

- elastic behaviour,
- plastic behaviour, and
- creep.

Real materials rarely show a behaviour which can be related to only one of these items. In many situations it is unavoidable to combine these simplified models, what leads to viscoelastic, viscoplastic, elastic-plastic, etc., models. Fig. 3.1a shows the stress-strain curve for a brittle material (1), which can be approximated by a linear elastic law, and for a ductile material (2), which can be approximately characterised as an elastic-plastic material with nonlinear hardening. Fig. 3.1b shows a rubber-like (1) material and the constitutive equation should be nonlinear elastic. Fig. 3.1b (2) shows the typical behaviour of a viscoelastic material, which partly creeps.

Another possibility of classification is the dependence on the kind of loading (Fig. 3.2). If we perform tension and compression tests, we observe that some materials show an identical behaviour (Fig. 3.2a). In other cases (Fig. 3.2b and c) the behaviour is different with respect to the microstructure of the material. The elastic range of the stress-strain curves is marked by thick lines.

In this paper the following classification of the material behaviour is used:

- classical material behaviour, and
- non–classical material behaviour.

For better understanding of the further discussions a definition of classical and non–classical material behaviour with respect to creep problems will be presented according to [3]:

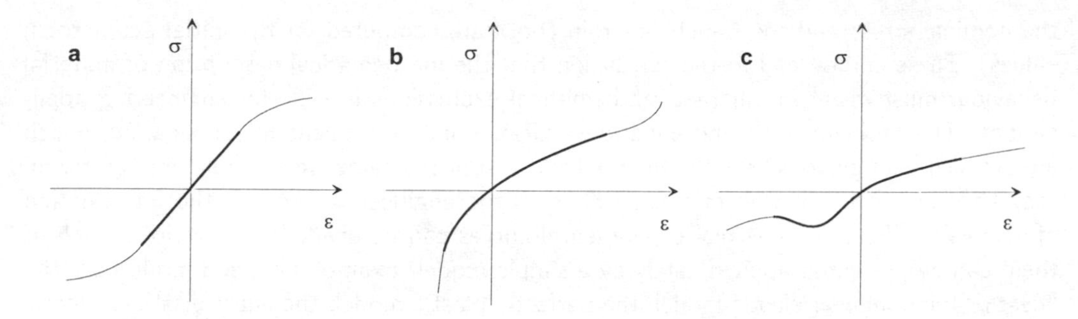

Figure 3.2. Stress-strain curves from tension and compression tests for different types of materials. a – crystalline material (identical behaviour), b – rubber-like material (different behaviour), c – material with a cellular structure (different behaviour)

- *Let us assume that the creep equation can be derived from a potential function*

$$\Phi = \Phi(\sigma_{\mathrm{eq}}) \tag{3.1}$$

σ_{eq} *is the equivalent stress, which is a function of the stress tensor* $\boldsymbol{\sigma}$ *and which can be defined in a suitable manner. With the help of this equivalent stress we can compare multiaxial and uniaxial stress states. The expression of the equivalent stress can be additionally dependent on some specific material parameters, which must be found by basic tests with standard specimens.*

- *Definition 1: All material models for isotropic materials of which the specific material parameters of the equivalent stress can be identified with help of one basic test (mostly uniaxial tension test) are called classical models. If more than one basic test is necessary for the parameter identification we have a non–classical or generalised model.*

- *Definition 2: All material models for anisotropic materials for which the identification of the specific material parameters is possible with help of one basic test, but in different directions, are defined as classical models, otherwise the models are called non–classical or generalised models.*

Remark: Both definitions are not restricted to creep models. Elasticity, plasticity or certain types of failure behaviour can be analysed in the same way [3].

3.3. Approaches in material modelling

Mathematical-mechanical equations which are able to describe the material behaviour should reflect the individual response of a certain material on external influences (for example, mechanical and thermal loadings) caused by its specific internal constitution. It is necessary to introduce a precise number of constitutive equations[1] and, if necessary, evolu-

[1] In this paper term "constitutive equation" is used in a weak sense, which leads to similarities with "state equation", etc. A strong definition of the term "constitutive equation" is presented in the contribution of

tion equations to supplement the material independent equations of structural mechanics. In such a way the mechanical problem becomes determined. The material models can be classified with respect to the observation scale: the physics scale is connected with the deformation mechanisms, the materials science scale - with structure-properties relations and the continuum mechanics scale - with the phenomenological behaviour (Table 3.1). For each case we have different types of material behaviour equations. Limiting our discussions to phenomenological equations three different possibilities for modelling material behaviour can be distinguished:

- the deductive derivation of constitutive equations based on complex mathematical expressions and on fundamental principles of material theory [4, 6, 9, 10],
- the inductive derivation based on experimental performances, simple mathematical relations and a step-by-step generalisation [3, 11], and
- the rheological modelling based on simple rheological basic models and their combinations [10, 12, 13].

The first approach is preferable from the point of view of mathematics and has to take into consideration so-called constitutive axioms (causality, determinism, objectivity, etc.) This approach is briefly discussed by P. Gummert in these Lecture Notes. In addition, the use of the deductive derivation is close to some mathematical techniques, especially the tensor function representation for isotropic and anisotropic materials, discussed, for example, by Betten in [14]. From the point of view of practical applications this first approach can be recommended only in some cases, if other possibilities do not work (solids with complex behaviour, mixtures, etc.). The method of rheological modelling, which is widely used in visco–elasticity for modelling plastics, is connected with an increasing development during the last years. The reasons for this is the simplicity and the easy formulation of complex models by some combination rules. The inductive approach is mostly used in practical applications because this approach is connected with general rules of engineering modelling. For the classical and non-classical creep models, which are presented here this approach will be used.

A lot of constitutive equations for creep processes are proposed in the literature. The mechanism-oriented description (microstructural level) leads to equations of the following form for the creep strain rate (see, e.g., [15])

$$\dot{\varepsilon}^{\mathsf{cr}} = A_i D_i \left(\frac{\tau}{G}\right)^m \left(\frac{\tau\Omega}{kT}\right) \left(\frac{b}{d}\right)^n . \tag{3.2}$$

$\tau\Omega/kT$ is the relation between the mechanical and thermal energy (τ - applied stress, k - Boltzmann's constant, Ω - atomic volume, T - absolute temperature), b/d describes the grain size dependence (b - Burger's vector, d -grain size), D_i is the diffusion coefficient and A_i - the "diffusion area measure". G denotes the shear modulus, m and n are exponents specific for each material and loading condition. The materials science-based models start

P. Gummert in these Lecture Notes.

from experimental observations (micrographs). By this way the type of the creep process, the nucleation and growth of damage and the factors, which have a significant influence on the creep process, can be classified. Equations related to this approach are presented, e.g., in [16, 17]. With respect to the different creep mechanisms (climb-controlled dislocation motion, diffusion-controlled creep, etc.) the mathematical expressions and the influence factors differ from one model to another. All these equations are inconvenient for engineering structural analysis, but lead to suitable results when they are applied to the description of deformation and damage processes in the microstructure of the material. It should be remarked that all these micromechanics-based equations are supporting the phenomenological creep curve describing equations. Successful use in engineering applications of mechanism-based equations is reported by D. Hayhurst in these Lecture Notes.

The phenomenological approach to creep processes is based on constitutive equations of the following type [18, 19, 20]

$$\varepsilon = f(\sigma, t, T), \tag{3.3}$$

where t denotes the time. The classical creep equations for metals were summarised, e.g., in [21]. The form of the constitutive equation differs from author to author. The reason for this is that these equations can be written for the strains ε, for the creep strains $\varepsilon^{\mathrm{cr}}$, for the strain rates $\dot{\varepsilon}$ and for the creep strain rates $\dot{\varepsilon}^{\mathrm{cr}}$. The problem is how to find suitable expressions for the creep equation. Additional difficulties are connected with the extension to the multiaxial case.

A possibility of extending the classical creep equations is the introduction of a damage parameter. The damage parameter allows a simple description of the tertiary creep. Creep-damage equations are discussed, e.g., in [18, 22, 23, 24, 25], but the number of necessary damage parameters, the mathematical nature of these variables (scalar-valued, vector-valued, tensor-valued) remains an open questions. The discussion of these problems is included in the contributions of J. Skrzypek and J.L. Chaboche to these Lecture Notes.

4. CLASSICAL CREEP MODELS

Creep is a time- and temperature-dependent phenomenon that occurs in engineering materials at elevated temperatures usually exceeding 0.4 times the melting temperature T_{m} in the case of metals [26]. Creep tests are performed at constant stress and fixed temperature. The creep strain vs. time curve can be separated into three stages (Fig. 4.1):

- primary or transient creep with hardening of the material,
- secondary or steady-state creep with hardening and softening equilibrium, and
- tertiary or accelerated creep with softening (degradation of the material)[2].

[2]This definition of tertiary creep is a simplification. A different approach is discussed in J.L. Chaboche's contribution to these Lecture Notes.

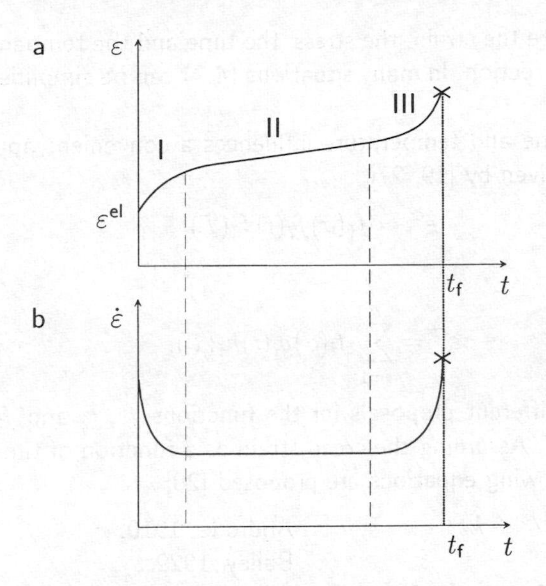

Figure 4.1. Characteristic of the creep behaviour. a - Typical creep curve (total strain vs. time) and b - Creep strain rate vs. time curve (t_f - failure time)

Each stage can be characterised by a typical strain rate curve shape (Fig. 4.1b). During the first stage the creep rate is decreasing, during the second stage we obtain an approximately constant creep rate and the third stage corresponds to increasing creep strain rate.

All three stages can be obtained in tests, but they are more or less developed. For example, if the temperature is moderate the tertiary creep does not occur. The significance of the creep stages also depends on the kind of material.

As shown in Fig. 4.1a in the case of moderate loads the deformation starts with an instantaneous elastic part $\varepsilon^{\mathrm{el}}$. If the load is removed from the specimen at the time t_1 the elastic part of the strains is recovered instantaneously, too. It is established by tests that the creep strain will be recovered incompletely over the time, hence a non-zero permanent strain should be taken into account. In the case of loads which result in plastic deformations the instantaneous deformations are composed of two parts: an elastic and a plastic part.

It must be underlined that the different stages of creep are connected with different creep mechanisms. This leads to different mathematical descriptions of creep processes. For engineering applications simplification must be introduced for a better handling of the model equations and for a reduction of the experimental efforts in the identification of material parameters.

4.1. Uniaxial creep models

Creep constitutive equations are more complicated than the elastic equations. In general, we presume in the uniaxial case

$$f(\varepsilon, \sigma, t, T) = 0, \tag{4.1}$$

where ε, σ, t and T denote the strain, the stress, the time and the temperature, respectively. ε and σ have the same direction. In many situations (4.1) can be simplified or approximated by simpler equations.

Separating stress, time and temperature influences a convenient approximation of the creep law (4.1) can be given by [19, 27]

$$\varepsilon^{\mathrm{cr}} = f_1(\sigma) f_2(t) f_3(T) \tag{4.2}$$

or [28]

$$\varepsilon^{\mathrm{cr}} = \sum_{i=1}^{n} f_i(\sigma) g_i(t) h_i(T). \tag{4.3}$$

$\varepsilon^{\mathrm{cr}}$ is the creep strain. Different proposals for the functions f_1, f_2 and f_3 are presented in the literature [18, 19, 20]. Assuming the creep strain as a function of time with fixed stress and temperature the following equations are proposed [20] ·

$$\begin{array}{ll}
\varepsilon^{\mathrm{cr}} = \beta t^{1/3} + kt & \text{Andrade, 1910,} \\
\varepsilon^{\mathrm{cr}} = F t^n & \text{Bailey, 1929,} \\
\varepsilon^{\mathrm{cr}} = G[1 - \exp(-qt)] + Ht & \text{McVetty, 1934,} \\
\varepsilon^{\mathrm{cr}} = \varepsilon_1 + A \lg t + Bt & \text{Leaderman, 1943} \\
\varepsilon^{\mathrm{cr}} = \varepsilon_1 + \varepsilon t^n \quad (n < 1) & \text{Findley, 1944,} \\
\varepsilon^{\mathrm{cr}} = \varepsilon_1 + A \lg t & \text{Philips, 1956,} \\
\varepsilon^{\mathrm{cr}} = \sum\limits_i a_i t^{n_i} & \text{Graham and Walles, 1955.}
\end{array} \tag{4.4}$$

For the creep strain rate-stress relation the following proposals exist

$$\begin{array}{ll}
\dot{\varepsilon}^{\mathrm{cr}} = K\sigma^m \quad (3 < m < 7) & \text{Norton, 1929, Bailey, 1929,} \\
\dot{\varepsilon}^{\mathrm{cr}} = B\left[\exp\dfrac{\sigma}{\sigma^*} - 1\right] & \text{Soderberg, 1936,} \\
\dot{\varepsilon}^{\mathrm{cr}} = A \sinh\dfrac{\sigma}{\sigma^*} & \text{Nadai, 1938, McVetty, 1943,} \\
\dot{\varepsilon}^{\mathrm{cr}} = D_1\sigma^{m_1} + D_2\sigma^{m_2} & \text{Johnson et al., 1963,} \\
\dot{\varepsilon}^{\mathrm{cr}} = A\left[\sinh\dfrac{\sigma}{\sigma^*}\right]^m & \text{Garofalo, 1965,} \\
\dot{\varepsilon}^{\mathrm{cr}} = \dfrac{d}{dt}\left(\dfrac{\sigma}{\sigma^*}\right)^{n_0} + \left(\dfrac{\sigma}{\sigma^*}\right)^n & \text{Odqvist, 1966.}
\end{array} \tag{4.5}$$

The approximation of creep data by the power function leads in some cases to mathematical difficulties, which can be solved, for example, if the approximation is realised by the sinh-function [29]. Other problems with the approximation of creep data are reported in [30].

The description of the temperature dependencies is more complicated. The reason for this can be seen in the temperature influence on the material parameters (temperature dependent parameters) and on the structural changes (the increase of the temperature yields different creep mechanisms). In dependence on the stress and the temperature level, the grain size, etc. different creep mechanisms are existing. So-called deformation maps (normalised shear stress vs. homologous temperature) show, which kind of creep mechanism can be assumed. Such deformation maps are derived from experimental observations for

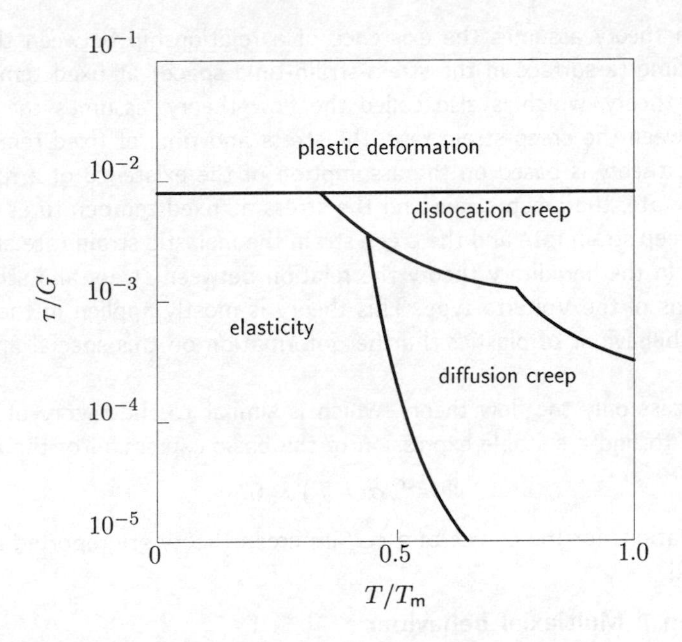

Figure 4.2: Typical deformation map: normalised shear stress vs. homologous temperature

many materials. A typical deformation map is shown in Fig. 4.2. Various creep deformation mechanisms are described, e.g., in [15]. Several examples of deformation maps are presented in [31]. Temperature-dependent creep equations can be expressed by

$$
\begin{aligned}
\dot{\varepsilon}^{\mathrm{cr}} &\approx \exp[-(Q-\gamma\sigma)/RT] && \text{Kauzman, 1941,} \\
\varepsilon^{\mathrm{cr}} &= f[t\exp(-Q/RT)]f_1(\sigma) && \text{Dorn and Tietz, 1949,} \\
\dot{\varepsilon}^{\mathrm{cr}} &\approx \frac{\sigma}{T}\exp(-Q/RT) && \text{Lifshiz, 1963,} \\
\varepsilon^{\mathrm{cr}} &= f[t\exp(-Q/RT)]^n f_1(\sigma) && \text{Penny and Marriott, 1971.}
\end{aligned}
\tag{4.6}
$$

Q, R, γ, n denote the activation energy, the gas constant and material parameters, respectively.

At varying stress the creep theories which include primary and secondary creep can be formulated by different approaches. The main directions are [18, 19, 20]

- the total strain theory,
- the time hardening theory,
- the strain hardening theory and
- the hereditary theory.

The total strain theory assumes the existence of a relationship between the total strain, the stress and time (a surface in the stress-strain-time space) at fixed temperatures. The time hardening theory, which is also called the flow theory, assumes the existence of a relationship between the creep strain rate, the stress and time at fixed temperatures. The strain hardening theory is based on the assumption of the existence of a relation between the creep strain rate, the creep strain and the stress at fixed temperatures. In some cases instead of the creep strain rate and the creep strain the inelastic strain rate and the inelastic strain are used. In the hereditary theory the relation between stress and strain is given by integral equations of the Volterra type. This theory is mostly applied in the description of the viscoelastic behaviour of plastics. Further information on this special approach can be taken from [18].

Here we discuss only the flow theory, which is similar to the theory of plasticity. The main problem is to find a suitable expression of the basic expression of the flow theory

$$\Phi(\dot{\boldsymbol{\varepsilon}}^{\mathrm{cr}}, \boldsymbol{\sigma}, t, T) = 0. \tag{4.7}$$

The recommendations for the choice of a certain creep theory are reported in detail in, for example, [20, 32].

4.2. Extension I: Multiaxial behaviour

All constitutive equations formulated here are restricted by the following assumptions:

- geometrically linear theory, that means infinitesimal strains,
- non–polar, homogeneous materials, that means symmetrical stress tensor, and
- isothermal processes, that means elevated, but constant temperatures.

From the first assumption, it follows that the coordinates of the stress tensor in the Eulerian and Lagrangian coordinates and the engineering stresses are identical [4]. The deformation state can be described by Cauchy's strain tensor. From the second assumption follows that only the symmetric part of the strain tensor is involved in the material behaviour equations [4]. In addition, the material characteristics are constant in all parts of the volume. The third assumption simplifies the formulations of the material model, because the material characteristics do not change with temperature. This assumption can be used in the case of fixed temperatures.

The classical multiaxial creep equations can be derived from the following creep potential [18]

$$\Phi = \Phi(\boldsymbol{\sigma}, q_n), \tag{4.8}$$

$\boldsymbol{\sigma}$ denotes the second rank symmetric stress tensor and the $q_n (n = 1, \ldots, m)$ are a set of inner (rheological) parameters characterising hardening, damage, etc. Assuming the equivalence between the uniaxial creep state (known from tests) and the three–dimensional creep state the potential should be a function of the equivalent stress σ_{eq}. This scalar-valued function of a tensor-valued argument can be restricted by some assumptions. In the case of isotropic materials the equivalent stress is a function of the stress tensor invariants. Neglecting the influence of the kind of loading a suitable approximation of the equivalent stress can be given by the quadratic invariant. Restricting the discussion to cases not influenced by the hydrostatic pressure we get with respect to the flow theory

$$\Phi(\sigma_{\text{eq}}) - \xi^2(q_n) = 0, \quad \sigma_{\text{eq}} = \sigma_{\text{vM}} = \sqrt{\frac{3}{2}\boldsymbol{s} \cdot\cdot \boldsymbol{s}}, \tag{4.9}$$

σ_{vM} denotes the von Mises equivalent stress and $\boldsymbol{s}$ - the stress deviator. The function ξ is a characteristic property of the creep surface which can be influenced by hardening and other phenomena. The creep strain rate tensor can be calculated by

$$\dot{\boldsymbol{\varepsilon}}^{\text{cr}} = \dot{\eta}\frac{\partial \Phi}{\partial \boldsymbol{\sigma}}. \tag{4.10}$$

The unknown factor $\dot{\eta}$ follows from the assumption that the dissipated power P^{diss} in the three–dimensional case is the same as in the equivalent uniaxial case

$$P^{\text{diss}} = \boldsymbol{\sigma} \cdot\cdot \dot{\boldsymbol{\varepsilon}}^{\text{cr}} = \sigma_{\text{eq}}\dot{\varepsilon}^{\text{cr}}_{\text{eq}}. \tag{4.11}$$

Taking into account some specifications of the creep potential Φ we can calculate $\dot{\eta}$. In the case of the so-called von Mises-type theory $\xi = \xi_0$ is constant and Φ depends on the second invariant of the stress deviator $\boldsymbol{s}$

$$\Phi(\sigma_{\text{eq}}) - \xi^2 = {\sigma_{\text{eq}}}^2 - \xi_0^2 = 0. \tag{4.12}$$

Following [21], we get for the creep strain rate

$$\dot{\boldsymbol{\varepsilon}}^{\text{cr}} = 3\dot{\eta}\boldsymbol{s}. \tag{4.13}$$

The equivalence relation (4.11) leads to

$$\dot{\eta} = \frac{\dot{\varepsilon}^{\text{cr}}_{\text{eq}}}{2\sigma_{\text{eq}}} \tag{4.14}$$

and we finally find [27].

$$\dot{\boldsymbol{\varepsilon}}^{\text{cr}} = \frac{3}{2}\frac{\dot{\varepsilon}^{\text{cr}}_{\text{eq}}}{\sigma_{\text{eq}}}\boldsymbol{s}. \tag{4.15}$$

The function $\dot{\varepsilon}^{\text{cr}}_{\text{eq}}$ must be determined by tests. Usual approximation is the Norton's creep law. With respect to (4.5) we assume

$$\dot{\varepsilon}^{\text{cr}}_{\text{eq}} = K{\sigma_{\text{eq}}}^m \tag{4.16}$$

which results in the well-known equation

$$\dot{\boldsymbol{\varepsilon}}^{\text{cr}} = \frac{3}{2}K{\sigma_{\text{eq}}}^{m-1}\boldsymbol{s} \tag{4.17}$$

presented in the classical textbooks. Additional discussions on classical creep equation can be found in [18, 19, 20, 21, 27, 28, 33, 34].

4.3. Extension II: Anisotropic behaviour

In tests many engineering materials are (more or less) anisotropic, which means that their constitutive behaviour is dependent on the given test direction. A typical example of this group of materials are single crystals, but anisotropic behaviour can also result from certain technological processes. For all these materials we need a suitable description by mathematical equations.

Limiting our discussions to the case of creep behaviour the introduction of a simple model which corresponds to the isotropic creep model (flow theory) is possible. The starting point is again a creep potential

$$\Phi = \Phi(\boldsymbol{\sigma}, q_n). \tag{4.18}$$

In the case of anisotropic behaviour we must introduce material tensors similar to the theory of plasticity

$$\Phi = \Phi(\boldsymbol{\sigma}, {}^{(4)}\boldsymbol{b}, q_n), \tag{4.19}$$

${}^{(4)}\boldsymbol{b}$ denotes the fourth rank material tensor. The dependence on the inner (rheological) parameters q_n can be dropped in some cases. A convenient proposal for the potential Φ is a quadratic function

$$\Phi = \boldsymbol{\sigma} \cdot\cdot {}^{(4)}\boldsymbol{b} \cdot\cdot \boldsymbol{\sigma}, \tag{4.20}$$

and the creep strain rate tensor can be calculated by the standard flow rule

$$\dot{\varepsilon}^{\mathsf{cr}} = \dot{\eta}\frac{\partial \Phi}{\partial \boldsymbol{\sigma}} = 2\dot{\eta}\,{}^{(4)}\boldsymbol{b} \cdot\cdot \boldsymbol{\sigma}. \tag{4.21}$$

The unknown factor $\dot{\eta}$ follows from the dissipated power P^{diss}. (4.20) is the generalisation of the isotropic creep potential, based on the second invariant of the stress deviator. This potential was suggested in [35]. The fourth rank tensor ${}^{(4)}\boldsymbol{b}$ satisfies the symmetry conditions of the Hookean tensor which means that 21 components of this tensor are independent.

Different additional proposals restricting the form of the presented anisotropic creep equation are discussed in literature. For example, as in the classical theory of isotropic creep, incompressibility can be presumed. Such restrictions result in a reduced number of material properties of the material tensor ${}^{(4)}\boldsymbol{b}$. Taking into account the incompressibility condition (no influence of the hydrostatic stress state) in the anisotropic case the number of independent components reduces to 15. Another possibility of reducing the experimental and computational efforts is given by the assumption of special kinds of anisotropy. For example, orthotropic behaviour, which can be established in thin sheets, !eads to a reduction of the number of independent components to 9. The additional introduction of the incompressibility condition reduces it to 6. This case was discussed in the theory of plasticity by Hill [36].

4.4. Extension III: Creep-damage coupling

Engineering materials operate under different mechanical and environmental conditions leading to microstructural changes which decrease their strength. These processes in the materials are irreversible and in the literature they are called damage. There are different kinds of damage (creep damage, ductile plastic damage, fatigue damage, embrittlement, environmental degradation, etc.), which are discussed in several textbooks, e.g., [23, 24, 37]. Here we restrict to creep-damage interaction. Additional information are presented in J.L. Chaboche's and J. Skrzypek's contributions to these Lecture Notes.

The classical creep-damage model introduced by Kachanov and Rabotnov is based on a constitutive law for the creep strain rate and on an evolution law for the scalar damage variable ω (isotropic damage). With respect to the classical creep equations connecting the creep strain rates with the stresses for the description of the creep-damage coupling, we substitute the stresses by the effective stresses (stresses divided by the continuity $\psi = 1-\omega$, which is the complementary variable to the damage variable) using the equivalent strain principle. In addition, we have to formulate an evolution equation for the damage variable. Thus, the starting point is the following system of equations

$$\dot{\boldsymbol{\varepsilon}}^{\mathsf{cr}} = \dot{\boldsymbol{\varepsilon}}^{\mathsf{cr}}(\boldsymbol{\sigma}, \omega, \ldots), \quad \dot{\omega} = \dot{\omega}(\boldsymbol{\sigma}, \omega, \ldots). \tag{4.22}$$

The concept of two coupled equations (one for the creep behaviour and one for damage) is working successfully, because we can give understandable interpretations for this model. The continuity (or damage) variable is connected with the changes of the cross-section area in a test specimen. If A denotes the actual area, which is calculated with respect to the "lost damaged" parts of the initial area A_0, the damage is

$$\omega = 1 - \frac{A}{A_0}, \quad 0 \leq \omega \leq 1. \tag{4.23}$$

This description is correct, if isotropic damage is presumed. It must be underlined that the damage is a monotonically increasing variable ($\dot{\omega} \geq 0$) and for real materials we obtain in many cases that an initial damage is existing (ω is different from zero at the starting moment) and that the maximum value is less than 1 (the rupture occurs on a lower level of damage).

In accordance to the classical flow theory a similar creep-damage model can be formulated. Neglecting instantaneous deformations a dependence between the creep strain rate and the stress deviator follows from the flow theory

$$\dot{\boldsymbol{\varepsilon}}^{\mathsf{cr}} = 3\dot{\eta}\boldsymbol{s}. \tag{4.24}$$

With respect to Odqvist's flow rule for steady state creep [21] and Rabotnov's scalar damage variable [18] the creep-damage equations can be established:

- the constitutive equation

$$\dot{\boldsymbol{\varepsilon}}^{\mathsf{cr}} = \frac{3}{2} F(\sigma_{\mathsf{vM}}) K(\omega) \frac{\boldsymbol{s}}{\sigma_{\mathsf{vM}}}, \tag{4.25}$$

- and a damage evolution equation

$$\dot{\omega} = R(\omega) H[\langle \chi(\boldsymbol{\sigma}) \rangle] \tag{4.26}$$

The functions F, K, R and H can be specified as

$$F(\sigma) = a\sigma^n, \quad K(\omega) = (1-\omega)^{-m}, \quad H(\sigma) = b\sigma^k, \quad R(\omega) = (1-\omega)^{-l}. \tag{4.27}$$

a, n, m, b, k and l are material parameters which must be determined by creep tests. In order to incorporate different damage mechanisms the following expression for $\chi(\boldsymbol{\sigma})$ can be used [38]

$$\chi(\boldsymbol{\sigma}) = \alpha\sigma_I + 3\beta\sigma_{\mathsf{H}} + (1 - \alpha - \beta)\sigma_{\mathsf{vM}}, \tag{4.28}$$

$\sigma_I, \sigma_{\mathsf{H}}$ denote the maximum principal stress and the hydrostatic stress, respectively, and

$$\langle\chi(\boldsymbol{\sigma})\rangle = \begin{cases} \chi(\boldsymbol{\sigma}), & \chi(\boldsymbol{\sigma}) > 0 \\ 0, & \chi(\boldsymbol{\sigma}) \leq 0 \end{cases} . \tag{4.29}$$

It should be underlined that the creep constitutive equation uses a different expression for the equivalent stress (von Mises-equivalent stress σ_{vM}) than the damage evolution equation (more general equivalent stress $\chi(\boldsymbol{\sigma})$) since these equations describe two different mechanisms.

5. NON–CLASSICAL CREEP MODELS

Creep behaviour, which is influenced by the kind of loading, can be obtained in tests for some geomaterials, some kind of ice, some light metals and their alloys, graphite, several ceramics, polymers etc. Such effects like different behaviour in tension and compression or the influence of the hydrostatic stresses on creep are ignored in the classical theories. Below the experimental results are discussed and a mathematical model for the description of non–classical behaviour is proposed.

5.1. Experimental observations

Some materials show a strong dependence of their behaviour on the kind of loading in tests. One of these effects, which is often reported (see [18, 39, 40, 41, 42, 43]), is the different behaviour under tensile or compressive load. In the case of the same absolute value of tensile and compressive stress and the same temperature, the first approximation of the creep behaviour (creep strain vs. time) gives identical curves (Fig. 5.1a). Such behaviour is assumed in classical theories. For some materials, different behaviour is obtained, which is schematically shown in Fig. 5.1b. This behaviour cannot be described by classical theories, which are not sensitive to the sign of the load.

A very simple, but often used creep law in engineering calculations is the Norton's law. Let us assume the following relationship in the case of tension

$$\dot{\varepsilon}^{\mathsf{cr}} = L_+\sigma^n, \quad L_+ > 0. \tag{5.1}$$

L_+ and n are similar to the material parameters in the previous section (Norton's law). In the case of identical behaviour in tension and compression we have an analogous equation for compression with the same material parameters (only the sign is opposite)

$$\dot{\varepsilon}^{\mathsf{cr}} = -L_+|\sigma|^n. \tag{5.2}$$

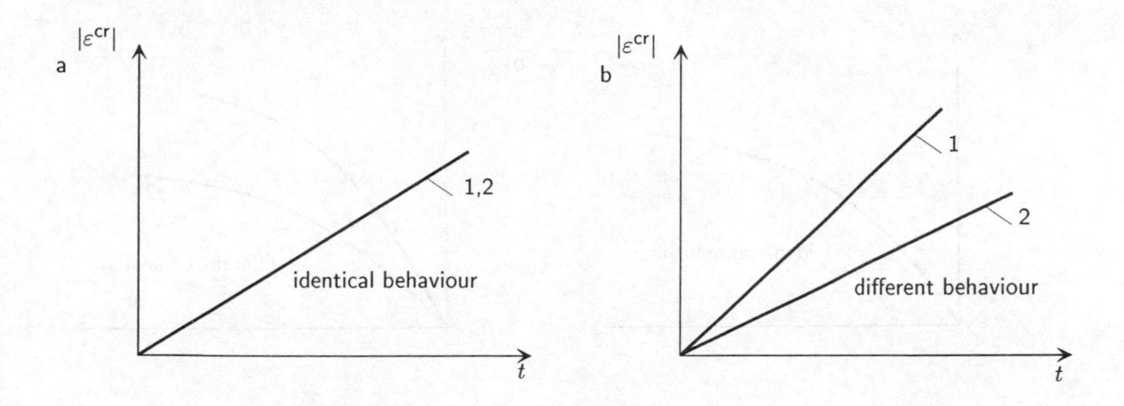

Figure 5.1: Identical and different behaviour in tension (1) and compression (2)

In the case of non–identical behaviour it is necessary to introduce a second equation for compression

$$\dot{\varepsilon}^{cr} = -L_{-}|\sigma|^n, \quad L_{-} > 0, \tag{5.3}$$

with $L_{-} \neq L_{+}$ as a new material parameter. The creep exponent n can be assumed independent on the kind of loading. This follows from experimental observations [34].

The ratio L_{+}/L_{-} can reach different values: for light alloys 2 ... 3 [44] and for polymers 1,5 ... 5 [39, 44]. Higher values can be obtained for ceramics: 39 for aluminium oxide based ceramics and 289 for a silicone nitride based ceramics [41].

We can also observe different behaviour in tension and compression in the case of nonlinear creep vs. time curves, which is schematically shown in Fig. 5.2. Examples of such behaviour are given in [3].

Another type of dependence of the material behaviour on the kind of loading can be

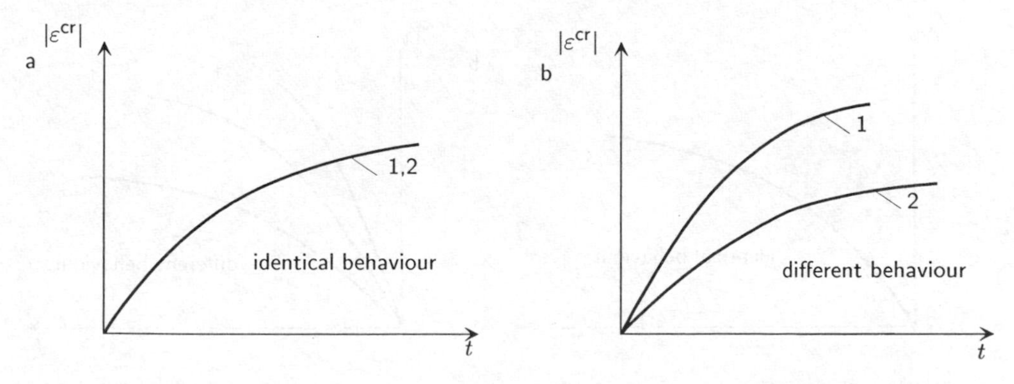

Figure 5.2. Identical (a) and different (b) primary creep behaviour in tension (1) and compression (2)

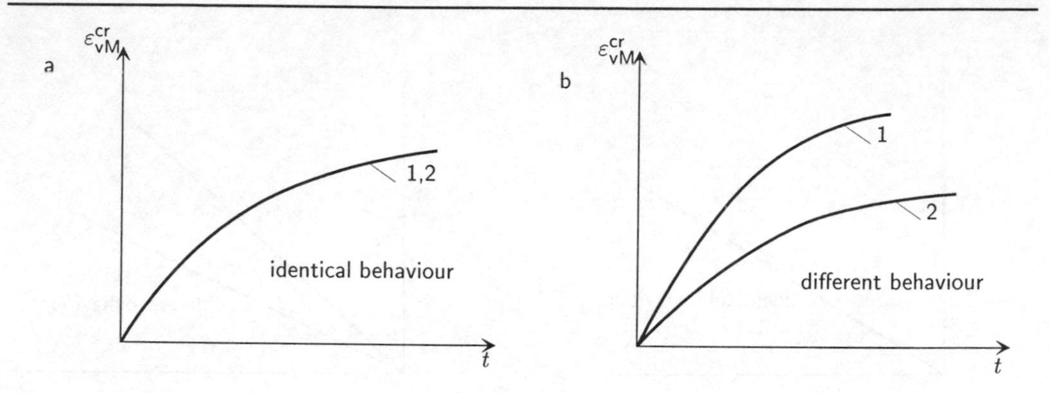

Figure 5.3. Equivalent strain vs. time curves for tension (1) and torsion (2). a - identical behaviour, b - different behaviour

obtained, if results of creep tests under tensile force and torsional momentum are compared. In the first case we get an axial creep strain vs. time curve, the second case leads to a shear creep strain vs. time curve. The comparison of both curves is possible, if we introduce equivalent stress and equivalent strain values. Suitable expressions were proposed by von Mises for the equivalent stress and strain

$$\begin{aligned} \sigma_{\text{vM}} &= \sqrt{\frac{3}{2}\boldsymbol{s} \cdot\cdot \boldsymbol{s}}, \qquad \boldsymbol{s} = \boldsymbol{\sigma} - \frac{1}{3}(\boldsymbol{\sigma} \cdot\cdot \boldsymbol{I})\boldsymbol{I}, \\ \varepsilon_{\text{vM}}^{\text{cr}} &= \sqrt{\frac{2}{3}\boldsymbol{e}^{\text{cr}} \cdot\cdot \boldsymbol{e}^{\text{cr}}}, \quad \boldsymbol{e}^{\text{cr}} = \boldsymbol{\varepsilon}^{\text{cr}} - \frac{1}{3}(\boldsymbol{\varepsilon}^{\text{cr}} \cdot\cdot \boldsymbol{I})\boldsymbol{I}. \end{aligned} \tag{5.4}$$

$\boldsymbol{\sigma}, \boldsymbol{s}$ denote the stress tensor and deviator, $\boldsymbol{\varepsilon}^{\text{cr}}, \boldsymbol{e}^{\text{cr}}$ - the creep strain tensor and deviator. The classical material behaviour is characterised by identical equivalent creep strain vs. time curves in tension and torsion (Fig. 5.3a). Some tests show a behaviour, which yields

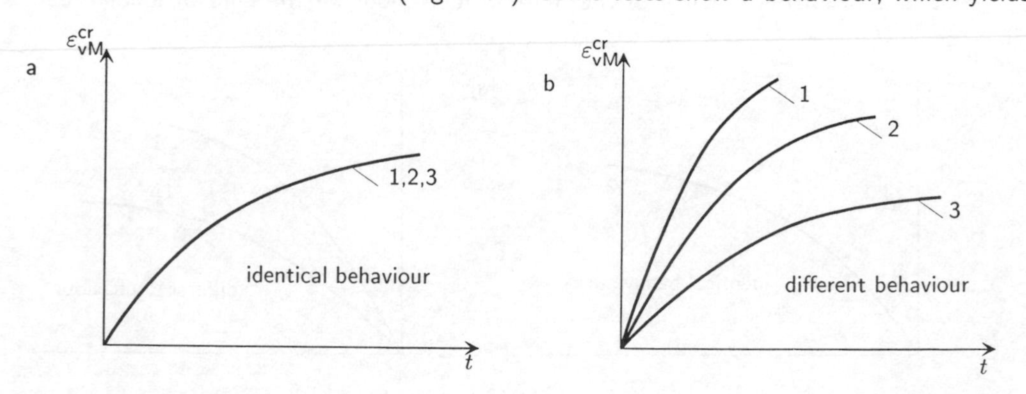

Figure 5.4. Identical (a) and different (b) equivalent strain vs. time curves for tension (1), compression (2) and torsion (3)

different curves (Fig. 5.3b). In both tests the equivalent stress has got the same value. Experimental results, connected with this effect, are published for some kind of austenitic steels, polymers and pure copper [45, 46, 47, 48, 49, 50].

By taking experimental test data from tension, compression and torsion tests, identical equivalent creep strain vs. time curves can be obtained for most engineering materials (Fig. 5.4a). Some polymers [42, 51] show different equivalent creep strain curves for tension, compression and torsion (the equivalent stress takes the same value in tension, compression and torsion). Such dependence on the kind of loading is shown in Fig. 5.4b.

In the literature other non–classical effects [1, 39, 52, 53, 54, 55, 56], obtained in creep tests are described. The classical models neglect the influence of the hydrostatic pressure on the creep and of pure torsion on the volumetric creep, etc. These experimental observations of non–classical behaviour are reported in [3]. Non–classical effects are obtained for primary, secondary, and also for tertiary creep.

5.2. Non–classical creep law

We presume that the deformation behaviour of a material can be derived from a potential Φ, which is a function of the stress tensor components, the temperature and, may be, some additional parameters. In analogy to the classical creep law assumptions, the potential is assumed to be temperature-independent, which means that all processes are performed at fixed temperatures. The dependence on the stress tensor components is substituted by a dependence on an equivalent stress σ_{eq}. If neither the temperature nor the additional parameters enter into the potential function, its most simple form is

$$\Phi = \Phi(\sigma_{\mathsf{eq}}). \tag{5.5}$$

For the equivalent stress we can find several definitions in the literature. In the case of strength or plasticity criteria some possibilities are discussed in [57].

Considering non–classical creep behaviour in the case of isotropic materials the equivalent stress itself is a function of the invariants of the stress tensor. The set of invariants can be defined in different ways (see, e.g., [58, 59]). For our derivations the following definition of the equivalent stress is useful

$$\sigma_{\mathsf{eq}} = \alpha\sigma_1 + \beta\sigma_2 + \gamma\sigma_3, \tag{5.6}$$

with the linear, quadratic and cubic invariants

$$\sigma_1 = \mu_1 I_1, \quad \sigma_2^2 = \mu_2 I_1^2 + \mu_3 I_2, \quad \sigma_3^3 = \mu_4 I_1^3 + \mu_5 I_1 I_2 + \mu_6 I_3, \tag{5.7}$$

and

$$I_1(\boldsymbol{\sigma}) = \boldsymbol{\sigma} \cdot\cdot\, \boldsymbol{I}, \quad I_2(\boldsymbol{\sigma}) = \boldsymbol{\sigma} \cdot\cdot\, \boldsymbol{\sigma}, \quad I_3(\boldsymbol{\sigma}) = (\boldsymbol{\sigma} \cdot \boldsymbol{\sigma}) \cdot\cdot\, \boldsymbol{\sigma}. \tag{5.8}$$

The I_i $(i = 1, 2, 3)$ are the linear, the quadratic and the cubic basic invariants, the μ_j $(j = 1, \ldots, 6)$ are parameters, which depend on the material properties. α, β, γ are numerical coefficients for weighting the different parts in the expression of the equivalent stress. Such weighting is also used for other material behaviour models. In [60] similar coefficients are introduced for characterising different failure modes.

The basic invariants are connected with the principal invariants (coefficients in the characteristic equation). Using the relations between the principal and the basic invariants (see subsection 2.3.) the principal invariants can be calculated from the basic invariants and vice versa.

The equivalent stress σ_{eq} introduced above contains the von Mises equivalent stress σ_{vM} as a special case. We compare the von Mises equivalent stress

$$\sigma_{\mathsf{vM}}^2 = \frac{3}{2}\boldsymbol{s}\cdot\cdot\,\boldsymbol{s}, \tag{5.9}$$

and the equivalent stress equation (5.6) with $\alpha = \gamma = 0,\ \beta = 1$

$$\begin{aligned}\sigma_{\mathsf{eq}}{}^2 = \sigma_2^2 &= \mu_2 I_1^2 + \mu_3 I_2 = \mu_2(\boldsymbol{\sigma}\cdot\cdot\,\boldsymbol{I})^2 + \mu_3\boldsymbol{\sigma}\cdot\cdot\,\boldsymbol{\sigma}\\ &= (\mu_2 + \frac{1}{3}\mu_3)(\boldsymbol{\sigma}\cdot\cdot\,\boldsymbol{I})^2 + \mu_3\boldsymbol{s}\cdot\cdot\,\boldsymbol{s}.\end{aligned}$$

The comparison of the coefficients in both expressions for the von Mises stress σ_{vM} and for the equivalent stress σ_{eq} leads to

$$\mu_3 = \frac{3}{2}, \quad \mu_2 = -\frac{1}{2}. \tag{5.10}$$

Therefore one can conclude: If $\alpha = \gamma = 0$, $\beta = 1$ and $\mu_3 = 1.5$, $\mu_2 = -0.5$ the von Mises equivalent stress expression results from (5.6). For a better comparison of the classical approaches and the generalised creep equation, we will set $\beta \equiv 1$ in the following derivations, which leads to the equivalent stress

$$\sigma_{\mathsf{eq}} = \alpha\sigma_1 + \sigma_2 + \gamma\sigma_3. \tag{5.11}$$

Introducing the creep strain rate tensor $\dot{\boldsymbol{\varepsilon}}^{\mathsf{cr}}$ with respect to the flow theory [18] we can propose the following creep law (see subsection 4.2.)

$$\dot{\boldsymbol{\varepsilon}}^{\mathsf{cr}} = \dot{\eta}\frac{\partial\Phi(\sigma_{\mathsf{eq}})}{\partial\boldsymbol{\sigma}}, \tag{5.12}$$

where $\dot{\eta}$ is an unknown scalar factor. Let us calculate the derivative of the potential in the generalised creep equation with respect to the introduced equivalent stress and the chain rule

$$\begin{aligned}\frac{\partial\Phi(\sigma_{\mathsf{eq}})}{\partial\boldsymbol{\sigma}} = \frac{\partial\Phi}{\partial\sigma_{\mathsf{eq}}}\frac{\partial\sigma_{\mathsf{eq}}}{\partial\boldsymbol{\sigma}} &= \frac{\partial\Phi}{\partial\sigma_{\mathsf{eq}}}\left(\frac{\partial\sigma_{\mathsf{eq}}}{\partial\sigma_1}\frac{\partial\sigma_1}{\partial\boldsymbol{\sigma}} + \frac{\partial\sigma_{\mathsf{eq}}}{\partial\sigma_2}\frac{\partial\sigma_2}{\partial\boldsymbol{\sigma}} + \frac{\partial\sigma_{\mathsf{eq}}}{\partial\sigma_3}\frac{\partial\sigma_3}{\partial\boldsymbol{\sigma}}\right)\\ &= \frac{\partial\Phi}{\partial\sigma_{\mathsf{eq}}}\left(\alpha\frac{\partial\sigma_1}{\partial\boldsymbol{\sigma}} + \frac{\partial\sigma_2}{\partial\boldsymbol{\sigma}} + \gamma\frac{\partial\sigma_3}{\partial\boldsymbol{\sigma}}\right).\end{aligned} \tag{5.13}$$

Taking into account the relations between the invariants σ_i $(i = 1,2,3)$ and the basic invariants I_i $(i = 1,2,3)$ we get

$$\begin{aligned}\frac{\partial\sigma_1}{\partial\boldsymbol{\sigma}} &= \mu_1\boldsymbol{I}, \quad \frac{\partial\sigma_2}{\partial\boldsymbol{\sigma}} = \frac{\mu_2 I_1\boldsymbol{I} + \mu_3\boldsymbol{\sigma}}{\sigma_2},\\ \frac{\partial\sigma_3}{\partial\boldsymbol{\sigma}} &= \frac{\mu_4 I_1^2\boldsymbol{I} + \frac{\mu_5}{3}I_2\boldsymbol{I} + \frac{2}{3}\mu_5 I_1\boldsymbol{\sigma} + \mu_6\boldsymbol{\sigma}\cdot\boldsymbol{\sigma}}{\sigma_3^2}.\end{aligned} \tag{5.14}$$

Finally we obtain the following generalised creep equation

$$\dot{\boldsymbol{\varepsilon}}^{\text{cr}} = \dot{\eta}\frac{\partial \Phi}{\partial \sigma_{\text{eq}}}\left[\alpha\mu_1\boldsymbol{I} + \frac{\mu_2 I_1 \boldsymbol{I} + \mu_3 \boldsymbol{\sigma}}{\sigma_2} + \gamma\frac{\left(\mu_4 I_1^2 + \frac{\mu_5}{3}I_2\right)\boldsymbol{I} + \frac{2}{3}\mu_5 I_1 \boldsymbol{\sigma} + \mu_6 \boldsymbol{\sigma}\cdot\boldsymbol{\sigma}}{\sigma_3^2}\right]. \qquad (5.15)$$

The creep law (5.15) is tensorial nonlinear.
Remark: The derivation is only correct for the case $\sigma_3 \neq 0$, otherwise the third term in the equivalent stress (5.11) vanishes and our derivation should be repeated with the consideration that

$$\sigma_{\text{eq}} = \alpha\sigma_1 + \sigma_2. \qquad (5.16)$$

In this case, we get a reduced tensorial linear creep equation

$$\dot{\boldsymbol{\varepsilon}}^{\text{cr}} = \dot{\eta}\frac{\partial \Phi}{\partial \sigma_{\text{eq}}}\left(\alpha\mu_1\boldsymbol{I} + \frac{\mu_2 I_1 \boldsymbol{I} + \mu_3 \boldsymbol{\sigma}}{\sigma_2}\right). \qquad (5.17)$$

The unknown scalar factor $\dot{\eta}$ can be determined by the assumption of the equivalence of the uniaxial and the three-dimensional states of stress and strain in the material. Introducing an invariant value P^{diss} (the dissipation power during the creep process), it can be presumed

$$P^{\text{diss}} = \boldsymbol{\sigma}\cdot\cdot\,\dot{\boldsymbol{\varepsilon}}^{\text{cr}} \equiv \sigma_{\text{eq}}\dot{\varepsilon}_{\text{eq}}^{\text{cr}}. \qquad (5.18)$$

The substitution of the creep strain rate tensor $\dot{\boldsymbol{\varepsilon}}^{\text{cr}}$ according to (5.15) leads to

$$P^{\text{diss}} = \boldsymbol{\sigma}\cdot\cdot\,\dot{\eta}\frac{\partial \Phi}{\partial \sigma_{\text{eq}}}\left[\alpha\mu_1\boldsymbol{I} + \frac{\mu_2 I_1 \boldsymbol{I} + \mu_3 \boldsymbol{\sigma}}{\sigma_2} + \gamma\frac{\left(\mu_4 I_1^2 + \frac{\mu_5}{3}I_2\right)\boldsymbol{I} + \frac{2}{3}\mu_5 I_1 \boldsymbol{\sigma} + \mu_6 \boldsymbol{\sigma}\cdot\boldsymbol{\sigma}}{\sigma_3^2}\right]. \qquad (5.19)$$

With respect to the definitions of the basic invariants and the equivalent stress the previous equations lead after reordering to

$$P^{\text{diss}} = \dot{\eta}\frac{\partial \Phi}{\partial \sigma_{\text{eq}}}(\alpha\sigma_1 + \sigma_2 + \gamma\sigma_3) = \dot{\eta}\frac{\partial \Phi}{\partial \sigma_{\text{eq}}}\sigma_{\text{eq}} \Longrightarrow \dot{\eta} = \frac{P^{\text{diss}}}{\sigma_{\text{eq}}\dfrac{\partial \Phi}{\partial \sigma_{\text{eq}}}}, \qquad (5.20)$$

which results in

$$\dot{\boldsymbol{\varepsilon}}^{\text{cr}} = \frac{P^{\text{diss}}}{\sigma_{\text{eq}}}\left[\alpha\mu_1\boldsymbol{I} + \frac{\mu_2 I_1 \boldsymbol{I} + \mu_3 \boldsymbol{\sigma}}{\sigma_2} + \gamma\frac{\left(\mu_4 I_1^2 + \frac{\mu_5}{3}I_2\right)\boldsymbol{I} + \frac{2}{3}\mu_5 I_1 \boldsymbol{\sigma} + \mu_6 \boldsymbol{\sigma}\cdot\boldsymbol{\sigma}}{\sigma_3^2}\right]. \qquad (5.21)$$

As in (4.11), we assume that $P^{\text{diss}} = \sigma_{\text{eq}}\dot{\varepsilon}_{\text{eq}}$ and it follows

$$\dot{\boldsymbol{\varepsilon}}^{\text{cr}} = \dot{\varepsilon}_{\text{eq}}^{\text{cr}}\left[\alpha\mu_1\boldsymbol{I} + \frac{\mu_2 I_1 \boldsymbol{I} + \mu_3 \boldsymbol{\sigma}}{\sigma_2} + \gamma\frac{\left(\mu_4 I_1^2 + \frac{\mu_5}{3}I_2\right)\boldsymbol{I} + \frac{2}{3}\mu_5 I_1 \boldsymbol{\sigma} + \mu_6 \boldsymbol{\sigma}\cdot\boldsymbol{\sigma}}{\sigma_3^2}\right]. \qquad (5.22)$$

The generalised creep equation can be used, if the parameters μ_i $(i = 1, \ldots, 6)$ are defined by basic tests and the function $\dot{\varepsilon}_{\text{eq}}^{\text{cr}} = \dot{\varepsilon}_{\text{eq}}^{\text{cr}}(\sigma_{\text{eq}})$ is known. As in the previous section, suitable

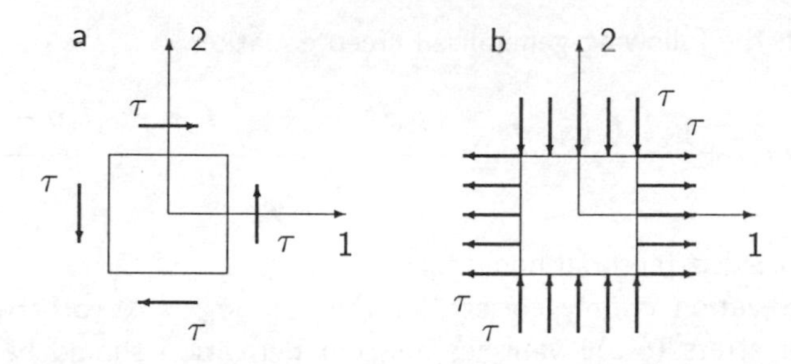

Figure 5.5: Different loading conditions leading to a shear stress state

approximations for the experimental creep data in the case of secondary creep processes are the power function, the hyperbolic sine function and the exponential function:

$$\phi(\sigma_{\mathsf{eq}}) = {\sigma_{\mathsf{eq}}}^n, \quad \phi(\sigma_{\mathsf{eq}}) = \sinh\left(\frac{\sigma_{\mathsf{eq}}}{a}\right), \quad \phi(\sigma_{\mathsf{eq}}) = \exp\left(\frac{\sigma_{\mathsf{eq}}}{b}\right), \tag{5.23}$$

n, a, b depend on the material, temperature, etc. Other proposals for approximations of the function $\dot{\varepsilon}_{\mathsf{eq}} = \phi(\sigma_{\mathsf{eq}})$ can be taken from the literature (see, e.g., [19]).
Remark: The proposed generalised creep equation has the form of the general tensorial nonlinear relation between two co–axial tensors [18, 61]

$$\dot{\boldsymbol{\varepsilon}}^{\mathsf{cr}} = H_0 \boldsymbol{I} + H_1 \boldsymbol{\sigma} + H_2 \boldsymbol{\sigma} \cdot \boldsymbol{\sigma}. \tag{5.24}$$

For isotropic materials the coefficients H_i $(i = 1, 2, 3)$ depend only on the invariants of the stress tensor $\boldsymbol{\sigma}$. The comparison of the general tensorial nonlinear relation (5.24) and the deduced constitutive equation (5.22) used here leads to

$$\begin{aligned} H_0 &= \dot{\varepsilon}^{\mathsf{cr}}_{\mathsf{eq}} \left(\alpha\mu_1 + \frac{\mu_2 I_1}{\sigma_2} + \gamma \frac{3\mu_4 I_1^2 + \mu_5 I_2}{3\sigma_3^2} \right), \\ H_1 &= \dot{\varepsilon}^{\mathsf{cr}}_{\mathsf{eq}} \left(\frac{\mu_3}{\sigma_2} + \gamma \frac{2\mu_5 I_1}{3\sigma_3^2} \right), \\ H_2 &= \dot{\varepsilon}^{\mathsf{cr}}_{\mathsf{eq}} \gamma \frac{\mu_6}{\sigma_3^2}. \end{aligned} \tag{5.25}$$

(5.22) is a tensorial nonlinear equation because quadratic terms of the stress tensor are included. Tensorial nonlinear constitutive equations lead to so-called second order effects [2, 14, 62]. For example, pure shear loadings lead to shear creep strain rates, and additional axial creep strain rates (Poynting-Swift effect) and volumetric creep strain rates (similar to the Kelvin effect in elasticity). Assuming tensorial linear constitutive equations ($\gamma = 0$) we get tensorial linear relations between the creep strain rate tensor and the stress tensor.

Let us discuss two short examples. In the first example a shear state is presumed. This state can be performed by different loading conditions. In the first case (Fig. 5.5a) this shear stress state is defined by pure shear loading

$$\boldsymbol{\sigma} = \tau(\boldsymbol{e}_1\boldsymbol{e}_2 + \boldsymbol{e}_2\boldsymbol{e}_1), \tag{5.26}$$

in the second case by bi–axial normal loading (Fig. 5.5b)

$$\boldsymbol{\sigma} = \tau(\boldsymbol{e}_1\boldsymbol{e}_1 - \boldsymbol{e}_2\boldsymbol{e}_2). \tag{5.27}$$

Let us calculate all coordinates of the creep strain rate tensor and its invariants for both cases:

- case a:
  basic invariants and equivalent stress

$$I_1 = \boldsymbol{\sigma} \cdot\cdot \boldsymbol{I} = 0, \quad I_2 = \boldsymbol{\sigma} \cdot\cdot \boldsymbol{\sigma} = 2\tau^2, \quad I_3 = (\boldsymbol{\sigma} \cdot \boldsymbol{\sigma}) \cdot\cdot \boldsymbol{\sigma} = 0, \tag{5.28}$$

$$\sigma_1 = \mu_1 I_1 = 0,\ \sigma_2^2 = \mu_2 I_1^2 + \mu_3 I_2 = 2\mu_3\tau^2,\ \sigma_3^3 = \mu_4 I_1^3 + \mu_5 I_1 I_2 + \mu_6 I_3 = 0, \tag{5.29}$$

$$\sigma_{\mathsf{eq}} = \sigma_2 = \sqrt{2\mu_3}\tau, \tag{5.30}$$

  creep strain rate tensor

$$\dot{\boldsymbol{\varepsilon}}^{\mathsf{cr}} = \dot{\varepsilon}^{\mathsf{cr}}_{\mathsf{eq}} \left(\alpha\mu_1 \boldsymbol{I} + \frac{\mu_3 \boldsymbol{\sigma}}{\sigma_2} \right) = \dot{\varepsilon}^{\mathsf{cr}}_{\mathsf{eq}} \left[\alpha\mu_1 \boldsymbol{I} + \frac{\sqrt{2\mu_3}(\boldsymbol{e}_1\boldsymbol{e}_2 + \boldsymbol{e}_2\boldsymbol{e}_1)}{2} \right], \tag{5.31}$$

$$\dot{\varepsilon}^{\mathsf{cr}}_{11} = \dot{\varepsilon}^{\mathsf{cr}}_{22} = \dot{\varepsilon}^{\mathsf{cr}}_{33} = \dot{\varepsilon}^{\mathsf{cr}}_{\mathsf{eq}}\alpha\mu_1, \quad \dot{\varepsilon}^{\mathsf{cr}}_{12} = \dot{\varepsilon}^{\mathsf{cr}}_{\mathsf{eq}} \frac{\sqrt{2\mu_3}}{2}, \quad \dot{\varepsilon}^{\mathsf{cr}}_{13} = \dot{\varepsilon}^{\mathsf{cr}}_{23} = 0, \tag{5.32}$$

  invariants of the creep strain rate tensor

$$\begin{aligned} I_1(\dot{\boldsymbol{\varepsilon}}^{\mathsf{cr}}) &= \dot{\boldsymbol{\varepsilon}}^{\mathsf{cr}} \cdot\cdot \boldsymbol{I} = \dot{\varepsilon}^{\mathsf{cr}}_{ii} = 3\dot{\varepsilon}^{\mathsf{cr}}_{\mathsf{eq}}\alpha\mu_1, \\ I_2(\dot{\boldsymbol{\varepsilon}}^{\mathsf{cr}}) &= \dot{\boldsymbol{\varepsilon}}^{\mathsf{cr}} \cdot\cdot \dot{\boldsymbol{\varepsilon}}^{\mathsf{cr}} = \left(\dot{\varepsilon}^{\mathsf{cr}}_{\mathsf{eq}}\right)^2 (3\alpha^2\mu_1^2 + \mu_3), \\ I_3(\dot{\boldsymbol{\varepsilon}}^{\mathsf{cr}}) &= (\dot{\boldsymbol{\varepsilon}}^{\mathsf{cr}} \cdot \dot{\boldsymbol{\varepsilon}}^{\mathsf{cr}}) \cdot\cdot \dot{\boldsymbol{\varepsilon}}^{\mathsf{cr}} = \left(\dot{\varepsilon}^{\mathsf{cr}}_{\mathsf{eq}}\right)^3 \alpha\mu_1 \left(\alpha^2\mu_1^2 - \frac{\mu_3}{2}\right); \end{aligned} \tag{5.33}$$

- case b:
  basic invariants and equivalent stress - see case a,
  creep strain rate tensor

$$\dot{\boldsymbol{\varepsilon}}^{\mathsf{cr}} = \dot{\varepsilon}^{\mathsf{cr}}_{\mathsf{eq}} \left[\alpha\mu_1 \boldsymbol{I} + \frac{\sqrt{2\mu_3}(\boldsymbol{e}_1\boldsymbol{e}_1 - \boldsymbol{e}_2\boldsymbol{e}_2)}{2} \right], \tag{5.34}$$

$$\dot{\varepsilon}^{\mathsf{cr}}_{11} = \dot{\varepsilon}^{\mathsf{cr}}_{\mathsf{eq}} \left(\alpha\mu_1 + \frac{\sqrt{2\mu_3}}{2} \right), \quad \dot{\varepsilon}^{\mathsf{cr}}_{22} = \dot{\varepsilon}^{\mathsf{cr}}_{\mathsf{eq}} \left(\alpha\mu_1 - \frac{\sqrt{2\mu_3}}{2} \right), \quad \dot{\varepsilon}^{\mathsf{cr}}_{33} = \dot{\varepsilon}^{\mathsf{cr}}_{\mathsf{eq}}\alpha\mu_1, \tag{5.35}$$

$$\dot{\varepsilon}^{\mathsf{cr}}_{12} = \dot{\varepsilon}^{\mathsf{cr}}_{13} = \dot{\varepsilon}^{\mathsf{cr}}_{23} = 0, \tag{5.36}$$

  invariants of the creep strain rate tensor - see case a.

The calculations show that different stress states result in different creep strain rate tensors. On the other hand, the invariant values are the same. The explanation can be given by calculation of the principal values and the principal directions for both cases. The principal values are the same, but in case 'a' the principal directions are rotated by an angle of 45^0 to the axis 1 and 2.

The second example is related to the hydrostatic pressure state

$$\boldsymbol{\sigma} = -p\boldsymbol{I}. \tag{5.37}$$

The calculations of the basic invariants, the equivalent stress and the creep strain rate tensor lead to:
basic invariants and equivalent stress

$$I_1 = -3p, \quad I_2 = 3p^2, \quad I_3 = -3p^3, \tag{5.38}$$

$$\sigma_1 = -3p\mu_1, \quad \sigma_2^2 = 9p^2\mu_2 + 3p^2\mu_3, \quad \sigma_3^3 = -27p^3\mu_4 - 9p^3\mu_5 - 3p^3\mu_6, \tag{5.39}$$

creep strain rate tensor

$$\dot{\boldsymbol{\varepsilon}}^{\mathsf{cr}} = \dot{\varepsilon}^{\mathsf{cr}}_{\mathsf{eq}} \left(\alpha\mu_1 - \frac{1}{\sqrt{3}}\sqrt{3\mu_2 + \mu_3} + \gamma\frac{1}{\sqrt[3]{9}}\sqrt[3]{9\mu_4 + 3\mu_5 + \mu_6} \right) \boldsymbol{I}, \tag{5.40}$$

$$\dot{\varepsilon}^{\mathsf{cr}}_{11} = \dot{\varepsilon}^{\mathsf{cr}}_{22} = \dot{\varepsilon}^{\mathsf{cr}}_{33}, \quad \dot{\varepsilon}^{\mathsf{cr}}_{ij} = 0 \quad (i \neq j). \tag{5.41}$$

The calculation of the invariants of the creep strain rate tensor is elementary.

5.3. Identification of the material dependent parameters

The generalised isotropic creep equation contains unknown parameters, which must be identified by tests. Several possibilities of identification procedures, based on static or dynamic tests, are known. In the case of creep problems creep tests with constant loads can be recommended[3]. The identification procedure is shown schematically in Fig. 5.6. The choice of the kind of tests depends at first on the experimental facilities. On the other hand this choice must correspond to the possibilities of receiving analytical solutions.

It is necessary to determine the parameters $\mu_1, \ldots, \mu_6$ from basic tests. In accordance to the identification procedure (Fig. 5.6) we can propose the following tests:

A. Physical tests

(a) uniaxial tension

$$\dot{\varepsilon}^{\mathsf{cr}}_{11} = L_+\sigma_{11}^n, \quad \dot{\varepsilon}^{\mathsf{cr}}_{22} = -Q\sigma_{11}^n, \tag{5.42}$$

(b) uniaxial compression

$$\dot{\varepsilon}^{\mathsf{cr}}_{11} = -L_-|\sigma_{11}|^n, \tag{5.43}$$

(c) torsion

$$\dot{\gamma}^{\mathsf{cr}}_{12} = 2\dot{\varepsilon}^{\mathsf{cr}}_{12} = N\sigma_{12}^n, \quad \dot{\varepsilon}^{\mathsf{cr}}_{11} = M\sigma_{12}^n, \tag{5.44}$$

[3]This approach is widely used in engineering applications, but should be carefully handled, because the material dependent parameters are determinated for a given constant load level. The use of state equations with these parameters is allowed only in a small range of stresses in the neighbourhood of the given load level [29]. Other possibilities (the use of functionals) are discussed in the contribution of P. Gummert in these Lecture Notes.

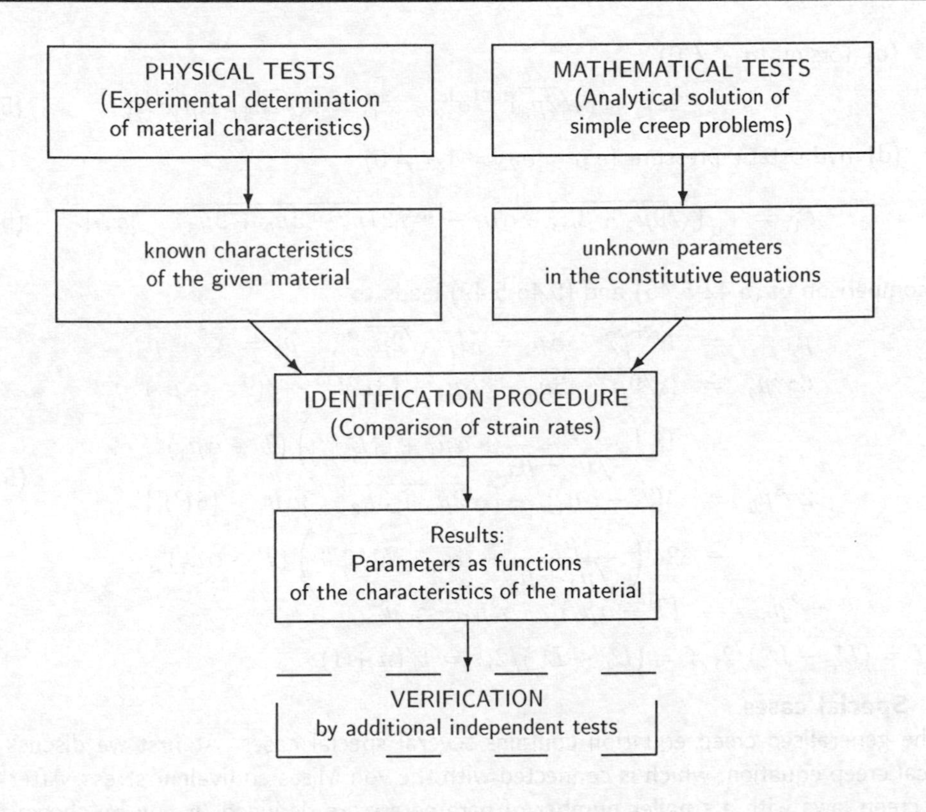

Figure 5.6: Identification procedure for the parameters in the creep equations

(d) hydrostatic pressure

$$\dot{\varepsilon}_{11}^{\mathsf{cr}} = \dot{\varepsilon}_{22}^{\mathsf{cr}} = \dot{\varepsilon}_{33}^{\mathsf{cr}} = -P|\sigma_{11}|^n, \tag{5.45}$$

L_+, L_-, Q, N, M, P, n are parameters from test data.

B. Mathematical tests

Let us assume the relation between $\dot{\varepsilon}_{\mathsf{eq}}^{\mathsf{cr}}$ and σ_{eq} in analogy to the Norton's creep law $\phi(\sigma_{\mathsf{eq}}) = \sigma_{\mathsf{eq}}{}^n$. From (5.22) we get the following results for the different loadings:

(a) uniaxial tension ($\sigma_{11} > 0$)

$$\begin{aligned} \dot{\varepsilon}_{11}^{\mathsf{cr}} &= (\alpha\mu_1 + \sqrt{\mu_2 + \mu_3} + \gamma\sqrt[3]{\mu_4 + \mu_6 + \mu_5})^{n+1}\sigma_{11}^n, \\ \dot{\varepsilon}_{22}^{\mathsf{cr}} &= (\alpha\mu_1 + \sqrt{\mu_2 + \mu_3} + \gamma\sqrt[3]{\mu_4 + \mu_6 + \mu_5})^n \\ &\times \left[\frac{\mu_2}{\sqrt{\mu_2 + \mu_3}} + \alpha\mu_1 + \gamma\frac{\mu_4 + \dfrac{\mu_5}{3}}{(\mu_4 + \mu_6 + \mu_5)^{2/3}}\right]\sigma_{11}^n, \end{aligned} \tag{5.46}$$

(b) uniaxial compression ($\sigma_{11} < 0$)

$$\dot{\varepsilon}_{11}^{\mathsf{cr}} = -(-\alpha\mu_1 + \sqrt{\mu_2 + \mu_3} - \gamma\sqrt[3]{\mu_4 + \mu_6 + \mu_5})^{n+1}|\sigma_{11}|^n, \tag{5.47}$$

(c) torsion ($\sigma_{12} \neq 0$)

$$2\dot{\varepsilon}_{12}^{\text{cr}} = (\sqrt{2\mu_3})^{n+1}\sigma_{12}^n, \quad \dot{\varepsilon}_{11}^{\text{cr}} = (\sqrt{2\mu_3})^n \alpha\mu_1\sigma_{12}^n, \tag{5.48}$$

(d) hydrostatic pressure ($\sigma_{11} = \sigma_{22} = \sigma_{33} \neq 0$)

$$\dot{\varepsilon}_{11}^{\text{cr}} = -\frac{1}{3}(\sqrt{9\mu_2 + 3\mu_3} - \alpha\mu_1 - \gamma\sqrt[3]{27\mu_4 + 3\mu_6 + 9\mu_5})^{n+1}|\sigma_{11}|^n. \tag{5.49}$$

The comparison of (5.42-5.45) and (5.46-5.49) leads to

$$\begin{aligned}
\mu_3 &= N^{2r}/2, \quad \alpha\mu_1 = M/(\sqrt{2\mu_3})^n, \quad \mu_2 = X^2 - \mu_3, \\
6\gamma^3\mu_4 &= [\sqrt{9\mu_2 + 3\mu_3} - 3\alpha\mu_1 - (3P)^r]^3 - 3(T - \alpha\mu_1)^3 \\
&+ 18\left(\frac{\mu_2}{\sqrt{\mu_2 + \mu_3}} + \alpha\mu_1 + QL_+^{-nr}\right)(T - \alpha\mu_1)^2, \\
2\gamma^3\mu_5 &= 3(T - \alpha\mu_1)^3 - [\sqrt{9\mu_2 + 3\mu_3} - 3\alpha\mu_1 - (3P)^r]^3 \\
&- 24\left(\frac{\mu_2}{\sqrt{\mu_2 + \mu_3}} + \alpha\mu_1 + QL_+^{-nr}\right)(T - \alpha\mu_1)^2, \\
\gamma^3\mu_6 &= (T - \alpha\mu_1)^3 - \gamma^3\mu_4 - \gamma^3\mu_5,
\end{aligned} \tag{5.50}$$

with $T = (L_+^r - L_-^r)/2, X = (L_+^r + L_-^r)/2, r = 1/(n+1)$.

5.4. Special cases

The generalised creep equation contains several special cases. At first we discuss the classical creep equation, which is connected with the von Mises equivalent stress. After this some creep laws with a smaller number of parameters are deduced. It can be shown that different models with three parameters can be introduced and we get tensorial linear or nonlinear equations.

The classical creep equation, based on the von Mises equivalent stress, can be derived from the generalised equation (5.22) in connection with the equivalent stress and the following parameters

$$\alpha = \gamma = 0, \quad \mu_2 = -1/2, \quad \mu_3 = 3/2, \tag{5.51}$$

$$\sigma_{\text{eq}} = \sigma_2 = \sqrt{-\frac{1}{2}I_1^2 + \frac{3}{2}I_2} = \sqrt{\frac{3}{2}\boldsymbol{s} \cdot\cdot \boldsymbol{s}} = \sigma_{\text{vM}}. \tag{5.52}$$

From this the creep rate strain tensor follows

$$\dot{\boldsymbol{\varepsilon}}^{\text{cr}} = \phi\left(\sqrt{\frac{3}{2}\boldsymbol{s} \cdot\cdot \boldsymbol{s}}\right)\frac{3\boldsymbol{\sigma} - I_1\boldsymbol{I}}{2\sqrt{\frac{3}{2}\boldsymbol{s} \cdot\cdot \boldsymbol{s}}} = \frac{3}{2}\frac{\phi(\sigma_{\text{vM}})}{\sigma_{\text{vM}}}\boldsymbol{s}, \tag{5.53}$$

which is identical to [21, 18] and the similar equation in subsection 4.2. The classical model can be recommended, if material constants, determined from basic tests, approximately fulfill the relations

$$L_+ = L_-, \quad 3L_+^{2r} = N^{2r}, \quad M = P = 0. \tag{5.54}$$

These conditions follow from (5.50) with respect to the values (5.51).

Let us assume identical behaviour in tension and compression and neglect the Poynting-Swift-effect and the influence of the hydrostatic pressure. Setting $\alpha\mu_1 = 0, \gamma = 0$ we get the equivalent stress as

$$\sigma_{\text{eq}} = \sigma_2 = \sqrt{\mu_2 I_1^2 + \mu_3 I_2}, \tag{5.55}$$

which leads to

$$\dot{\boldsymbol{\varepsilon}}^{\text{cr}} = \phi(\sigma_2)\left(\frac{\mu_2 I_1 \boldsymbol{I} + \mu_3 \boldsymbol{\sigma}}{\sigma_2}\right). \tag{5.56}$$

This is a tensorial linear equation with two parameters (μ_2, μ_3). They can be determined from uniaxial tension and torsion tests (L_+, N) or only from uniaxial tension test (L_+, Q). The constitutive equation (5.56) can be recommended, if the relations

$$L_+ = L_-, \quad M = 0, \quad 9L_+^{2r} - 3N^{2r} = (3P)^{2r} \tag{5.57}$$

are experimentally obtained.

Assuming no influence of the third invariant $(\gamma = 0)$, the strain rate can be expressed as

$$\dot{\boldsymbol{\varepsilon}}^{\text{cr}} = \phi(\sigma_{\text{eq}})\left(\alpha\mu_1 \boldsymbol{I} + \frac{\mu_2 I_1 \boldsymbol{I} + \mu_3 \boldsymbol{\sigma}}{\sigma_2}\right). \tag{5.58}$$

For this equation the relations

$$T = MN^{-nr}, \quad \sqrt{9X^2 - 3N^{2r}} = 3T + (3P)^r \tag{5.59}$$

between the characteristics of the material should be obtained from tests.

Including only the quadratic and cubic invariants $(\alpha\mu_1 = \mu_4 = \mu_5 = 0)$, we get a tensorial nonlinear equation

$$\dot{\boldsymbol{\varepsilon}}^{\text{cr}} = \phi(\sigma_{\text{eq}})\left(\frac{\mu_2 I_1 \boldsymbol{I} + \mu_3 \boldsymbol{\sigma}}{\sigma_2} + \gamma \frac{\mu_6 \boldsymbol{\sigma} \cdot \boldsymbol{\sigma}}{\sigma_3^2}\right). \tag{5.60}$$

For this special case we should obtain from tests

$$3T^3 - \left[\sqrt{9X^2 - 3N^{2r}} - (3P)^r\right]^3 = Y = M = 0 \tag{5.61}$$

with $Y = X = N^{2r}/(2X) + QL_+^{-nr}$. Another tensorial linear equation can be deduced, setting $\alpha\mu_1 = \mu_4 = \mu_6 = 0$

$$\dot{\boldsymbol{\varepsilon}}^{\text{cr}} = \phi(\sigma_{\text{eq}})\left[\frac{\mu_2 I_1 \boldsymbol{I} + \mu_3 \boldsymbol{\sigma}}{\sigma_2} + \gamma \frac{\mu_5 (I_2 \boldsymbol{I} + 2 I_1 \boldsymbol{\sigma})}{3\sigma_3^2}\right]. \tag{5.62}$$

This equation can be recommended for use, if

$$\left[\sqrt{9X^2 - 3N^{2r}} - (3P)^r\right]^3 - 9T^3 = T + 3Y = M = 0 \tag{5.63}$$

is obtained from tests.

6. EXTENSIONS OF THE NON–CLASSICAL CREEP LAW

Below we discuss possible extensions of the non–classical creep law. Two interesting cases are creep-damage coupling and anisotropic behaviour.

6.1. Extension I: Creep-damage coupling

For describing the tertiary creep behaviour with the primary creep neglected (cp. Fig. 6.1) we can use the generalised isotropic creep equation for secondary creep. A possible

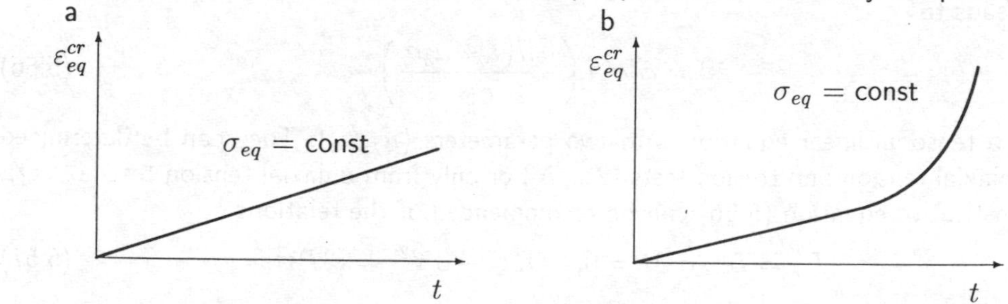

Figure 6.1: Creep curves: a - secondary creep, b - secondary and tertiary creep

extension of the generalised equation follows by introducing a damage variable and an evolution law for this variable. According to the papers of Kachanov, Rabotnov and others the specific dissipation power P^{diss} can be proposed as a measure of the intensity of creep φ

$$P^{\text{diss}} = \boldsymbol{\sigma} \cdot\cdot \dot{\boldsymbol{\varepsilon}}^{\text{cr}}, \quad \varphi = \int\limits_0^t P^{\text{diss}}\, dt \tag{6.1}$$

At $t \equiv 0$, we assume that no damage has occurred: $\varphi(t = 0) = \varphi_0 = 0$. If $t = t_*$ the dissipated energy is independent on the kind of loading

$$\varphi(t = t_*) = \varphi_* = \text{const.} \tag{6.2}$$

In real materials an ideal undamaged state is never obtained on the occasion of manufacturing, etc. The independence of the final state on the kind of loading can be observed in tests [63].

Let us assume that

$$P^{\text{diss}} = f(\sigma_{\text{eq}}, \varphi) \tag{6.3}$$

with

$$f(\sigma_{\text{eq}}, \varphi) = \kappa(\sigma_{\text{eq}}) \frac{\varphi_*^m}{(\varphi_* - \varphi)^m}, \tag{6.4}$$

$m, \kappa(\sigma_{\text{eq}})$ and φ_* should be determined by tests. For the introduced damage variable the relation to Rabotnov's damage parameter ω [18] can be shown

$$\omega = \frac{\varphi}{\varphi_*}, \quad 0 \le \omega \le 1. \tag{6.5}$$

Thus we get

$$P^{\text{diss}} = \kappa(\sigma_{\text{eq}})\frac{1}{(1-\omega)^m}. \tag{6.6}$$

On the other hand

$$P^{\text{diss}} = \sigma_{\text{eq}}\dot{\varepsilon}^{\text{cr}}_{\text{eq}}, \tag{6.7}$$

which leads to

$$\dot{\varepsilon}^{\text{cr}}_{\text{eq}} = \frac{P^{\text{diss}}}{\sigma_{\text{eq}}} = \frac{g(\sigma_{\text{eq}})}{(1-\omega)^m} \tag{6.8}$$

with

$$g(\sigma_{\text{eq}}) = \frac{\kappa(\sigma_{\text{eq}})}{\sigma_{\text{eq}}}. \tag{6.9}$$

Substituting ω by φ/φ_* we get

$$\dot{\varepsilon}^{\text{cr}}_{\text{eq}} = g(\sigma_{\text{eq}})\frac{\varphi_*^m}{(\varphi_* - \varphi)^m} \tag{6.10}$$

and the constitutive equation

$$\begin{aligned} \dot{\boldsymbol{\varepsilon}}^{\text{cr}} &= g(\sigma_{\text{eq}})\frac{\varphi_*^m}{(\varphi_* - \varphi)^m} \\ &\times \left[\alpha\mu_1 \boldsymbol{I} + \frac{\mu_2 I_1 \boldsymbol{I} + \mu_3 \boldsymbol{\sigma}}{\sigma_2} + \gamma\frac{\mu_4 I_1^2 \boldsymbol{I} + \mu_6 \boldsymbol{\sigma}\cdot\boldsymbol{\sigma} + \frac{\mu_5}{3}(I_2\boldsymbol{I} + 2I_1\boldsymbol{\sigma})}{\sigma_3^2}\right], \end{aligned} \tag{6.11}$$

which should be completed by an evolution equation

$$\dot{\varphi} = \kappa(\sigma_{\text{eq}})\frac{\varphi_*^m}{(\varphi_* - \varphi)^m}. \tag{6.12}$$

The unknown parameters μ_i must be determined from basic tests.

In accordance to the identification procedure (Fig. 5.6) we can propose now the following tests:

A. Physical Tests

(a) uniaxial tension

$$\begin{aligned} \dot{\varepsilon}^{\text{cr}}_{11} &= L_+\sigma_{11}^n\frac{\varphi_*^m}{(\varphi_* - \varphi)^m}, \\ \dot{\varepsilon}^{\text{cr}}_{22} &= -Q\sigma_{11}^n\frac{\varphi_*^m}{(\varphi_* - \varphi)^m}. \end{aligned} \tag{6.13}$$

The evolution equation for both strain rates is assumed to have the same form

$$\dot{\varphi} = L_+\sigma_{11}^{n+1}\frac{\varphi_*^m}{(\varphi_* - \varphi)^m}. \tag{6.14}$$

The evolution equation follows from the assumption that the damage can be characterised by a scalar-valued damage variable. The φ_* should be identified experimentally.

(b) uniaxial compression

$$\begin{aligned} \dot{\varepsilon}_{11}^{\text{cr}} &= -L_-|\sigma_{11}|^n \frac{\varphi_*^m}{(\varphi_* - \varphi)^m}, \\ \dot{\varphi} &= L_-|\sigma_{11}|^{n+1} \frac{\varphi_*^m}{(\varphi_* - \varphi)^m}. \end{aligned} \tag{6.15}$$

(c) torsion

$$\begin{aligned} \dot{\gamma}_{12}^{\text{cr}} &= 2\dot{\varepsilon}_{12}^{\text{cr}} = N\sigma_{12}^n \frac{\varphi_*^m}{(\varphi_* - \varphi)^m}, \\ \dot{\varepsilon}_{11}^{\text{cr}} &= M\sigma_{12}^n \frac{\varphi_*^m}{(\varphi_* - \varphi)^m}. \end{aligned} \tag{6.16}$$

The evolution equation for both strain rates is assumed to be the same

$$\dot{\varphi} = N\sigma_{12}^{n+1} \frac{\varphi_*^m}{(\varphi_* - \varphi)^m}. \tag{6.17}$$

(d) hydrostatic pressure

$$\begin{aligned} \dot{\varepsilon}_{11}^{\text{cr}} &= \dot{\varepsilon}_{22}^{\text{cr}} = \dot{\varepsilon}_{33}^{\text{cr}} = -P|\sigma_{11}|^n \frac{\varphi_*^m}{(\varphi_* - \varphi)^m}, \\ \dot{\varphi} &= 3P|\sigma_{11}|^{n+1} \frac{\varphi_*^m}{(\varphi_* - \varphi)^m}. \end{aligned} \tag{6.18}$$

$L_+, L_-, Q, N, M, P, n, m$ and φ_* characterise the material behaviour.

B. Mathematical Tests

Let us presume $g(\sigma_{\text{eq}}) = {\sigma_{\text{eq}}}^n$. From (6.11) we get the following results for the different loading cases:

(a) uniaxial tension ($\sigma_{11} > 0$)

$$\begin{aligned} \dot{\varepsilon}_{11}^{\text{cr}} &= (\sqrt{\mu_2 + \mu_3} + \alpha\mu_1 + \gamma\sqrt[3]{\mu_4 + \mu_5 + \mu_6})^{n+1} \sigma_{11}^n \frac{\varphi_*^m}{(\varphi_* - \varphi)^m}, \\ \dot{\varepsilon}_{22}^{\text{cr}} &= (\sqrt{\mu_2 + \mu_3} + \alpha\mu_1 + \gamma\sqrt[3]{\mu_4 + \mu_5 + \mu_6})^n \\ &\times \left[\frac{\mu_2}{\sqrt{\mu_2 + \mu_3}} + \alpha\mu_1 + \gamma \frac{\mu_4 + \dfrac{\mu_5}{3}}{(\mu_4 + \mu_5 + \mu_6)^{2/3}} \right] \sigma_{11}^n \frac{\varphi_*^m}{(\varphi_* - \varphi)^m}, \\ \dot{\varphi} &= (\sqrt{\mu_2 + \mu_3} + \alpha\mu_1 + \gamma\sqrt[3]{\mu_4 + \mu_5 + \mu_6})^{n+1} \sigma_{11}^{n+1} \frac{\varphi_*^m}{(\varphi_* - \varphi)^m}, \end{aligned} \tag{6.19}$$

(b) uniaxial compression ($\sigma_{11} < 0$)

$$\dot{\varepsilon}_{11}^{\mathrm{cr}} = -(\sqrt{\mu_2+\mu_3} - \alpha\mu_1 - \gamma\sqrt[3]{\mu_4+\mu_5+\mu_6})^{n+1}|\sigma_{11}|^n \frac{\varphi_*^m}{(\varphi_*-\varphi)^m},$$

$$\dot{\varphi} = (\sqrt{\mu_2+\mu_3} - \alpha\mu_1 - \gamma\sqrt[3]{\mu_4+\mu_5+\mu_6})^{n+1}|\sigma_{11}|^{n+1} \frac{\varphi_*^m}{(\varphi_*-\varphi)^m}, \tag{6.20}$$

(c) torsion ($\sigma_{12} \neq 0$)

$$2\dot{\varepsilon}_{12}^{\mathrm{cr}} = (\sqrt{2\mu_3})^{n+1}\sigma_{12}^n \frac{\varphi_*^m}{(\varphi_*-\varphi)^m},$$

$$\dot{\varepsilon}_{11}^{\mathrm{cr}} = (\sqrt{2\mu_3})^{n}\alpha\mu_1\sigma_{12}^n \frac{\varphi_*^m}{(\varphi_*-\varphi)^m}, \tag{6.21}$$

$$\dot{\varphi} = (\sqrt{2\mu_3})^{n+1}\sigma_{12}^{n+1} \frac{\varphi_*^m}{(\varphi_*-\varphi)^m},$$

(d) hydrostatic pressure ($\sigma_{11} = \sigma_{22} = \sigma_{33} \neq 0$)

$$\dot{\varepsilon}_{11}^{\mathrm{cr}} = -\frac{1}{3}(\sqrt{9\mu_2+3\mu_3} - \alpha\mu_1 - \gamma\sqrt[3]{27\mu_4+9\mu_5+3\mu_6})^{n+1}|\sigma_{11}|^n \frac{\varphi_*^m}{(\varphi_*-\varphi)^m},$$

$$\dot{\varphi} = (\sqrt{9\mu_2+3\mu_3} - \alpha\mu_1 - \gamma\sqrt[3]{27\mu_4+9\mu_5+3\mu_6})^{n+1}|\sigma_{11}|^{n+1} \frac{\varphi_*^m}{(\varphi_*-\varphi)^m}. \tag{6.22}$$

Comparing the results of the physical and mathematical tests, we receive the unknown material parameters μ_i as functions of the material characteristics. Due to the fact that the μ_i are identified for secondary creep behaviour only the non–classical material behaviour effects are included mainly into the secondary creep. The introduction of the damage variable leads to a material behaviour model for the tertiary creep similar to the Kachanov-Rabotnov approach.

In analogy to the previous section the classical creep theory follows from the general creep rate equation, if the following conditions for the characteristics are obtained in tests

$$L_+ = L_-, \quad M = P = 0, \quad N^{2r} = 3L_+^{2r}. \tag{6.23}$$

Considering

$$\alpha = \gamma = 0, \quad \mu_3 = -3\mu_2, \quad \mu_2 = -\frac{1}{2}, \tag{6.24}$$

we finally get the strain rate equation and the damage evolution equation [3]

$$\dot{\boldsymbol{\varepsilon}}^{\mathrm{cr}} = \frac{3}{2}\frac{g(\sigma_{\mathrm{vM}})\varphi_*^m}{(\varphi_*-\varphi)^m\sigma_{\mathrm{vM}}}\boldsymbol{s}, \quad \dot{\varphi} = \kappa(\sigma_{\mathrm{vM}})\frac{\varphi_*^m}{(\varphi_*-\varphi)^m}. \tag{6.25}$$

Taking into account $\omega = \varphi/\varphi_*$ the classical creep-damage equations follow:

$$\dot{\varepsilon}^{\text{cr}} = \frac{3}{2} g(\sigma_{\text{vM}}) \frac{1}{(1-\omega)^m} \frac{\boldsymbol{s}}{\sigma_{\text{vM}}}, \quad \dot{\omega} = \frac{\kappa(\sigma_{\text{vM}})}{\varphi_*} \frac{1}{(1-\omega)^m}. \tag{6.26}$$

These equations are similar to [27] or subsection 4.4.

6.2. Extension II: Anisotropic behaviour

Assuming again the existence of a creep potential

$$\Phi = \Phi(\sigma_{\text{eq}}), \tag{6.27}$$

the constitutive equation follows from

$$\dot{\boldsymbol{\varepsilon}}^{\text{cr}} = \dot{\eta} \frac{\partial \Phi}{\partial \boldsymbol{\sigma}}, \tag{6.28}$$

with $\dot{\eta}$ as a scalar and σ_{eq} as the equivalent stress, which is a function depending on

- the stress tensor and
- some material tensors (tensors of the material constants)

The physical state of an anisotropic continuum can be described with the help of different tensors of the material constants, e.g., $\boldsymbol{a}$, ${}^{(4)}\boldsymbol{b}$, ${}^{(6)}\boldsymbol{c}$. Using the mixed invariants

$$\sigma_1 = \boldsymbol{a} \cdot\cdot\, \boldsymbol{\sigma}, \quad \sigma_2^2 = \boldsymbol{\sigma} \cdot\cdot\, {}^{(4)}\boldsymbol{b} \cdot\cdot\, \boldsymbol{\sigma}, \quad \sigma_3^3 = \boldsymbol{\sigma} \cdot\cdot\, (\boldsymbol{\sigma} \cdot\cdot\, {}^{(6)}\boldsymbol{c} \cdot\cdot\, \boldsymbol{\sigma}), \tag{6.29}$$

the equivalent stress results in

$$\sigma_{\text{eq}} = \alpha\sigma_1 + \sigma_2 + \gamma\sigma_3. \tag{6.30}$$

This is a suitable generalisation of the non–classical creep law reported in section 5.. The equivalent stress expression, proposed in [64], can be deduced from (6.30) setting $\alpha \equiv \gamma \equiv 0$

$$\sigma_{\text{eq}} = \sigma_2. \tag{6.31}$$

Assuming $\alpha \neq \gamma \neq 0$ after some calculations we get the creep strain rate tensor

$$\dot{\boldsymbol{\varepsilon}}^{\text{cr}} = \dot{\eta} \frac{\partial \Phi(\sigma_{\text{eq}})}{\partial \sigma_{\text{eq}}} \left(\alpha \frac{\partial \sigma_1}{\partial \boldsymbol{\sigma}} + \frac{\partial \sigma_2}{\partial \boldsymbol{\sigma}} + \gamma \frac{\partial \sigma_3}{\partial \boldsymbol{\sigma}} \right) \tag{6.32}$$

and with respect to

$$\frac{\partial \sigma_1}{\partial \boldsymbol{\sigma}} = \boldsymbol{a}, \quad \frac{\partial \sigma_2}{\partial \boldsymbol{\sigma}} = \frac{{}^{(4)}\boldsymbol{b} \cdot\cdot\, \boldsymbol{\sigma}}{\sigma_2}, \quad \frac{\partial \sigma_3}{\partial \boldsymbol{\sigma}} = \frac{\boldsymbol{\sigma} \cdot\cdot\, {}^{(6)}\boldsymbol{c} \cdot\cdot\, \boldsymbol{\sigma}}{\sigma_3^2}, \tag{6.33}$$

the generalised anisotropic creep law can be obtained

$$\dot{\boldsymbol{\varepsilon}}^{\text{cr}} = \dot{\eta} \frac{\partial \Phi}{\partial \sigma_{\text{eq}}} \left(\alpha \boldsymbol{a} + \frac{{}^{(4)}\boldsymbol{b} \cdot\cdot\, \boldsymbol{\sigma}}{\sigma_2} + \gamma \frac{\boldsymbol{\sigma} \cdot\cdot\, {}^{(6)}\boldsymbol{c} \cdot\cdot\, \boldsymbol{\sigma}}{\sigma_3^2} \right). \tag{6.34}$$

The determination of $\dot{\eta}$ can be related to the invariant P^{diss} (the product of the creep strain rate tensor and the stress tensor in uniaxial or multiaxial cases). After some calculations (similar to section 5.) follows

$$\dot{\boldsymbol{\varepsilon}}^{\text{cr}} = \frac{P^{\text{diss}}}{\sigma_{\text{eq}}} \left(\alpha \boldsymbol{a} + \frac{{}^{(4)}\boldsymbol{b} \cdot\cdot\, \boldsymbol{\sigma}}{\sigma_2} + \gamma \frac{\boldsymbol{\sigma} \cdot\cdot\, {}^{(6)}\boldsymbol{c} \cdot\cdot\, \boldsymbol{\sigma}}{\sigma_3^2} \right), \tag{6.35}$$

or considering

$$P^{\mathsf{diss}} = \sigma_{\mathsf{eq}} \dot{\varepsilon}^{\mathsf{cr}}_{\mathsf{eq}}, \tag{6.36}$$

we finally get

$$\dot{\boldsymbol{\varepsilon}}^{\mathsf{cr}} = \dot{\varepsilon}^{\mathsf{cr}}_{\mathsf{eq}} \left(\alpha \boldsymbol{a} + \frac{{}^{(4)}\boldsymbol{b} \cdot\cdot \boldsymbol{\sigma}}{\sigma_2} + \gamma \frac{\boldsymbol{\sigma} \cdot\cdot {}^{(6)}\boldsymbol{c} \cdot\cdot \boldsymbol{\sigma}}{\sigma_3^2} \right). \tag{6.37}$$

The unified generalised anisotropic constitutive equation (6.37) corresponds to the generalised anisotropic creep equation proposed in [18, 65]

$$\dot{\boldsymbol{\varepsilon}}^{\mathsf{cr}} = \boldsymbol{H} + {}^{(4)}\boldsymbol{M} \cdot\cdot \boldsymbol{\sigma} + ({}^{(6)}\boldsymbol{L} \cdot\cdot \boldsymbol{\sigma}) \cdot\cdot \boldsymbol{\sigma}. \tag{6.38}$$

From the comparison of (6.37) and (6.38) the material tensors $\boldsymbol{H}$, ${}^{(4)}\boldsymbol{M}$ and ${}^{(6)}\boldsymbol{L}$ can be obtained.

Equations with a reduced number of parameters can also be derived:

- $\alpha = 1, \gamma = 0$ leads to

$$\sigma_{\mathsf{eq}} = \sigma_1 + \sigma_2, \quad \dot{\boldsymbol{\varepsilon}}^{\mathsf{cr}} = \dot{\varepsilon}^{\mathsf{cr}}_{\mathsf{eq}} \left(\boldsymbol{a} + \frac{{}^{(4)}\boldsymbol{b} \cdot\cdot \boldsymbol{\sigma}}{\sigma_2} \right), \tag{6.39}$$

- whereas for $\alpha = 0, \gamma = 1$ follows

$$\sigma_{\mathsf{eq}} = \sigma_2 + \sigma_3, \quad \dot{\boldsymbol{\varepsilon}}^{\mathsf{cr}} = \dot{\varepsilon}^{\mathsf{cr}}_{\mathsf{eq}} \left(\frac{{}^{(4)}\boldsymbol{b} \cdot\cdot \boldsymbol{\sigma}}{\sigma_2} + \frac{\boldsymbol{\sigma} \cdot\cdot {}^{(4)}\boldsymbol{b} \cdot\cdot \boldsymbol{\sigma}}{\sigma_3^2} \right). \tag{6.40}$$

In the first case we get a tensorial linear constitutive equation, in the second case a tensorial nonlinear one. The tensors $\boldsymbol{a}$, ${}^{(4)}\boldsymbol{b}$ and ${}^{(6)}\boldsymbol{c}$ contain 819 material parameters ($\boldsymbol{a}$ - 9, ${}^{(4)}\boldsymbol{b}$ - 81, ${}^{(6)}\boldsymbol{c}$ - 729), which should be determined from tests. The solution of this problem is impossible. Thus we need some simplification for practical use of the generalised anisotropic equation. The first simplification follows from the symmetry of the stress tensor and the kinematical tensor and from the assumption of the existence of a potential. From this we get a reduction to 83 characteristics ($\boldsymbol{a}$ - 6, ${}^{(4)}\boldsymbol{b}$ - 21, ${}^{(6)}\boldsymbol{c}$ - 56).

With the help of the transformation rules of tensors we can deduce simplified relations for special forms of anisotropy:

- Orthotropic material

$$\begin{aligned} \dot{\varepsilon}^{\mathsf{cr}}_{11} &= \dot{\varepsilon}^{\mathsf{cr}}_{\mathsf{eq}} \Bigg\{ \alpha a_{11} + \frac{b_{1111}\sigma_{11} + b_{1122}\sigma_{22} + b_{1133}\sigma_{33}}{\sigma_2} \\ &\quad + \gamma \Bigg[\frac{c_{111111}\sigma_{11}^2 + c_{112222}\sigma_{22}^2 + c_{113333}\sigma_{33}^2}{\sigma_3^2} \\ &\quad + \frac{2(c_{111122}\sigma_{11}\sigma_{22} + c_{112233}\sigma_{22}\sigma_{33} + c_{111133}\sigma_{11}\sigma_{33})}{\sigma_3^2} \\ &\quad + \frac{4(c_{111212}\sigma_{12}^2 + c_{112323}\sigma_{23}^2 + c_{111313}\sigma_{13}^2)}{\sigma_3^2} \Bigg] \Bigg\}, \end{aligned} \tag{6.41}$$

$$\dot{\varepsilon}_{12}^{\mathrm{cr}} = \dot{\varepsilon}_{\mathrm{eq}}^{\mathrm{cr}}\left[\frac{2b_{1212}\sigma_{12}}{\sigma_2} + \gamma\left(\frac{4c_{121211}\sigma_{12}\sigma_{11} + 4c_{121222}\sigma_{12}\sigma_{22}}{\sigma_3^2}\right.\right.$$
$$\left.\left. + \frac{4c_{121233}\sigma_{12}\sigma_{33} + 8c_{122313}\sigma_{23}\sigma_{13}}{\sigma_3^2}\right)\right].$$

The other elements of the creep strain rate tensor can be calculated by cyclic exchange of the indices 1, 2 and 3. The invariants can be computed to be:

$$\begin{aligned}
\sigma_1 &= a_{11}\sigma_{11} + a_{22}\sigma_{22} + a_{33}\sigma_{33},\\
\sigma_2^2 &= b_{1111}\sigma_{11}^2 + b_{2222}\sigma_{22}^2 + b_{3333}\sigma_{33}^2\\
&+ 2b_{1122}\sigma_{11}\sigma_{22} + 2b_{2233}\sigma_{22}\sigma_{33} + 2b_{1133}\sigma_{11}\sigma_{33}\\
&+ 4b_{1212}\sigma_{12}^2 + 4b_{2323}\sigma_{23}^2 + 4b_{1313}\sigma_{13}^2,\\
\sigma_3^3 &= c_{111111}\sigma_{11}^3 + c_{222222}\sigma_{22}^3 + c_{333333}\sigma_{33}^3 + 3c_{111122}\sigma_{11}^2\sigma_{22} + 3c_{111133}\sigma_{11}^2\sigma_{33}\\
&+ 3c_{222211}\sigma_{22}^2\sigma_{11} + 3c_{222233}\sigma_{22}^2\sigma_{33} + 3c_{333311}\sigma_{33}^2\sigma_{11} + 3c_{333322}\sigma_{33}^2\sigma_{22}\\
&+ 6c_{112233}\sigma_{11}\sigma_{22}\sigma_{33}\\
&+ 12c_{121211}\sigma_{12}^2\sigma_{11} + 12c_{121222}\sigma_{12}^2\sigma_{22} + 12c_{121233}\sigma_{12}^2\sigma_{33}\\
&+ 12c_{232311}\sigma_{23}^2\sigma_{11} + 12c_{232322}\sigma_{23}^2\sigma_{22} + 12c_{232333}\sigma_{23}^2\sigma_{33}\\
&+ 12c_{131311}\sigma_{13}^2\sigma_{11} + 12c_{131322}\sigma_{13}^2\sigma_{22} + 12c_{131333}\sigma_{13}^2\sigma_{33}\\
&+ 48c_{122313}\sigma_{12}\sigma_{23}\sigma_{13}.
\end{aligned} \tag{6.42}$$

In the case of orthotropic material behaviour the number of independent elements reduces to 32 ($\boldsymbol{a}$ - 3, ${}^{(4)}\boldsymbol{b}$ - 9, ${}^{(6)}\boldsymbol{c}$ - 20).

- Isotropic material behaviour

$$\begin{aligned}
a_{ij} &= \mu_1\delta_{ij},\\
b_{ijkl} &= \mu_2\delta_{ij}\delta_{kl} + \frac{1}{2}\mu_3(\delta_{ik}\delta_{jl} + \delta_{li}\delta_{jk}),\\
c_{ijklmn} &= \mu_4\delta_{ij}\delta_{kl}\delta_{mn}\\
&+ \frac{\mu_5}{6}(\delta_{ij}\delta_{km}\delta_{ln} + \delta_{ij}\delta_{kn}\delta_{lm} + \delta_{kl}\delta_{im}\delta_{jn}\\
&+ \delta_{kl}\delta_{in}\delta_{jm} + \delta_{mn}\delta_{ik}\delta_{jl} + \delta_{mn}\delta_{il}\delta_{jk})\\
&+ \frac{\mu_6}{8}(\delta_{ik}\delta_{jm}\delta_{ln} + \delta_{ik}\delta_{jn}\delta_{lm} + \delta_{il}\delta_{km}\delta_{jn} + \delta_{il}\delta_{kn}\delta_{jm}\\
&+ \delta_{im}\delta_{kj}\delta_{ln} + \delta_{im}\delta_{kn}\delta_{lj} + \delta_{in}\delta_{kj}\delta_{lm} + \delta_{in}\delta_{km}\delta_{lj}).
\end{aligned} \tag{6.43}$$

The number of linear-independent elements is reduced to 6 ($\boldsymbol{a}$ - 1, ${}^{(4)}\boldsymbol{b}$ - 2, ${}^{(6)}\boldsymbol{c}$ - 3). An additional reduction can be obtained if the incompressibility condition is used.

An anisotropic creep-damage model can be deduced analogously. The starting point is the following particular form of the potential

$$\Phi = \sigma_{\text{eq}} \tag{6.44}$$

and the generalised anisotropic creep rate equation (6.34)

$$\dot{\boldsymbol{\varepsilon}}^{\text{cr}} = 2\dot{\eta}\sigma_{\text{eq}}\left(\alpha\boldsymbol{a} + \frac{{}^{(4)}\boldsymbol{b}\cdot\cdot\boldsymbol{\sigma}}{\sigma_2} + \gamma\frac{\boldsymbol{\sigma}\cdot\cdot{}^{(6)}\boldsymbol{c}\cdot\cdot\boldsymbol{\sigma}}{\sigma_3^2}\right). \tag{6.45}$$

Introducing again the specific dissipation power:

$$2\dot{\eta}\sigma_{\text{eq}} = \frac{P^{\text{diss}}}{\sigma_{\text{eq}}}, \tag{6.46}$$

we can presume a relation between the creep and the damage processes

$$P^{\text{diss}} = f(\sigma_{\text{eq}},\varphi) = \kappa(\sigma_{\text{eq}})\frac{\varphi_*^m}{(\varphi_* - \varphi)^m} \tag{6.47}$$

The constitutive equation then leads to

$$\dot{\boldsymbol{\varepsilon}}^{\text{cr}} = g(\sigma_{\text{eq}})\frac{\varphi_*^m}{(\varphi_* - \varphi)^m}\left(\alpha\boldsymbol{a} + \frac{{}^{(4)}\boldsymbol{b}\cdot\cdot\boldsymbol{\sigma}}{\sigma_2} + \gamma\frac{\boldsymbol{\sigma}\cdot\cdot{}^{(6)}\boldsymbol{c}\cdot\cdot\boldsymbol{\sigma}}{\sigma_3^2}\right), \tag{6.48}$$

which should be completed by an evolution equation

$$\dot{\varphi} = \kappa(\sigma_{\text{eq}})\frac{\varphi_*^m}{(\varphi_* - \varphi)^m}, \tag{6.49}$$

with

$$\kappa(\sigma_{\text{eq}}) = g(\sigma_{\text{eq}})\sigma_{\text{eq}}. \tag{6.50}$$

The identification of the generalised anisotropic creep-damage law is very complicated, too. The material parameter tensors include a great number of parameters, which should be determined from tests. In this case reduced models can be introduced.

Let us assume a simplified expression for the equivalent stress, which results from the general case $\alpha \equiv 1$ and $\gamma \equiv 0$. It results in

$$\dot{\boldsymbol{\varepsilon}}^{\text{cr}} = (\sigma_1+\sigma_2)^n\left(\boldsymbol{a} + \frac{{}^{(4)}\boldsymbol{b}\cdot\cdot\boldsymbol{\sigma}}{\sigma_2}\right) = (\boldsymbol{a}\cdot\cdot\boldsymbol{\sigma}+\sqrt{\boldsymbol{\sigma}\cdot\cdot{}^{(4)}\boldsymbol{b}\cdot\cdot\boldsymbol{\sigma}})^n\left(\boldsymbol{a} + \frac{{}^{(4)}\boldsymbol{b}\cdot\cdot\boldsymbol{\sigma}}{\sigma_2}\right). \tag{6.51}$$

Presuming that the coordinate axes are identical to the principle directions of orthotropy and a plane stress state can be obtained in the material, the following basic tests for the identification of the material parameters can be proposed:

- uniaxial tension in direction 1

$$\begin{aligned} \dot{\varepsilon}_{11}^{\text{cr}} &= \left(a_{11} + \sqrt{b_{1111}}\right)^{n+1}\sigma_{11}^n, \\ \dot{\varepsilon}_{22}^{\text{cr}} &= \left(a_{11} + \sqrt{b_{1111}}\right)^n\left(a_{22} + \frac{b_{2211}}{\sqrt{b_{1111}}}\right)\sigma_{11}^n. \end{aligned} \tag{6.52}$$

From experimental data we get

$$\dot{\varepsilon}_{11}^{\text{cr}} = D_1^+\sigma_{11}^n,\ \dot{\varepsilon}_{22}^{\text{cr}} = -\mu_{21}\dot{\varepsilon}_{11}^{\text{cr}}; \tag{6.53}$$

- uniaxial compression in direction 1

$$\dot{\varepsilon}_{11}^{\mathrm{cr}} = -\left(-a_{11} + \sqrt{b_{1111}}\right)^{n+1} |\sigma_{11}|^n. \qquad (6.54)$$

From tests follows

$$\dot{\varepsilon}_{11}^{\mathrm{cr}} = -D_1^- |\sigma_{11}|^n; \qquad (6.55)$$

- uniaxial tension in direction 2

$$\dot{\varepsilon}_{22}^{\mathrm{cr}} = \left(a_{22} + \sqrt{b_{2222}}\right)^{n+1} \sigma_{22}^n. \qquad (6.56)$$

The experimental data lead to

$$\dot{\varepsilon}_{22}^{\mathrm{cr}} = D_2^+ \sigma_{22}^n; \qquad (6.57)$$

- uniaxial compression in direction 2

$$\dot{\varepsilon}_{22}^{\mathrm{cr}} = -\left(-a_{22} + \sqrt{b_{2222}}\right)^{n+1} |\sigma_{22}|^n. \qquad (6.58)$$

The experimental data result in

$$\dot{\varepsilon}_{22}^{\mathrm{cr}} = -D_2^- |\sigma_{22}|^n; \qquad (6.59)$$

- pure torsion

$$\dot{\gamma}_{12}^{\mathrm{cr}} = 2\dot{\varepsilon}_{12}^{\mathrm{cr}} = 2\left(\sqrt{4b_{1212}}\right)^n \sqrt{b_{1212}}\sigma_{12}^n. \qquad (6.60)$$

From experimental data we get

$$\dot{\gamma}_{12}^{\mathrm{cr}} = D_{12}\sigma_{12}^n. \qquad (6.61)$$

Thus we have 6 equations for the unknown parameters a_{11}, a_{22}, b_{1111}, b_{1122}, b_{2222}, b_{1212}, which are dependent on the material characteristics D_1^+, D_1^-, D_2^+, D_2^-, D_{12}, μ_{21}. The creep exponent n is assumed independent from the orientation.

7. APPLICATIONS

The models discussed in the previous sections are applied in different situations. Some solutions of practical problems have been reported in [3] and several articles. The main attention has been directed to

- the identification of some models with a reduced number of parameters and
- the use of reduced models in structural analysis of plates and shells

The first item is connected with the selection of suitable test results and the verification of the identified models by independent tests. The second item leads to additional discussions on the foundations of plate and shell theory.

The proposed creep equations are verified on the basis of different test data, published, for example, in [66, 67, 68]. These verifications are related to the identification of the parameters in the creep equations by basic tests and by the verification on the basis of independent multiaxial tests. Here only the main results are discussed. More details are given in the cited literature.

7.1. Identification and verification of reduced models

The first example is the identification and verification of a 2-parameter-model for isotropic creep. Experimental data for pure copper M1E (Cu 99,9 %, $T = 573$ K) are taken from [46, 47, 48]. The tests have been performed with tubular specimens. The identification of the creep law is based on the following basic tests

- uniaxial tension ($\sigma_{11} \neq 0$) and
- pure torsion ($\sigma_{12} \neq 0$).

The verification is performed by multiaxial tests (combined uniaxial tension and pure torsion). An adequate description is proposed using an assumed potential $\Phi(\sigma_{\text{eq}}) = \sigma_{\text{eq}}^n$ and constitutive equations with a reduced number of parameters

$$\dot{\varepsilon}_{11}^{\text{cr}} = \sigma_2^{(n-1)}(\mu_2 + \mu_3)\sigma_{11}, \quad 2\dot{\varepsilon}_{12}^{\text{cr}} = \dot{\gamma}_{12} = 2\sigma_2^{(n-1)}\mu_3\sigma_{12}. \tag{7.1}$$

Here σ_2 takes the expression

$$\sigma_2 = \sqrt{(\mu_2 + \mu_3)\sigma_{11}^2 + 2\mu_3\sigma_{12}^2} \tag{7.2}$$

The details of the model are published in [67]. From the basic tests L_+, N and the creep exponent n are obtained. At first, the creep exponent n was identified by minimizing

$$F = \left(\dot{\varepsilon}_{11}^{\text{cr theor}} - \dot{\varepsilon}_{11}^{\text{cr exp}}\right)^2 + \left(\dot{\gamma}_{12}^{\text{cr theor}} - \dot{\gamma}_{12}^{\text{cr exp}}\right)^2, \tag{7.3}$$

with

$$\dot{\varepsilon}_{11}^{\text{cr theor}} = L_+\sigma_{11\,\text{max}}^n, \quad \dot{\gamma}_{12}^{\text{cr theor}} = N\sigma_{12\,\text{max}}^n, \tag{7.4}$$

and

$$L_+ = \frac{1}{3}\sum_{i=1}^{3}\dot{\varepsilon}_{11}^{\text{cr }(i)}\left(\sigma_{11}^{(i)}\right)^{-n}, \quad N = \frac{1}{3}\sum_{i=1}^{3}\dot{\gamma}_{12}^{\text{cr }(i)}\left(\sigma_{12}^{(i)}\right)^{-n}, \tag{7.5}$$

i is the number of averaged creep curves. The minimum of F was obtained for the given material for $n = 5,09$. The characteristics of the material are as follows

$$L_+ = 1,39 \cdot 10^{-12}\ \text{MPa}^{-n}\text{h}^{-1}, \quad N = 1,61 \cdot 10^{-11}\ \text{MPa}^{-n}\text{h}^{-1}, \tag{7.6}$$

and the parameters μ_2, μ_3 can be computed from

$$L_+ = \left(\sqrt{\mu_2 + \mu_3}\right)^{(n+1)}, \quad N = \left(\sqrt{2\mu_3}\right)^{(n+1)}, \tag{7.7}$$

which lead to the solution

$$\begin{aligned} \mu_2 &= L_+^{2r} - \frac{1}{2}N^{2r} = -1,50 \cdot 10^{-5}\ \mathsf{MPa}^{-n}\ \mathsf{h}^{-1}, \\ \mu_3 &= \frac{1}{2}N^{2r} = 1,43 \cdot 10^{-4}\ \ \mathsf{MPa}^{-n}\ \mathsf{h}^{-1} \end{aligned} \tag{7.8}$$

with $r = 1/(n+1)$. The results of the verification are shown in Table 7.1. The variation

Table 7.1. Comparison of theoretical values and experimental values of the creep strain rates in the case of multiaxial loading (copper M1E, T = 573 K)

Nr.	Stresses, MPa			Creep strain rates $\cdot 10^5$ h^{-1}					
Experiment	σ_{11}	σ_{12}	σ_{vM}	$\dot{\varepsilon}_{11}^{cr}$		$\dot{\gamma}_{12}^{cr}$		$\dot{\varepsilon}_{vM}^{cr}$	
				Exp.	Theor.	Exp.	Theor.	Exp.	Theor.
1	26,8	8,9	31	3,2	4,14	2,4	3,08	3,5	4,51
2	15,5	15,5	31	1,8	1,77	4,4	3,96	3,1	2,89
3	35,5	11,8	41	14,5	17,20	12,4	12,82	16,0	18,73
4	20,5	20,5	41	7,5	7,36	19,6	16,44	13,5	12,01
5	39,0	13,0	45	19,2	27,60	14,6	20,58	21,0	30,09
6	22,5	22,5	45	10,5	11,82	26,4	26,42	18,5	19,30

of the stresses σ_{11} and σ_{12} has been done in such a way that σ_{vM} = const. The von Mises equivalent stress and strain can be calculated in the particular case of combined tension and pure shear to

$$\sigma_{vM} = \sqrt{\sigma_{11}^2 + 3\sigma_{12}^2}, \quad \dot{\varepsilon}_{vM}^{cr} = \sqrt{(\dot{\varepsilon}_{11}^{cr})^2 + \frac{(\dot{\gamma}_{12}^{cr})^2}{3}}. \tag{7.9}$$

From the verification we can conclude that the behaviour of copper differs in tension and torsion. The maximum difference between theoretical and experimental data was 18 %.

Now the comparison of different isotropic 3-parameter-models is discussed. The following 3-parameter-models for isotropic creep, which are particular cases of the generalised isotropic creep model with 6 parameters, can be introduced

- $\gamma = 0$

$$\sigma_{eq} = \alpha\sigma_1 + \sigma_2 = \alpha\mu_1 I_1 + \sqrt{\mu_2 I_1^2 + \mu_3 I_2}, \tag{7.10}$$

$$\dot{\boldsymbol{\varepsilon}}^{cr} = \phi(\sigma_{eq}) \left(\alpha\mu_1 \boldsymbol{I} + \frac{\mu_2 I_1 \boldsymbol{I} + \mu_3 \boldsymbol{\sigma}}{\sigma_2} \right); \tag{7.11}$$

- $\alpha = \mu_4 = \mu_5 = 0$

$$\sigma_{\mathsf{eq}} = \sigma_2 + \gamma\sigma_3 = \sqrt{\mu_2 I_1^2 + \mu_3 I_2} + \gamma\sqrt[3]{\mu_6 I_3}, \tag{7.12}$$

$$\dot{\boldsymbol{\varepsilon}}^{\mathsf{cr}} = \phi(\sigma_{\mathsf{eq}})\left(\frac{\mu_2 I_1 \boldsymbol{I} + \mu_3\boldsymbol{\sigma}}{\sigma_2} + \gamma\frac{\mu_6\boldsymbol{\sigma}\cdot\boldsymbol{\sigma}}{\sigma_3^2}\right); \tag{7.13}$$

- $\alpha = \mu_4 = \mu_6 = 0$

$$\sigma_{\mathsf{eq}} = \sigma_2 + \gamma\sigma_3 = \sqrt{\mu_2 I_1^2 + \mu_3 I_2} + \gamma\sqrt[3]{\mu_5 I_1 I_2}, \tag{7.14}$$

$$\dot{\boldsymbol{\varepsilon}}^{\mathsf{cr}} = \phi(\sigma_{\mathsf{eq}})\left[\frac{\mu_2 I_1 \boldsymbol{I} + \mu_3\boldsymbol{\sigma}}{\sigma_2} + \gamma\frac{\mu_6(I_2\boldsymbol{I} + 2I_1\boldsymbol{\sigma})}{3\sigma_3^2}\right]. \tag{7.15}$$

In Tables 7.2 and 7.3 the comparison of experimental data for plastics (PVC) at room temperature [39] and for an aluminum alloy (AK4-1T) at 473 K [69] and our calculations is presented. It shows that the more sophisticated models do not deliver significantly better

Table 7.2. Comparison of theoretical values and experimental values of the creep strains (P/C tubular specimens, inner pressure, tensile force, $t = 100$ h)

Stresses, MPa		Creep strains $\varepsilon_{11}^{\mathsf{cr}} \cdot 10^3$			
σ_{11}	σ_{22}	Experiment	(7.11)	(7.13)	(7.15)
-14,88	14,88	3,12	4,22	3,85	3,97
-17,10	17,10	4,99	6,76	6,17	6,37
9,93	9,93	0,18	0,15	0,16	0,16
22,05	22,05	2,10	2,29	2,43	2,43

results, so that the most simple model (the tensorial linear model) could be recommended. It also allows the description of the Poynting-Swift effect.

The next example is the verification of a reduced orthotropic model. Orthotropic creep can be described by a tensorial linear constitutive equation, deduced from the generalised anisotropic law with $\gamma = 0$ (for more details see [66])

$$\dot{\boldsymbol{\varepsilon}}^{\mathsf{cr}} = (\alpha\boldsymbol{a}\cdot\cdot\,\boldsymbol{\sigma} + \sqrt{\boldsymbol{\sigma}\cdot\cdot\,{}^{(4)}\boldsymbol{b}\cdot\cdot\,\boldsymbol{\sigma}})^n\left(\alpha\boldsymbol{a} + \frac{{}^{(4)}\boldsymbol{b}\cdot\cdot\,\boldsymbol{\sigma}}{\sqrt{\boldsymbol{\sigma}\cdot\cdot\,{}^{(4)}\boldsymbol{b}\cdot\cdot\,\boldsymbol{\sigma}}}\right). \tag{7.16}$$

For the aluminum alloy D16T at 523 K some experimental data are published in [70], which lead to

$$\begin{aligned} b_{1111} &= 1,58\cdot 10^s\ \mathsf{MPa}^{2k}\ \mathsf{h}^{2r},\ b_{2222} = 2,03\cdot 10^s\ \mathsf{MPa}^{2k}\ \mathsf{h}^{2r}, \\ b_{1122} &= 7,63\cdot 10^{s+k+r}\ \mathsf{MPa}^{2k}\ \mathsf{h}^{2r}, \\ \alpha a_{11} &= 1,74\cdot 10^{s-k}\ \mathsf{MPa}^{k}\ \mathsf{h}^{r},\ \alpha a_{22} = 9,57\cdot 10^{s+k}\ \mathsf{MPa}^{k}\ \mathsf{h}^{r}, \end{aligned} \tag{7.17}$$

Table 7.3. Comparison of theoretical values and experimental values of the creep strain rates (aluminum alloy AK4-1T, combined tension and torsion, T = 473 K)

Stresses, MPa		creep strain rates$\cdot 10^5$, h^{-1}							
σ_{11}	σ_{12}	$\dot{\varepsilon}_{11}^{cr}$	$2\dot{\varepsilon}_{12}^{cr}$	$\dot{\varepsilon}_{11}^{cr}$	$2\dot{\varepsilon}_{12}^{cr}$	$\dot{\varepsilon}_{11}^{cr}$	$2\dot{\varepsilon}_{12}^{cr}$	$\dot{\varepsilon}_{11}^{cr}$	$2\dot{\varepsilon}_{12}^{cr}$
		Experiment		(7.11)		(7.13)		(7.15)	
107,5	62,0	1,20	2,40	1,29	2,56	1,35	2,75	1,34	2,72
-60,6	84,4	-0,72	3,64	-0,75	4,25	-0,69	3,79	-0,75	3,89
-152,9	36,6	-1,74	1,42	-1,45	1,32	-1,43	1,27	-1,47	1,30
0,0	98,0	0,16	9,02	0,20	9,91	0,00	9,91	0,00	9,91

Table 7.4. Comparison of theoretical values ($\dot{\varepsilon}_{11}^{cr\ theor}, \dot{\varepsilon}_{22}^{cr\ theor}$) and experimental values ($\dot{\varepsilon}_{11}^{cr\ exp}, \dot{\varepsilon}_{22}^{cr\ exp}$) of the creep strain rate (aluminum alloy D16T, T=523 K)

stresses, MPa		creep strain rates $\cdot 10^3$, h^{-1}			
σ_{11}	σ_{22}	$\dot{\varepsilon}_{11}^{cr\ exp}$	$\dot{\varepsilon}_{11}^{cr\ theor}$	$\dot{\varepsilon}_{22}^{cr\ exp}$	$\dot{\varepsilon}_{22}^{cr\ theor}$
-109,8	54,9	-1,60	-0,93	1,60	0,91
-80,6	80,6	-0,92	-0,82	1,38	1,05
70,0	140,0	0,00	0,08	1,65	1,44
-37,6	112,8	-0,79	-0,76	1,59	1,49
124,0	124,0	0,59	0,71	1,18	1,04

with $r = -1/(n+1), k = nr, s = 5k + 3r$ and $n = 6,5$. The comparison of predictions, based on the theoretical model, and experimental data is presented in Table 7.4. The results of calculations are in a satisfying agreement with the experimental data.

The last example in this section is an isotropic creep-damage model, based on 3-parameter-equations and Norton's creep law and which was verified in [71]. The tests were performed for a titanium alloy OT-4 at 748 K [63]. The basic tests are uniaxial tension, uniaxial compression and pure torsion. A second series of tests were performed for an aluminum alloy AK4-1T at T = 473 K [63]. From the basic tests uniaxial tension, uniaxial compression and pure torsion the parameters in the creep equations were obtained. The prediction of the creep damage behaviour in both cases is in a good agreement with independent test data for secondary creep. For the tertiary creep some differences are obtained. The reason for these differences is that the material degradation is not only caused by creep, but also by other dissipative mechanisms. For more details see [71].

7.2. Applications to plates and shells

The special cases with a reduced number of material parameters discussed in the previous section are applied to plate and shell problems. Here the main results are summarised.

Further information can be found in the references.

In [68] the creep behaviour of thin shells of revolution made of anisotropic material with different behaviour in tension and compression was investigated. The simulations were based on different creep models reflecting the dependence on the kind of loading. For the isotropic and orthotropic cases models with a reduced number of parameters were introduced. All models were tensorial linear. Using different material models and the set of governing equations of the shell theory, the initial-boundary-value problem was formulated. The special numerical solution technique was based on a modified Kutta method for the initial value problem. The linearised boundary problem was solved by Godunov's orthogonalisation method.

For a cylinder with a clamped boundary and a free boundary, loaded by outside pressure and made of the isotropic aluminium alloy AK4-1T at 473 K, the calculation of the axial and transverse displacements and the meridian stresses has been performed. The creep calculations are based on three models: one classical and two non–classical. For the first non–classical model identical behaviour in tension and compression, but different equivalent behaviour in torsion was presumed, the second non–classical model - differences in tension and compression, but identical torsion behaviour. It has been shown that in these three cases the results are not the same. The calculations lead to the conclusion that the tendency in the distribution of the stresses is not identical for axial and transverse deflections. Therefore, it can be concluded that a model with three independent tests may lead to more satisfying results. The same example, but with an assumed anisotropic material behaviour was discussed, too. Similar conclusions as in the case of isotropic material behaviour can be drawn from these calculations.

In [72] the non–classical material model for isotropic creep considering different behaviour in tension and compression has been applied to the analysis of shells. The results show that the introduction of large transverse deflections leads to qualitatively similar curves, but a redistribution of stresses can be obtained. These conclusions are correct in the case of identical material behaviour with respect to tension and compression, as well as in the case of different material behaviour.

The creep-damage problems of thin rectangular plates, axisymmetrically loaded shells of revolution and circular plates are discussed in [73]. The creep-damage equations were formulated using the power law creep function and a scalar damage parameter. The corresponding initial-boundary value problems was defined using the nonlinear kinematics of shells considering finite (comparable with the shell thickness) deflections. The results show that in the case of the finite deflection approach, the deflection growth and damage evolution is substantially different from the geometrically linear one. The effects similar to "structure hardening" as the result of membrane forces in the initial state (due to the shell curvature) or their generation during the creep process (as a consequence of geometrical non–linearities) have been discussed. The geometrically linear approach overestimates the deflections and leads to a significant underestimation of the failure time.

ACKNOWLEDGMENTS

The manuscript is partly based on a graduate course "Mathematical modelling of creep processes in metals" held in August 1996 during the CEEPUS Summer School "Analysis of elastomers and creep and flow of glass and metals" at the University of Transport and Communications in Žilina (Slovak Republic). The author is thankful to Prof. Vlado Kompiš for the invitation and the opportunity to teach the course during the Summer School.

The author wishes to express his appreciation for the helpful discussions with the colleagues and the students during the course. Additional thanks to Dipl.-Ing. Johannes Meenen and Dr.-Ing. Konstantin Naumenko for help in preparation the manuscript and the critical remarks.

REFERENCES

1. Billington E.W. : The Poynting-Swift effect in relation to initial and post-yield deformation, Solids & Structures, 21(1985), pp. 355 – 372.
2. Truesdell C. : Second-Order Effects in the Mechanics of Materials, in: Second-Order Effects in Elasticity, Plasticity and Fluid Dynamics (Edited by M. Reiner and D. Abir), Pergamon Press, Oxford et al., 1964 pp. 228 – 251, pp. 228 – 251.
3. Altenbach H. , Altenbach J. and Zolochevsky A. : Erweiterte Deformationsmodelle und Versagenskriterien der Werkstoffmechanik, Deutscher Verlag für Grundstoffindustrie, Stuttgart, 1995.
4. Altenbach J. and Altenbach H. : Einführung in die Kontinuumsmechanik, Teubner Studienbücher Mechanik, Teubner, Stuttgart, 1994.
5. Lurie A.I. : Nonlinear Theory of Elasticity, North-Holland Publ. Company, Amsterdam, 1990.
6. Truesdell C. and Noll W. : The non-linear field theories of mechanics, in: Encyclopedia of Physics (Edited by S. Flügge), volume III/3, Springer, Heidelberg u.a., 1965 .
7. Ashby M.F. : Technology of the 1990's: advanced materials and predictive design, Phil. Trans. R. Soc. London, A322(1987), pp. 393 – 407.
8. Altenbach H. and Blumenauer H. : Grundlagen und Anwendungen der Schädigungsmechanik, Neue Hütte, 34(1989), pp. 214 – 219.
9. Haupt P. : On the mathematical modelling of material behaviour in continuum mechanics, Acta Mechanica, 100(1993), pp. 129 – 154.
10. Krawietz A. : Materialtheorie. Mathematische Beschreibung des phänomenologischen thermomechanischen Verhalten, Springer, Berlin et al., 1986.
11. Lemaitre J. and Chaboche J.-L. : Mechanics of Solid Materials, Cambridge University Press, Cambridge, 1990.
12. Giesekus H. : Phänomenologische Rheologie, Springer, Berlin, 1994.
13. Palmov V. : Vibrations in Elasto-Plastic Bodies, Springer, Berlin et al., 1998.
14. Betten J. : Anwendungen von Tensorfunktionen in der Kontinuumsmechanik anisotroper Materialien, ZAMM, 78(1998), pp. 507 – 521.

15. Riedel H. : Fracture at High Temperatures, Materials Research and Engineering, Springer, Berlin et al., 1987.
16. Kowalewski Z.L. , Hayhurst D.R. and Dyson B.F. : Mechanisms-based creep constitutive equations for an aluminium alloy, J. Strain Anal., 29(1994), pp. 309 – 316.
17. Tvergaard V. : Micromechanical modelling of creep rupture, ZAMM, 71(1991), pp. 23 – 32.
18. Rabotnov Yu.N. : Creep Problems in Structural Members, North Holland, Amsterdam, London, 1969.
19. Służalec A. : Introduction to Nonlinear Thermomechanics, Springer, Berlin et al., 1992.
20. Skrzypek J.J. : Plasticity and Creep, CRC Press, Boca Raton et al., 1993.
21. Odqvist F.K.G. and Hult J. : Kriechfestigkeit metallischer Werkstoffe, Springer, Berlin u.a., 1962.
22. Hayhurst D.R. : The use of continuum damage mechanics in creep analysis and design, J. Strain Anal., 29(1994), pp. 233 – 241.
23. Kachanov L.M. : Introduction to Continuum Damage Mechanics, Martinus Nijhoff Publishers, Dordrecht et al., 1986.
24. Lemaitre J. : A Course on Damage Mechanics, Springer, Berlin et al., 1992.
25. Murakami S. : Progress in continuum damage mechanics, JSME, 30(1987), pp. 701 – 710.
26. Besseling J.F. and van der Giessen E. : Mathematical Modelling of Inelastic Deformation, Chapman & Hall, London et al., 1993.
27. Penny R.K. and Mariott D.L. : Design for Creep, Chapman & Hall, London et al., 1995.
28. Kraus H. : Creep Analysis, Wiley, New York et al., 1980.
29. Gummert P. : Contributions and discussions (during the course), 1998, CISM Udine.
30. Rabotnov Yu.N. : Mekhanika deformiruemogo tverdogo tela (Mechanics of the deformable solid body, in Russ.), Nauka, Moskva, 1979.
31. Frost H.J. and Ashby M.F. : Deformation-Mechanism Maps, Pergamon Press, Oxford et al., 1982.
32. Krempl E. : Cyclic creep - An interpretive literature survey, Welding Research Council Bull., (1974), pp. 63 – 120.
33. Boyle J.T. and Spence J. : Stress Analysis for Creep, Butterworth, London, 1983.
34. Malinin N.N. : Raschet na polzuchest' konstrukcionnykh elementov (Creep calculations of structural elements, in Russ.), Mashinostroenie, Moskva, 1981.
35. Mises R. v. : Mechanik der plastischen Formänderung von Kristallen, ZAMM, 8(1928), pp. 161 – 185.
36. Hill R. : On the classical constitutive relations for elastic/plastic solids, in: Recent Progress in Applied Mechanics (Folke Odqvist Volume), Almqvist and Viksell, Stockholm, 1967 pp. 241 – 249, pp. 241 – 249.
37. Skrzypek J.J. and Ganczarski A. : Modelling of Material Damage and Failure of Structures, Springer, Berlin et al., 1998.
38. Leckie F.A. and Hayhurst D.R. : Constitutive equations for creep rupture, Acta Metall., 25(1977), pp. 1059 – 1070.

39. Lewin G. and Lehmann B. : Ergebnisse über das Spannungs-Verformungsverhalten von PVC, dargestellt an einem zylindrischen Bauelement, Wiss. Z. TH Magdeburg, 21(1977), pp. 415 – 422.
40. Nechtelberger E. : Raumtemperaturkriechen und Spannungsabhängigkeit des E-Moduls von Graugußwerkstoffen, Österr. Ing. Archit. Z., 130(1985), pp. 29 – 36.
41. Pintschovius L. , Gering E. , Munz D. , Fett T. and Soubeyroux J.L. : Determination of non-symmetric secondary creep behaviour of ceramics by residual stress measurements using neutron diffractometry, J. Mater. Sci. Lett., 8(1989), pp. 811 – 813.
42. Schlimmer M. : Zeitabhängiges mechanisches Werkstoffverhalten, Springer, Berlin et al., 1984.
43. Vandervoort R.R. and Barmore W.L. : Compressive creep of polycrystalline beryllium oxide, J. Amer. Ceram., 46(1963), pp. 180 – 184.
44. Zolochevskij A.A. : Kriechen von Konstruktionselementen aus Materialien mit von der Belastung abhängigen Charakteristiken, Technische Mechanik, 9(1988), pp. 177 – 184.
45. Betten J. and Waniewski M. : Einfluß der plastischen Anisotropie auf das sekundäre Kriechverhalten inkompressibler Werkstoffe, Rheologica Acta, 25(1986), pp. 166 – 174.
46. Kowalewski Z. : The surface of constant rate of energy dissipation under creep and its experimental determination, Arch. Mech., 39(1987), pp. 445 – 459.
47. Kowalewski Z. : Secondary creep surface and its evolution influenced by room temperature plastic deformation, Besançon, 1988.
48. Kowalewski Z. : Creep behaviour of copper under plane stress state, Int. J. Plasticity, 7(1991), pp. 387 – 400.
49. Murakami S. and Sawczuk A. : A unified approach to constitutive equations of inelasticity based on tensor function representations, Nucl. Eng. Des., 65(1981), pp. 33 – 47.
50. Waniewski M. : A simple law of steady-state creep for material with anisotropy introduced by plastic prestraining, Ingenieur-Archiv, 55(1985), pp. 368 – 375.
51. Sarabi B. : Anstrengungsverhalten von Kunststoffen unter biaxial-statischer Belastung, Kunststoffe, 76(1986), pp. 182 – 186.
52. Cristescu N. : Rock Rheology, Kluwer Academic Publ., Dordrecht, 1989.
53. Foux A. : An Experimental Investigation of the Poynting Effect, in: Second-Order Effects in Elasticity, Plasticity and Fluid Dynamics (Edited by M. Reiner and D. Abir), Pergamon Press, Oxford et al., 1964 pp. 228 – 251, pp. 228 – 251.
54. Hecker F.W. : Die Wirkung des Bauschinger-Effekts bei großen Torsions-Formänderungen, Ph.D. thesis, TH Hannover, 1967.
55. Nishitani T. : Mechanical behavior of nonlinear viscoelastic celluloid under superimposed hydrostatic pressure, Trans. ASME. J. Pressure Vessel Technol., 100(1978), pp. 271 – 276.
56. Swift H.W. : Length changes in metals under torsional overstrain, Engineering, 163(1947), pp. 253 – 257.
57. Altenbach H. and Zolochevsky A. : A generalized failure criterion for three-dimensional behaviour of isotropic materials, Engng Frac. Mech., 54(1996), pp. 75 – 90.

58. Chen W.F. and Zhang H. : Structural Plasticity, Springer, Berlin et al., 1991.
59. Życzkowski M. : Combined Loadings in the Theory of Plasticity, PWN-Polish Scientific Publisher, Warszawa, 1981.
60. Hayhurst D.R. : Creep rupture under multiaxial states of stress, J. Mech. Phys. Solids, 20(1972), pp. 381 – 390.
61. Betten J. : Zur Verallgemeinerung der Invariantentheorie in der Kriechmechanik, Rheologica Acta, 14(1975), pp. 715 – 720.
62. Backhaus G. : Deformationsgesetze, Akademie-Verlag, Berlin, 1983.
63. Gorev B.V. , Rubanov V.V. and Sosnin O.V. : O postroenii uravnenii polzuchesti dlya materialov s raznymi svoistvami na rastyazhenie i szhatie (On the formulation of creep equations for materials with different properties in tension and compression, in Russ.), Zhurnal prikladnoi mekhaniki i tekhnicheskoi fiziki, (1979), pp. 121 – 128.
64. Hill R. : The Mathematical Theory of Plasticity, Materials Research and Engineering, Oxford University Press, London, 1950.
65. Betten J. : Zur Aufstellung einer Integritätsbasis für Tensoren zweiter und vierter Stufe, ZAMM, 62(1982), pp. T 274 – T 275.
66. Altenbach H. , Dankert M. and Zoločevskij A. : Anisotrope mathematisch-mechanische Modelle für Werkstoffe mit von der Belastung abhängigen Eigenschaften, Technische Mechanik, 11(1990), pp. 5 – 13.
67. Altenbach H. , Schieße P. and Zolochevsky A. : Zum Kriechen isotroper Werkstoffe mit komplizierten Eigenschaften, Rheologica Acta, 30(1991), pp. 388 – 399.
68. Altenbach H. and Zolochevsky A. : Kriechen dünner Schalen aus anisotropen Werkstoffen mit unterschiedlichem Zug-Druck-Verhalten, Forschung im Ingenieurwesen, 57(1991), pp. 172 – 179.
69. Tsvelodub I.Yu. : Postulat ustoichivosti i ego prilozheniya v teorii polzuchesti metallicheskikh materialov (On the stability postulate and its applications in the creep theory of metallic materials, in Russ.), Institut gidrodinamiki, Novosibirsk, 1991.
70. Nikitenko A.F. and Tsvelodub I.Yu. : O polzuchesti anizotropnykh materialov s raznymi svoistvami na rastyazhenie i szhatie (On creep of anisotropic materials with different properties in tension and compression, in Russ.), Dinamika sploshnoi sredy, 43(1979), pp. 69 – 78.
71. Altenbach H. and Zolochevsky A.A. : Eine energetische Variante der Theorie des Kriechens und der Langzeitfestigkeit für isotrope Werkstoffe mit komplizierten Eigenschaften, ZAMM, 74(1994), pp. 189 – 199.
72. Altenbach H. , Morachkovsky O. , Naumenko K. and Sichov A. : Zum Kriechen dünner Rotationsschalen unter Einbeziehung geometrischer Nichtlinearität sowie der Asymmetrie der Werkstoffeigenschaften, Forschung im Ingenieurwesen, 62(1996), pp. 47 – 57.
73. Altenbach H. , Morachkovsky O. , Naumenko K. and Sychov A. : Geometrically nonlinear bending of thin-walled shells and plates under creep-damage conditions, Arch. Appl. Mech., 67(1997), pp. 339 – 352.

MATERIAL DAMAGE MODELS FOR CREEP FAILURE ANALYSIS AND DESIGN OF STRUCTURES

J.J. Skrzypek
Cracow University of Technology, Cracow, Poland

ABSTRACT

A concise review of one and three–dimensional theories of isotropic or anisotropic damage coupled constitutive equations of time–dependent elastic or inelastic materials is systematically presented. When damage is considered as isotropic phenomenon both phenomenologically–based damage–creep–plasticity models (Kachanov, Rabotnov, Hayhurst, Leckie, Kowalewski, Dunne, etc.) and unified irreversible thermodynamics formulation of coupled isotropic damage–thermo–elastic–creep–plastic materials (Lemaitre and Chaboche, Mou and Han, Saanouni, Foster and Ben Hatira) are reported. In case when anisotropic nature of damage is described in frame of the continuum damage mechanics (CDM) approach, a concept of the fourth–rank damage effect tensor $\mathbf{M}$ is introduced in order to define the constitutive tensors of damaged materials, stiffness or compliance $\widetilde{\mathbf{\Lambda}}$ or $\widetilde{\mathbf{\Lambda}}^{-1}$ in terms of those of virgin isotropic materials. Matrix representation of constitutive tensors is reviewed in case of energy based damage coupled constitutive model of elastic–brittle (Litewka, Murakami and Kamiya) or elastic–plastic engineering materials (Hayakawa and Murakami). Particular attention is paid to the orthotropic creep–damage model and its computer applications to the case of non–proportional loading conditions, when the objective damage rate is applied. A non–classical problem of thermo–damage coupling is developed, when the second–rank tensors of thermal conductivity $\widetilde{\mathbf{L}}$ and radiation $\widetilde{\mathbf{\Gamma}}$ in the extended heat transfer equation are defined for damaged material in terms of the damage tensor $\mathbf{D}$.

The CDM based finite difference method (FDM) and finite element method (FEM) computer applications to the analysis and design of simple engineering structures under damage conditions are developed. Structures of uniform creep damage strength are examined from the point of view of maximum lifetime prediction when the equality and inequality constraints are imposed, and the thickness and initial prestressing are chosen as design variables.

1. DAMAGE VARIABLES AND CDM EQUIVALENCE PRINCIPLES

1.1. State of damage and damage variables

State of material damage is identified as the existence of distributed microvoids, microcavities or microcracks in a volume of a material. Irreversible time–dependent microprocesses, when the microdefects nucleation, growth and coalescence cause a progressive degradation of the physical and thermomechanical properties through reduction of strength, elasticity modulae, microhardness, ultrasonic wave speed, heat conductivity, etc., is called the damage evolution. When the continuum damage mechanics CDM method is used the true distribution of microdeffects, their size, density and orientation, is homogenized by a selection of the set of internal variables of different nature, scalar D, vector D_α, second–rank tensor $\mathbf{D}$, fourth–rank tensor $\widehat{\mathbf{D}}$, etc., that measure the state of damage. These variables are called the damage variables which serve as internal variables $\mathcal{D} = \left\{D, D_\alpha, \mathbf{D}, \widehat{\mathbf{D}}, \ldots\right\}$ in the state and dissipation potential.

Damage variables have systematically been reviewed by Skrzypek and Ganczarski [1]. Scalar damage variables D or ω are applicable for description of isotropic damage, however they are also frequently used for description of anisotropic damage under creep–damage conditions (Kachanov [2, 3], Rabotnov [4], Martin and Leckie [5], Hayhurst and Leckie [6], Leckie and Hayhurst [7], Hayhurst [8, 9], Trąpczyński, Hayhurst and Leckie [10], Lemaitre and Chaboche [11, 12, 13], Chaboche [14], Dunne and Hayhurst [15, 16, 17, 18], Othman, Hayhurst and Dyson [19], Germain, Nguyen and Suquet [20], Dufailly and Lemaitre [21], Mou and Han [22], Saanouni, Forster and Ben Hatira [23], etc.).

Vector damage variables D_α or ω_α are applicable for description of the damage orthotropy or weak anisotropy (Davison and Stevens [24], Kachanov [3, 25], Krajcinovic and Fonseka [26], Krajcinovic [27, 28], Lubarda and Krajcinovic [29], etc.).

However, in order to describe fully anisotropic damage evolution, when effect of rotation of principal damage axes is allowed, the second–rank tensors $\mathbf{D}$ or $\boldsymbol{\Omega}$ must be used as the damage representation (Rabotnov [30], Vakulenko and Kachanov [31], Murakami and Ohno [32], Cordebois and Sidoroff [33, 34], Betten [35, 36], Litewka [37, 38, 39], Murakami [40, 41, 42, 43], Chow and Lu [44], Chaboche [14, 45, 46], Murakami and Kamiya [47], Hayakawa and Murakami [48, 49], Skrzypek and Ganczarski [1, 50, 51], etc.).

On the other hand, fourth–rank damage tensors are capable of describing strong damage anisotropy (Leckie and Onat [52], Chaboche [53], Simo and Ju [54], Krajcinovic [55, 28], Lubarda and Krajcinovic [29], Chen and Chow [56], Voyiadjis and Park [57, 58], Qi and Bertram [59], etc.). However, although the fourth–rank damage effect tensors can be used as a linear transformation tensors to define the effective stress and strain tensors $\widetilde{\boldsymbol{\sigma}}$, $\widetilde{\boldsymbol{\varepsilon}}$ in terms of the conventional stress and strain tensors $\boldsymbol{\sigma}$, $\boldsymbol{\varepsilon}$, $\widetilde{\boldsymbol{\sigma}} = \mathbf{M}(\mathcal{D}) : \boldsymbol{\sigma}$, $\widetilde{\boldsymbol{\varepsilon}}^{\mathrm{e}} = \mathbf{M}^{-1}(\mathcal{D}) : \boldsymbol{\varepsilon}^{\mathrm{e}}$, $\mathrm{d}\widetilde{\boldsymbol{\varepsilon}}^{\mathrm{p}} = \mathbf{M}^{-1}(\mathcal{D}) : \mathrm{d}\boldsymbol{\varepsilon}^{\mathrm{p}}$ (Chow and Lu [44]), it is not easy to identify physically the fourth–rank damage tensor compared to the second–rank damage tensor (Voyiadjis and Park [58]).

Scalar damage variable D, also called the damage parameter, is defined at the material point $\mathbf{X}$ of the surface element δA as the ratio of the damaged area δA_{D} to the total (undamaged or virgin) area δA, $D = \delta A_{\mathrm{D}}/\delta A$, such that $D = 0$ corresponds to the undamaged virgin state and it gradually grows up, to eventually reach the value $D = 1$ for the completely damaged element $\delta A_{\mathrm{D}} = \delta A$. Considering planes of various normals $\mathbf{n}_k$ we can define surface damage in an arbitrary direction $\mathbf{n}_k$ as $D(\mathbf{n}_k) = \delta A_{\mathrm{D}k}/\delta A_k$.

Second–rank damage tensor $\mathbf{D}$ defined by Murakami and Ohno [32] is represented as follows:

$$\mathbf{D} = \sum_{k=1}^{3} D_k \mathbf{n}_k \otimes \mathbf{n}_k \quad \text{or} \quad D_{ij} = \sum_{k=1}^{3} D_k n_i^k n_j^k \qquad (\text{no sum in } k), \tag{1.1}$$

where D_k are eigenvalues of the tensor $\mathbf{D}$ and $\mathbf{n}_k$ are the eigenvectors corresponding to eigenvalues D_k. Eigenvalues D_k may be interpreted here as the ratio of the area reduction in the plane perpendicular to $\mathbf{n}_k$, caused by development of damage components $D_k = \delta A_{\mathrm{D}k}/\delta A_k$.

On a ductile deformation process in crystalline materials the flow of mass through the lattice takes place, at which the lattice undergoes elastic reversible deformation only, whereas a total number of active atomic bonds remains approximately constant. Hence, none (or negligibly small) change of the effective material properties is assumed to occur. On the other hand, on a brittle deformation process the lattice itself is subjected to irreversible changes resulting from breaking of the atomic bonds and, hence, the progressive material degradation through the strength and stiffness reduction takes place. This fully coupled CDM approach to the elastic–brittle–damage or the creep–damage leads to the concept of fourth–rank elasticity tensors modified by damage $\mathcal{D}$, stiffness $\widetilde{\mathbf{\Lambda}}(\mathcal{D})$, or compliance $\widetilde{\mathbf{\Lambda}}^{-1}(\mathcal{D})$

$$\boldsymbol{\sigma} = \widetilde{\mathbf{\Lambda}}(\mathcal{D}) : \boldsymbol{\varepsilon}^{\mathrm{e}} \qquad \text{or} \qquad \boldsymbol{\varepsilon}^{\mathrm{e}} = \widetilde{\mathbf{\Lambda}}^{-1}(\mathcal{D}) : \boldsymbol{\sigma} \tag{1.2}$$

(cf. [37, 39, 56, 47, 48, 49] etc.). In general, the fourth–rank damage effect tensor $\mathbf{M}(\mathcal{D})$, that transforms the Cauchy stress tensor in a damaged configuration $\boldsymbol{\sigma}$ to the effective (conjugate) Cauchy stress tensor in an equivalent fictitious pseudo–undamaged solid $\widetilde{\boldsymbol{\sigma}}$, based on the appropriate damage equivalence hypothesis, takes into account the fully anisotropic nature of damage (cf. [44, 60])

$$\widetilde{\boldsymbol{\sigma}} = \mathbf{M}(\mathcal{D}) : \boldsymbol{\sigma}. \tag{1.3}$$

$\mathbf{M}(\mathcal{D})$ is an isotropic fourth–rank tensor–valued function of the damage state variable $\mathcal{D}$, and the effective stress tensor $\widetilde{\boldsymbol{\sigma}}(\boldsymbol{\sigma}, \mathcal{D})$ is an isotropic second–rank tensor–valued function of $\boldsymbol{\sigma}$ and $\mathcal{D}$ (damage isotropy principle), the representation of which depends on the equivalence principle adopted.

1.2. Strain, stress, and energy based CDM models

When the CDM approach is used the true discontinues and heterogeneous damaged material is approximated by the pseudo–undamaged continuum. The couples of state variables $(\boldsymbol{\varepsilon}, \boldsymbol{\sigma})$, (r, R) and $(\boldsymbol{\alpha}, \mathbf{X})$, representing strain and stress tensors, isotropic hardening variables and kinematic hardening tensors in the true (damaged) material, are replaced here by the effective state variables $(\widetilde{\boldsymbol{\varepsilon}}, \widetilde{\boldsymbol{\sigma}})$, $\left(\widetilde{r}, \widetilde{R}\right)$ and $\left(\widetilde{\boldsymbol{\alpha}}, \widetilde{\mathbf{X}}\right)$ referred to the pseudo–undamaged (fictitious) quasi–continuum. Definitions of effective variables depend on the equivalence principles used to define a quasi–continuum.

A. Principle of strain equivalence

For the isotropic damage described by the scalar D the following definitions of the effective variables hold (cf. Lemaitre and Chaboche [12]; Simo and Ju [54])

$$\widetilde{\boldsymbol{\varepsilon}}\left(\widetilde{\boldsymbol{\sigma}}, 0\right) = \boldsymbol{\varepsilon}\left(\boldsymbol{\sigma}, D\right), \qquad \widetilde{\boldsymbol{\sigma}} = \frac{\boldsymbol{\sigma}}{1 - D}. \tag{1.4}$$

In case of the damage anisotropy $\mathbf{D}$, the fourth–rank damage effect tensor $\mathbf{M}_{\mathrm{Ch}}\left(\mathbf{D}\right)$ is used in order to transform the Cauchy stress tensor $\boldsymbol{\sigma}$ into the effective stress tensor $\widetilde{\boldsymbol{\sigma}}$, e.g.:

$$\widetilde{\boldsymbol{\sigma}}\left(t\right) = \mathbf{M}_{\mathrm{Ch}}^{-1} : \boldsymbol{\sigma}\left(t\right), \tag{1.5}$$

whereas for the damage isotropy $\mathbf{M}_{\mathrm{Ch}}\left(D\right) = \left(1 - D\right)\mathbf{I}$,

$$\widetilde{\boldsymbol{\sigma}}\left(t\right) = \frac{\boldsymbol{\sigma}\left(t\right)}{1 - D\left(t\right)}, \tag{1.6}$$

where $\mathbf{I}$ denotes the fourth–rank identity tensor.

B. Principle of stress equivalence

For the isotropic damage characterized by the scalar D the following (dual) definitions of the effective variables are furnished (cf. Simo and Ju [54]):

$$\widetilde{\boldsymbol{\sigma}}\left(\widetilde{\boldsymbol{\varepsilon}}, 0\right) = \boldsymbol{\sigma}\left(\boldsymbol{\varepsilon}, D\right), \qquad \widetilde{\boldsymbol{\varepsilon}} = \left(1 - D\right)\boldsymbol{\varepsilon}. \tag{1.7}$$

In a more general case of the damage anisotropy characterized by the fourth–rank damage effect tensor $\mathbf{M}_{\mathrm{Ch}}\left(\mathbf{D}\right)$, the transformation from the damaged space to the pseudo–undamaged space is obtained:

$$\widetilde{\boldsymbol{\varepsilon}}\left(t\right) = \mathbf{M}_{\mathrm{Ch}} : \boldsymbol{\varepsilon}\left(t\right), \tag{1.8}$$

whereas for the damage isotropy $\mathbf{M}_{\mathrm{Ch}}\left(D\right) = \left(1 - D\right)\mathbf{I}$,

$$\widetilde{\boldsymbol{\varepsilon}}\left(t\right) = \left[1 - D\left(t\right)\right]\boldsymbol{\varepsilon}\left(t\right). \tag{1.9}$$

C. Generalized principle of strain equivalence

Three scalar generalized, total, elastic and plastic damage variables D^{t}, D^{e} and D^{p} are defined (cf. Taher et al. [61]) by the fourth–rank secant modulae degradation tensors $\widetilde{\boldsymbol{\Lambda}}\left(t\right)$, $\widetilde{\mathbf{E}}\left(t\right)$ and $\widetilde{\mathbf{P}}\left(t\right)$ in terms of damage evolution

$$\begin{aligned} \boldsymbol{\sigma} &= \widetilde{\boldsymbol{\Lambda}}\left(D^{\mathrm{t}}\right) : \boldsymbol{\varepsilon}, & \widetilde{\boldsymbol{\Lambda}}\left(t\right) &= \left[1 - D^{\mathrm{t}}\left(t\right)\right]\boldsymbol{\Lambda}, \\ \boldsymbol{\sigma} &= \widetilde{\mathbf{E}}\left(D^{\mathrm{e}}\right) : \boldsymbol{\varepsilon}^{\mathrm{e}}, & \widetilde{\mathbf{E}}\left(t\right) &= \left[1 - D^{\mathrm{e}}\left(t\right)\right]\mathbf{E}, \\ \boldsymbol{\sigma} &= \widetilde{\mathbf{P}}\left(D^{\mathrm{p}}\right) : \boldsymbol{\varepsilon}^{\mathrm{p}}, & \widetilde{\mathbf{P}}\left(t\right) &= \left[1 - D^{\mathrm{p}}\left(t\right)\right]\mathbf{P}, \end{aligned} \tag{1.10}$$

where $\boldsymbol{\Lambda}$, $\mathbf{E}$ and $\mathbf{P}$ denote the initial values of $\widetilde{\boldsymbol{\Lambda}}\left(t\right)$, $\widetilde{\mathbf{E}}\left(t\right)$ and $\widetilde{\mathbf{P}}\left(t\right)$, respectively (Fig. 1.1).

Inspection of the evolution of the generalized damage variables D^{t}, D^{e} and D^{p}, for two materials, a brittle (concrete) and a ductile (copper), leads to the conclusions that, in case of a brittle material under compression damage can be measured by the single damage variable D^{t}, whereas in case of a ductile material under tension a single damage variable is not capable of describing an uncoupled the total, elastic and plastic stiffness degradation as shown in Fig. 1.2

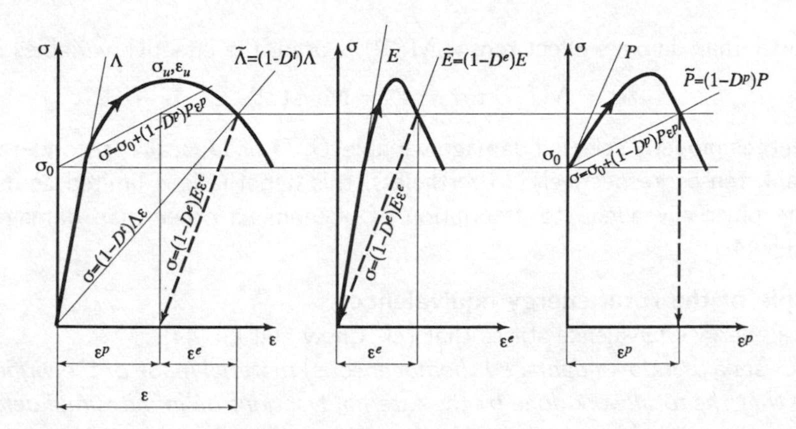

Figure 1.1. Total uniaxial strain split into the elastic and plastic components and the secant moduli degradation $\widetilde{\Lambda}$, $\widetilde{E}$, and $\widetilde{P}$ from damage D^{t}, D^{e} and D^{p} (after Taher et al. [61])

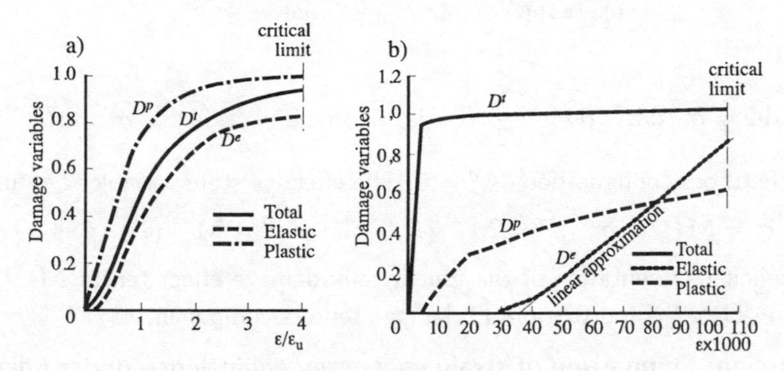

Figure 1.2. Evolution of generalized damage variables D^{t}, D^{e} and D^{p} in a) Concrete under compression and b) Copper under tension ($\varepsilon/\varepsilon_{\mathrm{u}}$ is the strain over the peak strain ratio), after Taher et al. [61]

D. Principle of the complementary elastic energy equivalence

The complementary elastic energy equivalence is postulated in order to define the fictitious pseudo–undamaged equivalent configuration and the corresponding effective variables $\widetilde{\boldsymbol{\sigma}}$ and $\widetilde{\boldsymbol{\varepsilon}}$ (cf. Cordebois and Sidoroff [33])

$$\Phi^{\mathrm{e}}(\boldsymbol{\sigma}, \mathcal{D}) = \widetilde{\Phi}^{\mathrm{e}}(\widetilde{\boldsymbol{\sigma}}, 0), \qquad \boldsymbol{\varepsilon}^{\mathrm{e}} = \frac{\partial \Phi^{\mathrm{e}}}{\partial \boldsymbol{\sigma}}, \tag{1.11}$$

$$\widetilde{\boldsymbol{\sigma}} = \left(\mathbf{I} - \widehat{\mathbf{D}}\right)^{-1} : \boldsymbol{\sigma}, \qquad \widetilde{\boldsymbol{\varepsilon}}^{\mathrm{e}} = \left(\mathbf{I} - \widehat{\mathbf{D}}\right) : \boldsymbol{\varepsilon}^{\mathrm{e}}, \tag{1.12}$$

where $\Phi^{\mathrm{e}} = (1/2)\,\boldsymbol{\sigma} : \boldsymbol{\varepsilon}^{\mathrm{e}}$ and $\widetilde{\Phi}^{\mathrm{e}} = (1/2)\,\widetilde{\boldsymbol{\sigma}} : \widetilde{\boldsymbol{\varepsilon}}^{\mathrm{e}}$, $\mathbf{I}$ and $\widehat{\mathbf{D}}$ are fourth–rank identity and damage tensors, whereas $\widehat{\mathbf{D}}$ is related to fourth–rank elasticity tensors $\mathbf{E}$ and $\widetilde{\mathbf{E}}$ of the damage equivalent (fictitious) and the current (physical) state of the material through

$$\widehat{\mathbf{D}} = \mathbf{I} - \widetilde{\mathbf{E}}^{1/2} : \mathbf{E}^{-1/2}. \tag{1.13}$$

When a fourth–rank damage effect tensor $\mathbf{M}(\mathcal{D})$ is used, the effective variables $\widetilde{\boldsymbol{\sigma}}$, $\widetilde{\boldsymbol{\varepsilon}}^{\mathrm{e}}$ are

$$\widetilde{\boldsymbol{\sigma}} = \mathbf{M}(\mathcal{D}) : \boldsymbol{\sigma}, \quad \widetilde{\boldsymbol{\varepsilon}}^{\mathrm{e}} = \mathbf{M}^{-1}(\mathcal{D}) : \boldsymbol{\varepsilon}^{\mathrm{e}}, \tag{1.14}$$

where $\mathcal{D}$ denotes properly selected damage variable D, $\mathbf{D}$ or $\widehat{\mathbf{D}}$, scalar, second–rank tensor or fourth–rank tensor, respectively. Nevertheless, this hypothesis is limited as it does not allow for the physically adequate description of phenomena other than damage coupled elasticity (cf. [44]).

E. Principle of the total energy equivalence

The total energy equivalence states that (cf. Chow and Lu [44]):

" *There exists a pseudo–undamaged (homogeneous) material made of the virgin material in the sense that the total work done by the external tractions on infinitesimal deformations during the same loading history as that for the real, damaged (heterogeneous) material is not changed* ":

$$\mathrm{d}\Phi^{\mathrm{e}} + \mathrm{d}\Phi^{\mathrm{d}} = \mathrm{d}\widetilde{\Phi}^{\mathrm{e}} \quad \text{and} \quad \mathrm{d}\Phi^{\mathrm{p}} = \mathrm{d}\widetilde{\Phi}^{\mathrm{p}}, \tag{1.15}$$

where

$$\mathrm{d}\widetilde{\Phi} = \widetilde{\boldsymbol{\sigma}} : \mathrm{d}\widetilde{\boldsymbol{\varepsilon}}, \quad \mathrm{d}\widetilde{\Phi}^{\mathrm{e}} = \frac{1}{2}\left(\widetilde{\boldsymbol{\sigma}} : \mathrm{d}\widetilde{\boldsymbol{\varepsilon}}^{\mathrm{e}} + \mathrm{d}\widetilde{\boldsymbol{\sigma}} : \widetilde{\boldsymbol{\varepsilon}}^{\mathrm{e}}\right), \quad \mathrm{d}\widetilde{\Phi}^{\mathrm{p}} = \widetilde{\boldsymbol{\sigma}} : \mathrm{d}\widetilde{\boldsymbol{\varepsilon}}^{\mathrm{p}}, \tag{1.16}$$

because in a fictitious configuration $\mathrm{d}\widetilde{\Phi}^{\mathrm{d}} = 0$. The effective state variables are furnished as

$$\widetilde{\boldsymbol{\sigma}} = \mathbf{M}(\mathcal{D}) : \boldsymbol{\sigma}, \quad \widetilde{\boldsymbol{\varepsilon}}^{\mathrm{e}} = \mathbf{M}^{-1}(\mathcal{D}) : \boldsymbol{\varepsilon}^{\mathrm{e}}, \quad \mathrm{d}\widetilde{\boldsymbol{\varepsilon}}^{\mathrm{p}} = \mathbf{M}^{-1}(\mathcal{D}) : \mathrm{d}\boldsymbol{\varepsilon}^{\mathrm{p}}, \tag{1.17}$$

where the explicit representation of the fourth–rank damage effect tensor $\mathbf{M}(\mathcal{D})$ depends on the second– $\mathbf{D}$ or the fourth–rank $\widehat{\mathbf{D}}$ damage tensors components.

1.3. Discussion: Comparison of strain vs. energy equivalence under uniaxial tension

A. 1D energy equivalence concept

In case of 1D elastic energy equivalence the following mapping holds:

$$\begin{gathered}\left\{\begin{matrix}\widetilde{\sigma}_1\\0\\0\end{matrix}\right\} = \begin{bmatrix}(1-D_1)^{-1} & 0 & 0\\ & (1-D_2)^{-1} & 0\\ & & (1-D_2)^{-1}\end{bmatrix}\left\{\begin{matrix}\sigma_1\\0\\0\end{matrix}\right\},\\ \left\{\begin{matrix}\widetilde{\varepsilon}^{\mathrm{e}}_1\\ \widetilde{\varepsilon}^{\mathrm{e}}_2\\ \widetilde{\varepsilon}^{\mathrm{e}}_3\end{matrix}\right\} = \begin{bmatrix}1-D_1 & 0 & 0\\ & 1-D_2 & 0\\ & & 1-D_2\end{bmatrix}\left\{\begin{matrix}\varepsilon^{\mathrm{e}}_1\\ \varepsilon^{\mathrm{e}}_2\\ \varepsilon^{\mathrm{e}}_3\end{matrix}\right\}.\end{gathered} \tag{1.18}$$

Hooke's law for the pseudo–undamaged continuum and for the damaged material is:

$$\left\{\begin{matrix}\widetilde{\sigma}_1\\0\\0\end{matrix}\right\} = \frac{E}{(1+\nu)(1-2\nu)}\begin{bmatrix}1-\nu & \nu & \nu\\ & 1-\nu & \nu\\ & & 1-\nu\end{bmatrix}\left\{\begin{matrix}\widetilde{\varepsilon}^{\mathrm{e}}_1\\ -\nu\widetilde{\varepsilon}^{\mathrm{e}}_1\\ -\nu\widetilde{\varepsilon}^{\mathrm{e}}_1\end{matrix}\right\},$$

$$\left\{\begin{array}{c}\sigma_1\\0\\0\end{array}\right\}=\frac{\widetilde{E}}{(1+\widetilde{\nu})(1-2\widetilde{\nu})}\left[\begin{array}{ccc}1-\widetilde{\nu} & \widetilde{\nu} & \widetilde{\nu}\\ & 1-\widetilde{\nu} & \widetilde{\nu}\\ & & 1-\widetilde{\nu}\end{array}\right]\left\{\begin{array}{c}\varepsilon_1^{\mathrm{e}}\\-\widetilde{\nu}\varepsilon_1^{\mathrm{e}}\\-\widetilde{\nu}\varepsilon_1^{\mathrm{e}}\end{array}\right\}. \tag{1.19}$$

Hence, two components describe damage evolution under uniaxial tension in the direction 1

$$D_1=1-\left(\frac{\widetilde{E}}{E}\right)^{1/2}, \qquad D_2=1-\frac{\nu}{\widetilde{\nu}}\left(\frac{\widetilde{E}}{E}\right)^{1/2}=1-\frac{\nu}{\widetilde{\nu}}(1-D_1). \tag{1.20}$$

B. 1D elastic strain equivalence concept

In case of 1D strain equivalence the mapping from $(\boldsymbol{\sigma},\boldsymbol{\varepsilon}^{\mathrm{e}})$ to $(\widetilde{\boldsymbol{\sigma}}^*,\widetilde{\boldsymbol{\varepsilon}}^{\mathrm{e}*})$ has a form:

$$\begin{gathered}\left\{\begin{array}{c}\widetilde{\sigma}_1^*\\0\\0\end{array}\right\}=\left[\begin{array}{ccc}(1-D_1^*)^{-1} & 0 & 0\\ & 1 & 0\\ & & 1\end{array}\right]\left\{\begin{array}{c}\sigma_1\\0\\0\end{array}\right\},\\ \left\{\begin{array}{c}\widetilde{\varepsilon}_1^{\mathrm{e}*}\\\widetilde{\varepsilon}_2^{\mathrm{e}*}\\\widetilde{\varepsilon}_3^{\mathrm{e}*}\end{array}\right\}=\left[\begin{array}{ccc}1 & 0 & 0\\ & 1 & 0\\ & & 1\end{array}\right]\left\{\begin{array}{c}\varepsilon_1^{\mathrm{e}}\\\varepsilon_2^{\mathrm{e}}\\\varepsilon_3^{\mathrm{e}}\end{array}\right\}.\end{gathered} \tag{1.21}$$

Hence, when Hooke's law analogous to (1.19) is used, after a simple rearrangement a single damage component D_1^* related to the Young's modulae ratio $\widetilde{E}/E$ is recovered, whereas Poisson's ratio $\widetilde{\nu}^*$ does not change,

$$D_1^*=1-\frac{\widetilde{E}}{E}, \qquad \widetilde{\nu}^*=\nu \quad (!). \tag{1.22}$$

This result contradicts a general observation that under uniaxial stress conditions microcracks of normals other than the main stress direction may appear (e.g. cylindrical transverse damages isotropy observed in rock–like materials under uniaxial compression, as mentioned by Chaboche [45]).

1.4. Exercise: Fourth–rank damage effect tensors

Legislation of the equivalence principles influences a particular form of the damage effects tensor representation in terms of the fourth–rank elasticity tensor change due to damage. Basic concepts of CDM based strain–, stress–, elastic energy or total energy equivalence, that result in constitutive tensor degradation with damage $\widetilde{\boldsymbol{\Lambda}}(D)$ and $\widetilde{\boldsymbol{\Lambda}}^{-1}(D)$, are sketched in Fig. 1.3. They will be discussed in details in what follows.

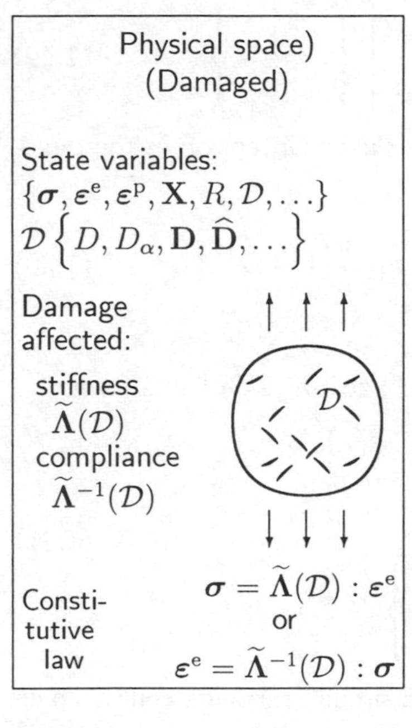

EQUIVALENCE PRINCIPLE

$\widetilde{\boldsymbol{\varepsilon}} = \boldsymbol{\varepsilon}$, or
$\widetilde{\boldsymbol{\sigma}} = \boldsymbol{\sigma}$, or
$\widetilde{\boldsymbol{\Phi}}^{\mathrm{e}} = \boldsymbol{\Phi}^{\mathrm{e}}$
$\mathrm{d}\widetilde{\boldsymbol{\Phi}}^{\mathrm{e}} = \mathrm{d}\boldsymbol{\Phi}^{\mathrm{e}} + \mathrm{d}\boldsymbol{\Phi}^{\mathrm{d}}$
$\mathrm{d}\widetilde{\boldsymbol{\Phi}}^{\mathrm{p}} = \mathrm{d}\boldsymbol{\Phi}^{\mathrm{p}}$

MAPPING

$\widetilde{\boldsymbol{\sigma}} = \mathbf{M}(\mathcal{D}) : \boldsymbol{\sigma}$
and
$\widetilde{\boldsymbol{\varepsilon}}^{\mathrm{e}} = \mathbf{M}^{-1}(\mathcal{D}) : \boldsymbol{\varepsilon}^{\mathrm{e}}$

$\mathrm{d}\widetilde{\boldsymbol{\varepsilon}}^{\mathrm{p}} = \mathbf{M}^{-1}(\mathcal{D}) : \mathrm{d}\boldsymbol{\varepsilon}^{\mathrm{p}}$

Effective (equivalent) space
(Pseudo-undamaged)

Effective state variables:
$\left\{\widetilde{\boldsymbol{\sigma}}, \widetilde{\boldsymbol{\varepsilon}}^{\mathrm{e}}, \widetilde{\boldsymbol{\varepsilon}}^{\mathrm{p}}, \widetilde{\mathbf{X}}, \widetilde{R}, \mathbf{O}, \ldots\right\}$

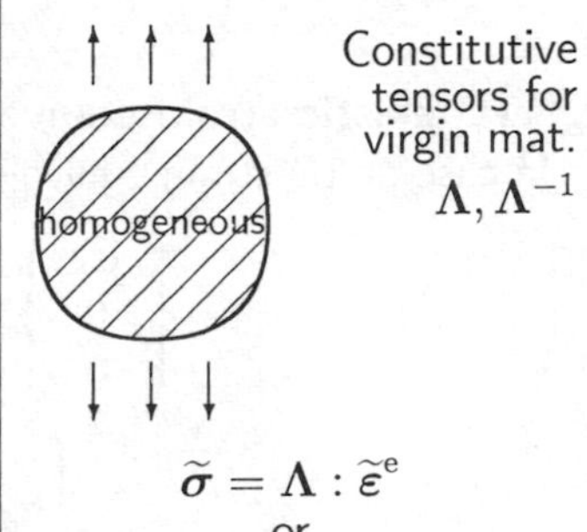

$\widetilde{\boldsymbol{\sigma}} = \boldsymbol{\Lambda} : \widetilde{\boldsymbol{\varepsilon}}^{\mathrm{e}}$
or
$\widetilde{\boldsymbol{\varepsilon}}^{\mathrm{e}} = \boldsymbol{\Lambda}^{-1} : \widetilde{\boldsymbol{\sigma}}$

damage effect tensor

$$\widetilde{\sigma}_{ij} = M_{ijkl}\left(D_{ijkl}\right)\sigma_{kl}$$

damaged constitutive tensor — initial constitutive tensor

$$\widetilde{\boldsymbol{\Lambda}}(\mathcal{D}) = \mathbf{M}^{-1}(\mathcal{D}) : \boldsymbol{\Lambda} : \mathbf{M}^{-\mathrm{T}}(\mathcal{D})$$
$$\widetilde{\boldsymbol{\Lambda}}^{-1}(\mathcal{D}) = \mathbf{M}^{\mathrm{T}}(\mathcal{D}) : \boldsymbol{\Lambda}^{-1} : \mathbf{M}(\mathcal{D})$$

Chaboche's notation

$$\underset{\approx}{\mathrm{M}} = \mathbf{M}^{-1}, \qquad \underset{\approx}{\Lambda} = \boldsymbol{\Lambda}, \qquad \underset{\approx}{\mathrm{S}} = \boldsymbol{\Lambda}^{-1},$$
$$\underset{\approx}{\widetilde{\Lambda}} = \underset{\approx}{\mathrm{M}} : \underset{\approx}{\Lambda} : \underset{\approx}{\mathrm{M}}^{\mathrm{T}}, \quad \underset{\approx}{\widetilde{\mathrm{S}}} = \underset{\approx}{\mathrm{M}}^{-\mathrm{T}} : \underset{\approx}{\mathrm{S}} : \underset{\approx}{\mathrm{M}}^{-1}$$

Matrix notation:

$$\{\widetilde{\boldsymbol{\sigma}}\} = [\mathrm{M}]\{\boldsymbol{\sigma}\} \qquad \{\boldsymbol{\sigma}\} = \left[\widetilde{\boldsymbol{\Lambda}}(\mathcal{D})\right] : \boldsymbol{\varepsilon}^{\mathrm{e}}$$
$$\{\boldsymbol{\sigma}\} = \{\sigma_{11}, \sigma_{22}, \sigma_{33}, \sigma_{23}, \sigma_{31}, \sigma_{12}\}^{\mathrm{T}} \qquad \{\boldsymbol{\varepsilon}^{\mathrm{e}}\} = \left[\widetilde{\boldsymbol{\Lambda}}^{-1}(\mathcal{D})\right] : \boldsymbol{\sigma}$$
$$\{\widetilde{\boldsymbol{\sigma}}\} = \{\widetilde{\sigma}_{11}, \widetilde{\sigma}_{22}, \widetilde{\sigma}_{33}, \widetilde{\sigma}_{23}, \widetilde{\sigma}_{31}, \widetilde{\sigma}_{12}\}^{\mathrm{T}}$$

(Symmetry)

Figure 1.3: Basic concepts of CDM

A. Principle of strain equivalence

When Hooke's law is written for the pseudo–undamaged (fictitious) and damaged (true) state we have:

$$
\begin{array}{ccc}
\text{(1D)} & \text{(3D)} & \text{(3D)} \\
\widetilde{\sigma} = E\varepsilon^{\mathrm{e}}, & \widetilde{\sigma}_{ij} = E_{ijkl}\varepsilon^{\mathrm{e}}_{kl}, & \widetilde{\boldsymbol{\sigma}} = \mathbf{E} : \boldsymbol{\varepsilon}^{\mathrm{e}}, \\
\sigma = \widetilde{E}\varepsilon^{\mathrm{e}}, & \sigma_{ij} = \widetilde{E}_{ijkl}\varepsilon^{\mathrm{e}}_{kl}, & \boldsymbol{\sigma} = \widetilde{\mathbf{E}} : \boldsymbol{\varepsilon}^{\mathrm{e}}, \\
\varepsilon^{\mathrm{e}} = \widetilde{E}^{-1}\sigma, & \varepsilon^{\mathrm{e}}_{kl} = \widetilde{E}^{-1}_{klij}\sigma_{ij}, & \boldsymbol{\varepsilon}^{\mathrm{e}} = \widetilde{\mathbf{E}}^{-1} : \boldsymbol{\sigma}.
\end{array}
\tag{1.23}
$$

$\widetilde{E}_{klij}$, $\widetilde{\mathbf{E}}$ denote the fourth–rank elasticity tensors modified by damage, and $\widetilde{\sigma}_{ij}$, $\widetilde{\boldsymbol{\sigma}}$ are the strain equivalent effective stress tensors, when both indices and absolute notation is used, hence:

$$
\begin{array}{ccc}
\text{(1D)} & \text{(3D)} & \text{(3D)} \\
\widetilde{\sigma} = \underbrace{E\widetilde{E}^{-1}}\sigma, & \widetilde{\sigma}_{ij} = \underbrace{E_{ijrs}\widetilde{E}^{-1}_{rskl}}\sigma_{kl}, & \widetilde{\boldsymbol{\sigma}} = \underbrace{\mathbf{E} : \widetilde{\mathbf{E}}^{-1}} : \boldsymbol{\sigma}, \\
\widetilde{\sigma} = (1-D)^{-1}\sigma, & \widetilde{\sigma}_{ij} = (I_{ijkl} - \widehat{D}_{ijkl})^{-1}\sigma_{kl}, & \widetilde{\boldsymbol{\sigma}} = (\mathbf{I} - \widehat{\mathbf{D}})^{-1} : \boldsymbol{\sigma}, \\
D = 1 - \widetilde{E}E^{-1}, & \widehat{D}_{ijkl} = I_{ijkl} - \widetilde{E}_{ijrs}E^{-1}_{rskl}, & \widehat{\mathbf{D}} = \mathbf{I} - \widetilde{\mathbf{E}} : \mathbf{E}^{-1}, \\
\widetilde{E}(D) = (1-D)E, & \widetilde{E}_{ijkl} = (I_{ijmn} - \widehat{D}_{ijmn})E_{mnkl}, & \widetilde{\mathbf{E}}(\widehat{\mathbf{D}}) = (\mathbf{I} - \widehat{\mathbf{D}}) : \mathbf{E}.
\end{array}
\tag{1.24}
$$

The 1D case was enclosed for a simple comparison.

B. Principle of elastic energy equivalence

A similar rearrangement based on the energy formulation yields:

$$
\begin{array}{ccc}
\text{(1D)} & \text{(3D)} & \text{(3D)} \\
\widetilde{\sigma} = E\widetilde{\varepsilon}^{\mathrm{e}}, & \widetilde{\sigma}_{ij} = E_{ijkl}\widetilde{\varepsilon}^{\mathrm{e}}_{kl}, & \widetilde{\boldsymbol{\sigma}} = \mathbf{E} : \widetilde{\boldsymbol{\varepsilon}}^{\mathrm{e}}, \\
\sigma = \widetilde{E}\varepsilon^{\mathrm{e}}, & \sigma_{ij} = \widetilde{E}_{ijkl}\varepsilon^{\mathrm{e}}_{kl}, & \boldsymbol{\sigma} = \widetilde{\mathbf{E}} : \boldsymbol{\varepsilon}^{\mathrm{e}}, \\
\Phi^{\mathrm{e}}(\sigma, \mathcal{D}) = \widetilde{\Phi}^{\mathrm{e}}(\widetilde{\sigma}, 0), & \Phi^{\mathrm{e}}(\sigma_{ij}, \mathcal{D}) = \widetilde{\Phi}^{\mathrm{e}}(\widetilde{\sigma}_{ij}, 0), & \Phi^{\mathrm{e}}(\boldsymbol{\sigma}, \mathcal{D}) = \widetilde{\Phi}^{\mathrm{e}}(\widetilde{\boldsymbol{\sigma}}, 0), \\
\widetilde{\Phi}^{\mathrm{e}}(\sigma, \mathcal{D}) = \frac{1}{2}\widetilde{\sigma}\widetilde{\varepsilon}^{\mathrm{e}}, & \widetilde{\Phi}^{\mathrm{e}} = \frac{1}{2}\widetilde{\sigma}_{ij}\widetilde{\varepsilon}^{\mathrm{e}}_{ij}, & \widetilde{\Phi}^{\mathrm{e}} = \frac{1}{2}\widetilde{\boldsymbol{\sigma}} : \widetilde{\boldsymbol{\varepsilon}}^{\mathrm{e}}, \\
\Phi^{\mathrm{e}} = \frac{1}{2}\sigma\varepsilon^{\mathrm{e}}, & \Phi^{\mathrm{e}} = \frac{1}{2}\sigma_{ij}\varepsilon^{\mathrm{e}}_{ij}, & \Phi^{\mathrm{e}} = \frac{1}{2}\boldsymbol{\sigma} : \boldsymbol{\varepsilon}^{\mathrm{e}}, \\
\widetilde{\sigma}\widetilde{\varepsilon}^{\mathrm{e}} = \sigma\varepsilon^{\mathrm{e}}, & \widetilde{\sigma}_{kl}\widetilde{\varepsilon}^{\mathrm{e}}_{kl} = \sigma_{rs}\varepsilon^{\mathrm{e}}_{rs}, & \widetilde{\boldsymbol{\sigma}} : \widetilde{\boldsymbol{\varepsilon}}^{\mathrm{e}} = \boldsymbol{\sigma} : \boldsymbol{\varepsilon}^{\mathrm{e}},
\end{array}
\tag{1.25}
$$

$$
\begin{array}{ccc}
\text{(1D)} & \text{(3D)} & \text{(3D)} \\
\widetilde{\sigma} = E\widetilde{\varepsilon}^{\mathrm{e}} & \widetilde{\sigma}_{ij} = E_{ijkl}\widetilde{\varepsilon}^{\mathrm{e}}_{kl} & \widetilde{\boldsymbol{\sigma}} = \mathbf{E} : \widetilde{\boldsymbol{\varepsilon}}^{\mathrm{e}} \\
= E\widetilde{\sigma}^{-1}\sigma\widetilde{E}^{-1}\sigma, & = E_{ijkl}\widetilde{\sigma}^{-1}_{kl}\sigma_{rs} & = \mathbf{E} : \widetilde{\boldsymbol{\sigma}}^{-1} : \boldsymbol{\sigma} \\
 & \times\widetilde{E}^{-1}_{rsmn}\sigma_{mn}, & : \widetilde{\mathbf{E}}^{-1} : \boldsymbol{\sigma}, \\
\widetilde{\sigma}^2 = E\widetilde{E}^{-1}\sigma^2, & \widetilde{\sigma}_{ik}\widetilde{\sigma}_{kj} = E_{ijrs}\widetilde{E}^{-1}_{rskl}\sigma_{kp}\sigma_{pl}, & \widetilde{\boldsymbol{\sigma}}^2 = \mathbf{E} : \widetilde{\mathbf{E}}^{-1} : \boldsymbol{\sigma}^2, \\
\widetilde{\sigma} = \underbrace{E^{1/2}\widetilde{E}^{-1/2}}\sigma, & \widetilde{\sigma}_{ij} = \underbrace{E^{1/2}_{ijrs}\widetilde{E}^{-1/2}_{rskl}}\sigma_{kl}, & \widetilde{\boldsymbol{\sigma}} = \underbrace{\mathbf{E}^{1/2} : \widetilde{\mathbf{E}}^{-1/2}} : \boldsymbol{\sigma}.
\end{array}
\tag{1.26}
$$

When a fourth–rank damage effect tensor $\mathbf{M}(\widehat{\mathbf{D}})$ is used, the mapping of the state variables $(\sigma, \varepsilon^{\mathrm{e}})$ or $(\sigma_{ij}, \varepsilon^{\mathrm{e}}_{ij})$ or $(\boldsymbol{\sigma}, \boldsymbol{\varepsilon}^{\mathrm{e}})$ from the physical (damaged) space to the fictitious (pseudo–undamaged) space $(\widetilde{\sigma}, \widetilde{\varepsilon}^{\mathrm{e}})$ or $(\widetilde{\sigma}_{ij}, \widetilde{\varepsilon}^{\mathrm{e}}_{ij})$ or $(\widetilde{\boldsymbol{\sigma}}, \widetilde{\boldsymbol{\varepsilon}}^{\mathrm{e}})$ is established

$$
\begin{array}{ccc}
\text{(1D)} & \text{(3D)} & \text{(3D)} \\
\widetilde{\sigma} = M(D)\sigma, & \widetilde{\sigma}_{ij} = M_{ijkl}(\widehat{D}_{ijkl})\sigma_{kl}, & \widetilde{\boldsymbol{\sigma}} = \mathbf{M}(\widehat{\mathbf{D}}) : \boldsymbol{\sigma}, \\
\widetilde{\varepsilon}^{\mathrm{e}} = M^{-1}(D)\varepsilon^{\mathrm{e}}, & \widetilde{\varepsilon}^{\mathrm{e}}_{ij} = M^{-1}_{ijkl}(\widehat{D}_{ijkl})\varepsilon^{\mathrm{e}}_{kl}, & \widetilde{\boldsymbol{\varepsilon}}^{\mathrm{e}} = \mathbf{M}^{-1}(\widehat{\mathbf{D}}) : \boldsymbol{\varepsilon}^{\mathrm{e}}, \\
M(D) = (1-D)^{-1} & M_{ijkl}(\widehat{D}_{ijkl}) = (I_{ijkl} - \widehat{D}_{ijkl})^{-1}, & \mathbf{M}(\widehat{\mathbf{D}}) = (\mathbf{I} - \widehat{\mathbf{D}})^{-1}.
\end{array}
\tag{1.27}
$$

2. MODELS OF ISOTROPIC DAMAGED (VISCO)PLASTIC MATERIALS

Damage evolution in virgin isotropic materials tested under creep conditions can often be considered as an isotropic phenomenon, for which scalar damage variables can be successfully used. In the following section isotropic damage models are considered, based both on the phenomenological observations as well as on the irreversible thermodynamics.

2.1. Phenomenological isotropic creep–damage models

Phenomenological isotropic damage models that account for a single damage mechanism, usually based either on the principal stress controlled grain boundary cavitation and growth mechanism (brittle damage) or the transgranular equivalent stress controlled slip-bands of plasticity mechanism (ductile damage), are known as the single state variable models (Robinson [62], Kachanov [2], Rabotnov [4], Hayhurst [8], Leckie and Hayhurst [7], Chrzanowski and Madej [63], Chaboche [64, 14], Othman and Hayhurst [65], Kowalewski, Hayhurst and Dyson [66], etc.). On the other hand, if two physical mechanisms of softening due to grain boundary cavitation on tertiary creep and multiplication of mobile dislocations are both considered, two damage state variables are necessary to describe this complex phenomenon (Othman, Hayhurst and Dyson [19], Dunne and Hayhurst [15, 16, 17], Kowalewski,

Hayhurst and Dyson [66], Kowalewski, Lin and Hayhurst [67], Hayhurst [68], etc.).

A. Single state variable creep–damage models

i. Uniaxial single state variable creep–damage models

In the simplest case, when the isotropic damage growth affects the uniaxial tertiary creep and the primary creep is ignored, the single damage variable D or ω introduces the creep–damage coupling (Kachanov [2], Rabotnov [4, 30]):

$$\frac{\dot{\varepsilon}^{\mathrm{c}}}{\dot{\varepsilon}_0} = \left(\frac{\sigma/\sigma_0}{1-\omega}\right)^n, \qquad \frac{\dot{\omega}}{\dot{\omega}_0} = \frac{(\sigma/\sigma_0)^\nu}{(1-\omega)^\varphi}, \tag{2.1}$$

where $\dot{\varepsilon}_0$, n and $\dot{\omega}_0$, ν, φ stand for the temperature dependent material constants in the creep law and the damage growth rule, respectively, whereas σ_0 is the reference stress. At $\sigma = \mathrm{const}$ and initial conditions for $t = 0 : \omega = \varepsilon^{\mathrm{c}} = 0$ the following holds:

$$\omega = 1 - \left(1 - \frac{t}{t_{\mathrm{f}}}\right)^{1/(1+\varphi)}, \qquad \frac{\varepsilon^{\mathrm{c}}}{\varepsilon_{\mathrm{f}}} = 1 - \left(1 - \frac{t}{t_{\mathrm{f}}}\right)^{\Delta}, \tag{2.2}$$

where $\Delta = (1+\varphi-n)\,/\,(1+\varphi)$, whereas symbols t_{f} and ε_{f} denote the time to failure $(\omega = 1)$ and the creep strain at failure, respectively:

$$t_{\mathrm{f}} = \frac{(\sigma_0/\sigma)^\nu}{(1+\varphi)\,\dot{\omega}_0}, \qquad \varepsilon_{\mathrm{f}} = \frac{\dot{\varepsilon}_0\,(\sigma/\sigma_0)^{n-\nu}}{\dot{\omega}_0\,(1+\varphi-n)} = \frac{\dot{\varepsilon}_{\mathrm{ss}} t_{\mathrm{f}}}{\Delta}, \tag{2.3}$$

and $\dot{\varepsilon}_{\mathrm{ss}} = \dot{\varepsilon}_0\,(\sigma/\sigma_0)^n$ stands for a steady–state or a minimum creep rate (no damage effect).

For the two batches of pure copper A and B tested to failure at temperature 300 C under the stress $\sigma = 32.4$ MPa, Rides et al. [69] obtained: $n = 6.56$, $\nu = 6.31$, $\varphi = 7.1$, $\sigma_0 = 300$ MPa; $\dot{\varepsilon}_0^{\mathrm{A}} = 11 \times 10^{-5}\ \mathrm{h}^{-1}$, $\dot{\omega}_0^{\mathrm{A}} = 6.68 \times 10^{-4}\ \mathrm{h}^{-1}$ and $\dot{\varepsilon}_0^{\mathrm{B}} = 2.54 \times 10^{-5}\ \mathrm{h}^{-1}$, $\dot{\omega}_0^{\mathrm{B}} = 2.74 \times 10^{-5}\ \mathrm{h}^{-1}$, however the model is often simplified by setting $\varphi = \nu$.

When the Kachanov's [2], and the Chaboche's [14] notation is used, the uniaxial damage growth rule is:

$$\frac{\mathrm{d}\psi}{\mathrm{d}t} = -C\left(\frac{\sigma_1}{\psi}\right)^r, \qquad \frac{\mathrm{d}D}{\mathrm{d}t} = \left(\frac{\sigma}{A}\right)^r (1-D)^{-k}, \tag{2.4}$$

where ψ and $D = \omega$ denote the continuity and the damage, respectively, if $\psi + D = 1$.

Earlier on, Robinson [62] established the life fraction rule:

$$\int_0^{t_{\mathrm{R}}} \frac{\mathrm{d}t}{t_{\mathrm{R}}(t)} = 1, \qquad t_{\mathrm{R}}(t) = \frac{1}{C\,(r+1)\,[\sigma_1(t)]^r}, \tag{2.5}$$

where $\sigma_1(t)$ is arbitrarily prescribed uniaxial stress function.

Generalization of the Kachanov's concept (2.1) to the case when both instantaneous damage (time independent) and creep–damage (time dependent) are accounted is due to Chrzanowski and Madej [63]:

$$\frac{\dot{\omega}}{\dot{\omega}_0} = \chi\frac{(\sigma/\sigma_0)^{\nu_0}}{(1+\omega)^{\varphi_0}}\left(\frac{\dot{\sigma}}{\sigma_0}\right) + \frac{(\sigma/\sigma_0)^\nu}{(1-\omega)^\varphi}. \tag{2.6}$$

At $\sigma = \text{const}$, integration of (2.6) twice, yields:

$$\omega = 1 - \left\{ \frac{t_{\mathrm{ff}}(\sigma)}{t_{\mathrm{f}}(\sigma)} \left[1 - \frac{t}{t_{\mathrm{ff}}(\sigma)} \right] \right\}^{1/(\varphi+1)}, \tag{2.7}$$

where

$$t_{\mathrm{f}}(\sigma) = \frac{1}{(1+\varphi)\,\dot{\omega}_0\,(\sigma/\sigma_0)^{\nu}}, \qquad t_{\mathrm{ff}}(\sigma) = t_{\mathrm{f}}(\sigma) \left[1 - \left(\frac{\sigma}{\sigma_{\mathrm{f}}} \right)^{1+\nu_0} \right]^{(1+\varphi)/(1+\varphi_0)} \tag{2.8}$$

$$\sigma_{\mathrm{f}} = \sigma_0 \left[\frac{1+\nu_0}{(1+\varphi_0)\,\chi\dot{\omega}_0} \right]^{1/(1+\nu_0)}. \tag{2.9}$$

$\dot{\omega}_0$, σ_0, ν_0, φ_0, ν, φ and χ are material constants for the combined instantaneous and subsequent damage mechanism.

In contrast to copper and aluminium alloy, where primary creep is negligible and the tertiary creep predominates, in case of stainless steel the primary creep manifests strongly. Hence, for 316 stainless steel tested at 210 C, 250 C and 550 C, Othman and Hayhurst [65], proposed to include the decaying time–function t^m $(m < 1)$ that accounts for primary creep:

$$\frac{\dot{\varepsilon}^{\mathrm{c}}}{\dot{\varepsilon}_0} = \left(\frac{\sigma/\sigma_0}{1-\omega} \right)^n t^m, \qquad \frac{\dot{\omega}}{\dot{\omega}_0} = \frac{(\sigma/\sigma_0)^{\nu}}{(1-\omega)^{\varphi}} t^m, \tag{2.10}$$

Integration of (2.10) at $\sigma = \text{const}$ yields:

$$\omega = 1 - \left[1 - \left(\frac{t}{t_{\mathrm{f}}} \right)^{m+1} \right]^{1/(1+\varphi)}, \qquad \frac{\varepsilon^{\mathrm{c}}}{\varepsilon_{\mathrm{f}}} = 1 - \left[1 - \left(\frac{t}{t_{\mathrm{f}}} \right)^{m+1} \right]^{\Delta}, \tag{2.11}$$

$$t_{\mathrm{f}} = \left[\frac{(1+m)\,(\sigma_0/\sigma)^{\nu}}{(1+\varphi)\,\dot{\omega}_0} \right]^{1/(1+m)}, \qquad \varepsilon_{\mathrm{f}} = \frac{\dot{\varepsilon}_0\,(\sigma/\sigma_0)^{n-\nu}}{\dot{\omega}_0\,(1+\varphi-n)}, \tag{2.12}$$

where Δ is defined in a similar way as in (2.2).

ii. Single state variable creep–damage models under multiaxial stress conditions

Multiaxial generalization of the single state variable creep–damage model is based on the concept of so called isochronous rupture curves when metallic materials are tested to failure (Johnson et al. [70, 71], Hayhurst [8], Trąpczyński, Hayhurst and Leckie [10], Kowalewski, Lin and Hayhurst [67], etc.). Roughly speaking, three classes of metallic materials with respect to rupture curves can be specified: principal stress controlled (copper–like), equivalent stress controlled (aluminium–like) and combined principal/equivalent stress controlled (steel–like). The above enables the following 3D generalization of (2.4), cf. Hayhurst [8], Chaboche [14]:

$$\frac{\mathrm{d}D}{\mathrm{d}t} = \left[\frac{\chi(\boldsymbol{\sigma})}{A} \right]^r (1-D)^{-k}, \tag{2.13}$$

where the damage equivalent stress $\chi(\boldsymbol{\sigma})$ is a linear combination of stress invariants

$$\chi(\boldsymbol{\sigma}) = \alpha J_0(\boldsymbol{\sigma}) + 3\beta J_1(\boldsymbol{\sigma}) + (1-\alpha-\beta)\,J_2(\boldsymbol{\sigma}) \tag{2.14}$$

or

$$\chi(\boldsymbol{\sigma}) = a\sigma_1 + 3b\sigma_{\mathrm{H}} + c\sigma_{\mathrm{eq}}, \tag{2.15}$$

where $J_0(\boldsymbol{\sigma}) = \sigma_1$, $J_1(\boldsymbol{\sigma}) = \sigma_{\mathrm{H}} = 1/3\,\mathrm{Tr}\,\boldsymbol{\sigma}$, $J_2(\boldsymbol{\sigma}) = \sigma_{\mathrm{eq}} = [3/2\,\mathrm{Tr}(\boldsymbol{\sigma}'^2)]^{1/2}$.

Multiaxial scalar creep–damage coupling with primary creep ignored (generalization of (2.1)) is due to Leckie and Hayhurst [7]:

$$\frac{\dot\varepsilon^{\mathrm{c}}_{ij}}{\dot\varepsilon_0} = \frac{1}{n+1}\frac{\partial\Omega^{n+1}(\sigma_{kl}/\sigma_0)}{\partial(\sigma_{ij}/\sigma_0)}\frac{1}{(1-\omega)^n}, \qquad \frac{\dot\omega}{\dot\omega_0} = \frac{\chi^\nu(\sigma_{ij}/\sigma_0)}{(1-\omega)^\varphi}. \tag{2.16}$$

$\Omega(\sigma_{kl}/\sigma_0) \equiv \sigma_{\mathrm{eq}}(\sigma_{kl}/\sigma_0)$ is a convex homogeneous potential function of degree 1 in stress, and $\chi(\sigma_{ij}/\sigma_0)$ is a properly defined damage equivalent stress determined by the isochronous rupture surface (2.15). When both the tertiary and the primary creep is accounted, Othman and Hayhurst [65] proposed:

$$\frac{\dot\varepsilon^{\mathrm{c}}_{ij}}{\dot\varepsilon_0} = \frac{1}{n+1}\frac{\partial\Omega^{n+1}(\sigma_{kl}/\sigma_0)}{\partial(\sigma_{ij}/\sigma_0)}\frac{f(t)}{(1-\omega)^n}, \qquad \frac{\dot\omega}{\dot\omega_0} = \frac{\chi^\nu(\sigma_{ij}/\sigma_0)\,f(t)}{(1-\omega)^\varphi}. \tag{2.17}$$

When the damage evolution is the equivalent stress controlled, and the Mises–type potential function is used, the following equations are furnished (Kowalewski, Hayhurst and Dyson [66]):

$$\dot\varepsilon^{\mathrm{c}}_{ij} = \frac{3}{2}A\frac{\sigma_{\mathrm{eq}}^{n-1}}{(1-\omega)^n}s_{ij}t^m, \qquad \dot\omega = B\frac{\sigma_{\mathrm{eq}}^{\nu}}{(1-\omega)^\varphi}t^m. \tag{2.18}$$

Calibration of the above model for aluminium alloy tested at 150 C yields: $A = 3.511\times10^{-29}$ (MPa)$^{-n}$/h^{m+1}, $B = 1.960\times10^{-23}$ (MPa)$^{-\nu}$/h^{m+1}, $n = 11.034$, $\nu = 8.220$, $\varphi = 12.107$, $m = -0.3099$, $E = 71.1\times10^3$ MPa.

B. Two state variables mechanisms–based damage models

i. Two–parameter multiaxial hyperbolic sinus models for nickel– and aluminium–based alloys

Othman, Hayhurst and Dyson [19] developed the model capable of describing nickel–based superalloy where the sinh function of the stress is used instead of the stress dependence in n powered in the single damage state variable model, whereas the primary creep is included through the additional variable $H(t)$ ranging from 0 to H^* (saturation)

$$\begin{aligned}
\frac{\mathrm{d}\varepsilon^{\mathrm{c}}_{ij}}{\mathrm{d}t} &= \frac{3}{2}A\frac{\sinh\{B\sigma_{\mathrm{eq}}[1-H(t)]\}}{(1-\omega_1)(1-\omega_2)^n}\left\{\frac{s_{ij}}{\sigma_{\mathrm{eq}}}\right\},\\
\frac{\mathrm{d}H}{\mathrm{d}t} &= \frac{h}{\sigma_{\mathrm{eq}}}A\frac{\sinh\{B\sigma_{\mathrm{eq}}[1-H(t)]\}}{(1-\omega_1)(1-\omega_2)^n}\left\{1-\frac{H(t)}{H^*}\right\},\\
\frac{\mathrm{d}\omega_1}{\mathrm{d}t} &= CA\frac{(1-\omega_1)}{(1-\omega_2)^n}\sinh\{B\sigma_{\mathrm{eq}}[1-H(t)]\},\\
\frac{\mathrm{d}\omega_2}{\mathrm{d}t} &= DA\left(\frac{\sigma_1}{\sigma_{\mathrm{eq}}}\right)^\nu N\frac{\sinh\{B\sigma_{\mathrm{eq}}[1-H(t)]\}}{(1-\omega_1)(1-\omega_2)^n},
\end{aligned} \tag{2.19}$$

where two physical mechanisms are included: the softening due to the multiplication of mobile dislocations ω_1 $(0\le\omega_1\le1)$ and to the creep constrained cavities nucleation and growth on the grain boundaries ω_2 $(0 \leq \omega_2 \leq 0.3)$. For aluminium alloy at temperature 150C the material constants are: $A = 2.96 \times 10^{-11}\mathrm{h}^{-1}$, $B = 7.17 \times 10^{-2}(\mathrm{MPa})^{-1}$, $C = 35$, $D = 6.63$, $H^* = 0.2032$, $h = 1.37 \times 10^5$ MPa, $\nu = 0$ (for more details see [66, 68]).

Another model based on a new mechanism of creep–damage in the particle hardened materials, namely tertiary creep softening due to both, ageing of microstructure Φ and grain boundary nucleation and growth ω_2 and primary creep effect using the variable H was developed by Othman, Hayhurst and Dyson [19]. More details about the model can be find in the Chapter 5 of this book (Hayhurst [68]).

ii. Creep–cyclic plasticity damage interaction model for copper

Dunne and Hayhurst [15, 16, 17] developed a model validated for creep–damage–cyclic plasticity interaction in copper specimens subjected to thermo–mechanical cyclic loadings at both room (20 C) and elevated (500 C) temperatures. Creep–cyclic plasticity–damage interaction is given by (2.20) where two variables ω_1 and ω_2 refer to the creep damage and the cyclic plasticity damage per cycle, each controlled by an independent damage evolution (2.21)

$$\begin{aligned} &D^{\mathrm{c}} = \omega_1 + \alpha_1 z(\omega_1)\,\omega_2, \quad D^{\mathrm{p}} = \omega_2 + \alpha_2 z(\omega_1)\,\omega_1, \\ &D = D^{\mathrm{c}} + D^{\mathrm{p}}, \qquad z(\omega_1) = \frac{1}{2} + \frac{1}{\pi}\arctan\mu\,(\omega_1 - \omega_0), \end{aligned} \tag{2.20}$$

$$\begin{aligned} \frac{\mathrm{d}\omega_1}{\mathrm{d}t} &= A\frac{\left[\delta\sigma_1 + (1-\delta)\,\sigma_{\mathrm{eq}}\right]^{\nu}}{(1-D^{\mathrm{c}})^{\varphi}}, \\ \frac{\mathrm{d}\omega_2}{\mathrm{d}N} &= \left[1 - (1-D^{\mathrm{p}})^{\beta+1}\right]^{\varrho}\left[\frac{A_{II}}{M\,(1-D^{\mathrm{p}})}\right]^{\beta}. \end{aligned} \tag{2.21}$$

The damage evolution model is combined, then, with the creep cyclic plasticity nonlinear kinematic hardening theory (extension of the Chaboche and Rousselier theory [72]), cf. also Skrzypek and Ganczarski [51], as follows:

$$\begin{aligned} &\dot{\boldsymbol{\varepsilon}}^{\mathrm{p}} = \frac{3}{2}\left\langle\frac{J_2(\boldsymbol{\sigma} - \mathbf{X})/(1-D) - \sigma_y}{K}\right\rangle^{n}\frac{\boldsymbol{\sigma}' - \mathbf{X}'}{J_2(\boldsymbol{\sigma} - \mathbf{X})}, \\ &\dot{\mathbf{X}}_{1,2} = \frac{2}{3}C_{1,2}\dot{\boldsymbol{\varepsilon}}^{\mathrm{p}}(1-D) - \gamma_{1,2}\mathbf{X}_{1,2}\dot{p} + \left(C'_{1,2}/C_{1,2}\right)\mathbf{X}_{1,2}\dot{T}, \\ &\mathbf{X} = \mathbf{X}_1 + \mathbf{X}_2, \quad \boldsymbol{\sigma} = E(1-D)\left(\boldsymbol{\varepsilon} - \boldsymbol{\varepsilon}^{\mathrm{p}} - \boldsymbol{\varepsilon}^{\mathrm{T}}\right), \quad \dot{p} = \left[\frac{2}{3}\dot{\boldsymbol{\varepsilon}}^{\mathrm{p}} : \dot{\boldsymbol{\varepsilon}}^{\mathrm{p}}\right]^{1/2}. \end{aligned} \tag{2.22}$$

2.2. Thermodynamics based formulation of the coupled isotropic damage–thermo-elastic–(visco)plasticity

A consistent way to obtain state and evolution equations of damaged materials is based on the thermodynamics of irreversible processes with internal variables (Germain, Nguyen and Suquet [20]). This approach is systematically introduced and discussed in Chapter 4 by Chaboche [46], when formulating basic thermodynamic concepts. In what follows some particular proposals of state and damage evolution equations, based on the simplified assumption that damage can be considered as isotropic phenomenon, are reviewed briefly.

A. Kinetic law of damage evolution

A simplified concept of elastic energy release due to isotropic damage cumulation in metallic materials, based on the strain equivalence and the effective stress concept, was developed by Lemaitre and Chaboche [13]. Assuming small strain theory, the total strain is written as the sum of the elastic and the inelastic part $\varepsilon = \varepsilon^{\mathrm{e}} + \varepsilon^{\mathrm{in}}$ and anisotropic elasticity law coupled with the isotropic damage is used for elastic strains

$$\sigma_{ij} = E_{ijkl}\varepsilon^{\mathrm{e}}_{kl}\,(1-D)\,, \tag{2.23}$$

to obtain the elastic strain energy density of damaged material

$$\Phi = \Phi^{\mathrm{e}} + \Phi^{\mathrm{in}}, \qquad \Phi^{\mathrm{e}} = \frac{1}{2}E_{ijkl}\varepsilon^{\mathrm{e}}_{ij}\varepsilon^{\mathrm{e}}_{kl}\,(1-D)\,. \tag{2.24}$$

Here the thermodynamic force Y^{e} associated with the internal scalar variable D contains the contribution of the elastic energy Φ^{e} only, whereas the inelastic part Φ^{in} is assumed not to be released by the damage growth

$$Y^{\mathrm{e}} \stackrel{\mathrm{def}}{=} \frac{1}{2}\frac{\mathrm{d}\Phi^{\mathrm{e}}}{\mathrm{d}D}\bigg|_{\sigma_{ij}=\mathrm{const}} = \frac{\Phi^{\mathrm{e}}}{1-D}. \tag{2.25}$$

The above model based on the anisotropic elasticity coupled with isotropic damage is inconsistent at all. So, confining ourselves to the isotropic elasticity law of isotropic damaged material

$$\begin{aligned} \varepsilon^{\mathrm{e}}_{ij} &= \frac{1+\nu}{E}\frac{\sigma_{ij}}{1-D} - \frac{\nu}{E}\frac{\sigma_{kk}}{1-D}\delta_{ij}, \\ \Phi^{\mathrm{e}} &= \frac{\sigma^2_{\mathrm{eq}}}{2E\,(1-D)}\left[\frac{2}{3}\,(1+\nu) + 3\,(1-2\nu)\left(\frac{\sigma_{\mathrm{H}}}{\sigma_{\mathrm{eq}}}\right)^2\right], \end{aligned} \tag{2.26}$$

the thermodynamic force Y^{e} is reduced to

$$Y^{\mathrm{e}} = \frac{\widetilde{\sigma}^2_{\mathrm{eq}}}{2E}R_\nu, \quad R_\nu = \frac{2}{3}\,(1+\nu) + 3\,(1-2\nu)\left(\frac{\sigma_{\mathrm{H}}}{\sigma_{\mathrm{eq}}}\right)^2, \quad \widetilde{\sigma}_{\mathrm{eq}} = \frac{\sigma_{\mathrm{eq}}}{1-D}. \tag{2.27}$$

i. Effective–stress based time–independent plasticity coupled with isotropic damage

Lemaitre [73, 74], and Lemaitre and Chaboche [13] introduced a single coupled dissipative potential in order to generalize the Chaboche–Rousselier nonlinear isotropic/kinematic hardening theory to obtain:

$$\begin{aligned} F^{\mathrm{L-Ch}}\,(\widetilde{\boldsymbol{\sigma}}, \mathbf{X}, R, D) &= f\,(\widetilde{\boldsymbol{\sigma}}, \mathbf{X}, R) + F^{\mathrm{d}}\,(Y^{\mathrm{e}}) = J_2\,(\widetilde{\boldsymbol{\sigma}} - \mathbf{X}) - R - \sigma_y + F^{\mathrm{d}}\,(Y^{\mathrm{e}})\,, \\ J_2\,(\widetilde{\boldsymbol{\sigma}} - \mathbf{X}) &= \left[\frac{3}{2}\left(\frac{\sigma'_{ij}}{1-D} - X'_{ij}\right)\left(\frac{\sigma'_{ij}}{1-D} - X'_{ij}\right)\right]^{1/2}. \end{aligned} \tag{2.28}$$

The generalized normality rule (associative theory) is governed by the single plastic multiplier $\dot{\lambda}$:

$$\dot{\boldsymbol{\varepsilon}}^{\mathrm{p}}=\dot{\lambda}\frac{\partial F^{\mathrm{L-Ch}}}{\partial \boldsymbol{\sigma}},\quad \dot{\boldsymbol{\alpha}}=-\dot{\lambda}\frac{\partial F^{\mathrm{L-Ch}}}{\partial \mathbf{X}},\quad \dot{r}=-\dot{\lambda}\frac{\partial F^{\mathrm{L-Ch}}}{\partial R},\quad \dot{D}=-\dot{\lambda}\frac{\partial F^{\mathrm{L-Ch}}}{\partial Y^{\mathrm{e}}},\tag{2.29}$$

where $f(\widetilde{\boldsymbol{\sigma}},\mathbf{X},R)=J_2(\widetilde{\boldsymbol{\sigma}}-\mathbf{X})-R-\sigma_y=0$ is a Mises–type partly coupled yield function and σ_y is the initial yield stress under uniaxial tension test. External state variables $(\boldsymbol{\varepsilon},\boldsymbol{\sigma})$ and internal state variables $(\boldsymbol{\alpha},\mathbf{X})$, (r,R), (D,Y^{e}) are introduced to represent kinematic hardening, isotropic hardening and isotropic damage. It is assumed in this simplified approach that hardening variables $\mathbf{X}$ and R are not amplified by damage and the effective hardening variables $\widetilde{\mathbf{X}}$ and $\widetilde{R}$ are not used. When (2.28) through (2.29) are applied, the state equations are obtained:

$$\begin{aligned}
\dot{\varepsilon}^{\mathrm{p}}_{ij}&=\frac{3}{2}\frac{\dot{\lambda}}{1-D}\frac{\widetilde{\sigma}'_{ij}-X'_{ij}}{J_2(\widetilde{\sigma}_{ij}-X_{ij})}=\frac{3}{2}\dot{\tilde{\lambda}}\frac{\widetilde{\sigma}'_{ij}-X'_{ij}}{J_2(\widetilde{\sigma}_{ij}-X_{ij})},\\
\dot{\alpha}_{ij}&=\frac{3}{2}\dot{\tilde{\lambda}}\frac{\widetilde{\sigma}'_{ij}-X'_{ij}}{J_2(\widetilde{\sigma}_{ij}-X_{ij})}=\dot{\varepsilon}^{\mathrm{p}}_{ij}(1-D),\\
\dot{r}&=\dot{\lambda}=\dot{p}(1-D),\quad \dot{D}=-\dot{\lambda}\frac{\partial F^{\mathrm{d}}(Y^{\mathrm{e}})}{\partial Y^{\mathrm{e}}}=-\frac{\partial F^{\mathrm{d}}(Y^{\mathrm{e}})}{\partial Y^{\mathrm{e}}}\dot{p}(1-D),
\end{aligned}\tag{2.30}$$

where $\dot{p}$ denotes the cumulative plastic strain $\dot{p}=\left[(2/3)\,\dot{\varepsilon}^{\mathrm{p}}_{ij}\dot{\varepsilon}^{\mathrm{p}}_{ij}\right]^{1/2}$ and $\dot{\tilde{\lambda}}=\dot{\lambda}/(1-D)$.

Assuming, after Lemaitre and Chaboche [13], that the damage term of the potential (2.28) is a quadratic function of Y^{e}, the kinetic law of damage evolution for rate–independent plasticity is furnished:

$$F^{\mathrm{d}}(Y^{\mathrm{e}})=-\frac{(Y^{\mathrm{e}})^2}{2S(1-D)},\qquad \dot{D}=\frac{Y^{\mathrm{e}}}{S}\dot{p}=\frac{\sigma_{\mathrm{eq}}^2R_\nu}{2ES(1-D)^2}\dot{p}.\tag{2.31}$$

If, in a more general case a damage potential is assumed as a power function of Y^{e} $(s>2)$ (Germain, Nguyen and Suquet [20], Dufailly and Lemaitre [21]), the generalized kinetic law of damage evolution holds

$$F^{\mathrm{d}}=-\left(\frac{Y^{\mathrm{e}}}{S}\right)^s\frac{Y^{\mathrm{e}}}{(s+1)(1-D)},\quad \dot{D}=\left(\frac{Y^{\mathrm{e}}}{S}\right)^s\dot{p}=\left[\frac{\sigma_{\mathrm{eq}}^2R_\nu}{2ES(1-D)^2}\right]^s\dot{p}.\tag{2.32}$$

ii. Kinetic law of damage of plastic materials

Assuming the Ramberg–Osgood isotropic power hardening function for damaged material (Lemaitre and Chaboche [13])

$$\widetilde{\sigma}_{\mathrm{eq}}=\frac{\sigma_{\mathrm{eq}}}{1-D}=\sigma_{\mathrm{s}}p^n\tag{2.33}$$

in the damage evolution equations (2.31) or (2.32), we obtain simplified rules

$$\mathrm{d}D=\frac{\sigma_{\mathrm{s}}^2}{2ES}R_\nu\left(\frac{\sigma_{\mathrm{H}}}{\sigma_{\mathrm{eq}}}\right)p^{2n}\mathrm{d}p\tag{2.34}$$

or

$$\mathrm{d}D=\left(\frac{\sigma_{\mathrm{s}}^2}{2ES}\right)^s\left[R_\nu\left(\frac{\sigma_{\mathrm{H}}}{\sigma_{\mathrm{eq}}}\right)\right]^s p^{2sn}\mathrm{d}p,\tag{2.35}$$

which describe the ductile damage growth by the scalar variable D that depends on the cumulative plastic strain p. In a particular case, when strain hardening is saturated $\dot{\mathbf{X}} = \mathbf{0}$, $\dot{R} = 0$, $f = \widetilde{\sigma}_{\mathrm{eq}} - \sigma_{\mathrm{s}} = 0$, the generalized kinetic law of damage evolution reduces to the simplified form

$$\dot{D} = \left(\frac{Y_{\mathrm{s}}^{\mathrm{e}}}{S}\right)^{s} \dot{p} = \left(\frac{\sigma_{\mathrm{s}}^{2} R_{\nu}}{2ES}\right)^{s} \dot{p}, \tag{2.36}$$

the integration of which in the case of proportional loadings, when the stress triaxiality ratio in R_ν is constant $\sigma_{\mathrm{H}}/\sigma_{\mathrm{eq}} = \mathrm{const}$, yields the linear damage growth with cumulative plastic strain p.

iii. Kinetic law of creep damage of metals and polymers

Lemaitre–Chaboche's kinetic law of damage evolution (2.31) combined with Norton's creep law yields:

$$\begin{gathered} \dot{D} = \frac{\sigma_{\mathrm{eq}}^{2} R_{\nu}}{2ES\,(1-D)^{2}} \dot{p} \mathrm{H}\,(p - p_0)\,, \quad \dot{p} = \left[\frac{\sigma_{\mathrm{eq}}}{K_v\,(1-D)}\right]^{N}, \\ \dot{D} = \frac{\sigma_{\mathrm{eq}}^{N+2} R_{\nu}}{2ESK_{v}^{N}\,(1-D)^{N+2}} \mathrm{H}\,(p - p_0)\,, \end{gathered} \tag{2.37}$$

where the Heaviside function $\mathrm{H}\,(p-p_0)$ is used for the initial damage at $p = p_0$.

In 1D case the classical Kachanov's equation (2.4) is recovered

$$\dot{D} = \left[\frac{\sigma}{A\,(1-D)}\right]^{r} \mathrm{H}\,(\varepsilon - \varepsilon_0)\,, \qquad A = \left(2ESK_{\nu}^{N}\right)^{1/(N+2)}, \qquad r = N+2. \tag{2.38}$$

B. Unified Helmholtz free energy–based CDM model of ductile isotropic damaged materials

A unified CDM model similar to the previously discussed Lemaitre and Chaboche's kinetic law of damage evolution is due to Mou and Han [22], who decomposed the total strain tensor into the elastic and inelastic tensors, $\boldsymbol{\varepsilon} = \boldsymbol{\varepsilon}^{\mathrm{e}} + \boldsymbol{\varepsilon}^{\mathrm{p}}$ and introduced, after Broberg [75], the new damage variable $D^{\mathrm{B}} = \ln\left(A/\widetilde{A}\right)$, as one of internal state variables which influences the Helmholtz free energy of the damaged material $\psi^{\mathrm{H}}\left(\boldsymbol{\varepsilon}^{\mathrm{e}}, \boldsymbol{\alpha}, r, D^{\mathrm{B}}, T\right)$

$$\psi^{\mathrm{H}}\left(\boldsymbol{\varepsilon}^{\mathrm{e}}, \boldsymbol{\alpha}, r, D^{\mathrm{B}}, T\right) = \psi^{\mathrm{e}}\left(\boldsymbol{\varepsilon}^{\mathrm{e}}, D^{\mathrm{B}}, T\right) + \psi^{\mathrm{in}}\left(\boldsymbol{\alpha}, r, T\right). \tag{2.39}$$

The generalized thermodynamic forces $(\boldsymbol{\sigma}, \mathbf{X}, R, Y^{\mathrm{e}})$ are associated with the elastic strain, kinematic hardening, isotropic hardening and damage $\left(\boldsymbol{\varepsilon}^{\mathrm{e}}, \boldsymbol{\alpha}, r, D^{\mathrm{B}}\right)$, respectively, through

$$\boldsymbol{\sigma} = \rho\frac{\partial\psi^{\mathrm{e}}}{\partial\boldsymbol{\varepsilon}^{\mathrm{e}}}, \quad \mathbf{X} = \rho\frac{\partial\psi^{\mathrm{in}}}{\partial\boldsymbol{\alpha}}, \quad R = \rho\frac{\partial\psi^{\mathrm{in}}}{\partial r}, \quad Y^{\mathrm{e}} = -\rho\frac{\partial\psi^{\mathrm{e}}}{\partial D^{\mathrm{B}}}, \tag{2.40}$$

whereas the specific entropy production rate can be expressed as

$$\boldsymbol{\sigma}\dot{\boldsymbol{\varepsilon}}^{\mathrm{p}} + R\dot{r} + \mathbf{X}\dot{\boldsymbol{\alpha}} + Y^{\mathrm{e}}\dot{D}^{\mathrm{B}} - \mathbf{q}\frac{1}{T}\mathbf{grad}T \geq 0, \tag{2.41}$$

where T denotes the absolute temperature, $\mathbf{q}$ is the heat flux vector and A and $\widetilde{A}$ denote the initial and the fictitious undamaged cross–sectional area, respectively. From the hypothesis of complementary energy equivalence the elastic energy affected by damage and the damage conjugate force $Y^{\mathrm{e}}\left(D^{\mathrm{B}}\right)$ are obtained:

$$\psi^{\mathrm{e}}=\frac{1}{2}\widetilde{E}_{ijkl}\left(D^{\mathrm{B}}\right)\varepsilon_{ij}^{\mathrm{e}}\varepsilon_{kl}^{\mathrm{e}}=\frac{1}{2}E_{ijkl}\varepsilon_{ij}^{\mathrm{e}}\varepsilon_{kl}^{\mathrm{e}}e^{-2D^{\mathrm{B}}},\quad \widetilde{\mathbf{E}}=\mathbf{E}\left(1-D\right)^{2}=\mathbf{E}\exp\left(-2D^{\mathrm{B}}\right), \tag{2.42}$$

$$Y^{\mathrm{e}}\left(D^{\mathrm{B}}\right)\stackrel{\mathrm{def}}{=}-\varrho\frac{\partial\psi^{\mathrm{e}}}{\partial D^{\mathrm{B}}}=-\frac{\sigma_{\mathrm{eq}}^{2}\exp\left(2D^{\mathrm{B}}\right)}{E}R_{\nu}\left(\frac{\sigma_{\mathrm{H}}}{\sigma_{\mathrm{eq}}}\right). \tag{2.43}$$

The dissipative potential for damage evolution is assumed as a quadratic function of Y^{e}

$$\psi^{\mathrm{d}}\left(Y^{\mathrm{e}},\dot{p},p,D^{\mathrm{B}},T\right)=CS\left(-\frac{Y^{\mathrm{e}}}{S}\right)^{2}\frac{\left(p_{\mathrm{cr}}-p\right)^{n-1}}{p^{2n}}e^{-2D^{\mathrm{B}}}\dot{p}, \tag{2.44}$$

$$\dot{D}^{\mathrm{B}}=\frac{\partial\psi^{\mathrm{d}}}{\partial Y^{\mathrm{e}}}=-\frac{2C}{ES}\sigma_{\mathrm{eq}}^{2}R_{\nu}\left(\frac{\sigma_{\mathrm{H}}}{\sigma_{\mathrm{eq}}}\right)\frac{\left(p_{\mathrm{cr}}-p\right)^{n-1}}{p^{2n}}\dot{p}. \tag{2.45}$$

For the Ramberg–Osgood hardening law $\sigma_{\mathrm{eq}}=Kp^{n}$ a particular form of the logarithmic damage evolution holds

$$\dot{D}^{\mathrm{B}}=-2\frac{CK^{2}}{ES}R_{\nu}\left(\frac{\sigma_{\mathrm{H}}}{\sigma_{\mathrm{eq}}}\right)\left(p_{\mathrm{cr}}-p\right)^{n-1}\dot{p}, \tag{2.46}$$

the integration of which under the assumption of constant R_{ν} (proportional loadings) results in the evolution of logarithmic damage variable D^{B} with cumulative plastic strain p (also Skrzypek and Ganczarski [51]).

C. Irreversible thermodynamics model of the coupled isotropic damage–thermo-elastic–(visco) plastic material

i. Helmholtz free energy representation and state equations

A consistent unified model, based on the assumptions that variable Y associated with the isotropic damage internal variable D contains both the classical elastic (reversible) energy Y^{e} and the inelastic (irreversible) energy Y^{in}, was developed by Saanouni, Forster and Ben Hatira [23]. Mechanical flux vector $\dot{\mathbf{J}}$ and its thermodynamic conjugate force vector $\mathbf{F}$ are defined as:

$$\dot{\mathbf{J}}=\left\{\dot{\boldsymbol{\varepsilon}}^{\mathrm{p}},\dot{\boldsymbol{\alpha}},\dot{r},\dot{D},\mathbf{q}\right\}^{T},\qquad \mathbf{F}=\left\{\boldsymbol{\sigma},\mathbf{X},R,Y,-\frac{1}{T}\mathbf{grad}T\right\}. \tag{2.47}$$

Total energy equivalence is applied to the elastic Φ^{e} and the inelastic Φ^{kh} and Φ^{ih} energy portions responsible for kinematic and isotropic hardening in the damaged and the fictitious pseudo–undamaged configurations:

$$\begin{gathered}\Phi^{\mathrm{e}}\left(\boldsymbol{\varepsilon}^{\mathrm{e}},D\right)=\frac{1}{2}\boldsymbol{\sigma}:\boldsymbol{\varepsilon}^{\mathrm{e}}=\frac{1}{2}\widetilde{\boldsymbol{\sigma}}:\widetilde{\boldsymbol{\varepsilon}}^{\mathrm{e}},\\ \Phi^{\mathrm{kh}}\left(\boldsymbol{\alpha},D\right)=\frac{1}{2}\mathbf{X}:\boldsymbol{\alpha}=\frac{1}{2}\widetilde{\mathbf{X}}:\widetilde{\boldsymbol{\alpha}},\qquad \Phi^{\mathrm{ih}}\left(r,D\right)=\frac{1}{2}rR=\frac{1}{2}\widetilde{r}\widetilde{R}.\end{gathered} \tag{2.48}$$

Couples of effective state variables are:

$$
\begin{aligned}
&\widetilde{\boldsymbol{\sigma}} = \frac{\boldsymbol{\sigma}}{g_e(D)}, \qquad \widetilde{\boldsymbol{\varepsilon}}^{\mathrm{e}} = g_e(D)\,\boldsymbol{\varepsilon}^{\mathrm{e}},\\
&\widetilde{\mathbf{X}} = \frac{\mathbf{X}}{h_\alpha(D)}, \quad \widetilde{\boldsymbol{\alpha}}^{\mathrm{e}} = h_\alpha(D)\,\boldsymbol{\alpha}, \quad \widetilde{R} = \frac{R}{h_r(D)}, \quad \widetilde{r} = h_r(D)\,r,
\end{aligned} \tag{2.49}
$$

where functions $g_e(D)$, $h_\alpha(D)$ and $h_r(D)$ are, generally, independently defined, however in what follows it is assumed for simplicity $g_e(D) = h_\alpha(D) = h_r(D) = (1-D)^{1/2}$.

Helmholtz free energy is taken as a state potential

$$
\psi^{\mathrm{H}}(\boldsymbol{\varepsilon}^{\mathrm{e}}, \boldsymbol{\alpha}, r, D, T) = \psi^{\mathrm{e}}(\widetilde{\boldsymbol{\varepsilon}}^{\mathrm{e}}, T) + \psi^{\mathrm{in}}(\widetilde{\boldsymbol{\alpha}}, \widetilde{r}), \tag{2.50}
$$

$$
\begin{aligned}
\rho\psi^{\mathrm{e}}(\widetilde{\boldsymbol{\varepsilon}}^{\mathrm{e}}, T) &= \frac{1}{2}\widetilde{\boldsymbol{\varepsilon}}^{\mathrm{e}} : \boldsymbol{\Lambda} : \widetilde{\boldsymbol{\varepsilon}}^{\mathrm{e}} - (T - T_0)\,\mathbf{k} : \widetilde{\boldsymbol{\varepsilon}}^{\mathrm{e}} - \rho c_v T\left[\log\left(\frac{T}{T_0}\right) - 1\right],\\
\rho\psi^{\mathrm{in}}(\widetilde{\boldsymbol{\alpha}}, \widetilde{r}, T) &= \frac{1}{3}C\widetilde{\boldsymbol{\alpha}} : \widetilde{\boldsymbol{\alpha}} + \frac{1}{2}Q\widetilde{r}^2.
\end{aligned} \tag{2.51}
$$

In this model damage affects both the elastic (reversible) and the inelastic (irreversible) energy portions and the effective state variables $\widetilde{\boldsymbol{\varepsilon}}^{\mathrm{e}}$, $\widetilde{\boldsymbol{\alpha}}$, $\widetilde{r}$ are consistently used for the state potential (free energy) of damaged material and $\mathbf{k}$ is second–rank tensor of thermal conductivity. Hence, the state equations are furnished from the state potential as follows:

$$
\begin{aligned}
&\boldsymbol{\sigma} = \rho\frac{\partial\psi^{\mathrm{H}}}{\partial\boldsymbol{\varepsilon}^{\mathrm{e}}} = \widetilde{\boldsymbol{\Lambda}} : \boldsymbol{\varepsilon}^{\mathrm{e}} - (T - T_0)\,\widetilde{\mathbf{k}}, \qquad \mathbf{X} = \rho\frac{\partial\psi^{\mathrm{H}}}{\partial\boldsymbol{\alpha}} = \frac{2}{3}\widetilde{C}\boldsymbol{\alpha}, \qquad R = \rho\frac{\partial\psi^{\mathrm{H}}}{\partial r} = \widetilde{Q}r,\\
&s = -\rho\frac{\partial\psi^{\mathrm{H}}}{\partial T} = \frac{1}{\rho}\widetilde{\mathbf{k}} : \boldsymbol{\varepsilon}^{\mathrm{e}} + c_v\log\left(\frac{T}{T_0}\right), \quad Y = -\rho\frac{\partial\psi^{\mathrm{H}}}{\partial D} = Y^{\mathrm{e}} + Y^{\mathrm{in}},
\end{aligned} \tag{2.52}
$$

where elastic and inelastic energy release rates are

$$
\begin{aligned}
Y^{\mathrm{e}} &= -\rho\frac{\partial\psi^{\mathrm{e}}}{\partial D} = \frac{1}{2}\boldsymbol{\varepsilon}^{\mathrm{e}} : \boldsymbol{\Lambda} : \boldsymbol{\varepsilon}^{\mathrm{e}} - \frac{1}{2}(T - T_0)\frac{\mathbf{k}}{(1-D)^{1/2}} : \boldsymbol{\varepsilon}^{\mathrm{e}},\\
Y^{\mathrm{in}} &= -\rho\frac{\partial\psi^{\mathrm{in}}}{\partial D} = \frac{1}{3}C\boldsymbol{\alpha} : \boldsymbol{\alpha} + \frac{1}{2}Qr^2,
\end{aligned} \tag{2.53}
$$

and the effective thermo–mechanical modulae are used

$$
\widetilde{\boldsymbol{\Lambda}} = (1-D)\,\boldsymbol{\Lambda}, \quad \widetilde{C} = (1-D)\,C, \quad \widetilde{Q} = (1-D)\,Q, \quad \widetilde{\mathbf{k}} = (1-D)^{1/2}\,\mathbf{k}. \tag{2.54}
$$

ii. Time–independent nonlinear plasticity model of isotropic/kinematic hardening coupled with isotropic damage

A consistent coupled Mises–type yield function is obtained when the classical state variables ($\boldsymbol{\sigma}$, $\mathbf{X}$, R) in the uncoupled yield function are replaced by the effective state variables ($\widetilde{\boldsymbol{\sigma}}$, $\widetilde{\mathbf{X}}$, $\widetilde{R}$) in the coupled one:

$$
\begin{aligned}
&f\left(\widetilde{\boldsymbol{\sigma}}, \widetilde{\mathbf{X}}, \widetilde{R}\right) = J_2\left(\widetilde{\boldsymbol{\sigma}} - \widetilde{\mathbf{X}}\right) - \widetilde{R} - \sigma_y = 0,\\
&J_2\left(\widetilde{\boldsymbol{\sigma}} - \widetilde{\mathbf{X}}\right) = \left[\frac{3}{2}\left(\widetilde{\boldsymbol{\sigma}} - \widetilde{\mathbf{X}}\right) : \left(\widetilde{\boldsymbol{\sigma}} - \widetilde{\mathbf{X}}\right)\right]^{1/2}.
\end{aligned} \tag{2.55}
$$

The fully coupled plastic potential, that generalizes the Lemaitre and Chaboche's (2.28), may be written as

$$F^{\mathrm{SFB}}\left(\widetilde{\boldsymbol{\sigma}}, \widetilde{\mathbf{X}}, \widetilde{R}, D, T\right) = f + \frac{1}{2}\frac{a}{C}J_2^2\left(\widetilde{\boldsymbol{\sigma}} - \widetilde{\mathbf{X}}\right) + \frac{1}{2}\frac{b}{Q}\widetilde{R}^2 + F^{\mathrm{d}}(Y), \tag{2.56}$$

where, following Germain, Nguyen and Suquet [20], damage evolution potential $F^{\mathrm{d}}(Y)$ is supposed to be a power function of the total (elastic and inelastic) energy release due to damage evolution $Y = Y^{\mathrm{e}} + Y^{\mathrm{in}}$ (extension of (2.32))

$$F^{\mathrm{d}}(Y) = -\frac{S}{(s+1)}\left(\frac{Y}{S}\right)^{s+1}\frac{1}{(1-D)^{\beta}}. \tag{2.57}$$

State equations for the nonlinear hardening theory are obtained by the generalized normality rule

$$\begin{aligned}
&\dot{\varepsilon}^{\mathrm{p}}_{ij} = \frac{3}{2}\dot{\widetilde{\lambda}}\frac{\sigma'_{ij} - X'_{ij}}{J_2(\sigma_{ij} - X_{ij})} = \frac{3}{2}\dot{p}\frac{\sigma'_{ij} - X'_{ij}}{J_2(\sigma_{ij} - X_{ij})},\\
&\dot{\alpha}_{ij} = \dot{\varepsilon}^{\mathrm{p}}_{ij} - a\dot{\lambda}\alpha_{ij} = \dot{\varepsilon}^{\mathrm{p}}_{ij} - \dot{p}\frac{3aX'_{ij}}{2C(1-D)^{1/2}}, \qquad \dot{r} = \dot{\widetilde{\lambda}}(1 - b\widetilde{r}) = \dot{p}\left(1 - \frac{b}{Q}\widetilde{R}\right),\\
&\dot{D} = -\dot{\lambda}\left(\frac{Y}{S}\right)^{s}\frac{1}{(1-D)^{\beta}} = -\dot{p}\left(\frac{Y}{S}\right)^{s}(1-D)^{(1-2\beta)/2},
\end{aligned} \tag{2.58}$$

where $\dot{\widetilde{\lambda}} = \dot{\lambda}(1-D)^{-1/2} = \dot{p} = \left[(2/3)\,\dot{\varepsilon}^{\mathrm{p}}_{ij}\dot{\varepsilon}^{\mathrm{p}}_{ij}\right]^{1/2}$. In case $a = b = 0$ the fully coupled linear hardening theory holds:

$$\begin{aligned}
&\dot{\varepsilon}^{\mathrm{p}}_{ij} = \frac{3}{2}\dot{p}\frac{\sigma'_{ij} - X'_{ij}}{J_2(\sigma_{ij} - X_{ij})}, \qquad \dot{\alpha}_{ij} = \dot{\varepsilon}^{\mathrm{p}}_{ij},\\
&\dot{r} = \dot{p}, \qquad \dot{D} = -\dot{p}(1-D)^{1/2}\left(\frac{Y}{S}\right)^{s}\frac{1}{(1-D)^{\beta}}.
\end{aligned} \tag{2.59}$$

iii. Time–dependent viscoplastic flow coupled with isotropic damage

In case of the time–dependent coupled damage–creep–isotropic/kinematic hardening material the single surface coupled visco–damage dissipation potential may be expressed as a sum of the viscoplastic and the creep–damage parts (cf. [23])

$$\Phi^{*}\left(\widetilde{\boldsymbol{\sigma}}, \widetilde{\mathbf{X}}, \widetilde{R}, D, T\right) = \Phi^{*^{\mathrm{vp}}}\left(\widetilde{\boldsymbol{\sigma}}, \widetilde{\mathbf{X}}, \widetilde{R}, T\right) + \Phi^{*^{\mathrm{d}}}(\boldsymbol{\sigma}, D, T), \tag{2.60}$$

where the viscoplastic term is represented by a following power function of f extended by the additional terms representing nonlinear hardening

$$\Phi^{*^{\mathrm{vp}}} = \frac{K}{n+1}\left\langle\frac{1}{K}\left[f + \frac{3}{4}\frac{a}{C}\widetilde{\mathbf{X}} : \widetilde{\mathbf{X}} - \frac{1}{3}aC\widetilde{\boldsymbol{\alpha}} : \widetilde{\boldsymbol{\alpha}} + \frac{1}{2}\frac{b}{Q}\widetilde{R}^2 - \frac{1}{2}bQ\widetilde{r}^2\right]\right\rangle^{n+1}, \tag{2.61}$$

whereas the creep–damage term is given by

$$\Phi^{*^{\mathrm{d}}} = -Y\left\langle\frac{\chi(\boldsymbol{\sigma})}{A}\right\rangle^{r}(1-D)^{-k}. \tag{2.62}$$

Symbol $f\left(\widetilde{\boldsymbol{\sigma}}, \widetilde{\mathbf{X}}, \widetilde{R}\right)$ denotes the fully coupled Mises–type isotropic/kinematic hardening yield function

$$f\left(\widetilde{\boldsymbol{\sigma}}, \widetilde{\mathbf{X}}, \widetilde{R}\right) = J_2\left(\widetilde{\boldsymbol{\sigma}} - \widetilde{\mathbf{X}}\right) - \widetilde{R} - \sigma_y > 0. \tag{2.63}$$

From the generalized normality rule the state equations are eventually obtained as:

$$\begin{aligned}
\dot{\varepsilon}_{ij}^{\mathrm{vp}} &= \frac{\partial \Phi^*}{\partial \sigma_{ij}} = \frac{3}{2}\frac{\langle f/K\rangle^n}{(1-D)^{1/2}}\frac{\sigma'_{ij} - X'_{ij}}{J_2(\sigma_{ij} - X_{ij})}, \\
\dot{\alpha}_{ij} &= -\frac{\partial \Phi^*}{\partial X_{ij}} = \frac{3}{2}\frac{\langle f/K\rangle^n}{(1-D)^{1/2}}\left[\frac{\sigma'_{ij} - X'_{ij}}{J_2(\sigma_{ij} - X_{ij})} - \frac{a}{C}\frac{X_{ij}}{(1-D)^{1/2}}\right], \\
\dot{r} &= -\frac{\partial \Phi^*}{\partial R} = \frac{\langle f/K\rangle^n}{(1-D)^{1/2}}\left[1 - \frac{b}{Q}\widetilde{R}\right], \qquad \dot{D} = -\frac{\partial \Phi^*}{\partial Y} = \left[\frac{\chi(\sigma_{ij})}{A}\right]^r (1-D)^{-k}.
\end{aligned} \tag{2.64}$$

3. ANISOTROPIC DAMAGE REPRESENTATION AND ACCUMULATION

3.1. Damage effect tensor

A. Fourth–rank damage effect tensor

When the simple principle of strain equivalence between the physical (damaged) and the fictitious (pseudo–undamaged) spaces is assumed, a tensorially linear transformation of the Cauchy stress $\boldsymbol{\sigma}$ to the effective Cauchy stress $\widetilde{\boldsymbol{\sigma}}$, through the fourth–rank damage effect tensor $\mathbf{M}$, is assumed (see Fig. 1.3)

$$\widetilde{\boldsymbol{\sigma}} = \mathbf{M}(\mathbf{D}) : \boldsymbol{\sigma} \qquad \text{or} \qquad \widetilde{\sigma}_{ij} = M_{ijkl}\sigma_{kj}\,. \tag{3.1}$$

A symmetric effective stress tensor $\widetilde{\sigma}_{ij}$ is used in (3.1) although the effective stress tensor $\widetilde{\boldsymbol{\sigma}} = \boldsymbol{\sigma} : (\mathbf{1} - \mathbf{D})$ needs not to be symmetric in a more general case under this transformation. Some proposal of various effective stress concepts are reviewed by Zheng and Betten [60]:

$$\begin{aligned}
\widetilde{\boldsymbol{\sigma}} &= \frac{1}{2}\left[\boldsymbol{\sigma} : (\mathbf{1} - \mathbf{D})^{-1} + (\mathbf{1} - \mathbf{D})^{-1} : \boldsymbol{\sigma}\right] && \text{Murakami [43]} \\
\widetilde{\boldsymbol{\sigma}} &= (\mathbf{1} - \mathbf{D})^{-1/2} : \boldsymbol{\sigma} : (\mathbf{1} - \mathbf{D})^{-1/2} && \text{Chow and Wang [76, 77]} \\
\widetilde{\boldsymbol{\sigma}} &= (\mathbf{1} - \mathbf{D})^{-1} : \boldsymbol{\sigma} : (\mathbf{1} - \mathbf{D})^{-1} && \text{Zheng and Betten [60].}
\end{aligned} \tag{3.2}$$

$\mathbf{1}$ and $\mathbf{D}$ denote here second–rank unit and damage tensors, respectively.

In a general case, when the fully anisotropic nature of damage is considered and the principle of total energy equivalence is used (Chow and Lu [44]) a concept of the fourth–rank damage effect tensor $\mathbf{M}(\mathcal{D})$ is introduced, that transforms the state variables in the physical space $\boldsymbol{\sigma}$, $\boldsymbol{\varepsilon}^{\mathrm{e}}$, $\boldsymbol{\varepsilon}^{\mathrm{p}}$ to the effective state variables in the fictitious space $\widetilde{\boldsymbol{\sigma}}$, $\widetilde{\boldsymbol{\varepsilon}}^{\mathrm{e}}$, $\widetilde{\boldsymbol{\varepsilon}}^{\mathrm{p}}$ (cf. Sec. 1, Fig. 1.3)

$$\widetilde{\boldsymbol{\sigma}} = \mathbf{M}(\mathcal{D}) : \boldsymbol{\sigma}, \qquad \widetilde{\boldsymbol{\varepsilon}}^{\mathrm{e}} = \mathbf{M}^{-1}(\mathcal{D}) : \boldsymbol{\varepsilon}^{\mathrm{e}}, \qquad \mathrm{d}\widetilde{\boldsymbol{\varepsilon}}^{\mathrm{p}} = \mathbf{M}^{-1}(\mathcal{D}) : \mathrm{d}\boldsymbol{\varepsilon}^{\mathrm{p}}. \tag{3.3}$$

$\mathcal{D}$ denotes a properly selected damage variable, scalar, second–rank tensor, fourth–rank tensors, D, $\mathbf{D}$, $\widehat{\mathbf{D}}$, etc., as argument of $\mathbf{M}(\mathcal{D})$.

The fourth–rank damage effect tensor $\mathbf{M}$ can also be explained in such a way that the following modification of the constitutive tensors of damaged material, the stiffness $\widetilde{\mathbf{\Lambda}}(\mathcal{D})$ or the compliance $\widetilde{\mathbf{\Lambda}}^{-1}(\mathcal{D})$, in terms of the constitutive tensors of virgin material $\mathbf{\Lambda}$ or $\mathbf{\Lambda}^{-1}$, holds:

$$\widetilde{\boldsymbol{\sigma}} = \mathbf{\Lambda} : \widetilde{\boldsymbol{\varepsilon}}^{\mathrm{e}}, \qquad \boldsymbol{\sigma} = \widetilde{\mathbf{\Lambda}}(\mathcal{D}) : \boldsymbol{\varepsilon}^{\mathrm{e}}, \qquad \widetilde{\mathbf{\Lambda}}(\mathcal{D}) = \mathbf{M}^{-1}(\mathcal{D}) : \mathbf{\Lambda} : \mathbf{M}^{-\mathrm{T}}(\mathcal{D}), \tag{3.4}$$

$$\widetilde{\boldsymbol{\varepsilon}}^{\mathrm{e}} = \mathbf{\Lambda}^{-1} : \widetilde{\boldsymbol{\sigma}}, \qquad \boldsymbol{\varepsilon}^{\mathrm{e}} = \widetilde{\mathbf{\Lambda}}^{-1}(\mathcal{D}) : \boldsymbol{\sigma}, \qquad \widetilde{\mathbf{\Lambda}}^{-1}(\mathcal{D}) = \mathbf{M}^{\mathrm{T}}(\mathcal{D}) : \mathbf{\Lambda}^{-1} : \mathbf{M}(\mathcal{D}). \tag{3.5}$$

Note that, when Chaboche's notation is used (cf. [46]), where $\underset{\approx}{\mathbf{M}} \equiv \mathbf{M}^{-1}$, $\underset{\approx}{\mathbf{\Lambda}} \equiv \mathbf{\Lambda}$, $\underset{\approx}{\mathbf{S}} \equiv \mathbf{\Lambda}^{-1}$, the equivalent formulae are furnished:

$$\underset{\approx}{\widetilde{\mathbf{\Lambda}}} = \underset{\approx}{\mathbf{M}} : \underset{\approx}{\mathbf{\Lambda}} : \underset{\approx}{\mathbf{M}}^{\mathrm{T}}, \qquad \underset{\approx}{\widetilde{\mathbf{S}}} = \underset{\approx}{\mathbf{M}}^{-\mathrm{T}} : \underset{\approx}{\mathbf{S}} : \underset{\approx}{\mathbf{M}}^{-1}. \tag{3.6}$$

B. Matrix representation of the damage effect tensors in terms of the second–rank damage tensors $[\mathbf{M}(\mathbf{D})]$

A tensorially linear transformation is assumed in Eq. (3.1) between the Cauchy stress tensor $\boldsymbol{\sigma}$ and the effective Cauchy stress tensor $\widetilde{\boldsymbol{\sigma}}$. Due to the symmetry assumed of both stress and effective stress tensors the fourth–rank tensor M_{ijkl} can be represented by a 6×6 matrix when the vector representation of $\{\boldsymbol{\sigma}\}$ and $\{\widetilde{\boldsymbol{\sigma}}\}$ is applied:

$$\{\widetilde{\sigma}_{11}\,\widetilde{\sigma}_{22}\,\widetilde{\sigma}_{33}\,\widetilde{\sigma}_{23}\,\widetilde{\sigma}_{31}\,\widetilde{\sigma}_{12}\}^{\mathrm{T}} = [M_{ijkl}]\,\{\sigma_{11}\,\sigma_{22}\,\sigma_{33}\,\sigma_{23}\,\sigma_{31}\,\sigma_{12}\}^{\mathrm{T}} \tag{3.7}$$

Three forms of $\mathbf{M}(\mathbf{D})$, expressed in terms of second–rank symmetric damage tensor $\mathbf{D}$, are discussed by Chen and Chow [56]:

$$\begin{aligned}
&\mathbf{M}^{2}_{\mathrm{C1}}(\mathbf{D}) = \widehat{\mathbf{P}}^{-1}(\mathbf{D}), && P_{ijkl} = \frac{1}{2}\left[(I_{ik} - D_{ik})(I_{jl} - D_{jl}) + (I_{il} - D_{il})(I_{jk} - D_{jk})\right], \\
&\mathbf{M}_{\mathrm{C2}}(\mathbf{D}) = \left[\mathbf{I} - \widehat{\mathbf{D}}(\mathbf{D})\right]^{-1}, && \widehat{D}_{ijkl} = \frac{1}{4}(I_{ik}D_{jl} + I_{il}D_{jk} + I_{jk}D_{il} + I_{jl}D_{ik}), \\
&\mathbf{M}_{\mathrm{C3}}(\mathbf{D}) = \widehat{\mathbf{\Phi}}(\mathbf{\Phi}), && \widehat{\Phi}_{ijkl} = \frac{1}{4}(I_{ik}\Phi_{jl} + I_{il}\Phi_{jk} + I_{jk}\Phi_{il} + I_{jl}\Phi_{ik}).
\end{aligned} \tag{3.8}$$

Symbols $\widehat{\mathbf{P}}(\mathbf{D})$, $\widehat{\mathbf{D}}(\mathbf{D})$ or $\widehat{\mathbf{\Phi}}(\mathbf{\Phi})$ denote fourth–rank damage tensors as expressed in terms of second–rank damage tensors $\mathbf{D}$ or $\mathbf{\Phi} = (\mathbf{1} - \mathbf{D})^{-1}$.

C. Matrix representation of damage effect tensors in the principal coordinates of the second–rank damage tensor $[\mathbf{M}(D_1, D_2, D_3)]$

Employing principal damage components D_1, D_2, D_3, $D_{23} = D_{31} = D_{12} = 0$, we obtain the diagonal forms for $\mathbf{M}_{\mathrm{C1}}$, $\mathbf{M}_{\mathrm{C2}}$, $\mathbf{M}_{\mathrm{C3}}$ (Voyiadjis and Kattan [78], and Voyiadjis and Park [57]), and $\mathbf{M}_{\mathrm{Ch}}$ (Chaboche, Lesne and Moire [79]) as shown in Table 3.1.

Table 3.1: Diagonal forms of matrix damage effect tensor representation in the principal coordinates of the second–rank damage tensor

$$\{\tilde{\sigma}_{11}\,\tilde{\sigma}_{22}\,\tilde{\sigma}_{33}\,\tilde{\sigma}_{23}\,\tilde{\sigma}_{31}\,\tilde{\sigma}_{12}\}^{\mathsf{T}} = [\mathbf{M}_{\mathrm{C1}}]\,\{\sigma_{11}\,\sigma_{22}\,\sigma_{33}\,\sigma_{23}\,\sigma_{31}\,\sigma_{12}\}^{\mathsf{T}},$$

$$[\mathbf{M}_{\mathrm{C1}}(D_1, D_2, D_3)] = \mathrm{diag}\left[\frac{1}{1-D_1}, \frac{1}{1-D_2}, \frac{1}{1-D_3}, \frac{1}{\sqrt{(1-D_2)(1-D_3)}}, \frac{1}{\sqrt{(1-D_1)(1-D_3)}}, \frac{1}{\sqrt{(1-D_1)(1-D_2)}}\right],$$

$$\{\tilde{\sigma}_{11}\,\tilde{\sigma}_{22}\,\tilde{\sigma}_{33}\,\tilde{\sigma}_{23}\,\tilde{\sigma}_{31}\,\tilde{\sigma}_{12}\}^{\mathsf{T}} = [\mathbf{M}_{\mathrm{C2}}]\,\{\sigma_{11}\,\sigma_{22}\,\sigma_{33}\,\sigma_{23}\,\sigma_{31}\,\sigma_{12}\}^{\mathsf{T}},$$

$$[\mathbf{M}_{\mathrm{C2}}(D_1, D_2, D_3)] = \mathrm{diag}\left[\frac{1}{1-D_1}, \frac{1}{1-D_2}, \frac{1}{1-D_3}, \frac{1}{1-\dfrac{D_2+D_3}{2}}, \frac{1}{1-\dfrac{D_1+D_3}{2}}, \frac{1}{1-\dfrac{D_1+D_2}{2}}\right],$$

$$\{\tilde{\sigma}_{11}\,\tilde{\sigma}_{22}\,\tilde{\sigma}_{33}\,\tilde{\sigma}_{23}\,\tilde{\sigma}_{31}\,\tilde{\sigma}_{12}\}^{\mathsf{T}} = [\mathbf{M}_{\mathrm{C3}}]\,\{\sigma_{11}\,\sigma_{22}\,\sigma_{33}\,\sigma_{23}\,\sigma_{31}\,\sigma_{12}\}^{\mathsf{T}},$$

$$[\mathbf{M}_{\mathrm{C3}}(D_1, D_2, D_3)], = \mathrm{diag}\left[\frac{1}{1-D_1}, \frac{1}{1-D_2}, \frac{1}{1-D_3}, \frac{1}{2}\left(\frac{1}{1-D_2}+\frac{1}{1-D_3}\right), \frac{1}{2}\left(\frac{1}{1-D_1}+\frac{1}{1-D_3}\right), \frac{1}{2}\left(\frac{1}{1-D_1}+\frac{1}{1-D_2}\right)\right],$$

$$\{\tilde{\varepsilon}_{11}\,\tilde{\varepsilon}_{22}\,\tilde{\varepsilon}_{33}\,\tilde{\varepsilon}_{23}\,\tilde{\varepsilon}_{31}\,\tilde{\varepsilon}_{12}\}^{\mathsf{T}} = [\mathbf{M}_{\mathrm{Ch}}]\,\{\varepsilon_{11}\,\varepsilon_{22}\,\varepsilon_{33}\,\varepsilon_{23}\,\varepsilon_{31}\,\varepsilon_{12}\}^{\mathsf{T}},$$

$$[\mathbf{M}_{\mathrm{Ch}}(d_1, d_2, d_3)] = \mathrm{diag}\left[(1-d_1), (1-d_2), (1-d_3), \sqrt{(1-d_2)(1-d_3)}, \sqrt{(1-d_1)(1-d_3)}, \sqrt{(1-d_1)(1-d_2)},\right].$$

3.2. Creep–damage models under non–proportional loadings

A. Orthotropic damage growth in case of constant principal directions of the stress tensor

Consider the simpler case when principal directions of the second–rank stress and damage tensors $\boldsymbol{\sigma}$, $\mathbf{D}$ coincide and do not change with time, such that the orthotropic theory of brittle damage coupled with the similarity of deviators of principal creep strain rates $\dot{\mathrm{e}}^{\mathrm{c}}$ and either the principal stress s (partly coupled) or the principal effective stress $\widetilde{\mathrm{s}}$ (fully coupled) is applicable (cf. Kachanov [25] and [80]).

The orthotropic creep–damage growth rule (direct extension of the Kachanov's concept (2.4) to principal continuity ψ_k or damage D_k components) holds

$$\dot{\psi}_k = \frac{\partial \psi_k}{\partial t} = -C_k \left\langle \frac{\sigma_k}{\psi_k} \right\rangle^{r_k} \quad \text{or} \quad \dot{D}_k = \frac{\partial D_k}{\partial t} = A_k \left\langle \frac{\sigma_k}{1 - D_k} \right\rangle^{r_k}. \tag{3.9}$$

B. Orthotropic damage cumulation in case of variable principal directions of the stress tensor

Consider after Skrzypek and Ganczarski [1] a more general case when principal directions $\alpha_k\,(1_\sigma, 2_\sigma, 3_\sigma)$ of the stress tensor $\boldsymbol{\sigma}$ rotate a small angle $\mathrm{d}\alpha_k$ from time t through $t+\mathrm{d}t$ to $\alpha'_k\,(1'_\sigma, 2_\sigma, 3'_\sigma)$. After damage has occurred the virgin isotropic material becomes orthotropic and the principal directions $\beta_k\,(1_{\mathrm{D}}, 2_{\mathrm{D}}, 3_{\mathrm{D}})$ follow the principal stress axes rotation, however the stress and damage tensors $\boldsymbol{\sigma}$ and $\mathbf{D}$ are no longer co–axial in their principal axes, $\alpha_k \neq \beta_k$, Fig. 3.1.

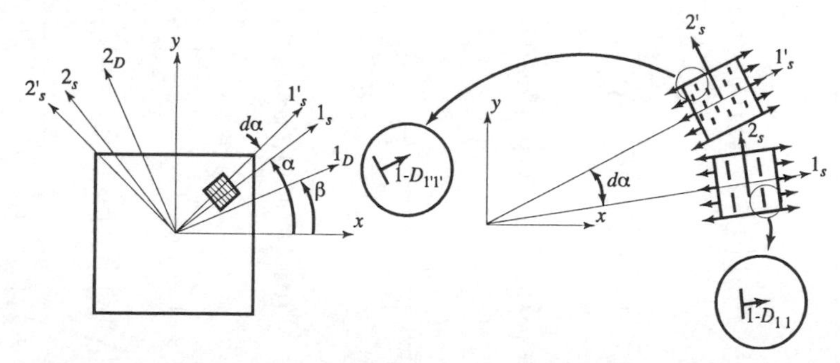

Figure 3.1. Schematic cumulation of several orthotropic damage increments of variable principal directions (after [1])

The Murakami–Ohno second–rank damage tensor (1.1) as represented through the principal values is used, whereas the non–objective damage rate tensor $\dot{\mathbf{D}}$ (rotation of principal axes ignored) is defined as

$$\dot{\mathbf{D}} = \sum_{k=1}^{3} \dot{D}_k \mathbf{n}^k \otimes \mathbf{n}^k. \tag{3.10}$$

The objective Zaremba–Jaumann derivative of the damage tensor $\mathbf{D}$ with respect to tensorial components D_k and the base vectors $\mathbf{n}^k$ (rotation of principal axes included) yields

$$\frac{\partial \mathbf{D}}{\partial t} = \sum_{k=1}^{3} \left(\dot{D}_k \mathbf{n}^k \otimes \mathbf{n}^k + D_k \dot{\mathbf{n}}^k \otimes \mathbf{n}^k + D_k \mathbf{n}^k \otimes \dot{\mathbf{n}}^k \right) \text{ or } \overset{\triangledown}{\mathbf{D}} = \dot{\mathbf{D}} - \mathbf{D}^T \mathbf{S} - \mathbf{S}^T \mathbf{D}, \quad (3.11)$$

where $\mathbf{S}$ is the skew–symmetric spin tensor due to rotation of principal stress directions $\mathrm{d}\alpha_k$, and $\overset{\triangledown}{\mathbf{D}}$ is the objective damage rate tensor. When the non–objective damage rate $\dot{\mathbf{D}}$ in current principal directions of the stress tensor α_k (effect of rotation of the base vector ignored) is assumed to be governed by the orthotropic damage growth rule (3.9), and the skew–symmetric spin tensor representation in terms of $\mathrm{d}\alpha_k$ is used, we obtain:

$$\overset{\triangledown}{D}_{IJ} = \dot{D}_{IJ} - D_{IJ}^{\mathrm{T}} \begin{bmatrix} 0 & \mathrm{d}\alpha_1 & -\mathrm{d}\alpha_2 \\ -\mathrm{d}\alpha_1 & 0 & \mathrm{d}\alpha_3 \\ \mathrm{d}\alpha_2 & -\mathrm{d}\alpha_3 & 0 \end{bmatrix} - \begin{bmatrix} 0 & -\mathrm{d}\alpha_1 & \mathrm{d}\alpha_2 \\ \mathrm{d}\alpha_1 & 0 & -\mathrm{d}\alpha_3 \\ -\mathrm{d}\alpha_2 & \mathrm{d}\alpha_3 & 0 \end{bmatrix} D_{IJ}, \quad (3.12)$$

where in the current base α_k it holds

$$\dot{D}_{IJ} = C_{IJ} \left\langle \frac{\sigma_{IJ}}{1 - D_{IJ}} \right\rangle^{r_{IJ}}, \quad (3.13)$$

$$D_{I'J'}(t + \Delta t) = D_{IJ}(t) + \overset{\triangledown}{D}_{IJ}(t)\Delta t, \qquad D_{I'J'}(t + \Delta t) \underset{\text{transforms}}{\longrightarrow} D_{ij}(t + \Delta t). \quad (3.14)$$

C. Creep–damage coupling formulations

Partly $(\varepsilon_{ij}^{\mathrm{c}}, s_{ij})$ or fully coupled $(\varepsilon_{ij}^{\mathrm{c}}, \widetilde{s}_{ij})$ approaches of creep–damage analysis are summarized in what follows for two different loading conditions: the stress and damage tensors are co–axial in their principal axes, $\alpha_k \equiv \beta_k$ (Table 3.1) or the stress and damage ten-

Table 3.1. Partly or fully coupled creep-orthotropic damage approach in case of constant principal directions

Partly coupled approach	Fully coupled approach
isotropic flow rule $\dot{\varepsilon}_{ij}^{\mathrm{c}} = \frac{3}{2}\frac{\dot{\varepsilon}_{\mathrm{eq}}^{\mathrm{c}}}{\sigma_{\mathrm{eq}}} s_{ij}$	modified orthotropic flow rule $\dot{\varepsilon}_{ij}^{\mathrm{c}} = \frac{3}{2}\frac{\dot{\varepsilon}_{\mathrm{eq}}^{\mathrm{c}}}{\widetilde{\sigma}_{\mathrm{eq}}} \widetilde{s}_{ij}$
multiaxial time–hardening $\dot{\varepsilon}_{\mathrm{eq}}^{\mathrm{c}} = (\widetilde{\sigma}_{\mathrm{eq}})^m \dot{\mathrm{f}}(t)$	
orthotropic damage growth rate $\dot{D}_\nu = C_\nu \left\langle \frac{\sigma_\nu}{1 - D_\nu} \right\rangle^{r_\nu}$	

Table 3.2. Partly or fully coupled creep–damage approaches applied to current principal stress axes

Partly coupled approach	Fully coupled approach
isotropic flow rule $\dot{\varepsilon}^{c}_{IJ}=\frac{3}{2}\frac{\dot{\varepsilon}^{c}_{eq}}{\sigma_{eq}}s_{IJ}$	current orthotropic flow rule $\dot{\varepsilon}^{c}_{IJ}=\frac{3}{2}\frac{\dot{\varepsilon}^{c}_{eq}}{\widetilde{\sigma}_{eq}}\widetilde{s}_{IJ}$
multiaxial time–hardening $\dot{\varepsilon}^{c}_{eq}=(\widetilde{\sigma}_{eq})^{m}\dot{\mathrm{f}}(t)$	
current non–objective damage growth rate $\dot{D}_{IJ}=C_{IJ}\left\langle\frac{\sigma_{IJ}}{1-D_{IJ}}\right\rangle^{r_{IJ}}$	
current objective damage growth rate $\overset{\triangledown}{\mathbf{D}}=\dot{\mathbf{D}}-\mathbf{D}^{\mathrm{T}}\mathbf{S}-\mathbf{S}^{\mathrm{T}}\mathbf{D}.$	

sors are no longer co–axial in the principal axis, $\alpha_k \neq \beta_k$ (Table 3.2) — cf. Skrzypek and Ganczarski [1].

In the first case constitutive equations are to be written in the fixed space (i,j) of principal stress and damage directions, whereas in the second case, when principal stress and damage directions change (independently) due to the shear effect, they have to be written in the current (rotating) principal stress space (I,J), and, on each time–step, a transformation from current to the global reference space (i,j) has to be done.

3.3. Orthotropic elastic–brittle damage in crystalline metallic solids

According to Litewka [37, 38] the anisotropic stress–strain law of elasticity affected by damage is assumed in the form

$$\varepsilon_{ij}=\widetilde{\Lambda}^{-1}_{ijkl}\sigma_{kl} \qquad \text{or} \qquad \boldsymbol{\varepsilon}=\widetilde{\boldsymbol{\Lambda}}^{-1}(\mathbf{D}^{*}):\boldsymbol{\sigma}, \tag{3.15}$$

$$\widetilde{\Lambda}^{-1}_{ijkl}=-\frac{\nu}{E}\delta_{ij}\delta_{kl}+\frac{1+\nu}{2E}(\delta_{ik}\delta_{jl}+\delta_{il}\delta_{jk})+\frac{D^{*}_{1}}{4(1+D^{*}_{1})E}(\delta_{ik}D^{*}_{jl}+\delta_{jl}D^{*}_{ik}+\delta_{il}D^{*}_{jk}+\delta_{jk}D^{*}_{il}) \tag{3.16}$$

or

$$\boldsymbol{\varepsilon}=-\frac{\nu}{E}\mathrm{Tr}\boldsymbol{\sigma}\,\mathbf{1}+\frac{1+\nu}{E}\boldsymbol{\sigma}+\frac{D^{*}_{1}}{2(1+D^{*}_{1})E}(\boldsymbol{\sigma}:\mathbf{D}^{*}+\mathbf{D}^{*}:\boldsymbol{\sigma}), \tag{3.17}$$

where E and ν denote Young's modulus and Poisson's ratio of the undamaged material, whereas D^{*}_{1} is the dominant principal component of the modified second–rank damage tensor $\mathbf{D}^*$, such that $D_k^* = D_k/(1-D_k)$, $D_k \in \langle 0,1\rangle$, $D_k^* \in \langle 0,\infty\rangle$ (cf. Vakulenko and Kachanov [31]).

As the corresponding Mises–type initial failure criterion the three–parameter damage affected isotropic scalar function of $\boldsymbol{\sigma}$ and $\mathbf{D}^*$ tensors is assumed:

$$F^{\mathrm{d}}(\boldsymbol{\sigma},\mathbf{D}^*) = C_1 \mathrm{Tr}^2\boldsymbol{\sigma} + C_2 \mathrm{Tr}\left(\boldsymbol{\sigma}'^2\right) + C_3 \mathrm{Tr}\left(\boldsymbol{\sigma}^2 : \mathbf{D}^*\right) - \sigma_{\mathrm{u}}^2 = 0. \tag{3.18}$$

σ_{u} denotes the ultimate strength of the undamaged material, whereas constants C_1, C_2 and C_3 are to be obtained from the uniaxial tension (direction 1), uniaxial tension (direction 2), and biaxial tension (1+2) tests (cf. Litewka and Hult [81]).

The damage evolution rule is formulated applying tensor function representation that accounts for both the isotropic and the anisotropic damage, Litewka [39],

$$\dot{\mathbf{D}} = B\left(\Phi^{\mathrm{e}}\right)^m \mathbf{1} + C\left(\Phi^{\mathrm{e}}\right)^n \boldsymbol{\sigma}^*, \tag{3.19}$$

$$\Phi^{\mathrm{e}}(\boldsymbol{\sigma},\mathbf{D}^*) = \frac{1-2\nu}{6E}\mathrm{Tr}^2\boldsymbol{\sigma} + \frac{1+\nu}{2E}\mathrm{Tr}(\boldsymbol{\sigma}'^2) + \frac{D_1^*}{2(1+D_1^*)E}\mathrm{Tr}(\boldsymbol{\sigma}^2 : \mathbf{D}^*), \tag{3.20}$$

where

$$\mathrm{Tr}\boldsymbol{\sigma} = 3\sigma_{\mathrm{H}} = 3\sigma_{ii}, \quad \mathrm{Tr}(\boldsymbol{\sigma}'^2) = \frac{2}{3}\sigma_{\mathrm{eq}}^2 = s_{ij}s_{ij}, \quad \mathrm{Tr}(\boldsymbol{\sigma}^2 : \mathbf{D}^*) = \sigma_{ik}\sigma_{kl}D_{li}^*. \tag{3.21}$$

$\boldsymbol{\sigma}^*$ is a modified stress tensor whose compressive principal components are replaced by zeros, whereas tensile ones are left unchanged. When the isotropic term is omitted $B=0$, and the exponent $n=2$ is set, (3.19) takes the simplified form (Litewka and Hult [81])

$$\dot{\mathbf{D}} = C\left(\Phi^{\mathrm{e}}\right)^2 \boldsymbol{\sigma}^* = C\left[\frac{1-2\nu}{6E}\mathrm{Tr}^2\boldsymbol{\sigma} + \frac{1+\nu}{2E}\mathrm{Tr}\left(\boldsymbol{\sigma}'^2\right) + \frac{D_1^*}{2E(1+D_1^*)}\mathrm{Tr}\left(\boldsymbol{\sigma}^2 : \mathbf{D}^*\right)\right]^2 \boldsymbol{\sigma}^*. \tag{3.22}$$

3.4. Unified constitutive and damage theory of anisotropic elastic–brittle materials by use of the Helmholtz free energy

Murakami and Kamiya [47] developed the model based on the Helmholtz free energy as a function of the elastic strain tensor $\boldsymbol{\varepsilon}^{\mathrm{e}}$, the second–rank damage tensor $\mathbf{D}$, and another scalar damage variable β responsible for isotropic damage:

$$\begin{aligned}\varrho\Psi(\boldsymbol{\varepsilon}^{\mathrm{e}},\mathbf{D},\beta) &= \varrho\Psi^{\mathrm{e}}(\boldsymbol{\varepsilon}^{\mathrm{e}},\mathbf{D}) + \varrho\Psi^{\mathrm{d}}(\beta) = \tfrac{1}{2}\lambda\mathrm{Tr}^2\boldsymbol{\varepsilon}^{\mathrm{e}} + \mu\mathrm{Tr}(\boldsymbol{\varepsilon}^{\mathrm{e}})^2 + \eta_1\mathrm{Tr}\mathbf{D}\mathrm{Tr}^2\boldsymbol{\varepsilon}^{\mathrm{e}}\\ &+\eta_2\mathrm{Tr}\mathbf{D}\mathrm{Tr}(\boldsymbol{\varepsilon}^{\mathrm{e}})^2 + \eta_3\mathrm{Tr}\boldsymbol{\varepsilon}^{\mathrm{e}}\mathrm{Tr}(\boldsymbol{\varepsilon}^{\mathrm{e}} : \mathbf{D}) + \eta_4\mathrm{Tr}[(\boldsymbol{\varepsilon}^{\mathrm{e}^*})^2 : \mathbf{D}] + \tfrac{1}{2}K^{\mathrm{d}}\beta^2,\end{aligned} \tag{3.23}$$

where $\lambda = E\nu/(1+\nu)(1-2\nu)$ and $\mu = E/2(1+\nu)$ are the Lamè constants for undamaged materials, η_1, η_2, η_3, η_4 and K^{d} are material constants, whereas $\boldsymbol{\varepsilon}^{\mathrm{e}^*}$ is a modified elastic strain tensor used to represent the unilateral damage response $\boldsymbol{\varepsilon}^{\mathrm{e}^*} = \langle\boldsymbol{\varepsilon}^{\mathrm{e}}\rangle - \zeta\langle-\boldsymbol{\varepsilon}^{\mathrm{e}}\rangle$:

$$[\langle\boldsymbol{\varepsilon}^{\mathrm{e}}\rangle] = \begin{bmatrix}\langle\varepsilon_1\rangle & 0 & 0\\ 0 & \langle\varepsilon_2\rangle & 0\\ 0 & 0 & \langle\varepsilon_3\rangle\end{bmatrix}, \qquad [\langle-\boldsymbol{\varepsilon}^{\mathrm{e}}\rangle] = \begin{bmatrix}\langle-\varepsilon_1^{\mathrm{e}}\rangle & 0 & 0\\ 0 & \langle-\varepsilon_2^{\mathrm{e}}\rangle & 0\\ 0 & 0 & \langle-\varepsilon_3^{\mathrm{e}}\rangle\end{bmatrix}. \tag{3.24}$$

Constitutive equations of anisotropic elasticity coupled with damage are:

$$\begin{aligned}\boldsymbol{\sigma} = & \frac{\partial(\varrho\Psi^{\mathrm{e}})}{\partial\boldsymbol{\varepsilon}^{\mathrm{e}}} = \widetilde{\boldsymbol{\Lambda}}(\mathbf{D}):\boldsymbol{\varepsilon}^{\mathrm{e}} = [\lambda\mathrm{Tr}\boldsymbol{\varepsilon}^{\mathrm{e}} + 2\eta_1\mathrm{Tr}\mathbf{D}\mathrm{Tr}\boldsymbol{\varepsilon}^{\mathrm{e}} + \eta_3\mathrm{Tr}(\boldsymbol{\varepsilon}^{\mathrm{e}}:\mathbf{D})]\,\mathbf{1} \\ & +2(\mu+\eta_2\mathrm{Tr}\mathbf{D})\,\boldsymbol{\varepsilon}^{\mathrm{e}} + \eta_3(\mathrm{Tr}\boldsymbol{\varepsilon}^{\mathrm{e}})\,\mathbf{D} + \eta_4\frac{\partial\boldsymbol{\varepsilon}^{\mathrm{e}^*}}{\partial\boldsymbol{\varepsilon}^{\mathrm{e}}}\left(\boldsymbol{\varepsilon}^{\mathrm{e}^*}:\mathbf{D}+\mathbf{D}:\boldsymbol{\varepsilon}^{\mathrm{e}^*}\right),\end{aligned} \tag{3.25}$$

whereas the thermodynamic damage conjugate forces of $\mathbf{D}$ and β are

$$\begin{aligned}\mathbf{Y} = & -\frac{\partial(\varrho\Psi^{\mathrm{e}})}{\partial\mathbf{D}} = -\left[\eta_1(\mathrm{Tr}\boldsymbol{\varepsilon}^{\mathrm{e}})^2 + \eta_2\mathrm{Tr}(\boldsymbol{\varepsilon}^{\mathrm{e}})^2\right]\mathbf{1} - \eta_3(\mathrm{Tr}\boldsymbol{\varepsilon}^{\mathrm{e}})\,\boldsymbol{\varepsilon}^{\mathrm{e}} - \eta_4\boldsymbol{\varepsilon}^{\mathrm{e}^*}:\boldsymbol{\varepsilon}^{\mathrm{e}^*}, \\ B = & \frac{\partial(\varrho\Psi^{\mathrm{d}})}{\partial\beta} = K^{\mathrm{d}}\beta.\end{aligned} \tag{3.26}$$

$\widetilde{\boldsymbol{\Lambda}}(\mathbf{D})$ is a fourth–rank symmetric secant stiffness tensor, as a function of the second–rank damage tensor $\mathbf{D}$. Thermodynamic conjugate force $\mathbf{Y}$, associated with $\mathbf{D}$, is known as the damage strain energy release rate, that is the derivative of strain energy with respect to $\mathbf{D}$ (the mechanical flux vector component). In case of the second rank–damage tensor $\mathbf{D}$, force $\mathbf{Y}$ is the second–rank tensor as well.

The damage criterion in the space $\{\mathbf{Y}, -B\}$ is assumed as:

$$\begin{aligned}F^{\mathrm{d}}(\mathbf{Y},B) = & \ Y_{\mathrm{eq}} - (B_0 + B) = 0, \\ Y_{\mathrm{eq}} = & \left(\frac{1}{2}\mathbf{Y}:\mathbf{L}:\mathbf{Y}\right)^{1/2}, \qquad L_{ijkl} = \frac{1}{2}\left(\delta_{ik}\delta_{jl} + \delta_{il}\delta_{jk}\right).\end{aligned} \tag{3.27}$$

The evolution equations for damage are furnished as follows:

$$\begin{aligned}\dot{\mathbf{D}} = & -\dot{\lambda}^{\mathrm{d}}\frac{\partial F^{\mathrm{d}}}{\partial\mathbf{Y}}, \qquad \dot{\beta} = \dot{\lambda}^{\mathrm{d}}\frac{\partial F^{\mathrm{d}}}{\partial(-B)} = \dot{\lambda}^{\mathrm{d}}, \\ \dot{\lambda}^{\mathrm{d}} = & \frac{(\partial F^{\mathrm{d}}/\partial\mathbf{Y}):\dot{\mathbf{Y}}}{(\partial B/\partial\beta)} = \alpha\frac{\mathbf{L}:\mathbf{Y}}{2K^{\mathrm{d}}\mathbf{Y}_{\mathrm{eq}}}:\dot{\mathbf{Y}},\end{aligned} \tag{3.28}$$

where $\alpha = 1$ if $F^{\mathrm{d}} = 0$ and $\partial F^{\mathrm{d}}/\partial\mathbf{Y}:\dot{\mathbf{Y}} > 0$ or $\alpha = 0$ if $F^{\mathrm{d}} < 0$ and $\partial F^{\mathrm{d}}/\partial\mathbf{Y}:\dot{\mathbf{Y}} \leq 0$.

3.5. Constitutive and evolution equations for anisotropic damage of initially isotropic elastic–plastic materials by use of the Gibbs thermodynamic potential

It is sometimes more convenient to define the damage conjugate forces as functions of stress tensor by use of the Gibbs thermodynamic potential Γ that consists of the complementary energy Γ^{e} due to the elastic deformation, the potential related to the plastic deformation Γ^{p} and the damage potential related to the free surface energy due to the microcavities nucleation Γ^{d} (cf. Hayakawa and Murakami [49])

$$\begin{aligned}\Gamma(\boldsymbol{\sigma},r,\mathbf{D},\beta) = & \ \Gamma^{\mathrm{e}}(\boldsymbol{\sigma},\mathbf{D}) + \Gamma^{\mathrm{p}}(r) + \Gamma^{\mathrm{d}}(\beta) \\ = & -\frac{\nu}{2E}(\mathrm{Tr}\boldsymbol{\sigma})^2 + \frac{1+\nu}{2E}\mathrm{Tr}\boldsymbol{\sigma}^2 + \vartheta_1\mathrm{Tr}\mathbf{D}(\mathrm{Tr}\boldsymbol{\sigma})^2 + \vartheta_2\mathrm{Tr}\mathbf{D}\mathrm{Tr}\boldsymbol{\sigma}^{*2}\end{aligned} \tag{3.29}$$

$$+\vartheta_3 \mathrm{Tr}\boldsymbol{\sigma}\left(\mathrm{Tr}\boldsymbol{\sigma}\mathbf{D}\right) + \vartheta_4 \mathrm{Tr}\left(\boldsymbol{\sigma}^{*2}\mathbf{D}\right) + R_\infty \left[r + \frac{1}{b}\exp\left(-br\right)\right] + \frac{1}{2}K^{\mathrm{d}}\beta^2,$$

where ϑ_1, ϑ_2, ϑ_3, ϑ_4 and R_∞, b and K^{d} are material constants and $\boldsymbol{\sigma}^*$ is the modified stress tensor responsible for the opening/closure effect $\boldsymbol{\sigma}^* = \langle\boldsymbol{\sigma}\rangle - \zeta\langle-\boldsymbol{\sigma}\rangle$.

The constitutive equation for elastic strain $\boldsymbol{\varepsilon}^{\mathrm{e}}$ is furnished as:

$$\begin{aligned}\boldsymbol{\varepsilon}^{\mathrm{e}} = \frac{\partial \Gamma^{\mathrm{e}}}{\partial \boldsymbol{\sigma}} = -\frac{\nu}{E}\left(\mathrm{Tr}\boldsymbol{\sigma}\right)\mathbf{1} + \frac{1+\nu}{E}\boldsymbol{\sigma} + 2\vartheta_1\left(\mathrm{Tr}\mathbf{D}\,\mathrm{Tr}\boldsymbol{\sigma}\right)\mathbf{1} + 2\vartheta_2\left(\mathrm{Tr}\mathbf{D}\right)\boldsymbol{\sigma}^* : \frac{\partial \boldsymbol{\sigma}^*}{\partial \boldsymbol{\sigma}} \\ +\vartheta_3\left[\mathrm{Tr}\left(\boldsymbol{\sigma}\mathbf{D}\right)\mathbf{1} + \left(\mathrm{Tr}\boldsymbol{\sigma}\right)\mathbf{D}\right] + \vartheta_4\left(\boldsymbol{\sigma}^*\mathbf{D} + \mathbf{D}\boldsymbol{\sigma}^*\right) : \frac{\partial \boldsymbol{\sigma}^*}{\partial \boldsymbol{\sigma}}\end{aligned} \qquad (3.30)$$

and the forces conjugate to internal variables $\mathbf{D}$, r and β are

$$\begin{aligned}\mathbf{Y} &= \frac{\partial \Gamma^{\mathrm{e}}}{\partial \mathbf{D}} = \left[\vartheta_1\left(\mathrm{Tr}\boldsymbol{\sigma}\right)^2 + \vartheta_2 \mathrm{Tr}\boldsymbol{\sigma}^{*2}\right]\mathbf{1} + \vartheta_3\left(\mathrm{Tr}\boldsymbol{\sigma}\right)\boldsymbol{\sigma} + \vartheta_4\boldsymbol{\sigma}^{*2}, \\ R &= \frac{\partial \Gamma^{\mathrm{p}}}{\partial r} = R_\infty\left[1 - \exp\left(-br\right)\right], \qquad B = \frac{\partial \Gamma^{\mathrm{d}}}{\partial \beta} = K^{\mathrm{d}}\beta.\end{aligned} \qquad (3.31)$$

Assuming also the Mises–type isotropic strain hardening yield condition of a damaged ductile material in the form

$$F^{\mathrm{p}}\left(\boldsymbol{\sigma}, R, \mathbf{D}\right) = \widetilde{\sigma}^{\mathrm{M}}_{\mathrm{eq}} - \left(\sigma_y + R\right) = 0 \qquad (3.32)$$

the constitutive equations for plastic strain rate $\dot{\varepsilon}^{\mathrm{p}}_{ij}$ and the rate of isotropic hardening $\dot{r}$ are obtained

$$\dot{\varepsilon}^{\mathrm{p}}_{ij} = \dot{\Lambda}^{\mathrm{p}}\frac{\partial F^{\mathrm{p}}}{\partial \sigma'_{ij}} = \frac{3}{2}\dot{\Lambda}^{\mathrm{p}}\frac{\widetilde{H}^{\mathrm{M}}_{ijkl}\sigma'_{kl}}{\widetilde{\sigma}_{\mathrm{eq}}}, \qquad \dot{r} = \dot{\Lambda}^{\mathrm{p}}\frac{\partial F^{\mathrm{p}}}{\partial\left(-R\right)} = \dot{\Lambda}^{\mathrm{p}}, \qquad (3.33)$$

where $\widetilde{H}^{\mathrm{M}}_{ijkl} = \widetilde{\mathbf{H}}^{\mathrm{M}}\left(\mathbf{D}\right)$ is a fourth–rank effective plastic characteristic tensor with $\mathbf{D}$ as an argument

$$\widetilde{\mathbf{H}}^{\mathrm{M}}\left(\mathbf{D}\right) = \frac{1}{2}\left(\delta_{ik}\delta_{jl} + \delta_{il}\delta_{jk}\right) + \frac{1}{2}c^{\mathrm{p}}\left(\delta_{ik}D_{jl} + D_{ik}\delta_{jl} + \delta_{il}D_{jk} + D_{il}\delta_{jk}\right), \qquad (3.34)$$

and the effective Mises–type equivalent stress $\widetilde{\sigma}^{\mathrm{M}}_{\mathrm{eq}}$ is

$$\widetilde{\sigma}^{\mathrm{M}}_{\mathrm{eq}} = \left[\frac{3}{2}\boldsymbol{\sigma}' : \widetilde{\mathbf{H}}^{\mathrm{M}}\left(\mathbf{D}\right) : \boldsymbol{\sigma}'\right]^{1/2} = \left[\frac{3}{2}\widetilde{\boldsymbol{\sigma}}' : \mathbf{H}^{\mathrm{M}} : \widetilde{\boldsymbol{\sigma}}'\right]^{1/2} \qquad \widetilde{\mathbf{H}}^{\mathrm{M}} = \mathbf{M}^{\mathrm{T}} : \mathbf{H}^{\mathrm{M}} : \mathbf{M}. \qquad (3.35)$$

The initial failure criterion (damage surface) is assumed in the form

$$F^{\mathrm{d}}\left(\mathbf{Y}, B, \mathbf{D}, r\right) = Y_{\mathrm{eq}} + c^{\mathrm{r}}r\mathrm{Tr}\mathbf{D}\mathrm{Tr}\mathbf{Y} - \left(B_0 + B\right) = 0, \quad Y_{\mathrm{eq}} = \left[\frac{1}{2}\mathbf{Y} : \widehat{\mathbf{L}}\left(\mathbf{D}\right) : \mathbf{Y}\right]^{1/2} \qquad (3.36)$$

that extends Eq. (3.27) by the additional damage–plasticity term corresponding to the isotropic hardening r. Eventually, the evolution equations are furnished as follows:

$$\dot{\mathbf{D}} = \dot{\Lambda}^{\mathrm{d}}\frac{\partial F^{\mathrm{d}}}{\partial \mathbf{Y}} = \dot{\Lambda}^{\mathrm{d}}\left[\frac{\widehat{\mathbf{L}} : \mathbf{Y}}{2Y_{\mathrm{eq}}} + c^{\mathrm{r}}r\left(\mathrm{Tr}\mathbf{D}\right)\mathbf{1}\right], \qquad \dot{\beta} = \dot{\Lambda}^{\mathrm{d}}\frac{\partial F^{\mathrm{d}}}{\partial\left(-B\right)} = \dot{\Lambda}^{\mathrm{d}}. \qquad (3.37)$$

Table 3.4: Constitutive elasticity tensors of initially isotropic materials in the principal axes of damage tensor

$$\{\varepsilon^e_{11}\,\varepsilon^e_{22}\,\varepsilon^e_{33}\,\varepsilon^e_{23}\,\varepsilon^e_{31}\,\varepsilon^e_{12}\}^{\mathsf{T}} = \left[\widetilde{\boldsymbol{\Lambda}}_1^{-1}(D_1,D_2,D_3)\right]\{\sigma_{11}\,\sigma_{22}\,\sigma_{33}\,\sigma_{23}\,\sigma_{31}\,\sigma_{12}\}^{\mathsf{T}},$$

$$\widetilde{\boldsymbol{\Lambda}}_{C1}^{-1}(D_\alpha) = \mathbf{M}_{C1}(D_\alpha) : \boldsymbol{\Lambda}^{-1} : \mathbf{M}_{C1}(D_\alpha)$$

$$= \frac{1}{E}\begin{bmatrix} \frac{1}{(1-D_1)^2} & -\frac{\nu}{(1-D_1)(1-D_2)} & -\frac{\nu}{(1-D_1)(1-D_3)} & 0 & 0 & 0 \\ -\frac{\nu}{(1-D_2)(1-D_1)} & \frac{1}{(1-D_2)^2} & -\frac{\nu}{(1-D_2)(1-D_3)} & 0 & 0 & 0 \\ -\frac{\nu}{(1-D_3)(1-D_1)} & -\frac{\nu}{(1-D_3)(1-D_2)} & \frac{1}{(1-D_3)^2} & 0 & 0 & 0 \\ 0 & 0 & 0 & \frac{1+\nu}{(1-D_2)(1-D_3)} & 0 & 0 \\ 0 & 0 & 0 & 0 & \frac{1+\nu}{(1-D_1)(1-D_3)} & 0 \\ 0 & 0 & 0 & 0 & 0 & \frac{1+\nu}{(1-D_1)(1-D_2)} \end{bmatrix};$$

$$\{\varepsilon^e_{11}\,\varepsilon^e_{22}\,\varepsilon^e_{33}\,\varepsilon^e_{23}\,\varepsilon^e_{31}\,\varepsilon^e_{12}\}^{\mathsf{T}} = \left[\widetilde{\boldsymbol{\Lambda}}_2^{-1}(D_1,D_2,D_3)\right]\{\sigma_{11}\,\sigma_{22}\,\sigma_{33}\,\sigma_{23}\,\sigma_{31}\,\sigma_{12}\}^{\mathsf{T}},$$

$$\widetilde{\boldsymbol{\Lambda}}_{C2}^{-1}(D_\alpha) = \mathbf{M}_{C2}(D_\alpha) : \boldsymbol{\Lambda}^{-1} : \mathbf{M}_{C2}(D_\alpha)$$

$$= \frac{1}{E}\begin{bmatrix} \frac{1}{(1-D_1)^2} & -\frac{\nu}{(1-D_1)(1-D_2)} & -\frac{\nu}{(1-D_1)(1-D_3)} & 0 & 0 & 0 \\ -\frac{\nu}{(1-D_2)(1-D_1)} & \frac{1}{(1-D_2)^2} & -\frac{\nu}{(1-D_2)(1-D_3)} & 0 & 0 & 0 \\ -\frac{\nu}{(1-D_3)(1-D_1)} & -\frac{\nu}{(1-D_3)(1-D_2)} & \frac{1}{(1-D_3)^2} & 0 & 0 & 0 \\ 0 & 0 & 0 & \frac{1+\nu}{(1-\frac{D_2+D_3}{2})^2} & 0 & 0 \\ 0 & 0 & 0 & 0 & \frac{1+\nu}{(1-\frac{D_3+D_1}{2})^2} & 0 \\ 0 & 0 & 0 & 0 & 0 & \frac{1+\nu}{(1-\frac{D_1+D_2}{2})^2} \end{bmatrix};$$

$$\{\varepsilon^e_{11}\,\varepsilon^e_{22}\,\varepsilon^e_{33}\,\varepsilon^e_{23}\,\varepsilon^e_{31}\,\varepsilon^e_{12}\}^{\mathsf{T}} = \left[\widetilde{\boldsymbol{\Lambda}}_3^{-1}(D_1,D_2,D_3)\right]\{\sigma_{11}\,\sigma_{22}\,\sigma_{33}\,\sigma_{23}\,\sigma_{31}\,\sigma_{12}\}^{\mathsf{T}},$$

$$\widetilde{\boldsymbol{\Lambda}}_{C3}^{-1}(D_\alpha) = \mathbf{M}_{C3}(D_\alpha) : \boldsymbol{\Lambda}^{-1} : \mathbf{M}_{C3}(D_\alpha)$$

$$= \frac{1}{E}\begin{bmatrix} \frac{1}{(1-D_1)^2} & -\frac{\nu}{(1-D_1)(1-D_2)} & -\frac{\nu}{(1-D_1)(1-D_3)} & 0 & 0 & 0 \\ -\frac{\nu}{(1-D_2)(1-D_1)} & \frac{1}{(1-D_2)^2} & -\frac{\nu}{(1-D_2)(1-D_3)} & 0 & 0 & 0 \\ -\frac{\nu}{(1-D_3)(1-D_1)} & -\frac{\nu}{(1-D_3)(1-D_2)} & \frac{1}{(1-D_3)^2} & 0 & 0 & 0 \\ 0 & 0 & 0 & \frac{1+\nu}{4}\left(\frac{1}{1-D_2}+\frac{1}{1-D_3}\right)^2 & 0 & 0 \\ 0 & 0 & 0 & 0 & \frac{1+\nu}{4}\left(\frac{1}{1-D_1}+\frac{1}{1-D_3}\right)^2 & 0 \\ 0 & 0 & 0 & 0 & 0 & \frac{1+\nu}{4}\left(\frac{1}{1-D_1}+\frac{1}{1-D_2}\right)^2 \end{bmatrix};$$

continued on next page

continued from previous page

$$\{\sigma_{11}\,\sigma_{22}\,\sigma_{33}\,\sigma_{23}\,\sigma_{31}\,\sigma_{12}\}^{\mathsf{T}} = \left[\widetilde{\mathbf{\Lambda}}_{\mathrm{Ch}}\,(d_1, d_2, d_3)\right] \{\varepsilon^{\mathrm{e}}_{11}\,\varepsilon^{\mathrm{e}}_{22}\,\varepsilon^{\mathrm{e}}_{33}\,\varepsilon^{\mathrm{e}}_{23}\,\varepsilon^{\mathrm{e}}_{31}\,\varepsilon^{\mathrm{e}}_{12}\}^{\mathsf{T}},$$

$$\widetilde{\mathbf{\Lambda}}_{\mathrm{Ch}}\,(d_\alpha) = \mathbf{M}_{\mathrm{Ch}}\,(d_\alpha) : \mathbf{\Lambda} : \mathbf{M}_{\mathrm{Ch}}\,(d_\alpha)$$

$$= \begin{bmatrix} (\lambda+2\mu)(1-d_1)^2 & \lambda(1-d_1)(1-d_2) & \lambda(1-d_1)(1-d_3) & 0 & 0 & 0 \\ \lambda(1-d_1)(1-d_2) & (\lambda+2\mu)(1-d_2)^2 & \lambda(1-d_2)(1-d_3) & 0 & 0 & 0 \\ \lambda(1-d_1)(1-d_3) & \lambda(1-d_2)(1-d_3) & (\lambda+2\mu)(1-d_3)^2 & 0 & 0 & 0 \\ 0 & 0 & 0 & 2\mu(1-d_2)(1-d_3) & 0 & 0 \\ 0 & 0 & 0 & 0 & 2\mu(1-d_3)(1-d_1) & 0 \\ 0 & 0 & 0 & 0 & 0 & 2\mu(1-d_1)(1-d_2) \end{bmatrix};$$

$$\{\varepsilon^{\mathrm{e}}_{11}\,\varepsilon^{\mathrm{e}}_{22}\,\varepsilon^{\mathrm{e}}_{33}\,\varepsilon^{\mathrm{e}}_{23}\,\varepsilon^{\mathrm{e}}_{31}\,\varepsilon^{\mathrm{e}}_{12}\}^{\mathsf{T}} = \left[\widetilde{\mathbf{\Lambda}}^{-1}_{\mathrm{L}}\,(D^\star_1, D^\star_2, D^\star_3)\right] \{\sigma_{11}\,\sigma_{22}\,\sigma_{33}\,\sigma_{23}\,\sigma_{31}\,\sigma_{12}\}^{\mathsf{T}},$$

$$\widetilde{\mathbf{\Lambda}}^{-1}_{\mathrm{L}}\,(D^\star_\alpha) = \mathbf{M}_{\mathrm{L}}\,(D^\star_\alpha) : \mathbf{\Lambda}^{-1} : \mathbf{M}_{\mathrm{L}}\,(D^\star_\alpha)$$

$$= \frac{1}{E}\begin{bmatrix} 1+\frac{D^{\star 2}_1}{1+D^\star_1} & -\nu & -\nu & 0 & 0 & 0 \\ -\nu & 1+\frac{D^\star_1 D^\star_2}{1+D^\star_1} & -\nu & 0 & 0 & 0 \\ -\nu & -\nu & 1+\frac{D^\star_1 D^\star_3}{1+D^\star_1} & 0 & 0 & 0 \\ 0 & 0 & 0 & (1+\nu)\left[1+\frac{D^\star_1(D^\star_2+D^\star_3)}{2(1+\nu)(1+D^\star_1)}\right] & 0 & 0 \\ 0 & 0 & 0 & 0 & (1+\nu)\left[1+\frac{D^\star_1(D^\star_1+D^\star_3)}{2(1+\nu)(1+D^\star_1)}\right] & 0 \\ 0 & 0 & 0 & 0 & 0 & (1+\nu)\left[1+\frac{D^\star_1(D^\star_1+D^\star_2)}{2(1+\nu)(1+D^\star_1)}\right] \end{bmatrix};$$

continued on next page

continued from previous page

$$\{\sigma_{11}\,\sigma_{22}\,\sigma_{33}\,\sigma_{23}\,\sigma_{31}\,\sigma_{12}\}^{\mathsf{T}} = \left[\widetilde{\mathbf{\Lambda}}_{\mathrm{MK}}\left(D_{11}, D_{22}, D_{33}\right)\right]\{\varepsilon^{\mathrm{e}}_{11}\,\varepsilon^{\mathrm{e}}_{22}\,\varepsilon^{\mathrm{e}}_{33}\,\varepsilon^{\mathrm{e}}_{23}\,\varepsilon^{\mathrm{e}}_{31}\,\varepsilon^{\mathrm{e}}_{12}\}^{\mathsf{T}},$$

$$\widetilde{\mathbf{\Lambda}}_{\mathrm{MK}}\left(D_{\alpha\alpha}\right) = \mathbf{M}_{\mathrm{MK}}\left(D_{\alpha\alpha}\right) : \mathbf{\Lambda} : \mathbf{M}_{\mathrm{MK}}\left(D_{\alpha\alpha}\right)$$

$$= \begin{bmatrix} \lambda+2\mu+2(\eta_1+\eta_2)\,\mathrm{Tr}\mathbf{D}+2(\eta_3+\eta_4)\,D_{11} & \lambda+2\eta_1\,\mathrm{Tr}\mathbf{D}+\eta_3\,(D_{11}+D_{22}) & \lambda+2\eta_1\,\mathrm{Tr}\mathbf{D}+\eta_3\,(D_{11}+D_{33}) & 0 & 0 & 0 \\ \lambda+2\eta_1\,\mathrm{Tr}\mathbf{D}+\eta_3\,(D_{11}+D_{22}) & \lambda+2\mu+2(\eta_1+\eta_2)\,\mathrm{Tr}\mathbf{D}+2(\eta_3+\eta_4)\,D_{22} & \lambda+2\eta_1\,\mathrm{Tr}\mathbf{D}+\eta_3\,(D_{33}+D_{22}) & 0 & 0 & 0 \\ \lambda+2\eta_1\,\mathrm{Tr}\mathbf{D}+\eta_3\,(D_{11}+D_{33}) & \lambda+2\eta_1\,\mathrm{Tr}\mathbf{D}+\eta_3\,(D_{22}+D_{33}) & \lambda+2\mu+2(\eta_1+\eta_2)\,\mathrm{Tr}\mathbf{D}+2(\eta_3+\eta_4)\,D_{33} & 0 & 0 & 0 \\ 0 & 0 & 0 & 2\mu+2\eta_2\,\mathrm{Tr}\mathbf{D}+\eta_4\,(D_{33}+D_{22}) & 0 & 0 \\ 0 & 0 & 0 & 0 & 2\mu+2\eta_2\,\mathrm{Tr}\mathbf{D}+\eta_4\,(D_{11}+D_{33}) & 0 \\ 0 & 0 & 0 & 0 & 0 & 2\mu+2\eta_2\,\mathrm{Tr}\mathbf{D}+\eta_4\,(D_{11}+D_{22}) \end{bmatrix};$$

$$\{\varepsilon^{\mathrm{e}}_{11}\,\varepsilon^{\mathrm{e}}_{22}\,\varepsilon^{\mathrm{e}}_{33}\,\varepsilon^{\mathrm{e}}_{23}\,\varepsilon^{\mathrm{e}}_{31}\,\varepsilon^{\mathrm{e}}_{12}\}^{\mathsf{T}} = \left[\widetilde{\mathbf{\Lambda}}^{-1}_{\mathrm{HM}}\left(D_{11}, D_{22}, D_{33}\right)\right]\{\sigma_{11}\,\sigma_{22}\,\sigma_{33}\,\sigma_{23}\,\sigma_{31}\,\sigma_{12}\}^{\mathsf{T}},$$

$$\widetilde{\mathbf{\Lambda}}^{-1}_{\mathrm{HM}}\left(D_{\alpha\alpha}\right) = \mathbf{M}_{\mathrm{HM}}\left(D_{\alpha\alpha}\right) : \mathbf{\Lambda}^{-1} : \mathbf{M}_{\mathrm{HM}}\left(D_{\alpha\alpha}\right)$$

$$= \begin{bmatrix} \frac{1}{E}+2\mathrm{Tr}\mathbf{D}(\vartheta_1+\vartheta_2)+2D_{11}(\vartheta_3+\vartheta_4) & -\frac{\nu}{E}+2\vartheta_1\mathrm{Tr}\mathbf{D}+\vartheta_3(D_{11}+D_{22}) & -\frac{\nu}{E}+2\vartheta_1\mathrm{Tr}\mathbf{D}+\vartheta_3(D_{11}+D_{33}) & 0 & 0 & 0 \\ -\frac{\nu}{E}+2\vartheta_1\mathrm{Tr}\mathbf{D}+\vartheta_3(D_{11}+D_{22}) & \frac{1}{E}+2\mathrm{Tr}\mathbf{D}(\vartheta_1+\vartheta_2)+2D_{22}(\vartheta_3+\vartheta_4) & -\frac{\nu}{E}+2\vartheta_1\mathrm{Tr}\mathbf{D}+\vartheta_3(D_{22}+D_{33}) & 0 & 0 & 0 \\ -\frac{\nu}{E}+2\vartheta_1\mathrm{Tr}\mathbf{D}+\vartheta_3(D_{11}+D_{33}) & -\frac{\nu}{E}+2\vartheta_1\mathrm{Tr}\mathbf{D}+\vartheta_3(D_{22}+D_{33}) & \frac{1}{E}+2\mathrm{Tr}\mathbf{D}(\vartheta_1+\vartheta_2)+2D_{33}(\vartheta_3+\vartheta_4) & 0 & 0 & 0 \\ 0 & 0 & 0 & \frac{1+\nu}{E}+2\vartheta_2\mathrm{Tr}\mathbf{D}+\vartheta_4(D_{22}+D_{33}) & 0 & 0 \\ 0 & 0 & 0 & 0 & \frac{1+\nu}{E}+2\vartheta_2\mathrm{Tr}\mathbf{D}+\vartheta_4(D_{11}+D_{33}) & 0 \\ 0 & 0 & 0 & 0 & 0 & \frac{1+\nu}{E}+2\vartheta_2\mathrm{Tr}\mathbf{D}+\vartheta_4(D_{11}+D_{22}) \end{bmatrix}.$$

The plasticity and damage multipliers $\dot{\Lambda}^{\mathrm{p}}$ and $\dot{\Lambda}^{\mathrm{d}}$ must be derived from the consistency conditions for plastic yield surface, and the fourth–rank tensor $\widehat{\mathbf{L}}(\mathbf{D})$ is represented by the formula analogous to (3.34) with the constant c^{p} replaced by the new constant c^{d}. This fourth–rank tensor function of the second–rank damage tensor $\mathbf{D}$ describes the damage induced change of the damage surface, such that for the initial damage $\mathbf{D} = \mathbf{0}$ and $c^r = 0$ (3.36) is reduced to (3.27).

A more general approach for thermodynamically admissible plasticity damage coupling, when initial material is assumed to be anisotropic and described by the Hill–type criterion, is discussed by Chow and Lu [44] and Chaboche [46].

3.6. Examples of matrix representation of fourth–rank elasticity tensors for damaged materials in the principal axes of damage tensor

Applying transformation formulae (3.3–3.5) we obtain matrix representations for constitutive elasticity tensors of damaged material $\widetilde{\mathbf{\Lambda}}$ or $\widetilde{\mathbf{\Lambda}}^{-1}$ and damaged effect tensors $\mathbf{M}^{-1}$ or $\mathbf{M}$. Assume for simplicity, that material is initially isotropic. When the definitions of $\mathbf{M}_{\mathrm{C1}}$, $\mathbf{M}_{\mathrm{C2}}$ or $\mathbf{M}_{\mathrm{C3}}$ (3.8) are used in terms of the principal damage components, $D_1 = D_{11}$, $D_2 = D_{22}$, $D_3 = D_{33}$, $D_{23} = D_{31} = D_{12} = 0$, the following symmetric compliance and stiffness matrices modified by damage are obtained (Table 3.4).

4. LIFETIME PREDICTION AND OPTIMAL DESIGN OF STRUCTURES BY USE OF CDM METHOD FOR CREEP–DAMAGE AND OTHER CONDITIONS

4.1. Structural optimization under damage conditions

When elastic structures are designed for minimum weight or maximum load, structures of uniform strength are optimal in most cases. In general, the condition of uniform strength is neither a necessary nor a sufficient optimality condition if the static indeterminacy of a structure or the geometric changes are taken into account (Gallagher [82]).

O.. the other hand, when inelastic structures made of time–dependent damaged material are subject to optimal design, the minimum weight or the maximum load remains the design objective, similar as in the elastic case. Essential changes occur in the state and evolution equations as well as in the constraints, where a new independent time variable plays an important rule. The new optimization constrains may be imposed not only on the strength, stiffness, and stability, like in the elastic case, but additionally on a limited stress relaxation or residual characteristic displacement, and lifetime prediction of the initial failure, $t_{\mathrm{I}} = t_{\mathrm{R}}$ or the complete structural failure, $t_{\mathrm{II}} = t_{\mathrm{F}}$.

General classification of the optimization problems, originally proposed by Życzkowski [83, 84, 85, 86] for creep conditions, is shown in Table 4.1. First two approaches are inconvenient in most cases since the lifetime, t_{I} or t_{II}, is usually not explicitly given, but it results from the constrains imposed on the internal damage variables $\mathcal{D} = \left\{D, D_\alpha, \mathbf{D}, \widehat{\mathbf{D}}, \ldots\right\}$ when a critical state of damage is reached. Time of failure initiation t_{I} is defined here in such a way that either the scalar damage variable D (isotropic damage), the dominant

Table 4.1. Classification of the global optimization problems of structures for creep–damage conditions (after Skrzypek and Ganczarski [51])

Type of the formulation	Optimality criteria	Global constraints
i	$V = \min$	$P = \mathrm{const}$ and t_{I} or $t_{\mathrm{II}} = \mathrm{const}$
ii	$P = \max$	$V = \mathrm{const}$ and t_{I} or $t_{\mathrm{II}} = \mathrm{const}$
iii	t_{I} or $t_{\mathrm{II}} = \max$	$P = \mathrm{const}$ and $V = \mathrm{const}$

damage component $\max\{D_\alpha\}$ (orthotropic damage) and $\sup\{D_{ij}\}$ (anisotropic damage) reaches a critical value D_{cr} or a corresponding initial failure criterion $F^{\mathrm{d}}(\boldsymbol{\sigma}, \mathbf{D}) = 0$, (3.18) for Litewka's model, $F^{\mathrm{d}}(\mathbf{Y}, B) = 0$ (3.27) for Murakami and Kamiya's model, $F^{\mathrm{d}}(\mathbf{Y}, B, R, \mathbf{D}, \beta, r) = 0$ (3.36) for Hayakawa and Murakami's model etc., is satisfied when a more general space of damage and plasticity internal variables $\mathbf{D}$, β, r, and their thermodynamically conjugate forces $\mathbf{Y}$, B, R is used for critical damage (initial failure) criterion. Time of complete failure under continuum damage conditions t_{II} is obtained when a global failure mechanism of the structure occurs, such that the fractured structure becomes unserviceable. Local CDM approach to design analysis, when the influence of all other microdefects within the RVE, in the neighbourhood of $\mathbf{x}$, $\mathcal{D}(\boldsymbol{\xi})$ is measured only through the change of effective mechanical properties, stiffness $\widetilde{\mathbf{\Lambda}}(\mathbf{x}, t)$ and compliance $\widetilde{\mathbf{\Lambda}}^{-1}(\mathbf{x}, t)$, effective thermal properties, conductivity $\widetilde{\mathbf{L}}(\mathbf{x}, t)$ and emissivity $\widetilde{\mathbf{\Gamma}}(\mathbf{x}, t)$, etc., all defined at point $\mathbf{x}$, was shown to be capable of predicting not only lifetime of crack initiation t_{I} with the stress redistribution due to damage, but also crack length growth and time of complete failure t_{II} ([87, 88, 89, 90, 91, 92, 93, 51, 94]).

Recently, a number of optimal solutions for surface structures with respect to creep–brittle–damage have been obtained: axisymmetric prestressed disk [[95, 96, 97, 98, 99, 100, 101]; axisymmetric thin plates optimally prestressed [102, 103, 80, 51]; Reissner's plates optimally prestressed [104]. It is worth to mention that both thickness and initial prestressing optimization of the membrane– or the bending–type occur to be an efficient tool for lifetime improvement of structures under creep–damage conditions.

A. Optimality criteria for time–dependent materials

Structures optimal with respect to lifetime $t_{\mathrm{I}} = t_{\mathrm{R}} \to \max$, may often be found among the class of structures of Uniform Creep Strength (UCS), cf. Życzkowski [83, 84, 85, 86]. Structures of uniform creep strength with respect to brittle rupture are defined in such a way that macrocracks initiate simultaneously either in every material point $\mathbf{x} \in V$ or along certain characteristic lines or surfaces. Hence, e.g. when the simple, scalar Kachanov–Hayhurst's isotropic damage growth rule (2.13) is used with $k = r$ and the integration is performed from the damage initiation $D(t_0) = 0$ up to initiation of the first macrocrack $D(t_I) = D_{\mathrm{cr}}$, the condition of Uniform Isotropic Damage Strength (UIDS) takes the following representation:

$$1-(1-D_{\rm cr})^{r+1}=C(r+1)\int_{t_0}^{t_I}\{\chi[\boldsymbol{\sigma}(\mathbf{x},t)]\}^r\,\mathrm{d}t,\qquad \forall\mathbf{x}\in V.$$

$$\dot{D}=C\left\langle\frac{\chi[\boldsymbol{\sigma}(\mathbf{x},t)]}{1-D}\right\rangle^r,\ \chi=a\sigma_1+3b\sigma_{\rm H}+c\sigma_{\rm eq},\tag{4.1}$$

For orthotropic damage (3.9) the condition of Uniform Orthotropic Damage Strength (UODS) can be written as:

$$\dot{D}_k=C_k\left\langle\frac{\sigma_k(\mathbf{x},t)}{1-D_k}\right\rangle^{r_k},\qquad \sup_{(1,2,3)}\left[\frac{D_k(\mathbf{x},t_I)}{D_{k\rm cr}}\right]\equiv 1,\qquad \forall\mathbf{x}\in V.\tag{4.2}$$

where three independent critical damage values $D_{k\rm cr}$ $(k=1,2,3)$ are used for damage orthotropy. In a more general case of damage anisotropy the isotropic scalar function of stress and damage tensors $\boldsymbol{\sigma}$ and $\mathbf{D}$ may be postulated as the failure criterion (damage surface) at the point $\mathbf{x}$ (3.18), the scalar function of damage conjugate forces $\mathbf{Y}$ and B (3.27), the scalar function of both the damage and plasticity internal variables and their thermodynamic conjugates (3.36), etc.:

$$\begin{aligned}F^{\rm d}\left[\boldsymbol{\sigma}(\mathbf{x},t_I),\mathbf{D}^*(\mathbf{x},t_I)\right]&=0,\\ F^{\rm d}\left\{\mathbf{Y}\left[\boldsymbol{\varepsilon}^{\rm e}(\mathbf{x},t_{\rm I})\right],B\left[\beta(\mathbf{x},t_{\rm I})\right]\right\}&=0,\\ F^{\rm d}\left\{\mathbf{Y}\left[\boldsymbol{\varepsilon}^{\rm e}(\mathbf{x},t_{\rm I})\right],B\left[\beta(\mathbf{x},t_{\rm I})\right],\mathbf{D}(\mathbf{x},t_{\rm I}),r(\mathbf{x},t_{\rm I})\right\}&=0,\end{aligned}\tag{4.3}$$

all satisfied at each point $\mathbf{x}$ of the volume V at $t=t_{\rm I}$.

If, for instance, the Litewka's model is applied the condition of Uniform Anisotropic Damage Strength (UADS) may be furnished as follows:

$$\begin{aligned}\dot{\mathbf{D}}&=C\left\{\Phi^{\rm e}\left[\boldsymbol{\sigma}(\mathbf{x},t),\mathbf{D}^*(\mathbf{x},t)\right]\right\}^2\boldsymbol{\sigma}^*,\\ F^{\rm d}(\boldsymbol{\sigma},\mathbf{D}^*)&=C_1{\rm Tr}^2\boldsymbol{\sigma}(\mathbf{x},t_I)+C_2{\rm Tr}\left[\boldsymbol{\sigma}'(\mathbf{x},t_I)\right]^2\\ &\quad+C_3{\rm Tr}\left[\boldsymbol{\sigma}^2(\mathbf{x},t_I):\mathbf{D}^*(\mathbf{x},t_I)\right]-\sigma_{\rm u}^2=0,\qquad\forall\mathbf{x}\in V,\end{aligned}\tag{4.4}$$

where $\Phi^{\rm e}\left[\boldsymbol{\sigma},\mathbf{D}^*\right]$ denotes the elastic energy affected by damage (3.20), $\mathbf{D}$ and $\mathbf{D}^*$ denote the second–rank damage tensors, classical Murakami–Ohno's (1.1) and the modified $D_i^*=D_i/(1-D_i)$, whereas $\boldsymbol{\sigma}^*$ is a modified stress tensor (Sec. 3.3).

B. Constraints

In general, both inequality constraints and equality constraints are to be imposed when an optimal solution (in a various sense) is sought for.

i. Inequality constraints:

Strength constraints may be imposed on the effective Mises–type equivalent stress, when strain hardening is saturated and damage isotropy is assumed (Sec. 2.2):

$$\widetilde{\sigma}_{\rm eq}(\boldsymbol{\sigma},D)=\frac{\sigma_{\rm eq}}{1-D}\le\frac{\sigma_{\rm cr}}{j},\tag{4.5}$$

where $\sigma_{\rm cr}$ denotes the critical effective stress for the material and j is the safety factor.

In a more general case of strain–hardening initially isotropic material (damage induced anisotropy) the following inequality constraint can be used (3.35):

$$\widetilde{\sigma}^{\mathrm{M}}_{\mathrm{eq}}\left[\boldsymbol{\sigma},\widetilde{\mathbf{H}}^{\mathrm{M}}(\mathbf{D})\right]-R=\left[\frac{3}{2}\boldsymbol{\sigma}':\widetilde{\mathbf{H}}^{\mathrm{M}}(\mathbf{D}):\boldsymbol{\sigma}'\right]^{1/2}-R\leq\frac{\sigma_{\mathrm{cr}}}{j}. \tag{4.6}$$

Eventually, when the initial material anisotropy is described by Hill's type fourth–rank characteristic tensor $\mathbf{H}^{\mathrm{H}}$ and kinematic hardening is included (cf. Chaboche [46]) the Hill's type inequality constraint holds:

$$\widetilde{\sigma}^{\mathrm{H}}_{\mathrm{eq}}\left[\boldsymbol{\sigma}-\mathbf{X},\widetilde{\mathbf{H}}^{\mathrm{H}}(\mathbf{D})\right]-R=\left[\frac{3}{2}(\boldsymbol{\sigma}'-\mathbf{X}'):\widetilde{\mathbf{H}}^{\mathrm{H}}(\mathbf{D}):(\boldsymbol{\sigma}'-\mathbf{X}')\right]^{1/2}-R\leq\frac{\sigma_{\mathrm{cr}}}{j}. \tag{4.7}$$

Note that strength inequality constraints (4.5–4.7) are physically based limitation of elasticity domains when damage coupled elastic–(visco)plastic material is considered ([76, 77]). On the other hand, when the CDM based local approach to fracture is used, the so called stress limitation method is sometimes incorporated to FEM code in order to suppress a mesh–dependence of the crack growth prediction, (e.g. [91, 92, 94]):

$$\sigma^{\mathrm{L}}_{ij}=\begin{cases}\sigma_{ij}, & \sigma_{\mathrm{eq}}\leq\sigma_{\mathrm{cr}}\\ k\sigma_{ij}, & \sigma_{\mathrm{eq}}\geq\sigma_{\mathrm{cr}}\end{cases},\qquad k\sqrt{\frac{3}{2}\sigma'_{ij}\sigma'_{ij}}=\sigma_{\mathrm{cr}}. \tag{4.8}$$

This numerically induced equivalent stress constraint, which is not physically induced, has to be applied as an additional inequality constraint to obtain convergence.

Initial stability constraints (elastic stability condition) should also be imposed, especially when initial prestressing of a structure is used to improve the lifetime t_{I} or t_{II}, (cf. [101, 95, 80, 104, 51]):

$$n_0<n_{\mathrm{E}}, \tag{4.9}$$

where n_{E} denotes the basic Eulerian force (if a possibility of creep buckling is not included into the analysis).

Often geometric constraints for thickness of the structure $h(\mathbf{x})$ and the in–plane prestressing eccentricity $e_{\max}$ have also to be imposed

$$h_{\min}<h(\mathbf{x})<h_{\max},\quad e_{\max}\leq h/2, \tag{4.10}$$

such that uniform damage strength as defined by conditions (4.1–4.3) is met in the part of a structure $\left(V^{\mathrm{UCS}}<V\right)$ only, where the thickness geometric constraint is not active.

ii. Equality constraints

Depending on the optimization problem, as classified in Table 4.1, global equality constraints may be furnished. Condition of constant volume (weight) for a uniform cross–section of axisymmetric structures is written as

$$V=2\pi\int_0^R h(r)r\mathrm{d}r=\mathrm{const}, \tag{4.11}$$

and for a two–point sandwich cross–section

$$V = 2\pi \int_0^R \left[\alpha\left(h_{\mathrm{s}} - g_{\mathrm{s}}\right) + 2\beta g_{\mathrm{s}}\right] r\mathrm{d}r = \mathrm{const}, \tag{4.12}$$

where $h\,(r)$, $h_{\mathrm{s}}\,(r)$ and $g_{\mathrm{s}}\,(r)$ denote thickness of the uniform cross–section, the sandwich cross–section and the sandwich working layers, respectively, whereas α and β are arbitrary weight factors for the core and layers materials.

Condition of constant lifetime prediction of macrocracks initiation t_{I} or the complete failure (fragmentation) t_{II} are sometimes used as global equality constraints, although, in general, these quantities are not explicitly given, but they result from a combined time–dependent process of damage evolution in a material, as governed by the constitutive and evolution equations on each step of the optimization procedure through the geometry and prestressing changes:

$$t_{\mathrm{I}} = t_{\mathrm{R}} = \mathrm{const}, \quad \text{or} \quad t_{\mathrm{II}} = t_{\mathrm{F}} = \mathrm{const}\,. \tag{4.13}$$

Condition of constant intensity of surface loadings (prestressing force excluded) may also be applied as a global equality constraint

$$q(\mathbf{x}, t) = q(\mathbf{x}). \tag{4.14}$$

C. Decision variables

When problems of optimization are formulated for prestressed structures under damage or damage/fracture conditions, vectors of control variables involve not only the thickness of a structure $h(\mathbf{x})$ or $h_{\mathrm{s}}(\mathbf{x})$ and $g_{\mathrm{s}}(\mathbf{x})$ for an uniform or sandwich cross–section, respectively, but also parameters of prestressing n_0 or Δ_0 in case of in–plane membrane–type prestressing (a force or a membrane distortion), and m_0 or φ_0 in case of bending–type prestressing (a bending moment or an initial bending distortion). Hence, the corresponding membrane–type or bending–type vectors of decision variables $\mathbf{c}_{\mathrm{m}}$ or $\mathbf{c}_{\mathrm{b}}$ are:

$$\{\mathbf{c}_{\mathrm{m}}^{\mathrm{u}}\} = \{n_0 \ \text{ or } \ \Delta_0, h(\mathbf{x})\} \ \text{ or } \ \{\mathbf{c}_{\mathrm{m}}^{\mathrm{s}}\} = \{n_0 \ \text{ or } \ \Delta_0, h_{\mathrm{s}}(\mathbf{x}), g_{\mathrm{s}}\,(\mathbf{x})\} \tag{4.15}$$

or

$$\{\mathbf{c}_{\mathrm{b}}^{\mathrm{u}}\} = \{m_0 \ \text{ or } \ \varphi_0, h(\mathbf{x})\} \ \text{ or } \ \{\mathbf{c}_{\mathrm{b}}^{\mathrm{s}}\} = \{m_0 \ \text{ or } \ \varphi_0, h_{\mathrm{s}}(\mathbf{x}), g_{\mathrm{s}}\,(\mathbf{x})\}\,, \tag{4.16}$$

in case of a uniform cross–section or a sandwich cross–section, respectively, Fig. 4.1.

Apart from the order of the theory, which may include or not the coupling between the membrane and bending effects (cf. Ganczarski and Skrzypek [80]), both states, membrane and bending, may additionally be coupled by the boundary conditions. Generally, such a coupling can be described by a function $\mathcal{F}$ which depends on the excitation parameters:

$$\mathcal{F}(n_0, m_0, \Delta_0, \varphi_0) = 0, \tag{4.17}$$

where n_0 is the initial prestressing force, Δ_0 the initial (membrane) distortion, m_0 the initial prestressing moment and φ_0 the initial angular distortion (curvature).

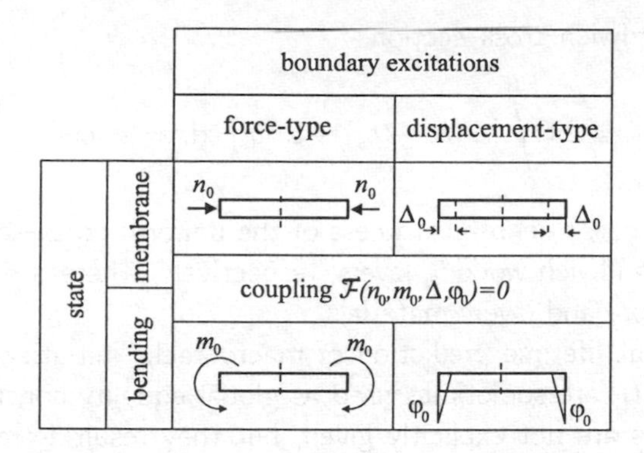

Figure 4.1: Boundary excitations in axisymmetric plates

4.2. Example: Optimal design of rotating disks in creep–damage conditions

A. State and evolution equations of rotationally–symmetric annular disks of variable thickness in coupled creep–orthotropic damage conditions

Consider an annular disk of variable thickness $h\,(r)$ and radii a and b, clamped at the inner edge, subjected to steady rotation with angular velocity ω and in–plane loading due to peripheral tension or prestressing (cf. Skrzypek and Egner [100]). Assuming plane stress $\sigma_z = 0$, creep incompressibility $\varepsilon^{\mathrm{c}}_m = 0$, the Mises–type flow rule associated with the multiaxial, Kachanov–Hayhurst time – hardening law and the orthotropic damage growth rule (Table 3.1, partly coupled approach) the following displacement–type state and damage evolution equations are obtained:

$$\begin{gathered}\frac{\mathrm{d}^2 u}{\mathrm{d}r^2} + \left(\frac{1}{h}\frac{\mathrm{d}h}{\mathrm{d}r} + \frac{1}{r}\right)\frac{\mathrm{d}u}{\mathrm{d}r} + \left(\frac{\nu}{h}\frac{\mathrm{d}h}{\mathrm{d}r} - \frac{1}{r}\right)\frac{u}{r} = \frac{f}{r} + \frac{\mathrm{d}g}{\mathrm{d}r} + \frac{1}{h}\frac{\mathrm{d}h}{\mathrm{d}r}g - kr, \\ \mathrm{d}\varepsilon^{\mathrm{c}}_{r/\vartheta} = \frac{\sigma_{\mathrm{eq}}^{m-1}}{(1-D)^m}\left(\sigma_{r/\vartheta} - \frac{\sigma_{\vartheta/r}}{2}\right)\dot{\mathrm{f}}(t)\mathrm{d}t, \quad \mathrm{d}\varepsilon^{\mathrm{c}}_z = -\left(\mathrm{d}\varepsilon^{\mathrm{c}}_r + \mathrm{d}\varepsilon^{\mathrm{c}}_\vartheta\right), \\ \mathrm{d}D = C\left\langle\frac{\chi(\boldsymbol{\sigma})}{1-D}\right\rangle^r \mathrm{d}t, \qquad \chi(\boldsymbol{\sigma}) = \delta\sigma_1 + (1-\delta)\sigma_{\mathrm{eq}},\end{gathered} \tag{4.18}$$

where $f = (1-\nu)(\varepsilon^{\mathrm{c}}_r - \varepsilon^{\mathrm{c}}_\vartheta), \quad g = \varepsilon^{\mathrm{c}}_r + \nu\varepsilon^{\mathrm{c}}_\vartheta, \quad k = \dfrac{1-\nu^2}{E}\rho\omega^2.$

The dimensionless form of governing equations for disk of a variable thickness at $t = 0$ (initial condition), and $t > 0$ (creep–damage conditions) is furnished as:

$$\begin{gathered}R^2\frac{\mathrm{d}^2 U}{\mathrm{d}R^2} + \left(\frac{R^2}{H}\frac{\mathrm{d}H}{\mathrm{d}R} + R\right)\frac{\mathrm{d}U}{\mathrm{d}R} + \left(\frac{\nu R}{H}\frac{\mathrm{d}H}{\mathrm{d}R} - 1\right)U = -KR^3 \quad (t=0), \\ R^2\frac{\mathrm{d}^2 \dot{U}}{\mathrm{d}R^2} + \left(\frac{R^2}{H}\frac{\mathrm{d}H}{\mathrm{d}R} + R\right)\frac{\mathrm{d}\dot{U}}{\mathrm{d}R} + \left(\frac{\nu R}{H}\frac{\mathrm{d}H}{\mathrm{d}R} - 1\right)\dot{U} = \dot{F}R + R^2\frac{\mathrm{d}\dot{G}}{\mathrm{d}R} + \frac{1}{H}\frac{\mathrm{d}H}{\mathrm{d}R}\dot{G} \quad (t>0), \\ \dot{E}^{\mathrm{c}}_r = \frac{S_{\mathrm{eq}}^{m-1}}{(1-D)^m}\left(S_r - \frac{S_\vartheta}{2}\right), \qquad \dot{E}^{\mathrm{c}}_\vartheta = \frac{S_{\mathrm{eq}}^{m-1}}{(1-D)^m}\left(S_\vartheta - \frac{S_r}{2}\right), \\ \mathrm{d}D = C\sigma_0^n t_{\mathrm{I}}\left\langle\frac{\chi\,(\boldsymbol{\sigma})}{1-D}\right\rangle^r \mathrm{d}\bar{t},\end{gathered} \tag{4.19}$$

where

$$\bar{t} = \frac{t}{t_{\mathrm{I}}}, \quad \frac{1}{t_{\mathrm{I}}(t)} = E\sigma_0^{m-1}\dot{\mathrm{f}}(t), \quad \dot{F} = (1-\nu)\left(\dot{E}_r^{\mathrm{c}} - \dot{E}_\vartheta^{\mathrm{c}}\right), \quad \dot{G} = \dot{E}_r^{\mathrm{c}} + \nu\dot{E}_\vartheta^{\mathrm{c}},$$

$$\varepsilon_0 = \frac{\sigma_0}{E}, \quad U = \frac{u}{a\varepsilon_0}, \quad R = \frac{r}{a}, \quad F = \frac{f}{\varepsilon_0}, \quad G = \frac{g}{\varepsilon_0}, \quad K = \frac{ka^2}{\varepsilon_0},$$

$$S_r = \frac{\sigma_r}{\sigma_0}, \quad S_\vartheta = \frac{\sigma_\vartheta}{\sigma_0}, \quad S_{\mathrm{eq}} = \frac{\sigma_{\mathrm{eq}}}{\sigma_0}, \quad E_r^{\mathrm{c}} = \frac{\varepsilon_r^{\mathrm{c}}}{\varepsilon_0}, \quad E_\vartheta^{\mathrm{c}} = \frac{\varepsilon_\vartheta^{\mathrm{c}}}{\varepsilon_0}, \tag{4.20}$$

$$T = tE\sigma_0^{m-1}\dot{\mathrm{f}}(t), \quad H = \frac{h}{a}, \quad P = \frac{p}{\sigma_0}, \quad R_0 = \frac{b}{a} = 5.$$

B. Boundary value problems

Two boundary value problems of disk optimization by use of CDM based FDM approach are solved when either the local optimality criterion of the UODS (4.2) or the global optimality criterion of maximum lifetime $t_{\mathrm{I}} \to \max$, are solved (Fig. 4.2).

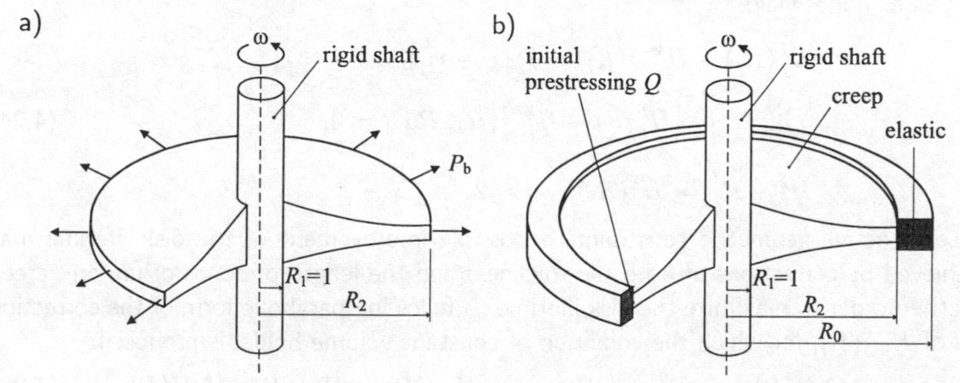

Figure 4.2. Schematics of clamped annular disks of a variable thickness subjected to: a) steady rotation and radial tension , b) steady rotation under initial prestressing constraint

Example A: Boundary conditions for clamped annular disk subjected to steady rotation and tension are:

$$\begin{aligned} &U(1) = 0, \quad H(R_2)S_r(R_2) = H_0P_b \quad (\bar{t} = 0), \\ &\dot{U}(1) = 0, \quad \dot{S}_r(R_2) = 0 \qquad\qquad\quad (\bar{t} > 0). \end{aligned} \tag{4.21}$$

Example B: Initial boundary and continuity conditions for clamped annular disk subjected to rotation under prestressing conditions are:

$$\begin{aligned} &\left.\begin{aligned} &U(1) = 0, && H(R_2)S_r(R_2) = -H_0Q \\ &S_r^{\mathrm{ring}}(R_0) = 0 && S_r^{\mathrm{ring}}(R_2) = -Q, \end{aligned}\right\} \quad (\bar{t} = 0), \\ &\left.\begin{aligned} &\dot{U}(1) = 0, && H(R_0)\dot{S}_r(R_2)\mathrm{d}\bar{t} = H_0\mathrm{d}S_r^{\mathrm{ring}}(R_2) \\ &S_r^{\mathrm{ring}}(R_0) = 0, && \dot{U}(R_2)\mathrm{d}\bar{t} = \mathrm{d}U^{\mathrm{ring}}(R_2) \end{aligned}\right\} \quad (\bar{t} > 0). \end{aligned} \tag{4.22}$$

The calculations are done for the following data:
$E = 1.77 \times 10^5$ MPa, $\nu = 0.3$, $a = 0.02$ m, $b = 5a$, $h_0 = 0.004$ m, $\sigma_0 = 118$ MPa, $P_b = 0.1$, $\rho = 7.9 \times 10^3$ kg/m^3, $\omega = 100$ s^{-1}(A) or 240 s^{-1}(B), $C = 2.13 \times 10^{-42}$ Pa^{-r}s^{-1}, $m = 5.6$, $r = 3.9$, $\delta = 0.5$ (B) or 1.0 (A).

C. Two step optimization approach by use of FDM

For the prescribed loading conditions the optimal distribution of disk thickness $H(R)$ and the initial prestressing Q that maximize the time of failure initiation t_{I} (first macro-cracks) under the condition of constant loads and volume and the additional geometric constraints are sought:

$$t_{\mathrm{I}}[H(R),Q] = \max; \quad \omega, P_{a,b} = \mathrm{const}, \qquad V = \mathrm{const}, \qquad H_{\mathrm{inf}} \leq H(R) \leq H_{\mathrm{sup}}. \tag{4.23}$$

As the first step of optimal design the shape of a disk of Uniform Creep Strength (UCS) is determined, $H_{\mathrm{ucs}}(R)$. Next, the corrections of the disk thickness are imposed with the constant volume condition applied, as long as the lifetime is maximized. The nodal correction of disk thickness is assumed to be proportional to the power function of the residual value of the nodal continuity function $\psi_j = 1 - D_j$ at rupture time $\bar{t}_{\mathrm{R}}$. Hence, the thickness correction rule, the constant volume condition (for the corrections) and the continuity of thickness at nodes yield:

$$\begin{aligned} &H_j^k(R_j) - H_j^{k-1}(R_j) = \mathcal{P}\left(\bar{\psi} - \psi_j\right)^{\alpha}, \quad j = 1,\ldots,N, \\ &\textstyle\sum_{j=1}^{N-1} \int_{R_j}^{R_{j+1}} \left[H_j^k(R) - H_j^{k-1}(R)\right] R\mathrm{d}R = 0, \\ &H_{I-1}^k(R_j) = H_I^k(R_j), \quad j = 2,\ldots,N-1. \end{aligned} \tag{4.24}$$

In case of active geometric constraints a possible improvement of the disk lifetime may be achieved by corrections of both the thickness and the length of zones of uniform creep strength in order to maximize the disk lifetime. A following parabolic form of the correction terms of $H_{\mathrm{ucs}}(R)$, for which the condition of constant volume holds, is proposed:

$$\Delta H(R) = a_1R^2 + a_2R + a_3, \quad H_{\mathrm{opt}}(R) = H_{\mathrm{ucs}}(R) + \Delta H(R); \tag{4.25}$$

$$\int_{R_1}^{R_2} \left(a_1R^2 + a_2R + a_3\right) R\mathrm{d}R = 0, \tag{4.26}$$

hence, two parameters are free to be optimized.

D. Results

Comparison of the rotating non–prestressed disks of the UES versus the UCS, when a continuous or a jump–like variable thickness is allowed, is shown in Fig. 4.3a.

The proposed design method allows to significantly elongate the disk lifetime when compared to the disk of a constant thickness, as summarized in Table 4.2

Lifetime of a rotating disk can essentially be improved when the initial prestressing is imposed on the disk by the elastic ring (Fig. 4.2b). Both, the initial prestressing Q and the distribution of thickness $H(R)$ are subjected to optimization. The initial prestressing considered as an additional decision variable causes non–uniqueness of the uniform creep strength solution. Hence, the optimization procedure consists in two steps: first, effect of prestressing force on the lifetime of disk of constant thickness is examined, second, additional lifetime elongation is met when thickness optimization $H_i(R)$ under the constant

Table 4.2. Comparison of lifetimes of disks of the uniform initial strength $H_{\rm ues}$ and the uniform creep strength $H_{\rm ucs}$ (Fig. 4.3)

	Lifetime			
	Constant thickness	Uniform Elastic	Uniform Creep Strength	
			jump-like	variable
Temp [K]	$\bar{t}_{\rm I}^{\rm ref}$	Strength	variable thickness	thickness
773	74.2	$2.9\bar{t}_{\rm I}^{\rm ref}$	$3.5\bar{t}_{\rm I}^{\rm ref}$	$3.9\bar{t}_{\rm I}^{\rm ref}$
873	79.0	$2.0\bar{t}_{\rm I}^{\rm ref}$	$2.7\bar{t}_{\rm I}^{\rm ref}$	$2.9\bar{t}_{\rm I}^{\rm ref}$

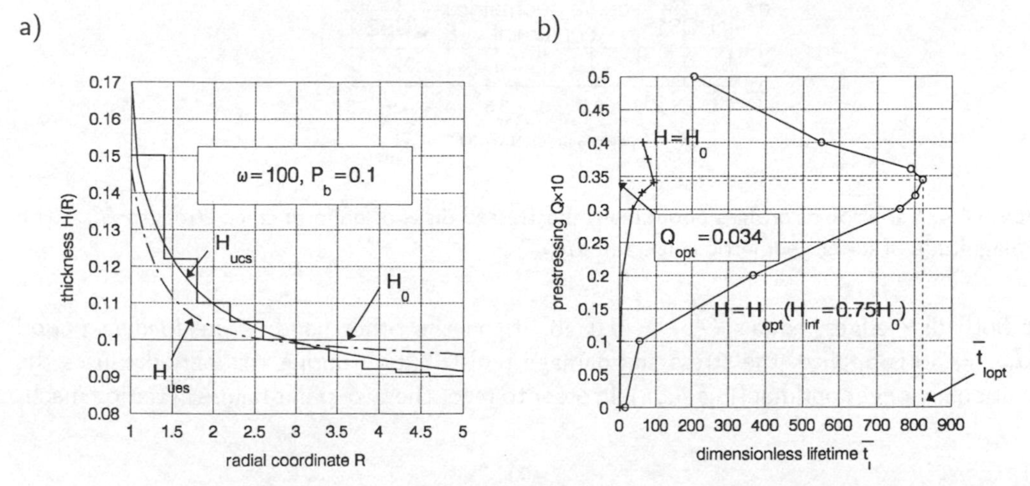

Figure 4.3. a) Disk of UCS versus disk of UES and disk of a jump–like variable thickness, b) Effect of initial prestressing Q on time to macrocrack initiation $\bar{t}_I$ of a disk of uniform creep strength $H_{\rm ucs}$ versus disk of a constant thickness H_0 (after Skrzypek and Egner [100])

volume (4.24) and given prestressing force Q_i is performed, to eventually yield the optimal shape and the corresponding prestressing force for which the lifetime is maximized $t_{\rm I}\left[H_{\rm ucs}\left(R\right), Q_{\rm opt}\right] = \max$, (Fig. 4.3b).

Optimal profiles of optimally prestressed disks without or with lower geometric constraint imposed are shown in Fig. 4.4.

The pre–loading evolution of damage due to prestressing may essentially influence the net–lifetime after the loading is imposed. It was examined by Egner and Skrzypek [101] in a simpler case of disk of constant thickness. Depending on the duration of the pre–loading period and the magnitude of the prestressing force, it may occur that the first macrocracks appear during the pre–loading stage. Therefore, a constraint must be imposed on both the magnitude of the prestressing force (mainly with respect to stability) and the duration of the pre–loading prestressing period. When the pre–loading damage caused by prestressing is disregarded or, in other words, the rotation is instantaneously applied at the instant of prestressing $\bar{t}_{00} = \bar{t}_0$, the optimal prestressing, for which damage simultaneously occurs

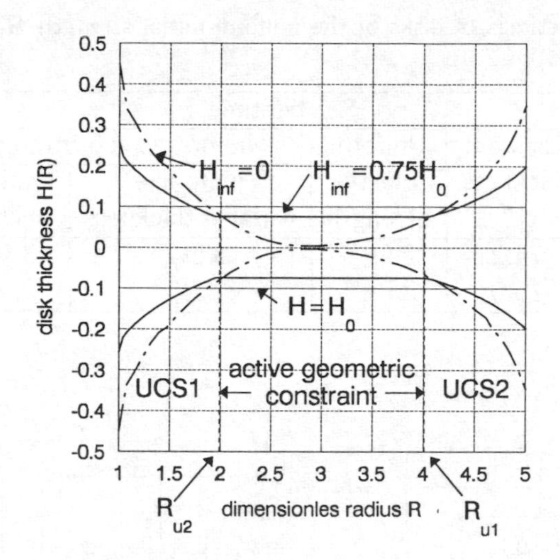

Figure 4.4. Variation of profiles of optimally prestressed disks of uniform creep strength $H_{\rm ucs}$ with a magnitude of lower geometric constraint $H_{\rm inf}$

at both disk edges, equals $Q^0_{\rm opt} = 0.0340$. If, on the other hand, a pre–loading period $\Delta\bar{t}_{\rm pre} = 50$ is applied, the stress and damage redistribution during this period causes the solution no longer optimal (Fig. 4.5a). In order to meet the two–point failure initiation mech-

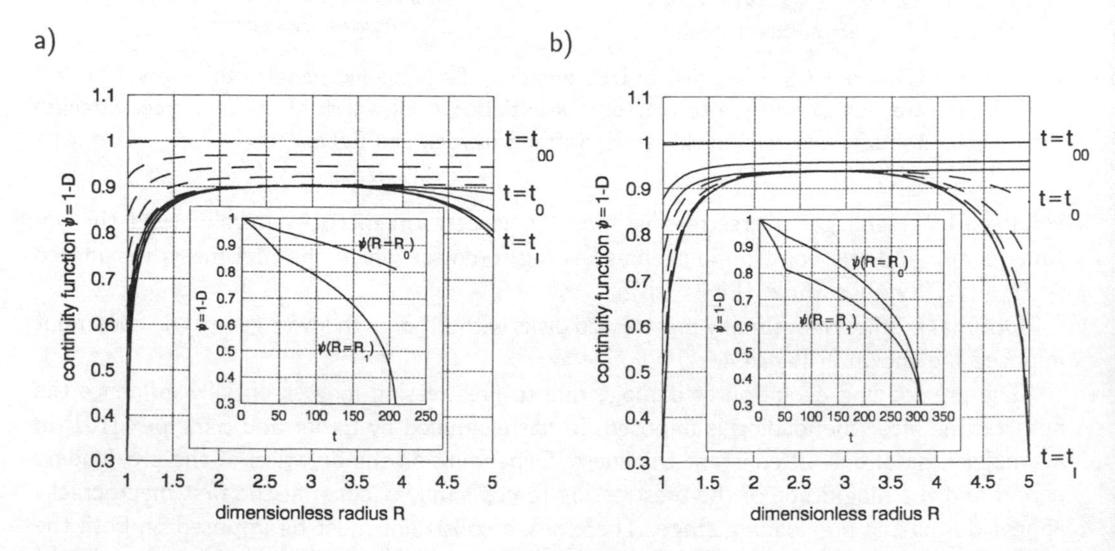

Figure 4.5. Evolution of the continuity ψ, during the pre–loading $\Delta t_{\rm pre} = t_0 - t_{00} = 50$ and the working loading $\Delta t_{\rm wor} = t_{\rm I} - t_0$ periods: a) non–optimal prestressing $Q^0_{\rm opt} = 0.0340$ b) optimal initial prestressing $Q^{50}_{\rm opt} = 0.0359$

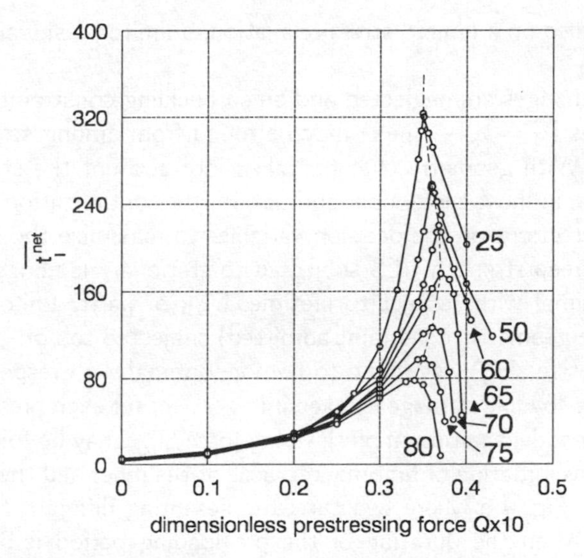

Figure 4.6. A family of net–lifetimes versus initial prestressing period (after Egner and Skrzypek [101])

anism that corresponds to the pre–loading period considered, the higher initial prestressing $Q^{50}_{\rm opt} = 0.0359$ has to be imposed (Fig. 4.5b). In the range of two–point optimization (simultaneous initiation of macrocracks at both disk edges) the optimum prestressing force increases when the duration of the pre–loading period increases. Additionally, the sharp net–lifetime corners in Fig. 4.6 are here highly sensitive to a possible imperfection of the optimal initial prestressing force. On the other hand, when the prestressing period is longer than approximately 15% of the net–lifetime, the maximum net–lifetime corresponds to the initiation of first macrocracks at a single disk edge (inner) with the analytical maximum in Fig. 4.6. In this range the optimum prestressing force decreases when the duration of the pre–loading period increases.

E. Conclusions: Non/optimality of disks of uniform creep strength with respect to lifetime

When elastic structures are designed for either the minimum weight or the maximum load under the constraint of strength, the structures of uniform elastic strength UES, also called the fully stressed design, is in most cases optimal. In general, when static indeterminacy of structure or when geometry changes are taken into account, the condition of uniform strength is neither a necessary nor a sufficient condition of optimality. Hence, the fully stressed design method is, in most cases, a first step towards the exact optimal solution when more rigorous optimization approaches are used (Gallagher [82]).

When optimization of structures under creep–damage conditions is formulated, the minimum weight (volume) or the maximum load remains the typical design objective, whereas constraints may be imposed not only on the strength (failure), stiffness and stability as in the elastic case, but also on a limited stress relaxation, a limited residual displacement or a given lifetime t_{I} or t_{II}.

When geometry changes are neglected and creep buckling constraints are not involved, the optimal structures ($t_{\mathrm{I}} = t_{\mathrm{R}} \rightarrow \max$) may be found from among structures of uniform creep strength UCS. With geometry changes taken into account the structure of uniform creep strength UCS is, in most cases, non–optimal. Further optimization may be performed by superimposing corrections on the decision variables to maximize the lifetime.

Disk of uniform creep strength UCS subjected to stationary loadings (not prestrained) was found to be optimal with respect to lifetime. Disk of partly uniform creep strength PUCS (zone of active geometric constraint admitted) subjected to non–stationary loadings (due to the initial prestressing) was found to be non–optimal with respect to lifetime.

When effect of pre–loading damage is taken into account for each prescribed pre–loading period $\Delta t^i_{\mathrm{pre}}$ the independent optimum prestressing force Q^i_{opt} may be found. It corresponds usually to simultaneous initiation of first macrocracks at the inner and the outer fibres of the disk (switch points in Fig. 4.6 where two curves representing different failure mechanisms intersect). However, when the duration of the pre–loading period is sufficiently long, it may happen that the initial damage during pre–loading at the inner fibre is rapid enough to reach the maximum net–lifetime without the switching effect. In this case the optimal prestressing is determined by the smooth extremum point on the curve $t^{\mathrm{net}}_I(Q)$ as shown in Fig. 4.6.

4.3. Example: Creep–damage and failure analysis of axisymmetric disks with shear effect included

A. Basic mechanical state and evolution equations of plane stress–rotationally symmetric creep–damage process

Geometrically linear theory is applied when small total strains are decomposed into the elastic and creep portions: $\varepsilon_{r/\vartheta} = \varepsilon^{\mathrm{e}}_{r/\vartheta} + \varepsilon^{\mathrm{c}}_{r/\vartheta}$, $\gamma_{r\vartheta} = \gamma^{\mathrm{e}}_{r\vartheta} + \gamma^{\mathrm{c}}_{r\vartheta}$. Elastic part is governed by the Hooke law (isotropic) and none additional effect of the material deterioration on elastic properties is taken into account. Creep part is governed by either the isotropic or the modified orthotropic flow theory and by the time–hardening hypothesis applied to current principal stress axes (Table 3.2). The Murakami–Ohno damage tensor $\mathbf{D}$ and its objective time–derivative $\overset{\nabla}{\mathbf{D}}$ are used, (3.11–3.12). Non–objective damage rate $\dot{\mathbf{D}}$ is governed by the orthotropic void growth rule (3.13) applied to current principal directions of stresses when rotation of principal axes of damage and stress tensors on creep–damage cumulation process in disks is accounted for, as shown in Fig. 4.7. Plane stress–rotationally symmetric problems are considered.

Reduced displacement–type mechanical state equations for coupled plane–stress creep–damage problem are (Skrzypek and Ganczarski [1]):

$$\left.\begin{aligned} \frac{\mathrm{d}^2 u_r}{\mathrm{d}r^2} + \frac{1}{r}\frac{\mathrm{d}u_r}{\mathrm{d}r} - \frac{u_r}{r^2} &= -\frac{1-\nu^2}{E}\rho\omega_0^2 r \\ \frac{\mathrm{d}^2 u_\vartheta}{\mathrm{d}r^2} + \frac{1}{r}\frac{\mathrm{d}u_\vartheta}{\mathrm{d}r} - \frac{u_\vartheta}{r^2} &= \frac{\rho\varepsilon_0}{G} r \end{aligned}\right\} (t=0),$$

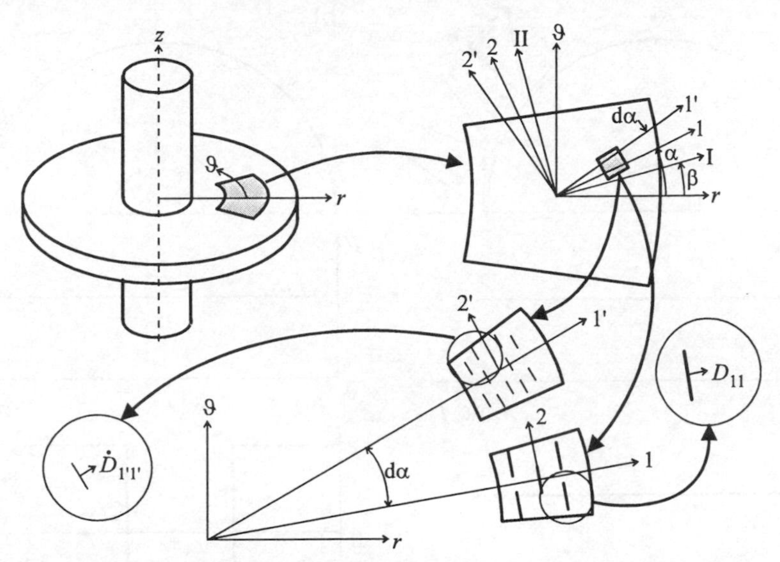

Figure 4.7. Schematic creep damage accumulation of several orthotropic increments coincided with current principal stress axes (1, 2) and resulting rotation of current principal damage axes (I, II) in case of a disk (after Skrzypek and Ganczarski [1])

$$\left.\begin{aligned}&\frac{\mathrm{d}^2\dot{u}_r}{\mathrm{d}r^2}+\frac{1}{r}\frac{\mathrm{d}\dot{u}_r}{\mathrm{d}r}-\frac{\dot{u}_r}{r^2}=\frac{\mathrm{d}(2\dot{g}-\dot{f})}{\mathrm{d}r}+\frac{\dot{f}}{r}-2\frac{(1-\nu^2)}{E}\rho\omega(t)\varepsilon(t)r\\&\frac{\mathrm{d}^2\dot{u}_\vartheta}{\mathrm{d}r^2}+\frac{1}{r}\frac{\mathrm{d}\dot{u}_\vartheta}{\mathrm{d}r}-\frac{\dot{u}_\vartheta}{r^2}=\frac{\mathrm{d}\dot{\gamma}^{\mathrm{c}}_{r\vartheta}}{\mathrm{d}r}+2\frac{\dot{\gamma}^{\mathrm{c}}_{r\vartheta}}{r}+\frac{\rho\dot{\varepsilon}(t)}{G}r\end{aligned}\right\}(t>0),$$

$$\begin{aligned}&\dot{\varepsilon}^{\mathrm{c}}_{1/2}=\frac{(\widetilde{\sigma}_{\mathrm{eq}})^m}{\sigma_{\mathrm{eq}}}\left(\sigma_{1/2}-\frac{\sigma_{2/1}}{2}\right)\dot{\mathsf{f}}(t)\qquad\text{(partly coupled)},\\&\dot{\varepsilon}^{\mathrm{c}}_{1/2}=(\widetilde{\sigma}_{\mathrm{eq}})^{m-1}\left(\frac{\sigma_{1/2}}{1-D_{1/2}}-\frac{1}{2}\frac{\sigma_{2/1}}{1-D_{2/1}}\right)\dot{\mathsf{f}}(t)\qquad\text{(fully coupled)},\end{aligned}\tag{4.27}$$

where, due to creep incompressibility $\dot{\varepsilon}^{\mathrm{c}}_3=-\dot{\varepsilon}^{\mathrm{c}}_1-\dot{\varepsilon}^{\mathrm{c}}_2$, the auxiliary symbols $\dot{f}$, $\dot{g}$ denote $\dot{f}=(1-\nu)\left(\dot{\varepsilon}^{\mathrm{c}}_r-\dot{\varepsilon}^{\mathrm{c}}_\vartheta\right)$, $\dot{g}=\dot{\varepsilon}^{\mathrm{c}}_r+\nu\dot{\varepsilon}^{\mathrm{c}}_\vartheta$, and $\varepsilon(t)=\dot{\omega}(t)$ is the angular acceleration, whereas

$$\sigma_{\mathrm{eq}}=\sqrt{\sigma_1^2+\sigma_2^2-\sigma_1\sigma_2},\qquad\widetilde{\sigma}_{\mathrm{eq}}=\sqrt{\left(\frac{\sigma_1}{1-D_1}\right)^2+\left(\frac{\sigma_2}{1-D_2}\right)^2-\frac{\sigma_1}{(1-D_1)}\frac{\sigma_2}{(1-D_2)}}.\tag{4.28}$$

The plane stress 2D objective derivative of the damage tensor takes the form

$$\begin{aligned}\begin{bmatrix}\overset{\nabla}{D}_{1'1'}&\overset{\nabla}{D}_{1'2'}\\\overset{\nabla}{D}_{2'1'}&\overset{\nabla}{D}_{2'2'}\end{bmatrix}&=\begin{bmatrix}\dot{D}_{11}&0\\0&\dot{D}_{22}\end{bmatrix}-\begin{bmatrix}D_{11}&D_{21}\\D_{12}&D_{22}\end{bmatrix}\begin{bmatrix}0&\mathrm{d}\alpha\\-\mathrm{d}\alpha&0\end{bmatrix}\\&\quad+\begin{bmatrix}0&-\mathrm{d}\alpha\\\mathrm{d}\alpha&0\end{bmatrix}\begin{bmatrix}D_{11}&D_{12}\\D_{21}&D_{22}\end{bmatrix},\end{aligned}\tag{4.29}$$

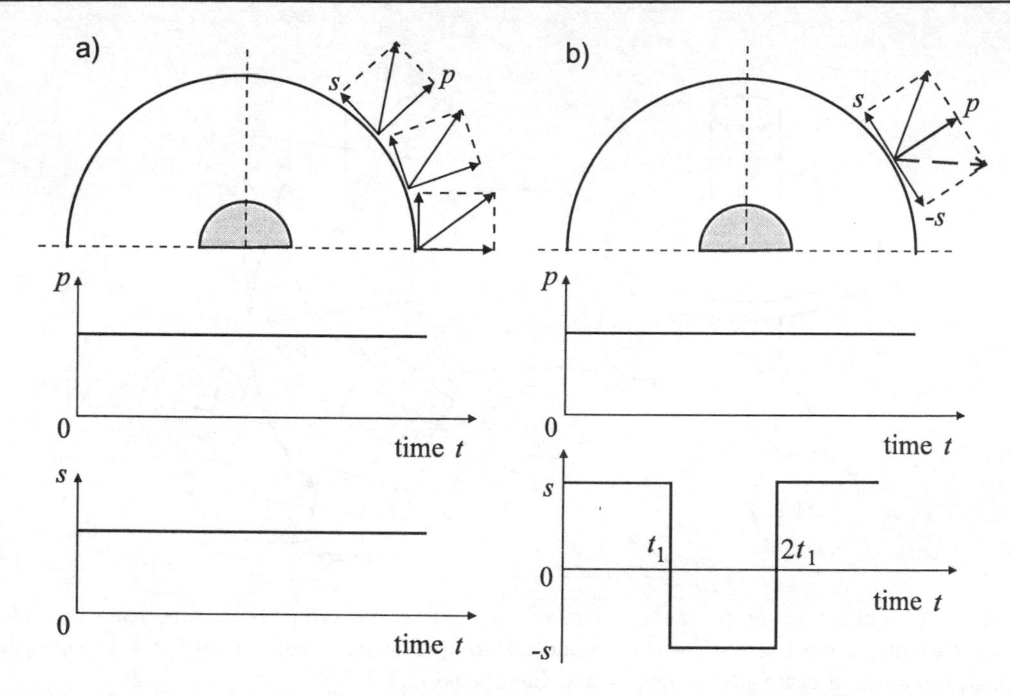

Figure 4.8. Layout of a circular disk subject to creep damage under: a) steady peripheral tension p and torsion s; b) steady peripheral tension p and multiple reverse torsion $\pm s$

where the non–objective damage rates associated with the current principal stress axes are

$$\dot{D}_{11} = C_1 \left\langle \frac{\sigma_{11}}{1 - D_{11}} \right\rangle^{r_1}, \qquad \dot{D}_{22} = C_2 \left\langle \frac{\sigma_{22}}{1 - D_{22}} \right\rangle^{r_2}. \tag{4.30}$$

When the objective damage rate tensor $\overset{\nabla}{D}_{IJ}$ (4.29) is transformed from current principal directions of the stress tensor (IJ) to the sampling coordinates (ij) $\overset{\nabla}{D}_{ij}$, the new damage tensor $D_{ij}(t + \Delta t)$ is achieved

$$\overset{\nabla}{D}_{IJ} \overset{\text{transf.}}{\longrightarrow} \overset{\nabla}{D}_{ij}, \qquad D_{ij}(t + \Delta t) = D_{ij}(t) + \overset{\nabla}{D}_{ij}(t)\Delta t, \tag{4.31}$$

and the creep strain rates (4.27) referring to the global coordinate system (ij) are:

$$\dot{\varepsilon}^{\mathrm{c}}_{r/\vartheta} = \frac{\dot{\varepsilon}^{\mathrm{c}}_1 + \dot{\varepsilon}^{\mathrm{c}}_2}{2} \pm \frac{\dot{\varepsilon}^{\mathrm{c}}_1 - \dot{\varepsilon}^{\mathrm{c}}_2}{2} \cos 2\alpha, \qquad \dot{\gamma}^{\mathrm{c}}_{r\vartheta} = (\dot{\varepsilon}^{\mathrm{c}}_1 - \dot{\varepsilon}^{\mathrm{c}}_2) \sin 2\alpha. \tag{4.32}$$

B. Boundary value problems

Consider annular disk of constant thickness h and radii a and b clamped at the inner edge, subjected to: (a) steady tension and torsion or (b) steady tension and multiple reverse torsion (Fig. 4.8).

The corresponding boundary conditions hold:

$$\left.\begin{array}{l} u^{\mathrm{e}}_r(a) = 0, \;\; u^{\mathrm{e}}_\vartheta(a) = 0, \\ \sigma^{\mathrm{e}}_r(b) = p, \;\; \tau^{\mathrm{e}}_{r\vartheta}(b) = s \end{array}\right\} \; (t = 0), \qquad \left.\begin{array}{l} \dot{u}_r(a) = 0, \;\; \dot{u}_\vartheta(a) = 0 \\ \dot{\sigma}_r(b) = 0, \;\; \dot{\tau}_{r\vartheta}(b) = 0 \end{array}\right\} \; (t > 0), \tag{4.33}$$

and

$$\left.\begin{array}{ll} u_r^{\rm e}(a)=0, & u_\vartheta^{\rm e}(a)=0, \\ \sigma_r^{\rm e}(b)=p, & \tau_{r\vartheta}^{\rm e}(b)=\pm s \end{array}\right\} \ (t=0), \qquad \left.\begin{array}{ll} \dot{u}_r(a)=0, & \dot{u}_\vartheta(a)=0 \\ \dot{\sigma}_r(b)=0, & \dot{\tau}_{r\vartheta}(b)=0 \end{array}\right\} \ (t>0). \quad (4.34)$$

Computer simulation is done for the ASTM 321 stainless steel of the following data at 500 C: $E = 180$ GPa, $\sigma_{0.2} = 120$ MPa, $\nu = 0.3$, $p = 0.2\times\sigma_{0.2}$, $s = p/20$ (a) or $s = \pm p/20$ (b).

C. Results

In the case of steady loading conditions first macrocrack appears at the inner edge ($r = a$), and the damage zone is limited to the closest neighbourhood of the fixed disk edge (Fig. 4.9a). Evolution of the principal directions of stress (α) and damage (β) depends on the partly or the fully coupled creep–damage approach (4.27). When the partly coupled approach is used, the isotropic flow rule requires similarity of stress and creep strain rate deviators s_{ij} and $e_{ij}^{\rm c}$. The angles α and β slightly differ from one another during the primary creep (Fig. 4.9c). When the fully coupled approach is used, the orthotropic flow rule requires similarity of the effective stress and creep strain rate deviators, $\widetilde{s}_{ij}$ and $\dot{e}_{ij}^{\rm c}$. Changes of both angles are slower than in the previously discussed case. However, lifetime prediction is hardly 0.3% longer when compared to the lifetime obtained for isotropic (partly coupled)

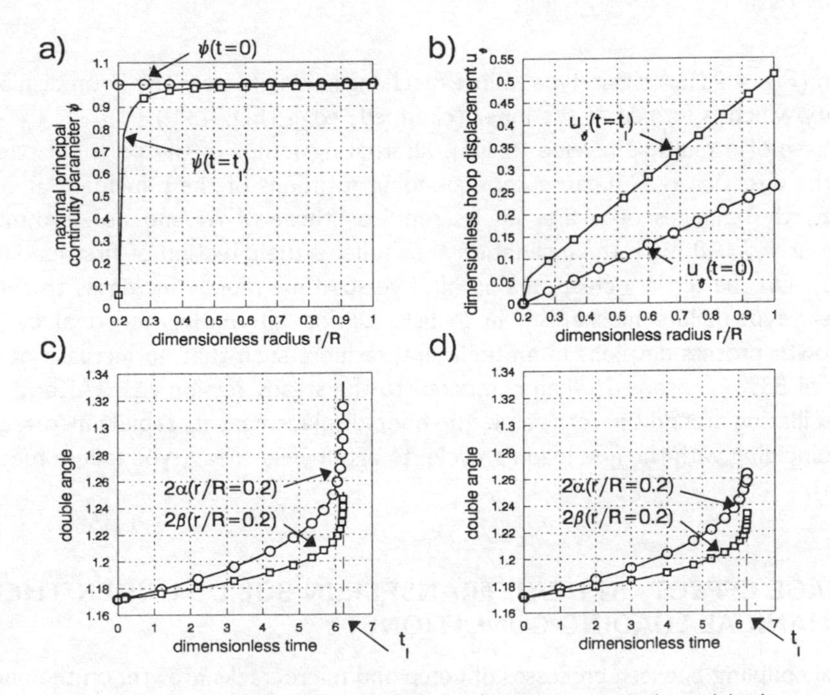

Figure 4.9. Disk under steady tension and torsion: a) damage evolution with time to failure, b) formation of the hoop displacement discontinuity, c) and d) rotation of principal stress axes α and principal damage axes β in case of scalar and tensorial creep-damage coupling, respectively (at inner disk edge $r/R = 0.2$) (after Skrzypek and Ganczarski [1])

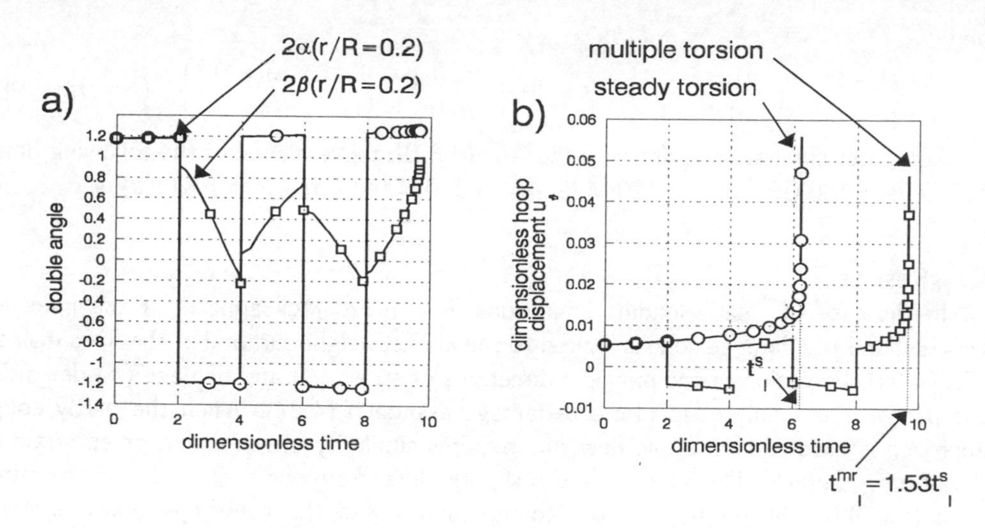

Figure 4.10. Disk subject to steady tension and multiple reverse torsion: a) principal damage axes rotation β resulting from principal stress axes oscilation α with time to failure, b) formation of bilateral hoop displacement discontinuity with time to failure in case of multiple reverse torsion, versus steady torsion

formulation (Fig. 4.9d). A shear–type failure mechanism results from the hoop displacement discontinuity which is formed at the inner (clamped) edge (Fig. 4.9b).

In the case of a multiple reverse torsion, alternating jumps of the principal stress axes α around the direction $\alpha = 0$ cause corresponding rotations of the principal damage axes β. However, the changes of β are not as rapid as those of α, and non–symmetrically oscillate around $\beta = 0$, with the inclination that follows the direction of first loading cycle (Fig. 4.10a). On the tertiary creep, a slop of β versus time rapidly increases to eventually yield a shear–type failure mechanism in a disk. During alternating torsional cycles, the damage growth process develops in an unilateral fashion, such that an increase of lifetime by amount of 53% is observed, when compared to the steady torsion case. After a number of cycles oscillating around the zero value, the hoop displacement u_ϑ rapidly increases in the direction coinciding with the first loading cycle, to again yield shear–type failure mechanism (Fig. 4.10b).

5. DAMAGE EFFECT ON HEAT TRANSFER IN SOLIDS UNDER THERMO–MECHANICAL LOADING CONDITIONS

Bilateral coupling between processes of creep and microcracks growth, on the one hand, and redistribution of temperature field, on the other hand, is considered. To this order the thermal conductivity function of a virgin material $\lambda_0(x, y, z)$ in the heat transfer equation is replaced by a new time–dependent, in general anisotropic, tensor function $\widetilde{\mathbf{L}}\,(x, y, z, t)$ that characterizes the thermal properties of a partly damaged material. Hence, when the isotropic damage is assumed as governed by a single scalar variable $D(\mathbf{x}, t)$, for non–steady states with internal heat sources $\dot{q}_v$, the extended heat conduction equation takes a form:

$$\begin{aligned}&\frac{\partial}{\partial x}\left\{\widetilde{\lambda}\left[\mathbf{x}, D(\mathbf{x}, t)\right] \frac{\partial T(\mathbf{x}, t)}{\partial x}\right\}+\frac{\partial}{\partial y}\left\{\widetilde{\lambda}\left[\mathbf{x}, D(\mathbf{x}, t)\right] \frac{\partial T(\mathbf{x}, t)}{\partial y}\right\} \\ &+\frac{\partial}{\partial z}\left\{\widetilde{\lambda}\left[\mathbf{x}, D(\mathbf{x}, t)\right] \frac{\partial T(\mathbf{x}, t)}{\partial z}\right\}+\frac{\partial q_v}{\partial t}=c_v \varrho \frac{\partial T(\mathbf{x}, t)}{\partial t},\end{aligned} \tag{5.1}$$

or

$$\operatorname{div}\left\{\widetilde{\lambda}\left[\mathbf{x}, D(\mathbf{x}, t)\right] \mathbf{grad} T\right\}+\dot{q}_v=c_v \varrho \dot{T}. \tag{5.2}$$

Damage effect on heat conduction is described here by the single scalar variable $\widetilde{\lambda}\left[\mathbf{x}, D(\mathbf{x}, t)\right]$. The mass density and the specific heat ϱ and c_v, are assumed to be time–independent constants.

5.1. Thermo–damage coupling models

A. Direct extension of the equation of thermal conductivity for damaged material

A linear heat conductivity drop with damage was assumed by Ganczarski and Skrzypek [105, 106]:

$$\widetilde{\lambda}\left[\mathbf{x}, D(\mathbf{x}, t)\right]=\lambda_0(\mathbf{x})\left[1-D(\mathbf{x}, t)\right], \tag{5.3}$$

where $\lambda_0(\mathbf{x})$ denotes a non–homogeneous, in general, distribution of the thermal conductivity in a virgin (undamaged) material. In this model, when material is locally completely damaged $D(\mathbf{x}, t) \equiv 1$, the thermal conductivity coefficient drops at this point to zero $\widetilde{\lambda}(D=1)=0$ and, hence, local heat conductivity through the completely damaged surface element must also drop to zero. In other words, the fully damaged RVE is assumed to be free from any kind of stress and unable to support heat conduction. When the energy based equivalence principle is used the other formula, instead of the linear conductivity drop, is derived from the state potential (cf. [23])

$$\widetilde{\lambda}\left[\mathbf{x}, D(\mathbf{x}, t)\right]=\lambda_0(\mathbf{x})\left[1-D(\mathbf{x}, t)\right]^{1/2}. \tag{5.4}$$

B. Concept of a combined change of thermal conduction and radiation through partly damaged material

An extension of the model A accounts for an additional term of heat flow through the damaged surface element portion by application of the Stefan–Boltzmann radiation law:

$$\begin{aligned}&\frac{\partial}{\partial x}\left\{\lambda_0(\mathbf{x})\left[1-D(\mathbf{x}, t)\right] \frac{\partial T(\mathbf{x}, t)}{\partial x}-\sigma \epsilon_0(\mathbf{x}, t) D(\mathbf{x}, t) T^4(\mathbf{x}, t)\right\} \\ &+\frac{\partial}{\partial y}\left\{\lambda_0(\mathbf{x})\left[1-D(\mathbf{x}, t)\right] \frac{\partial T(\mathbf{x}, t)}{\partial y}-\sigma \epsilon_0(\mathbf{x}, t) D(\mathbf{x}, t) T^4(\mathbf{x}, t)\right\} \\ &+\frac{\partial}{\partial z}\left\{\lambda_0(\mathbf{x})\left[1-D(\mathbf{x}, t)\right] \frac{\partial T(\mathbf{x}, t)}{\partial z}-\sigma \epsilon_0(\mathbf{x}, t) D(\mathbf{x}, t) T^4(\mathbf{x}, t)\right\} \\ &+\frac{\partial q_v}{\partial t}=c_v \varrho \frac{\partial T(\mathbf{x}, t)}{\partial t}.\end{aligned} \tag{5.5}$$

In the model B under consideration a combined conduction/radiation mechanism allows for a heat flux even though the damage at a point reaches level 1 (due to radiation across the microcracks).

C. Concept of the equivalent coefficient of thermal conductivity for a combined conduction/radiation heat flux through partly damaged material

A combined heat flux is characterized by the substitutive coefficient of thermal conductivity, modified in order to take into account a simultaneous influence of the conductivity $\widetilde{\lambda}$ through the RVE at the point $\mathbf{x}$, and the radiation from $\mathbf{x}$ to $\mathbf{x} + \mathrm{d}\mathbf{x}$. The equivalent coefficient of thermal conductivity λ^{eq} is expressed, therefore, by the equation:

$$\lambda^{\mathrm{eq}}\left[\mathbf{x}, D\left(\mathbf{x},t\right), T\left(\mathbf{x},t\right)\right] = \widetilde{\lambda}\left[\mathbf{x}, D\left(\mathbf{x},t\right)\right] + \Delta\widetilde{\lambda}^{\mathrm{rad}}\left[\mathrm{d}\mathbf{x}, D\left(\mathbf{x},t\right), T\left(\mathbf{x},t\right)\right]. \tag{5.6}$$

Consequently, the equation of heat transfer (5.1) may be extended to the following form:

$$\begin{aligned}&\frac{\partial}{\partial x}\left\{\lambda^{\mathrm{eq}}\left[\mathbf{x}, D\left(\mathbf{x},t\right), T\left(\mathbf{x},t\right)\right]\frac{\partial T\left(\mathbf{x},t\right)}{\partial x}\right\} + \frac{\partial}{\partial y}\left\{\lambda^{\mathrm{eq}}\left[\mathbf{x}, D\left(\mathbf{x},t\right), T\left(\mathbf{x},t\right)\right]\frac{\partial T\left(\mathbf{x},t\right)}{\partial y}\right\}\\ &+\frac{\partial}{\partial z}\left\{\lambda^{\mathrm{eq}}\left[\mathbf{x}, D\left(\mathbf{x},t\right), T\left(\mathbf{x},t\right)\right]\frac{\partial T\left(\mathbf{x},t\right)}{\partial z}\right\} + \frac{\partial q_v}{\partial t} = c_v \varrho \frac{\partial T\left(\mathbf{x},t\right)}{\partial t}.\end{aligned} \tag{5.7}$$

The equivalent (substitutive) coefficient of thermal conductivity λ^{eq} is obtained by equating the heat flux due to conduction and radiation through partly damaged cross section and the heat flux due to the corresponding conduction through the fictitious pseudo–undamaged cross section (Fig. 5.1). The specific formulae for $\Delta\widetilde{\lambda}^{\mathrm{rad}}$ will be discussed in the following section.

D. Axisymmetric heat flow in cylinders or disks under thermo–creep–damage coupling conditions

The heat transfer equations in partly damaged materials in the case of axisymmetric heat flow in cylinders or disks of constant thickness are:

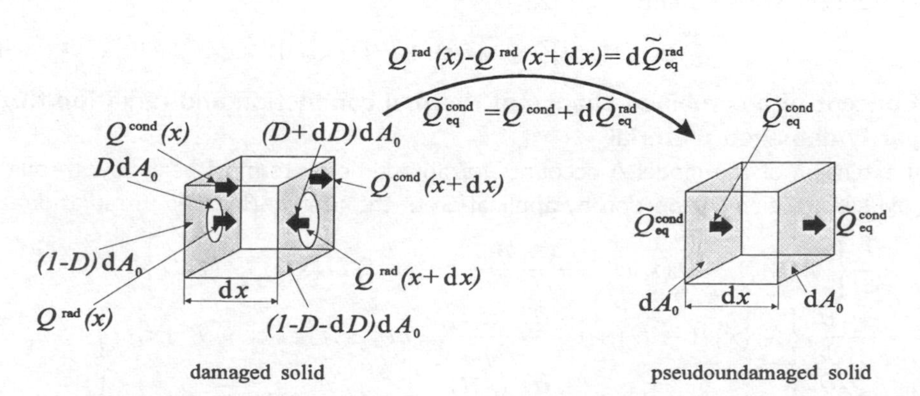

Figure 5.1: One–dimensional concept of the equivalent coefficient of thermal conductivity

Model A

$$\frac{1}{r}\frac{\mathrm{d}}{\mathrm{d}r}\left\{r\left[\lambda_0\left(1-D\left(r,t\right)\right)\frac{\mathrm{d}T\left(r,t\right)}{\mathrm{d}r}\right]\right\}+\dot{q}_v=c_v\varrho\dot{T}. \tag{5.8}$$

Model B

$$\frac{1}{r}\frac{\mathrm{d}}{\mathrm{d}r}\left\{r\left[\lambda_0\left(1-D\left(r,t\right)\right)\frac{\mathrm{d}T\left(r,t\right)}{\mathrm{d}r}-\sigma\epsilon_0 D\left(r,t\right)T^4\left(r,t\right)\right]\right\}+\dot{q}_v=c_v\varrho\dot{T}. \tag{5.9}$$

Model C

$$\begin{aligned}
&\frac{1}{r}\frac{\mathrm{d}}{\mathrm{d}r}\left\{r\left[\lambda^{\mathrm{eq}}\left(r,t,T\right)\frac{\mathrm{d}T\left(r,t\right)}{\mathrm{d}r}\right]\right\}+\dot{q}_v=c_v\varrho\dot{T},\\
&\lambda^{\mathrm{eq}}\left(r,t,T\right)=\widetilde{\lambda}\left(r,t\right)+\sigma\epsilon_0\left[4D+\frac{\mathrm{d}D/\mathrm{d}r}{\mathrm{d}T/\mathrm{d}r}T\right]T^3\Delta,\\
&\widetilde{\lambda}\left(r,t\right)=\lambda_0\left(1-D\left(r,t\right)\right)\quad\text{or}\quad\widetilde{\lambda}\left(r,t\right)=\lambda_0\left(1-D\left(r,t\right)\right)^{1/2}.
\end{aligned} \tag{5.10}$$

Extension of (5.9) and (5.10) to the rotationally symmetric disks of a variable thickness $h\left(r\right)$ yields (Δ stands for the average defect size):

Model B

$$\begin{aligned}
&\frac{1}{r}\frac{\mathrm{d}}{\mathrm{d}r}\left\{r\left[\lambda_0\psi\left(r,t\right)\frac{\mathrm{d}T\left(r,t\right)}{\mathrm{d}r}-\sigma\epsilon_0 D\left(r,t\right)T^4\left(r,t\right)\right]\right\}\\
&+\frac{1}{h}\frac{\mathrm{d}h}{\mathrm{d}r}\left[\lambda_0\psi\left(r,t\right)\frac{\mathrm{d}T\left(r,t\right)}{\mathrm{d}r}-\sigma\epsilon_0 D\left(r,t\right)T^4\left(r,t\right)\right]+\dot{q}_v=c_v\varrho\dot{T}.
\end{aligned} \tag{5.11}$$

Model C

$$\begin{aligned}
&\frac{1}{r}\frac{\mathrm{d}}{\mathrm{d}r}\left(r\lambda^{\mathrm{eq}}\frac{\mathrm{d}T\left(r,t\right)}{\mathrm{d}r}\right)+\frac{1}{h}\frac{\mathrm{d}h}{\mathrm{d}r}\lambda^{\mathrm{eq}}\frac{\mathrm{d}T\left(r,t\right)}{\mathrm{d}r}+\dot{q}_v=c_v\varrho\dot{T},\\
&\lambda^{\mathrm{eq}}\left(r,t,T\right)=\lambda_0\left(1-D\left(r,t\right)\right)+\sigma\epsilon_0\left[4D+\frac{\mathrm{d}D/\mathrm{d}r}{\mathrm{d}T/\mathrm{d}r}T\right]T^3\Delta.
\end{aligned} \tag{5.12}$$

5.2. Mechanical state equations of axisymmetric deformation under unsteady temperature field

Applying the geometrically linear theory of small displacements and decomposing the total strains into elastic, creep, and thermal parts $\boldsymbol{\varepsilon}=\boldsymbol{\varepsilon}^{\mathrm{e}}+\boldsymbol{\varepsilon}^{\mathrm{c}}+\boldsymbol{\varepsilon}^{\mathrm{th}}$, the problem may be expressed by the system of displacement and rate equations as follows:

$$\begin{aligned}
&\frac{\mathrm{d}^2u}{\mathrm{d}r^2}+\frac{1}{r}\frac{\mathrm{d}u}{\mathrm{d}r}-\frac{u}{r^2}=h(1+\nu)\alpha\frac{\mathrm{d}T}{\mathrm{d}r} && (t=0),\\
&\frac{\mathrm{d}^2\dot{u}}{\mathrm{d}r^2}+\frac{1}{r}\frac{\mathrm{d}\dot{u}}{\mathrm{d}r}-\frac{\dot{u}}{r^2}=\frac{\dot{f}}{r}+\frac{\mathrm{d}\dot{g}}{\mathrm{d}r}+h(1+\nu)\alpha\frac{\mathrm{d}\dot{T}}{\mathrm{d}r} && (t>0).
\end{aligned} \tag{5.13}$$

The solution of (5.13) yields the following formulae for displacements and stresses

$$\begin{aligned}
u &= \frac{1}{2}C_1 r+\frac{1}{2}C_2+h\left(1+\nu\right)I_0,\\
\sigma_r &= \frac{E}{1+\nu}\left(\frac{1}{2}kC_1-\frac{1}{r^2}C_2\right)-h\frac{E}{r}I_0,
\end{aligned} \tag{5.14}$$

$$\sigma_\vartheta = \frac{E}{1+\nu}\left(\frac{1}{2}kC_1 + \frac{1}{r^2}C_2\right) + h\frac{E}{r}\left(I_0 - \alpha r T\right),$$

and their rates ($t > 0$):

$$\begin{aligned}
\dot{u} &= \frac{1}{2}C_3 r + \frac{1}{2}C_4 + \dot{I}_1 + \dot{I}_2 + h\,(1+\nu)\,\dot{I}_0,\\
\dot{\sigma}_r &= \frac{E}{1+\nu}\left(\frac{1}{2}kC_3 - \frac{1}{r^2}C_4 + \frac{k\dot{I}_1 - \dot{I}_2}{r}\right) - h\frac{E}{r}\dot{I}_0,\\
\dot{\sigma}_\vartheta &= \frac{E}{1+\nu}\left(\frac{1}{2}kC_3 + \frac{1}{r^2}C_4 + \frac{k\dot{I}_1 + \dot{I}_2}{r} - \dot{g} + \dot{\varepsilon}_r^{\mathrm{c}} - \dot{\varepsilon}_\vartheta^{\mathrm{c}}\right) + h\frac{E}{r}\dot{I}_0 - hE\alpha\dot{T}.
\end{aligned} \tag{5.15}$$

In the case of plane strain and creep incompressibility additionally holds:

$$\begin{aligned}
\dot{\varepsilon}_z &= \frac{1}{E}\left[\dot{\sigma}_z - \nu\left(\dot{\sigma}_r + \dot{\sigma}_\vartheta\right)\right] + \alpha\dot{T} - \dot{\varepsilon}_r^{\mathrm{c}} - \dot{\varepsilon}_\vartheta^{\mathrm{c}} = 0,\\
\sigma_z &= \frac{E}{1+\nu}\nu k C_1 - hE\alpha T,\\
\dot{\sigma}_z &= \frac{E}{1+\nu}\left(\nu k C_3 + 2\nu k\frac{\dot{I}_1}{r} + h\dot{\varepsilon}_r^{\mathrm{c}} + \dot{\varepsilon}_\vartheta^{\mathrm{c}}\right) - hE\alpha\dot{T},
\end{aligned} \tag{5.16}$$

where auxiliary functions are defined in Table 5.1, and I_0, $\dot{I}_0$, $\dot{I}_1$, $\dot{I}_2$ denote

$$I_0 = \frac{\alpha}{r}\int_0^r T\xi\,\mathrm{d}\xi\,, \quad \dot{I}_0 = \frac{\alpha}{r}\int_0^r \dot{T}\xi\,\mathrm{d}\xi\,, \quad \dot{I}_1 = \frac{r}{2}\int_0^r \frac{\dot{f}}{\xi}\,\mathrm{d}\xi\,, \quad \dot{I}_2 = \frac{1}{2r}\int_0^r \left(2\dot{g} - \dot{f}\right)\xi\,\mathrm{d}\xi\,. \tag{5.17}$$

Table 5.1. Auxiliary functions for basic rotationally symmetric deformation under unsteady temperature field (after Ganczarski and Skrzypek [105])

Quantity	Plane stress	Plane strain
$\dot{f}$	$(1-\nu)\left(\dot{\varepsilon}_r^{\mathrm{c}} - \dot{\varepsilon}_\vartheta^{\mathrm{c}}\right)$	$\frac{1-2\nu}{1-\nu}\left(\dot{\varepsilon}_r^{\mathrm{c}} - \dot{\varepsilon}_\vartheta^{\mathrm{c}}\right)$
$\dot{g}$	$\dot{\varepsilon}_r^{\mathrm{c}} + \nu\dot{\varepsilon}_\vartheta^{\mathrm{c}}$	$\frac{1-2\nu}{1-\nu}\dot{\varepsilon}_r^{\mathrm{c}}$
h	1	$\frac{1}{1-\nu}$
k	$\frac{1+\nu}{1-\nu}$	$\frac{1}{1-2\nu}$

Axisymmetric constitutive coupled creep–damage equations are formulated as:

i) Plane stress, creep incompressibility conditions (orthotropic damage)

$$\dot{D}_k = C_k(T)\,\langle\widetilde{\sigma}_k\rangle^{r_k(T)}, \tag{5.18}$$

- partly coupled case

$$\dot{\varepsilon}^{\mathrm{c}}_{r/\vartheta} = \frac{(\widetilde{\sigma}_{\mathrm{eq}})^{m(T)}}{\sigma_{\mathrm{eq}}}\left(\sigma_{r/\vartheta} - \frac{\sigma_{\vartheta/r}}{2}\right)\dot{\mathsf{f}}(t), \qquad \dot{\varepsilon}^{\mathrm{c}}_z = -\left(\dot{\varepsilon}^{\mathrm{c}}_r + \dot{\varepsilon}^{\mathrm{c}}_\vartheta\right), \tag{5.19}$$

- fully coupled case

$$\dot{\varepsilon}^{\mathrm{c}}_{r/\vartheta} = (\widetilde{\sigma}_{\mathrm{eq}})^{m(T)-1}\left[\frac{\sigma_{r/\vartheta}}{1-D_{r/\vartheta}} - \frac{\sigma_{\vartheta/r}}{2(1-D_{\vartheta/r})}\right]\dot{\mathsf{f}}(t), \qquad \dot{\varepsilon}^{\mathrm{c}}_z = -\left(\dot{\varepsilon}^{\mathrm{c}}_r + \dot{\varepsilon}^{\mathrm{c}}_\vartheta\right). \tag{5.20}$$

ii) Plane strain conditions (isotropic damage)

- partly coupled case $\dot{D} = C(T)\left\langle\dfrac{\sigma_I}{1-D}\right\rangle^{r(T)}$,

$$\dot{\varepsilon}^{\mathrm{c}}_{r/\vartheta} = \frac{(\sigma_{\mathrm{eq}})^{m(T)-1}}{(1-D)^{m(T)}}\left(\sigma_{r/\vartheta} - \frac{\sigma_{\vartheta/r}+\sigma_z}{2}\right)\dot{\mathsf{f}}(t); \tag{5.21}$$

where $\widetilde{\sigma}_k = \dfrac{\sigma_k}{1-D_k}$ and $\widetilde{\sigma}_{\mathrm{eq}} = \dfrac{\sigma_{\mathrm{eq}}}{1-D}$.

5.3. Example: Cylinder subject to a non–stationary radial temperature field in–plane strain conditions

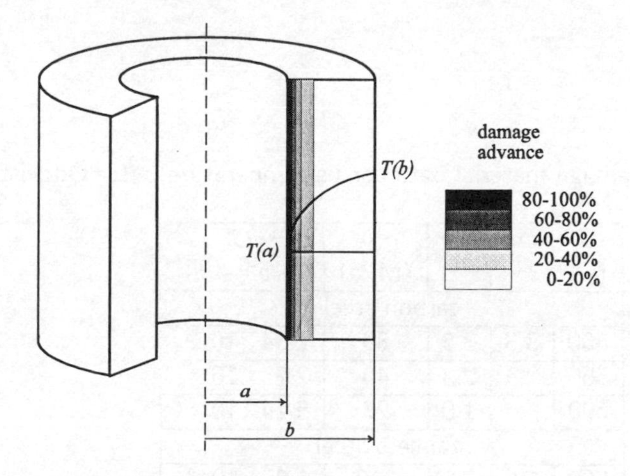

Figure 5.2. Long cylindrical thick–walled tube subject to a non–stationary radial temperature gradient under plane strain conditions (after Skrzypek and Ganczarski[50])

Consider a cylinder of inner and outer radii a and b, under the plane strain conditions, subject to a non–stationary radial temperature field (Fig. 5.2). Stresses and their rates satisfy (5.14–5.16) and boundary conditions for stress and temperature field hold:

$$\begin{array}{lll} \sigma_r(a) = 0 & \sigma_r(b) = 0 & (t = 0)\,, \\ \dot{\sigma}_r(a) = 0 & \dot{\sigma}_r(b) = 0 & (t > 0)\,, \end{array} \tag{5.22}$$

$$T(a) = T_a, \qquad T(b) = T_b. \tag{5.23}$$

Numerical simulation is done for two structural materials, carbon steel and stainless steel, the thermo–mechanical properties of which are (Holman [107]):

1. Carbon steel (rolled, 0.40 Mn, 0.25 Si, 0.12 C, normalized, annealed at 850C): $E =$ 150 GPa, $\sigma_{0.2} =$ 120 MPa, $\nu = 0.3$, $\alpha = 1.4 \times 10^{-5}$ K^{-1}, $\lambda_0 =$ 43 Wm^{-1}K^{-1}, $\varepsilon_0 = 0.60$.

2. ASTM 321 stainless steel (rolled, 18 Cr, 0.45 Si, 0.4 Mn, 0.1 C, Ti, Nb, stabilized, austenitic, annealed at 1070 C): $E =$ 150 GPa, $\sigma_{0.2} =$ 120 MPa, $\nu = 0.3$, $\alpha =$ 1.85×10^{-5} K^{-1}, $\lambda_0 =$ 23 Wm^{-1}K^{-1}, $\varepsilon_0 = 0.50$.

In both cases radii ratio $a/b = 0.5$, and the Stefan–Boltzmann constant $\sigma = 5.609 \times 10^{-8}$ Wm^{-1}K^{-4} are assumed. Temperature affected creep/damage material constants are listed in Table 5.2.

Table 5.2: Creep/damage material data versus temperature, after Odqvist [108]

T (C)	m	r	σ^5_{CB} (MPa)	C (Pa^{-r}s^{-1})
carbon steel				
500	3.3	3.5	80	1.34×10^{-37}
550	2.5	2.3	40	2.75×10^{-27}
600	—	1.0	27	5.14×10^{-17}
stainless steel				
500	5.6	3.9	210	1.98×10^{-42}
600	4.5	3.1	100	1.07×10^{-34}
650	4.0	2.8	60	1.21×10^{-31}
700	3.5	2.5	38	8.91×10^{-29}

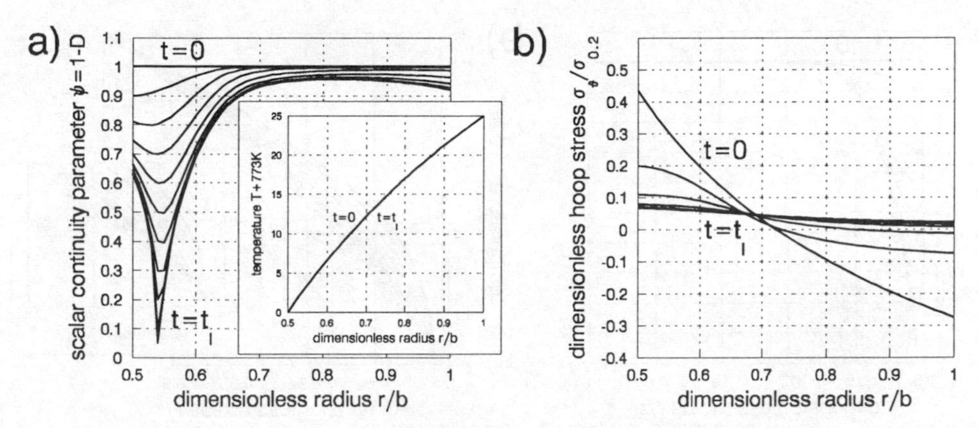

Figure 5.3. A tube subjected to creep under stationary temperature field (effect of thermo–damage coupling disregarded): a) scalar continuity parameter evolution, b) hoop stress redistribution (after Skrzypek and Ganczarski [50])

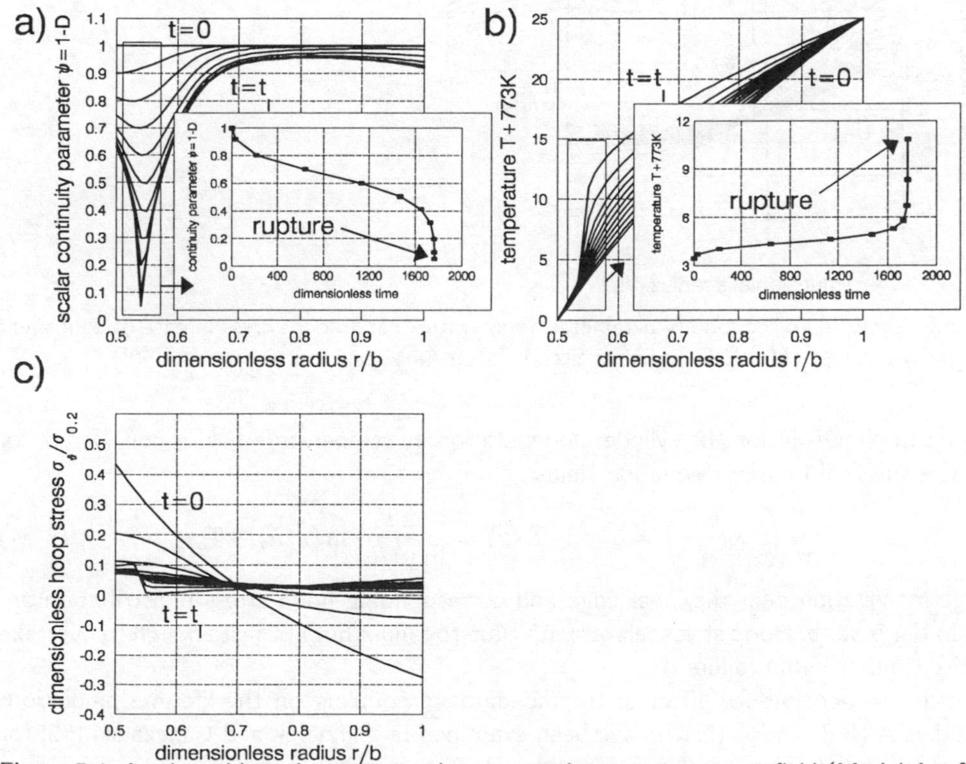

Figure 5.4. A tube subjected to creep under non–stationary temperature field (Model A: effect of thermo–damage coupling incorporated, pure conductivity $\epsilon = 0$): a) scalar continuity parameter evolution, b) temperature field evolution resulting from damage accumulation, c) hoop stress redistribution (after Skrzypek and Ganczarski [50])

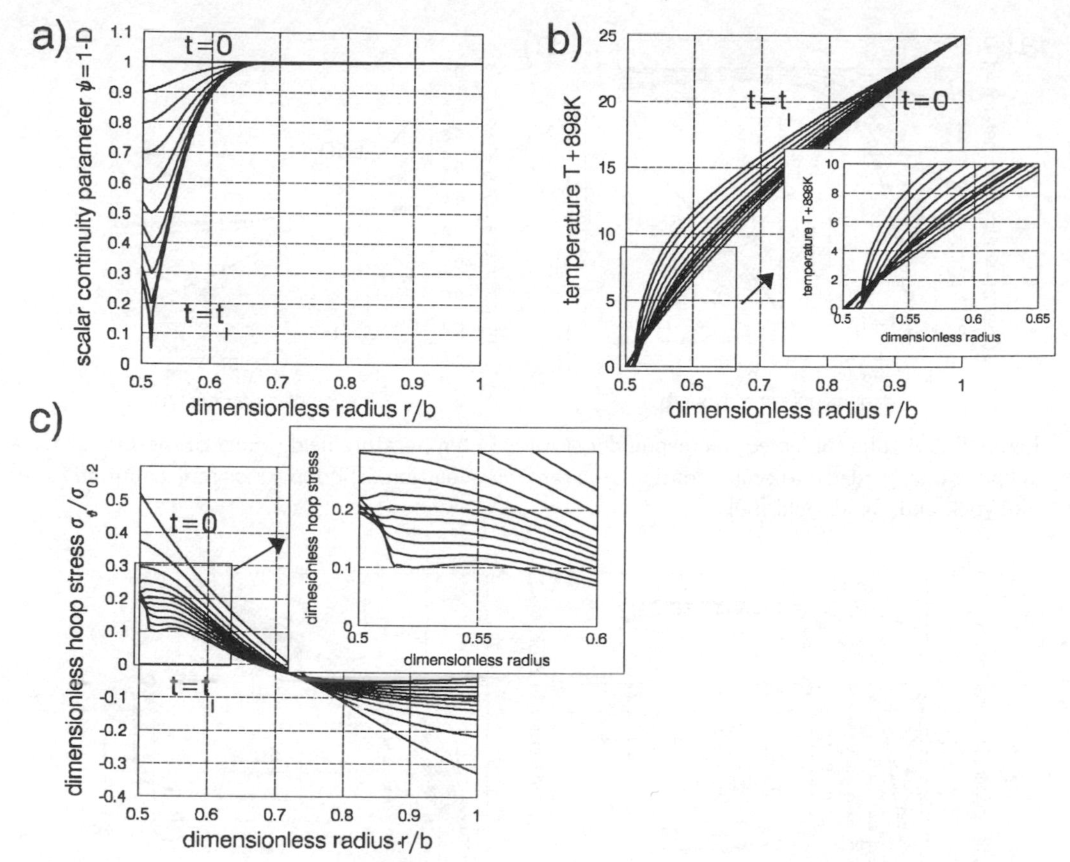

Figure 5.5. Evolution of continuity parameter, temperature, and hoop stress in case of equivalent conductivity concept (Model C, Stainless Steel), (after Skrzypek and Ganczarski [50])

As a sampling solution the cylinder under stationary temperature field is considered, for which the classical Fourier's equation yields:

$$\frac{1}{r}\frac{\mathrm{d}}{\mathrm{d}r}\left(r\lambda_0\frac{\mathrm{d}T}{\mathrm{d}r}\right)=0, \qquad T\left(r\right)=\frac{\Delta T}{\ln\left(a/b\right)}\ln\left(r/a\right)+T_a. \tag{5.24}$$

Damage localization near the inner edge and corresponding hoop stress redistribution are shown in Fig.5.3a, b. Hoop stress relaxes with time to failure but not fast enough to overtake the finite time of initial failure $t_{\mathrm{I}}^{\mathrm{c}}$.

In order to account for effect of thermo–damage coupling on the lifetime prediction, two models A (5.8) and C (5.10) have been examined by Skrzypek and Ganczarski [50] for carbon steel and stainless steel cylinder material, respectively (Table 5.2).

When the linear conductivity drop is used (model A), if time increases the conduction across the damaged surface in the failure zone asymptotically approaches zero, and, as a result, both temperature and hoop stress jumps are produced in this zone (Fig. 5.4b, c). An accompanying stress relaxation is not fast enough to prevent the structure from fracture. The corresponding lifetime $t_{\mathrm{I}}^{\mathrm{A}}$ is finite and shorter by amount of 15% when compared to the sampling solution if thermo–damage coupling is disregarded, $t_{\mathrm{I}}^{\mathrm{A}} = 0.85 t_{\mathrm{I}}^{\mathrm{c}}$.

In case of a more advanced model C (5.10), when equivalent coefficient of thermal conductivity accounts for both the conduction and radiation terms in stainless steel cylinder under thermo–damage conditions, damage localization close to the inner cylinder edge results in a corresponding redistribution of the temperature field which is, however, not as quick as in the case of model A, since even the completely damaged surface is capable of head transferring due to the residual radiation equivalent conduction (Fig. 5.5b). A hoop stress redistribution that eventually yields a discontinuity formed across the completely damaged surface, results in 22% shorter lifetime prediction when compared to the corresponding sampling solution, $t_{\mathrm{I}}^{\mathrm{c}} = 0.78 t_{\mathrm{I}}^{\mathrm{ss}}$.

5.4. Example: Optimal design of rotationally–symmetric disks in thermo–damage coupling conditions

A. Assumptions

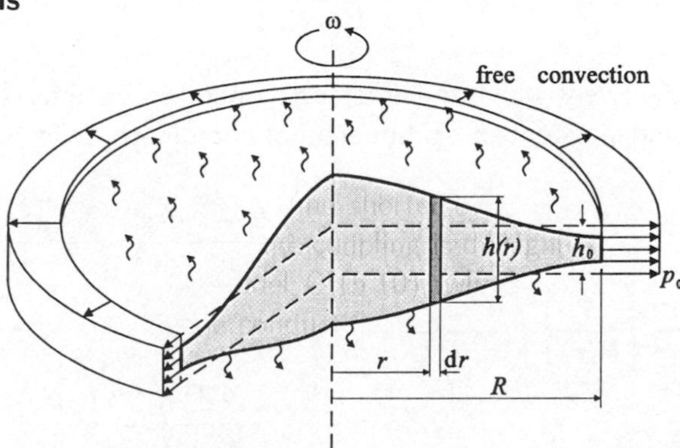

Figure 5.6. Rotating disk of variable thickness (vs. constant thickness disk of the same volume) stretched at periphery and cooled through faces (after Ganczarski and Skrzypek [98])

Thin axisymmetric disk of a variable thickness under plane stress conditions is considered (cf. Fig. 5.6). The geometrically linear theory of small displacements and the additive decomposition of strains are applied: $\varepsilon = \varepsilon^{\mathrm{e}} + \varepsilon^{\mathrm{c}} + \varepsilon^{\mathrm{th}}$. The fully coupled orthotropic creep–damage approach is used (Table 3.1). The coupled thermo–damage problem is solved (model C (5.12)) by the use of equivalent conduction concept. 1D non–stationary temperature field is assumed $T\left[r, D\left(r,t\right)\right]$ (temperature homogenization through the disk thickness) but only quasi–static changes of temperature are allowed ($\dot{T} = 0$). 1D volumetric inner heat sources are assumed

$$q_v = q_v \left\{ h\left(r\right), \frac{\mathrm{d}h\left(r\right)}{\mathrm{d}r}, T\left[r, D\left(r,t\right)\right] \right\}. \tag{5.25}$$

A uniform constant temperature along the periphery $T_0 = \mathrm{const}$ and the constant temperature of the cooling fluid stream (through the disk faces) $T_\infty = \mathrm{const}$ are assumed as the thermal boundary conditions. The body forces due to steady rotation with angular velocity

ω and the uniform peripheral tension, in a sense of constant force per unit length of the periphery p_0, are assumed as the mechanical loadings.

B. General equations of the mechanical state

A general mixed approach originally derived for a plate under membrane–bending state by Ganczarski and Skrzypek [80], where the equation of membrane state is written by use of the Airy function F whereas the equation of bending state is expressed by the appropriate deflection function, is reduced to the case of a disk. Hence, $n_r = (F'/r) + U$, $n_\vartheta = F'' + U$ where a potential of body forces is defined as $U' = -\varrho\omega^2 rh$, whereas symbol prime stands for the derivative with respect to r. Finally, the fundamental mechanical state equations are furnished:

$$\begin{aligned}
&\mathcal{F}[F] + (1-\nu)\mathcal{B}(r)\nabla^2\left[\frac{U}{\mathcal{B}(r)}\right] + (1-\nu^2)\mathcal{B}(r)\alpha\nabla^2 T = 0,\\
&\mathcal{F}[\dot{F}] + (1-\nu^2)\mathcal{B}(r)\alpha\nabla^2\dot{T} + \mathcal{B}(r)\nabla^2\left[\frac{\dot{n}^{\mathrm{c}}_\vartheta - \nu\dot{n}^{\mathrm{c}}_r}{\mathcal{B}(r)}\right] + \frac{1+\nu}{r}\mathcal{B}(r)\frac{\mathrm{d}}{\mathrm{d}r}\left[\frac{\dot{n}^{\mathrm{c}}_\vartheta - \dot{n}^{\mathrm{c}}_r}{\mathcal{B}(r)}\right] = 0,
\end{aligned} \tag{5.26}$$

for $t = 0$ and $t > 0$ respectively, where the differential operator $\mathcal{F}[...]$ as well as the auxiliary operators ∇^2 and ∇^4, independent of circumferential coordinate, take the form ($k = 0$):

$$\begin{aligned}
\mathcal{F}[...] &= \nabla^4 + \mathcal{B}(r)\frac{\mathrm{d}}{\mathrm{d}r}\left[\frac{1}{\mathcal{B}(r)}\right]\left(2\frac{\mathrm{d}^3...}{\mathrm{d}r^3} + \frac{2-\nu}{r}\frac{\mathrm{d}^2...}{\mathrm{d}r^2} - \frac{1}{r^2}\frac{\mathrm{d}...}{\mathrm{d}r}\right)\\
&+\mathcal{B}(r)\frac{\mathrm{d}^2}{\mathrm{d}r^2}\left[\frac{1}{\mathcal{B}(r)}\right]\left(\frac{\mathrm{d}^2...}{\mathrm{d}r^2} - \frac{\nu}{r}\frac{\mathrm{d}...}{\mathrm{d}r}\right),\\
\nabla^2... &= \frac{\mathrm{d}^2...}{\mathrm{d}r^2} + \frac{1}{r}\frac{\mathrm{d}...}{\mathrm{d}r}, \qquad \nabla^4... = \frac{\mathrm{d}^4...}{\mathrm{d}r^4} + \frac{2}{r}\frac{\mathrm{d}^3...}{\mathrm{d}r^3} - \frac{1}{r^2}\frac{\mathrm{d}^2...}{\mathrm{d}r^2} + \frac{1}{r^3}\frac{\mathrm{d}...}{\mathrm{d}r}.
\end{aligned} \tag{5.27}$$

The inelastic membrane forces expressed in terms of inelastic strains and the membrane stiffness are defined as follows:

$$n^{\mathrm{c}}_{r/\vartheta} = \mathcal{B}(r)(\varepsilon^{\mathrm{c}}_{r/\vartheta} + \nu\varepsilon^{\mathrm{c}}_{\vartheta/r}), \quad \mathcal{B}(r) = \frac{E(r)h(r)}{1-\nu^2}, \tag{5.28}$$

and constitutive equations for coupled creep–damage problem hold:

$$\begin{aligned}
\dot{\varepsilon}^{\mathrm{c}}_{kl} &= \frac{3}{2}\frac{\dot{\varepsilon}^{\mathrm{c}}_{eq}}{\widetilde{\sigma}_{\mathrm{eq}}}\widetilde{s}_{kl}, \quad k,l = r,\vartheta,\\
\dot{\varepsilon}^{\mathrm{c}}_{\mathrm{eq}} &= (\widetilde{\sigma}_{\mathrm{eq}})^{m(T)}\dot{\mathrm{f}}(t), \quad \widetilde{s}_{r/\vartheta} = \frac{2}{3}\left(\frac{\sigma_{r/\vartheta}}{1-D_{r/\vartheta}} - \frac{\sigma_{\vartheta/r}}{2(1-D_{\vartheta/r})}\right),\\
\widetilde{\sigma}_{\mathrm{eq}} &= \sqrt{\left(\frac{\sigma_r}{1-D_r}\right)^2 + \left(\frac{\sigma_\vartheta}{1-D_\vartheta}\right)^2 - \frac{\sigma_r\sigma_\vartheta}{(1-D_r)(1-D_\vartheta)}},
\end{aligned} \tag{5.29}$$

$$\dot{D}_\nu = C_\nu(T)\left\langle\frac{\sigma_\nu}{1-D_\nu}\right\rangle^{r_\nu(T)}, \quad \nu = r, \vartheta.$$

C. Coupled thermo–mechanical boundary problem

The mechanical state fulfills (5.26) and the following mechanical boundary conditions (see Fig. 5.6):

$$\begin{aligned} n_r(0) &= n_\vartheta(0), \quad n_r(R) = p_0 h_0 \quad (t=0), \\ \dot{n}_r(0) &= \dot{n}_\vartheta(0), \quad \dot{n}_r(R) = 0 \quad (t>0). \end{aligned} \tag{5.30}$$

The inner heat source intensity is defined:

$$\dot{q}_v \overset{\text{def}}{=} -\frac{\dot{Q}_v}{\mathrm{d}V} = -\frac{\dot{Q}_v}{r\mathrm{d}\vartheta h\mathrm{d}r}, \tag{5.31}$$

where an overall effect of convection through both disk faces is expressed by the classical Newton law of cooling (cf. Holman [107]):

$$\dot{Q}_v = 2\beta\mathrm{d}A(T-T_\infty), \quad \mathrm{d}A = \frac{r\mathrm{d}\vartheta\mathrm{d}r}{\cos\Theta}, \quad \cos\Theta = \frac{1}{\sqrt{1+\tan^2\Theta}} = \frac{1}{\sqrt{1+(\mathrm{d}h/\mathrm{d}r)^2}}, \tag{5.32}$$

where T_∞ is the temperature of the cooling fluid, hence:

$$\dot{q}_v = -2\beta\frac{\sqrt{1+(\mathrm{d}h/\mathrm{d}r)^2}}{h}(T-T_\infty). \tag{5.33}$$

The thermal boundary conditions are:

$$\begin{aligned} \mathrm{d}T/\mathrm{d}r\big|_{r=0} &= 0, \quad T(R) = T_0 \quad (t=0), \\ \mathrm{d}\dot{T}/\mathrm{d}r\big|_{r=0} &= 0, \quad \dot{T}(R) = 0 \quad (t>0). \end{aligned} \tag{5.34}$$

D. Optimization

As the optimality criterion the local condition for structures of uniform orthotropic damage strength UODS (4.2) is used. The distribution of the disk thickness $h\,(r)$ is considered as the decision variable. Two inequality constraints of a limited thickness and limited temperature gradients are checked

$$h_{\min} \le h\,(r) \le h_{\max}, \qquad \max\{\mathrm{d}T/\mathrm{d}r\} \le (\mathrm{d}T/\mathrm{d}r)_{\max}, \tag{5.35}$$

and the condition of constant volume (4.11) is used as the equality constraint.

A numerical procedure of optimization, based on the iterative corrections of the decision variable, is used. When optimization with respect to uniform orthotropic creep damage under constant volume is performed, increments of thickness are chosen proportionally to the level of dominant damage tensor components

$$\Delta h_j = P\Delta D_j - \Delta h_m, \qquad \Delta D_j = \sup\{D_{r,\vartheta}\}_j, \tag{5.36}$$

Table 5.3: Comparison of lifetimes for optimally designed disks (Model C)

	constant thickness $h(r) = h_0$; thermo-damage coupling	
	no $\lambda_{\mathrm{eq}} = \lambda_0$	yes $\lambda_{\mathrm{eq}} = \lambda_{\mathrm{eq}}(\lambda_0, \epsilon_0)$
lifetime	$t_{no}^{(\lambda_{\mathrm{eq}}=\lambda_0)} = t_{\mathrm{ref}}$	$t_{no}^{(\lambda_{\mathrm{eq}}=\lambda_{\mathrm{eq}}(\lambda_0,\epsilon_0))} = 1.01 t_{\mathrm{ref}}$
	uniform elastic strength $h_{\mathrm{UES}}(r)$; thermo-damage coupling	
	no $\lambda_{\mathrm{eq}} = \lambda_0$	yes $\lambda_{\mathrm{eq}} = \lambda_{\mathrm{eq}}(\lambda_0, \epsilon_0)$
lifetime		$t_{\mathrm{UES}}^{(\lambda_{\mathrm{eq}}=\lambda_{\mathrm{eq}}(\lambda_0,\epsilon_0))} = 1.03 t_{\mathrm{ref}}$
	uniform creep strength $h_{\mathrm{UCS}}(x)$; thermo-damage coupling	
	no $\lambda_{\mathrm{eq}} = \lambda_0$	yes $\lambda_{\mathrm{eq}} = \lambda_{\mathrm{eq}}(\lambda_0, \epsilon_0)$
lifetime	$t_{\mathrm{UCS}}^{(\lambda_{\mathrm{eq}}=\lambda_0)} = 4.43 t_{\mathrm{ref}}$	$t_{\mathrm{UCS}}^{(\lambda_{\mathrm{eq}}=\lambda_{\mathrm{eq}}(\lambda_0,\epsilon_0))} = 4.70 t_{\mathrm{ref}}$

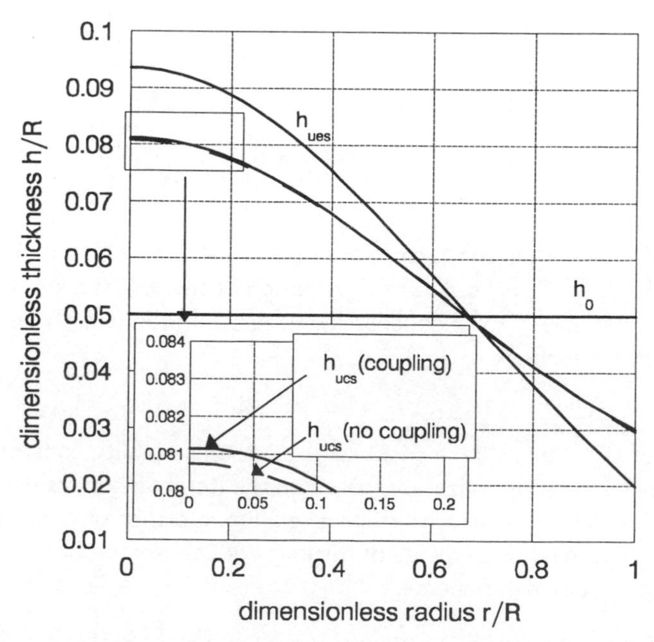

Figure 5.7: Optimal profiles of disks (after Ganczarski and Skrzypek [98])

where the average thickness correction satisfies the constant volume condition

$$\Delta h_m = \frac{\sum_j P\Delta D_j r_j}{\sum_j r_j}, \tag{5.37}$$

whereas the step factor P is experimentally chosen. The process of damage equalization is continued until the following condition is met:

$$\sup\{D_{r,\vartheta}\}_j \leq \mathrm{EPS} = 1. \tag{5.38}$$

To solve the complete coupled initial–boundary problem the FDM is applied where (5.12) and differential operators (5.27) are rewritten in terms of finite differences of T_i, F_i, h_i, λ_i^{eq} with respect to r_i coordinate at each time step t_k (cf. [98]). Numerical procedure begins when the elastic solution of the thermal and mechanical problems is known. Next, the creep problem is entered that requires the vector of equivalent effective stress as well as damage and strain tensors components are computed. The thermal problem is non–linear, hence, by inserting the previous solution for temperature $[T^*]_j$ to the equivalent thermal conductivity λ^{eq}, the solution of (5.12) in terms of finite differences provides the new temperature distribution $[T]_j$, considered as an approximate solution for λ^{eq} and temperature subiteration, until the calculated $[T]_j$ differs from $[T^*]_j$ with a given accuracy. Eventually, when rates of temperature $[\dot{T}]_j$ and inelastic forces $\left[\dot{n}^{\mathrm{c}}_{r,\vartheta}\right]_j$ are known, rates of Airy functions $[\dot{F}]_j$ are found and, finally, the vector of state is determined $[\dot{T}, \dot{n}^{\mathrm{c}}_{r,\vartheta}, \dot{\sigma}_{r,\vartheta}]_j$. In the next time step, the Runge–Kutta II is applied for the thermal and mechanical states, and when the new vector of state is computed the program jumps at the beginning of the creep loop. Numerical procedure is repeated until the conditional statement for highest damage component (5.38) is met.

Numerical simulation is done for ASTM 321 stainless steel at temperature ranging from 500C to 650C (Table 5.2). Profiles of disks of Uniform Elastic Strength (UES) and Uniform Creep Strength (UCS) are shown in Fig. 5.7. None essential difference in shape in cases of thermo–damage coupling disregarded ($\lambda^{\mathrm{eq}} = \lambda_0$) or accounted for $\lambda^{\mathrm{eq}}(\lambda_0, \epsilon_0)$ is observed. However, the essential differences in lifetime predictions occur (Table 5.3).

5.5. Three–dimensional thermo–damage coupling in initially isotropic material

In order to extend the previously discussed 1D thermo–damage coupling models to the 3D case, consider the heat flux decomposition into the conduction and the radiation vectors $\{\mathbf{q}^{\mathrm{cond}}\} = \{q^{\mathrm{cond}}_{x1}, q^{\mathrm{cond}}_{x2}, q^{\mathrm{cond}}_{x3}\}^{\mathrm{T}}$ and $\{\mathbf{q}^{\mathrm{rad}}\} = \{q^{\mathrm{rad}}_{x1}, q^{\mathrm{rad}}_{x2}, q^{\mathrm{rad}}_{x3}\}^{\mathrm{T}}$, as controlled by the tensors of thermal conduction $\widetilde{L}_{ij}$ and thermal radiation $\widetilde{\Gamma}_{ij}$ of damaged material defined as follows:

$$\widetilde{L}_{ij} = \lambda_0 (I_{ij} - D_{ij}), \qquad \widetilde{\Gamma}_{ij} = \sigma\varepsilon_0 D_{ij}. \tag{5.39}$$

Hence, the extension of the model B is written as:

$$\operatorname{div}\left[\lambda_0 (\mathbf{1} - \mathbf{D})\, \mathbf{grad} T - \sigma\varepsilon_0 \mathbf{D}\mathbf{n} T^4\right] + \dot{q}_v = c_v \rho \dot{T}, \tag{5.40}$$

or

$$\frac{\partial}{\partial x_i}\left[\lambda_0 (I_{ij} - D_{ij}) \frac{\partial T}{\partial x_j} - \sigma\varepsilon_0 D_{ij} n_j T^4\right] + \dot{q}_v = c_v \rho \dot{T}. \tag{5.41}$$

When the explicit representation is used we obtain

$$\begin{aligned}
&\frac{\partial}{\partial x}\left\{\lambda_0\left[(1-D_{xx})\frac{\partial T}{\partial x}+(1-D_{xy})\frac{\partial T}{\partial y}+(1-D_{xz})\frac{\partial T}{\partial z}\right]\right.\\
&\qquad\left.-\,\sigma\varepsilon_0\left(D_{xx}n_x+D_{xy}n_y+D_{xz}n_z\right)T^4\right\}\\
&+\frac{\partial}{\partial y}\left\{\lambda_0\left[(1-D_{yx})\frac{\partial T}{\partial x}+(1-D_{yy})\frac{\partial T}{\partial y}+(1-D_{yz})\frac{\partial T}{\partial z}\right]\right.\\
&\qquad\left.-\,\sigma\varepsilon_0\left(D_{yx}n_x+D_{yy}n_y+D_{yz}n_z\right)T^4\right\}\\
&+\frac{\partial}{\partial z}\left\{\lambda_0\left[(1-D_{zx})\frac{\partial T}{\partial x}+(1-D_{zy})\frac{\partial T}{\partial y}+(1-D_{zz})\frac{\partial T}{\partial z}\right]\right.\\
&\qquad\left.-\,\sigma\varepsilon_0\left(D_{zx}n_x+D_{zy}n_y+D_{zz}n_z\right)T^4\right\}+\dot{q}_v=c_v\rho\dot{T}.
\end{aligned}\tag{5.42}$$

The off–diagonal components of the corresponding matrices play a role of the diffusional conduction and radiation portions due to the transverse temperature gradients. However, both tensors $\widetilde{\mathbf{L}}$ and $\widetilde{\mathbf{\Gamma}}$ are co–axial in their principal axes with the damage tensor $\mathbf{D}$, therefore, there exists a locally orthogonal frame coinciding with directions of damage orthotropy such that (5.39) can be written as:

$$\widetilde{L}_\nu=\lambda_0\left(1-D_\nu\right),\qquad \widetilde{\Gamma}_\nu=\sigma\varepsilon_0 D_\nu,\qquad \nu=1,2,3. \tag{5.43}$$

Consequently, the heat flux rates expressed in terms of damage eigenvalues take the form (Fig. 5.8)

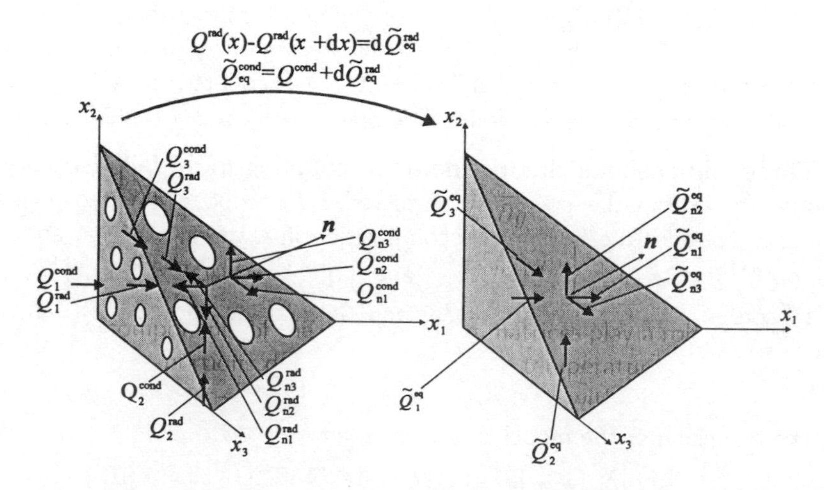

Figure 5.8: Three-dimensional concept of the equivalent heat conductivity

$$
\begin{Bmatrix} q_1^{\rm cond} \\ q_2^{\rm cond} \\ q_3^{\rm cond} \end{Bmatrix} = -\lambda_0 \begin{bmatrix} 1-D_1 & 0 & 0 \\ 0 & 1-D_2 & 0 \\ 0 & 0 & 1-D_3 \end{bmatrix} \begin{Bmatrix} \partial T/\partial x_1 \\ \partial T/\partial x_2 \\ \partial T/\partial x_3 \end{Bmatrix},
$$
$$
\begin{Bmatrix} q_1^{\rm rad} \\ q_2^{\rm rad} \\ q_3^{\rm rad} \end{Bmatrix} = \sigma\varepsilon_0 \begin{bmatrix} D_1 & 0 & 0 \\ 0 & D_2 & 0 \\ 0 & 0 & D_3 \end{bmatrix} \begin{Bmatrix} n_1 \\ n_2 \\ n_3 \end{Bmatrix} T^4. \tag{5.44}
$$

A heat flux equation reduced to the case of thermo–mechanical orthotropy (model B) is furnished now as follows:

$$
\begin{aligned}
&\frac{\partial}{\partial x_1}\left[\lambda_0 (1-D_1)\frac{\partial T}{\partial x_1} - \sigma\varepsilon_0 D_1 T^4\right] + \frac{\partial}{\partial x_2}\left[\lambda_0 (1-D_2)\frac{\partial T}{\partial x_2} - \sigma\varepsilon_0 D_2 T^4\right] \\
&+\frac{\partial}{\partial x_3}\left[\lambda_0 (1-D_3)\frac{\partial T}{\partial x_3} - \sigma\varepsilon_0 D_3 T^4\right] + \frac{\partial q_v}{\partial t} = c_v\rho\frac{\partial T}{\partial t}.
\end{aligned} \tag{5.45}
$$

Model A is recovered from (5.45) when the radiation terms are omitted ($\varepsilon_0 = 0$).

Extension of the equivalent thermal conductivity model to the case of thermo–mechanical orthotropy (model C) consists in introducing diagonal substitutive conductivity tensor $\Delta\widetilde{L}_\nu^{\rm rad}$ in pseudo–undamaged material that corresponds to the equivalent radiation in damaged material, such that:

$$
\Delta\widetilde{L}_\nu^{\rm rad} = \sigma\varepsilon_0\left(4D_\nu T^3 + \frac{\partial D_\nu/\partial x_\nu}{\partial T/\partial x_\nu}T^4\right)\Delta_\nu, \qquad \nu = 1,2,3. \tag{5.46}
$$

Δ_ν denote average dimensions ($\nu = 1,2,3$) of the defects (microcracks) in damaged material. Eventually, the 3D equivalent heat flux equation in terms of three components of the diagonal substitutive conductivity tensor $\widetilde{L}_\nu^{\rm eq} = \widetilde{L}_\nu + \Delta\widetilde{L}_\nu^{\rm rad}$, is furnished:

$$
\frac{\partial}{\partial x_1}\left(\widetilde{L}_1^{\rm eq}\frac{\partial T}{\partial x_1}\right) + \frac{\partial}{\partial x_2}\left(\widetilde{L}_2^{\rm eq}\frac{\partial T}{\partial x_2}\right) + \frac{\partial}{\partial x_3}\left(\widetilde{L}_3^{\rm eq}\frac{\partial T}{\partial x_3}\right) + \frac{\partial q_v}{\partial t} = c_v\rho\frac{\partial T}{\partial t}. \tag{5.47}
$$

In a general case, when the damaged material is anisotropic, complete representation of thermal conductivity and radiation tensor L_{ij} and Γ_{ij} must be used instead of their diagonal representation, such that additional terms connected with a diffusion due to the transverse temperature gradients must appear. On the other hand, when principal directions of the stress and damage change, a combined thermo–damage equations may be considered at current principal damage directions to yield current heat flux anisotropy, though in a reference space a general heat orthotropy occurs.

ACKNOWLEDGMENTS

Much material in the present chapter is originated from the book *Modeling of Material Damage and Failure of Structures* by Jacek Skrzypek and Artur Ganczarski, published within the series *Foundations of Engineering Mechanics* by Springer–Berlin, 1999. It is author's pleasure to acknowledge the great amount of thorough work of dr. A. Ganczarski when preparing examples of engineering applications, numerical procedures and computer codes to illustrate the material.

Thanks are addressed also to professor Vlado Kompiš from University of Žilina, Slovakia, principal organizer of CEEPUS Summer School held in Žilina, August 19–30, 1996, where some parts of this material were presented by the author.

The camera–ready manuscript was prepared by dr. A. Wróblewski, technical editor of this volume, and graphics by dr. A. Ganczarski. Special appreciation is addressed to both of them for their assistance.

The author acknowledges the support from the Committee for Scientific Research KBN Poland, Grant No. 7 T07A 02811 (421/T07/96/11).

REFERENCES

1. Skrzypek J. and Ganczarski A. : Application of the orthotropic damage growth rule to variable principal directions, Int. J. Damage Mech., 7(1998), pp. 180–206.
2. Kachanov L.M. : Time of the rupture process under creep conditions, Izv. AN SSR, Otd. Tekh. Nauk, 8(1958), pp. 26–31.
3. Kachanov L.M. : Foundations of Fracture Mechanics, Nauka, Moscow, 1974, in Russian.
4. Rabotnov Ju.N. : Creep rupture, in: Proc. 12 Int. Congr. Appl. Mech., Stanford, Calif., 1968 pp. 342–349.
5. Martin J.B. and Leckie F.A. : J. Mech. Phys. Solids, 20(1972), pp. 223.
6. Hayhurst D.R. and Leckie F.A. : The effect of creep constitutive and damage relationships upon rupture time of solid circular torsion bar, J. Mech. Phys. Solids, 21(1973), pp. 431–446.
7. Leckie F.A. and Hayhurst D.R. : Creep rupture of structures, Proc. Roy. Soc. London, A 340(1974), pp. 323–347.
8. Hayhurst D.R. : Creep rupture under multiaxial state of stress, J. Mech. Phys. Solids, 20(1972), pp. 381–390.
9. Hayhurst D.R. : On the role of creep continuum damage in structural mechanics, in: Engineering Approaches to High Temperature Design (Edited by Wilshire and D. Owen), Pineridge Press, Swansea, 1983 .
10. Trąpczyński W.A. , Hayhurst D.R. and Leckie F.A. : Creep rupture of copper and aluminium under non–proportional loading, J. Mech. Phys. Solids, 29(1981), pp. 353–374.
11. Lemaitre J. and Chaboche J.L. : A non–linear model of creep fatigue damage cumulation and interaction, in: Proc. of IUTAM Symp. Mechanics of Visco–Elastic Media and Bodies (Edited by J. Hult), Springer, Gothenburg, Sweden, 1975 pp. 291–301.
12. Lemaitre J. and Chaboche J.L. : Aspect phenomenologique de la rapture per endommagement, J. de Méchanique applique, 2(1978), pp. 317–365.
13. Lemaitre J. and Chaboche J.L. : Méchanique des Matériaux Solides, Dunod Publ., Paris, 1985.

14. Chaboche J.L. : Continuum damage mechanics: Part I: General concepts, Part II: Damage growth, crack initiation, and crack growth, J. Appl. Mech., 55(1988), pp. 59–71.
15. Dunne F.P.E. and Hayhurst D.R. : Continuum damage based constitutive eqaution for copper under high temperature creep and cyclic plasticity, Proc. R. Soc. Lond., A 437(1992), pp. 545–566.
16. Dunne F.P.E. and Hayhurst D.R. : Modelling of combined high–temperature creep and cyclic plasticity in components using Continuum Damage Mechanics, Proc. R. Soc. Lond., A 437(1992), pp. 567–589.
17. Dunne F.P.E. and Hayhurst D.R. : Efficient cycle jumping techniques for the modelling of materials and structures under cyclic mechanical and thermal loadings, Eur. J. Mech., A/Solids, 13(1994), pp. 639–660.
18. Dunne F.P.E. and Hayhurst D.R. : Physically based temperature dependence of elastic–viscoplastic constitutive equations for copper between 20 and 500°C, Philosophical Mag., A 74(1995), pp. 359–382.
19. Othman A.M. , Hayhurst D.R. and Dyson B.F. : Skeletal point stresses in circumferentially notched tension bars undergoing tertiary creep modelled with physically based constitutive equations, Proc. R. Soc. London, 441(1993), pp. 343–358.
20. Germain P. , Nguyen Q.S. and Suquet P. : Continuum Thermodynamics, ASME J. Appl. Mech., 50(1983), pp. 1010–1020.
21. Dufailly J. and Lemaitre J. : Modeling very low cycle fatigue, Int. J. Damage Mech., 4(1995), pp. 153–170.
22. Mou Y.H. and Han R.P.S. : Damage evolution in ductile materials, Int. J. Damage Mech., 5(1996), pp. 241–258.
23. Saanouni K. , Forster C.H. and Hatira F. Ben : On the anelastic flow with damage, Int. J. Damage Mech., 3(1994), pp. 140–169.
24. Davison L. and Stevens A.L. : Thermodynamical constitution of spalling elastic bodies, J. Appl. Phys., 44(1973), pp. 668–674.
25. Kachanov L.M. : Introduction to Continuum Damage Mechanics, Martinus Nijhoff, The Netherlands, 1986.
26. Krajcinovic D. and Fonseka G.U. : The continuous damage theory of brittle materials, Part I and II: General theory, J. Appl. Mech., Trans ASME, 48(1981), pp. 809–824.
27. Krajcinovic D. : Constitutive theory of damaging materials, J. Appl. Mech., Trans. ASME, 50(1983), pp. 355–360.
28. Krajcinovic D. : Damage Mechanics, North Holland Series in Appl. Math. and Mech., Elsevier, Amsterdam, 1996.
29. Lubarda V.A. and Krajcinovic D. : Damage tensors and the crack density distribution, Int. J. Solids Struct., 30(1993), pp. 2859–2877.
30. Rabotnov Ju.N. : Creep Problems in Structural Members, North-Holland, Amsterdam, 1969, engl. trans. by F.A. Leckie.
31. Vakulenko A.A. and Kachanov M. L. : Continuum theory of medium with cracks, Izv. A.N. SSSR, M.T.T., 4(1971), pp. 159–166, in Russian.

32. Murakami S. and Ohno N. : A continuum theory of creep and creep damage, in: Creep in Structures (Edited by A. Ponter and D. Hayhurst), Springer, Berlin, 1980 pp. 422–444.
33. Cordebois J.P. and Sidoroff F. : Damage induced elastic anisotropy, in: Col. EUROMECH 115, Villard de Lans, 1979 Also in Mechanical Behavior of Anisotropic Solids (Ed. Boehler, J. P.), Martinus Nijhoff, Boston 1983, 761–774.
34. Cordebois J.P. and Sidoroff F. : Endommagement anisotrope an élasticité at plasticité, J. Méc. Théor. Appl., Numero Spécial, (1982), pp. 45–60.
35. Betten J. : Damage tensors in continuum mechanics, J. Méc. Théor. Appl., 1(1983), pp. 13–32.
36. Betten J. : Application of tensor functions in continuum damage mechanics, Int. J. Damage Mech., 1(1992), pp. 47–59.
37. Litewka A. : Effective material constants for orthotropically damaged elastic solid, Arch. Mech., 6(1985), pp. 631–642.
38. Litewka A. : Analytical and experimental study of fracture of damaging solids, in: Proc. of IUTAM/ICM Symp. Yielding, Damage, and Failure of Anisotropic Solids (Edited by J. Boehler), Mech. Eng. Publ, London, 1987 pp. 655–665.
39. Litewka A. : Creep rupture of metals under multi-axial state of stress, Arch. Mech., 41(1989), pp. 3–23.
40. Murakami S. : Notion of continuum damage mechanics and its applications to anisotropic creep damage theory, J. Eng. Mater. Technol., 105(1983), pp. 99–105.
41. Murakami S. : Failure Criterion of Structural Media, Balkema, 1986 .
42. Murakami S. : Progress of continuum mechanics, JSME, Int. J., 30(1987), pp. 701–710.
43. Murakami S. : Mechanical modelling of material damage, J. Appl. Mech., Trans. ASME, 55(1988), pp. 280–286.
44. Chow C.L. and Lu T.J. : An analytical and experimental study of mixed-mode ductile fracture under nonproportional loading, Int. J. Damage Mech., 1(1992), pp. 191–236.
45. Chaboche J.L. : Development of continuum damage mechanics for elastic solids sustaining anisotropic and unilateral damage, Int. J. Damage Mech., 2(1993), pp. 311–329.
46. Chaboche J.L. : Thermodynamically founded CDM models for creep and other conditions, in: Creep and Damage in Materials and Structures (Edited by H. Altenbach and J. Skrzypek), Advanced School No. 187, Udine, Sept. 7–11, 1998, Springer Vienna, 1999 .
47. Murakami S. and Kamiya K. : Constitutive and damage evolution equations of elastic–brittle materials based on irreversible thermodynamics, Int. J. Solids Struct., 39(1997), pp. 473–486.
48. Hayakawa K. and Murakami S. : Thermodynamical modeling of elastic–plastic damage and experimental validation of damage potential, Int. J. Damage Mech., 6(1997), pp. 333–362.
49. Hayakawa K. and Murakami S. : Space of damage conjugate force and damage

potential of elastic–plastic damage materials, in: Damage Mechanics in Engineering Materials (Edited by G. Z. Voyiadjis, J.-W. Ju and J.-L. Chaboche), Elsevier Science, Amsterdam, 1998 pp. 27–44.
50. Skrzypek J. and Ganczarski A. : Modeling of damage effect on heat transfer in time–dependent non–homogeneous solids, J. Thermal Stresses, 21(1998), pp. 205–231.
51. Skrzypek J. and Ganczarski A. : Modeling of Material Damage and Failure of Structures, Springer, Berlin–Heidelberg, 1999.
52. Leckie F.A. and Onat E.T. : Tensorial nature of damage measuring internal variables, in: Proc. of IUTAM Symp. Physical Non–linearities in Structural Analysis (Edited by J. Hult and J. Lemaitre), Springer, Berlin, 1981 .
53. Chaboche J.L. : Le concept de contraine appliqué à l'élasticité et la viscoplasticité en présence d'un endommagement anisotrope, in: Mechanical Behaviour of Anisotropic Solids (Edited by J. Boehler), Col. EUROMECH 115, Grenoble 1979, Editions du CNRS No. 295, Paris, 1982 pp. 737–760.
54. Simo J.C. and Ju J.W. : Strain– and stress–based continuum damage models. I — Formulation, II — Computational aspects, Int. J. Solids Struct., 23(1987), pp. 821–869.
55. Krajcinovic D. : Damage mechanics, Mech. Mater., 8(1989), pp. 117–197.
56. Chen X.F. and Chow C.L. : On damage strain energy release rate Y, Int. J. Damage Mech., 4(1995), pp. 251–236.
57. Voyiadjis G.Z. and Park T. : Anisotropic damage for the characterization of the onset of macro–crack initiation in metals, Int. J. Damage Mech., 5(1996), pp. 68–92.
58. Voyiadjis G.Z. and Park T. : Kinematics of large elastoplastic damage deformation, in: Damage Mechanics in Engineering Materials (Edited by G. Voyiadjis, J.-W. Ju and J.-L. Chaboche), Elsevier Science, Amsterdam, 1998 pp. 45–64.
59. Qi W. and Bertram A. : Anisotropic creep damage modeling of single crystal super-alloys, Techn. Mechanik, 17(1997), pp. 313–332.
60. Zheng Q.-S. and Betten J. : On damage effective stress and equivalence hypothesis, Int. J. Damage Mech., 5(1996), pp. 219–240.
61. Taher S.F. , Baluch M.H. and Al-Gadhib A.H. : Towards a canonical elastoplastic damage model, Eng. Fracture Mechanics, 48(1994), pp. 151–166.
62. Robinson E.L. : Effect of temperature variation on the long time rupture strength of steel, Trans. ASME, 74(1952), pp. 777–780.
63. Chrzanowski M. and Madej J. : Construction of the failure curves based on the damage parameter concept, Mech. Teor. Stos., 4(1980), pp. 587–601, (in Polish).
64. Chaboche J.L. : Une Loi Différentielle d'Endommagement de Fatigue avec Cumulation non Linéaire, Revue Française de Mecanique, (1974), pp. 50–51, english trans. in : Annales de l'IBTP, HS 39, (1977).
65. Othman A.M. and Hayhurst D.R. : Multi–axial creep rupture of a model structure using a two–parameter material model, Int. J. Mech. Sci., 32(1990), pp. 35–48.
66. Kowalewski Z.L. , Hayhurst D.R. and Dyson B.F. : Mechanisms–based creep constitutive equations for an aluminium alloy, J. Strain Analysis, 29(1994), pp. 309–316.

67. Kowalewski Z.L. , Lin J. and Hayhurst D.R. : Experimental and theoretical evaluation of a high–accuracy uni–axial creep testpiece with slit extensometers ridges, Int. J. Mech. Sci., 36(1994), pp. 751–769.
68. Hayhurst D.R. : Material data bases and mechanisms–based constitutive equations for use in design, in: Creep and Damage in Materials and Structures (Edited by H. Altenbach and J. Skrzypek), Advanced School No. 187, Udine, Sept. 7–11, 1998, Springer Vienna, 1999 .
69. Rides M. , Cocks A.C. and Hayhurst D.R. : The elastic response of damaged materials, J. Appl. Mech., 56(1989), pp. 493–498.
70. Johnson A.E. , Henderson J. and Mathur V.D. : Combined stress creep fracture of commercial copper at 250°, The Engineer, 24(1956), pp. 261–265.
71. Johnson A.E. , Henderson J. and Khan B. : Complex-stress creep, relaxation and fracture of metallic alloys, HMSO, Edinbourgh, 1962.
72. Chaboche J.L. and Rousselier G. : On the plastic and viscoplastic constitutive equations – P.1: Rules developed with internal variable concept, P.2: Application of internal variable concepts to the 316 stainless steel, J. Pressure Vessel Technol, 105(1983), pp. 153–164.
73. Lemaitre J. : A continuum damage mechanics model for ductile fracture, ASME J. Engng. Mat. and Technology, 107(1985), pp. 83–89.
74. Lemaitre J. : Formulation and identification of damage kinetic constitutive equations, in: Continuum damage mechanics — theory and application (Edited by D. Krajcinovic and J. Lemaitre), CISM Courses and Lectures, 295, Springer, Berlin, 1987 pp. 37–89.
75. Broberg H. : Damage measures in creep deformation and rupture, Swedish Solid Mechanics Report, 8(1974), pp. 100–104.
76. Chow C.L. and Wang L. : An anisotropic theory of elasticity for continuum damage mechanics, Int. J. Fracture, 33(1987), pp. 3–16.
77. Chow C.L. and Wang L. : An anisotropic theory of continuum damage mechanics for ductile materials, Eng. Fract. Mech., 27(1987), pp. 547–558.
78. Voyiadjis G.Z. and Kattan P.I. : A plasticity–damage theory for large deformation of solids, Part I: Theoretical formulation, Int. J. Eng. Sci., 30(1992), pp. 1089–1108.
79. Chaboche J.L. , Lesne P.M. and Moire J.F. : Continuum damage mechanics, anisotropy and damage deactivation for brittle materials like concrete and ceramic composites, Int. J. Damage Mech., 4(1995), pp. 5–22.
80. Ganczarski A. and Skrzypek J. : Effect of initial prestressing on the optimal design of plates with respect to orthotropic brittle rupture, Arch. Mech., 46(1994), pp. 463–483.
81. Litewka A. and Hult J. : One parameter CDM model for creep rupture prediction, Eur. J. Mech., A/Solids, 8(1989), pp. 185–200.
82. Gallagher R.H. : Fully stressed design, in: Optimal Structural Design (Edited by R. Gallagher and O. Zienkiewicz), John Willey, New York, 1973 pp. 19–23.
83. Życzkowski M. : Optimal structural design in rheology, J. Appl. Mech., 38(1971), pp. 39–46, proc. 12 Int. Cong. Theor. Appl. Mech., Standford 1968.
84. Życzkowski M. : Optimal structural design under creep conditions (1), Appl. Mech.

Reviews, 41(1988), pp. 453–461.
85. Życzkowski M. : Problems of structural optimization under creep conditions, in: Proc. of IUTAM Symp. Creep in Structures IV, 1990 (Edited by M. Życzkowski), Springer, Berlin, 1991 pp. 519–530.
86. Życzkowski M. : Optimal structural design under creep conditions (2), Appl. Mech. Reviews, 49(1996), pp. 433–446.
87. Hayhurst D.R. , Dimmer P.R. and Chernuka M.W. : Estimates of the creep rupture lifetimes of structures using the finite element methods, J. Mech. Phys. Solids, 23(1975), pp. 335–355.
88. Hayhurst D.R. , Dimmer P.R. and Morrison C.J. : Development of continuum damage in the creep rupture of notched bars, Phil. Trans. R. Soc. London, A 311(1984), pp. 103–129.
89. Saanouni K. , Chaboche J.L. and Bathias C. : On the creep crack growth prediction by a non–local damage formulation, Eur. J. Mech., A/Solids, 8(1986), pp. 677–691.
90. Liu Y. , Murakami S. and Kanagawa Y. : Mesh–dependence and stress singularity in finite element analysis of creep crack growth by Continuum Damage Mechanics, Eur. J. Mech. A/Solids, 13(1994), pp. 395–417.
91. Murakami S. , Kawai M. and Rong H. : Finite element analysis of creep crack growth by a local approach, Int. J. Mech. Sci., 30(1988), pp. 491–502.
92. Murakami S. and Liu Y. : Mesh–dependence in local approach to creep fracture, Int. J. Damage Mech., 4(1995), pp. 230–250.
93. Skrzypek J. , Kuna-Ciskał H. and Ganczarski A. : On CDM modelling of pre– and post–critical failure modes in the elastic–brittle structures, in: Beiträge zur Festschrift zum 60 Geburstag von Prof. Dr.–Ing. Peter Gummert, Mechanik, Berlin, 1998 pp. 203–228.
94. Skrzypek J. , Kuna-Ciskał H. and Ganczarski A. : Continuum damage mechanics modeling of creep–damage and elastic–damage–fracture in materials and structures, in: Proc. Workshop on Modeling Damage, Localization and Fracture Process in engineering Materials, Kazimierz Dolny, 1999 (to be published).
95. Ganczarski A. and Skrzypek J. : Optimal prestressing and design of rotating disks against brittle rupture under unsteady creep conditions, Eng. Trans., 37(1989), pp. 627–649, (in Polish).
96. Ganczarski A. and Skrzypek J. : On optimal design of disks with respect to creep rupture, in: Proc. of IUTAM Symp. Creep in Structures (Edited by M. Życzkowski), Cracow, 1990 pp. 571–577.
97. Ganczarski A. and Skrzypek J. : Optimal shape of prestressed disks against brittle rupture under unsteady creep conditions, Struct. Optim., 4(1992), pp. 47–54.
98. Ganczarski A. and Skrzypek J. : Optimal design of rotationally symmetric disks in thermo–damage coupling conditions, Techn. Mechanik, 17(1997), pp. 365–378.
99. Skrzypek J.J. : Plasticity and Creep, Theory, Examples, and Problems, Begell House – CRC Press, Boca Raton, 1993, ed. R. B. Hetnarski.
100. Skrzypek J. and Egner W. : On the optimality of disks of uniform creep strength

against brittle rupture, Eng. Opt., 21(1993), pp. 243–264.
101. Egner W. and Skrzypek J. : Effect of pre-loading damage on the net–lifetime of optimally prestressed rotating disks, Arch. Appl. Mech., 64(1994), pp. 447–456.
102. Ganczarski A. and Skrzypek J. : Axisymmetric plates optimally designed against brittle rupture, in: Proc. World Congr. on Optimal Design of Structural Systems, Structural Optimization 93 (Edited by J. Herskovits), Rio de Janeiro 1993, 1993 pp. 197–204.
103. Ganczarski A. and Skrzypek J. : Brittle-rupture mechanisms of axisymmetric plates subject to creep under surface and thermal loadings, in: Proc. of SMIRT–12 (Edited by K. Kussmaul), Stuttgart 1993, 1993 pp. 263–268.
104. Ganczarski A. , Freindl L. and Skrzypek J. : Orthotropic brittle rupture of Reissner's prestressed plates, in: Proc. 5 Int. Conf. On Computational Plasticity (Edited by D. Owen, E. O. nate and E. Hinton), Barcelona, 1997 pp. 1904–1909.
105. Ganczarski A. and Skrzypek J. : Concept of thermo–damage coupling in continuum damage mechanics, in: Proc. First Int. Symp. Thermal Stresses '95 (Edited by R. Hetnarski and N. Noda), Hamamatsu, Japan, Act City, 1995 pp. 83–86.
106. Ganczarski A. and Skrzypek J. : Modeling of damage effect of heat transfer in solids, in: Proc. Second Int. Symp. Thermal Stresses '97 (Edited by R. Hetnarski and N. Noda), Rochester, NY, 1997 pp. 213–216.
107. Holman J.P. : Heat Transfer, McGraw-Hill, 1990.
108. Odqvist F.K.G. : Mathematical Theory of Creep and Creep Rupture, Oxford Mathematical Monographs, Clarendon Press, Oxford, 1966.

MATERIALS DATA BASES AND MECHANISMS-BASED CONSTITUTIVE EQUATIONS FOR USE IN DESIGN

...ayhurst

...ester, Manchester, UK

...omputer simulation as a wealth creational tool in ...he route is highlighted from laboratory testing of ...nisms-based constitutive equations to the super- ...f high-temperature engineering components. The ...echanics (CDM) is used as an example of a tool that ca... the damage/rupture behaviour of a wide range of engineering compon... ...erated... high-temperatures. The importance is stressed of using mechanisms-based constitutive equations in order to achieve accurate predictions/extrapolations. Procedures are discussed for the selection of the dominant mechanisms from laboratory data, and hence the relevant constitutive equations. It is argued that the barrier to progress in the use of these techniques to achieve wealth creation will be a paucity of good materials data.

1. INTRODUCTION

Design and manufacture of new products are the key to sustaining a healthy wealth creational base. However possession of a design and manufacturing facility is no longer a passport to success. There are many other factors which are decisive in the determination of successful wealth creation. Some of these may be appreciated by reference to Fig. 1 in which a top-down systems approach is given to the design process. The diagram is presented not as a definitive methodology but as a discussion aid, and certainly it is recognised that many countries carry out such functions in a highly parallel or concurrent mode, Preiss [1].

Marketing research, sometimes referred to as marketing but which is very different from the sales function, is the key to the identification of which market sectors, and of their size, that a particular product should be aimed at. Having done that, market research can then be used to create the design specification. Traditionally technologists have felt able, and qualified, to "know the minds" of their clients and customers, and to be able to bypass the first two stages in the process i.e. marketing research and design specification. Current activity in world, and in national markets is such that large databases on market information have to be established and continually updated. Quick access, ahead of the competition, is an important aspect of maintaining and of extending ones market share. It is recognised that data and information management associated with marketing are an increasingly important part of design management. However, it will not be covered in this paper explicitly.

Having identified a product and a market sector and from this information written an unprejudiced design specification, which can be cast in basic or abstract terms, the engineering design process can commence. It does so with the formation of concept designs by encouraging the design team to think laterally; to innovate, often using new technological breakthroughs; to use new materials; and, to use new manufacturing techniques which will get the product to market quicker, at the required quality, and at lower cost. This process requires large databases, not just involving the lead company, but those of other companies such as design out-sources, component suppliers, manufacturers and fabricators. This leads to the requirements of shared data between companies, often competing, with a need to maintain confidentiality and ones competitive edge. The very process of sharing data, whilst exercising bounded control over accessibility, can slow down the very process which one needs to speed up in order to be successful. In this environment the correct hierarchical structuring of data, and the level of accessibility is vital to success. Hence, data structures and robust software are an essential part of the competitive process.

At the concept design stage the principal function is to propose design variants, and to carry out assessments of the extent to which they satisfy the design specification. It will be necessary to carry out design calculations operating with lower levels of information for which the degree of precision need not necessarily be high. The coarse sorting type of assessments carried out at the concept phase lead to the rejection of inadmissible concept variants and to the identification of a reduced sub-set; which, in turn, requires assessment at higher levels of information to enable the final concept selection to take place. This further process is often referred to as embodiment design. It repeats the concept level feedback - feed forward; and interactions with the manufacturing - assembly - packaging functions, i.e. feed forward - feedback. Many of these functions are materials selection dependent, and require the use of materials databases containing different data sets each possessing a range of information qualities. The embodiment phase naturally leads to a chosen design preference with clearly delineated options and fall back strategies. However, in what follows in this paper attention will be focussed on the materials database requirements.

The detailed design phase then follows in which a single design is examined in great detail. Traditionally the role of this phase has been to satisfy the designer that the technical function and the risk of unacceptable failure, as perceived by the user and society, are acceptable. The phase naturally leads on to the manufacture of the product or system, and to its infeed to the marketplace.

The manufacturing function is key to product quality and reliability. The subject of manufacturing systems impacts on the cost and time to market, whilst the technological aspects of manufacture relate directly to product integrity, quality, cost and technological function. The role of materials processing in manufacture is the technological aspect which is considered here. This has been selected since it involves an upstream - downstream coupling with the design stage of the process, which is brought about by the common material thread.

The major technological change which has taken place in recent years is the availability of lower cost computer workstations and supercomputers, with access to large, low cost, fast-access data storage facilities. At the same time, numerical techniques and software for the solution of combined boundary-initial value problems, often involving the finite element method, are becoming available in robust forms. Hence it is possible, given the appropriate materials data and models, to simulate complex physical processes in design and manufacture. In parallel with these advances computer visualisation techniques and the development of video animation facilities have reached a state of the art where real time simulation, sometimes labelled virtual reality, is an accessible tool for use in decision making in design and manufacture.

In this paper these aspects are first covered in more detail, then the materials data requirements for supercomputer simulation in high-temperature design are examined. The generation of materials data from laboratory testing is addressed, and its subsequent conversion into higher levels of data and information are discussed in the context of several applications which are related to high-temperature design. Firstly, the theory and principles of creep continuum damage mechanics are reviewed for a single damage state variable theory. Secondly, the multi-state variable modelling of an aluminium alloy, and the relationship of the models to the physics of damage evolution are discussed. Thirdly, the model is used to analyse the behaviour of a simple uni-axial creep testpiece to assess the accuracy of the measurement process. Fourthly, the paper considers general formulations of multi-damage state variable theories and addresses the problems associated with mechanism identification and model calibration. Finally, conclusions are made on general methodologies for use of materials databases to select mechanisms-based constitutive equations, and to calibrate them against laboratory data for use in detailed design/analysis.

Firstly, before these detailed aspects are addressed, an overview is now presented of data requirements throughout engineering.

2. REQUIREMENTS FOR DATA THROUGHOUT ENGINEERING

2.1. The wealth creation process

A top-down sequential approach to the wealth creation process [2] is shown in Fig. 1. The serial approach from market research to the delivery of the product in the marketplace is not always enacted in industry. This is largely dependent upon the type of design to be carried out. For example if the design is totally original then market research is essential and the entire process is enacted. If the design is an adaptive one then only part of it is original, and for that part the entire process is utilised as shown in Fig. 1; for the remaining part of the design

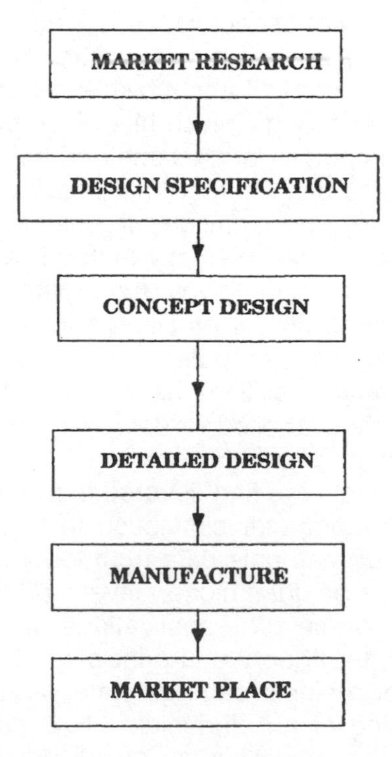

Figure 1. Top-down serial approach to Marketing, Design, and Manufacture, within the Wealth creation process.

activity, which could involve repetition of a previous design, only the latter phases of detailed design and manufacture etc. are implemented. If the design is variant in nature then by definition this involves the scaling of a previous design without any major conceptual or procedural changes; and in this case only sub-sets of the entire process of Fig. 1 are enacted, usually involving parametric computer techniques. A cost effective solution is then simply achieved by implementing what has been done before using the same workforce, without any innovation.

In practice the serial approach shown in Fig. 1 is increasingly less used as the benefits of simultaneous or parallel operation become more generally accepted [1]. The degree of parallelisation depends upon many factors which can include: company size; the divisibility of the product into sub-units; the multi-disciplinary nature of the design; and, whether the design is original, adaptive or variant. What is clear, however, is that the shift to parallelism highlights the need to plan, execute and control the design using a software based system. The difficulties associated with the management of task forces or design teams, and also the rewards to be gained are well known. They highlight the need for data which defines the current design status, the level of input from several disciplinary groups, the degrees of interaction, and project management; and, they all require dynamic databases for use with appropriate levels of interaction, accessibility and user constraints.

The main industrial drivers are to shorten:

(a) the time from concept initiation to the commissioning of the first prototype;
(b) the time from concept initiation to product launch;
(c) increase product quality, reduce cost;
(d) maximise reliability and minimise the risk of failure.

Since the use of materials is central to the creation of new products, their selection and utilisation are essential to success. In the following sections the materials related aspects are considered with particular reference to concept design and to detailed design and manufacture.

2.2. Materials data requirements for concept design

Implied in Fig. 1 are the information flow and material requirements for concept design. The first three boxes may include solid-surface modelling, rapid prototyping, and the need to consider company manufacturing methods, constraints and costs, and customer liaison. The need for materials data encompassing basic physical properties, availability and cost is required prior to the formal concept design stage. These needs are well known, Ashby [3] has discussed the type and level of precision of the data required at this stage. At the conceptual stage there is a need to technically innovate, and to utilise methods of market forecasting. Both of these features are crucial to the identification of new market opportunities; they each require a considerable company-industry database which is progressively updated, and which is seen as a major competitive edge for the particular company involved. It is therefore a database for which a high level of confidentiality must be maintained, and therefore access and level of usage has to be carefully controlled. At the bottom of Fig. 1 connectivity is required with company manufacturing methods etc. It is here that the manufacturing systems aspects of the company are logged in a large database, and opportunities for using spare manufacturing capacity within given time windows can be identified. The availability of manufacturing resource, and the potential to schedule new work within given timeframes, and hence maximising plant utilisation can be achieved through the use of this database. It is in this way that the interaction between concept design and the manufacturing function can develop synergy and be used to seize new opportunities. This can only be fully realised provided that the companies involved have a policy of continuous updating of their databases, and of reconfiguration to provide access to the right quality of information to make the necessary decisions; and, provided that the software can maintain a competitive edge for the company through appropriate access and style of database usage.

Whilst the aspect of management of materials data within suitably shortened timeframes at the concept stage of design is an essential feature of the successful company, this paper will not consider these aspects further. Instead it will concentrate on the subsequent detailed design stage, and its interaction with the manufacturing process.

2.3. Materials data requirements for detailed design

The latter part of the diagram given in Fig. 1 is considered in more detail in the wealth creation column of Fig. 2. It starts following embodiment design and progresses through the detailed design-analysis-simulation stage to manufacturing simulation, and on to fabrication, assembly, and quality assurance testing. The types of material data required to underpin this activity is accessed from a materials database linked to the process of finite element analysis coupled with pre- and post-processing for the analysis. This can be regarded as the first detailed design analysis of the behaviour of the component or system when subjected to in-service conditions. Coupled with this is the requirement for further information on constitutive equations. These are the equations which relate the deformation characteristics of the material with stress, temperature and strain rate conditions. The nature of these constitutive equations will be discussed in later sections of the paper, and the need will be highlighted for them to be mechanisms-based. Further down the sequence is the requirement for materials data regarding the flow of materials in manufacture, the associated heat transfer, the evolution of microstructure and the consequent mechanical properties. It is through the supercomputer modelling of processes such as forging, extrusion, pressing, casting, and plastic injection moulding that the design engineer can shape the component and design the process to achieve defect-free components entering service. Also included is the need for information on materials processes, surface durability and constitutive equations required to model fabrication processes such as: welding, joining and adhesives etc. These relate to the assembly of discrete smaller sub-components and units. The latter operations usually involve processes which bring together different materials into one operation, they also define a set

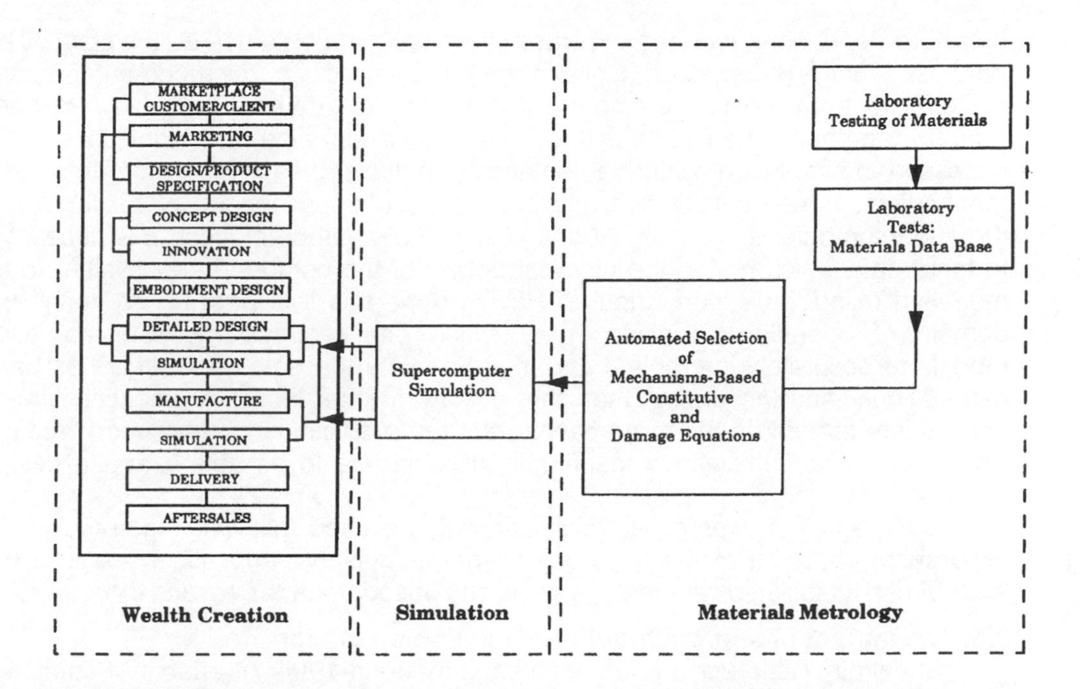

Figure 2. Diagrammatic representation of the route leading from Materials Testing, through Materials Metrology and Supercomputer Simulation, to the Wealth Creation Process.

of multi-materials problems for which detailed and specialised information is needed for analysis and subsequent design.

In this paper, the detailed design/simulation link, given in the wealth creation column of Fig. 2, will be considered together with the need: to select materials; to identify the mechanisms by which they deform, damage and fail; and to select and calibrate the appropriate mechanisms-based constitutive equations. The data requirements associated with this process will be examined through selected examples.

2.4. From laboratory testing of materials to supercomputer simulation

Shown in Fig. 2 is a diagrammatic representation of the route leading from laboratory materials testing, through materials metrology, and supercomputer simulation to wealth creation. On the far left of the figure is reproduced an extended version of Fig. 1. The coupling to the region of interest, namely detailed design and manufacture/simulation is shown by the arrows entering on the right-hand side of the wealth creation column. On the right-hand side of the figure, contained within the broken lined rectangle headed Materials Metrology, is shown Laboratory Testing of Materials. This is the source of materials data necessary for use in supercomputer simulation. The accuracy of the supercomputer simulation is therefore dependent upon the accuracy of the test data obtained at this stage. This data is usually collected in digital form and stored on microcomputer based systems to form the first elements of the materials data. When this has been brought together to constitute a database of sufficient size and significance, it is accessed by software which automatically selects the type and form of the mechanisms-based constitutive and damage equations necessary for input into the supercomputer simulation. At this point of automated selection it is necessary to input data which define the physical details of the component or system to be modelled using the supercomputer. Having done this it is possible to define the domain of operating conditions over which one requires mechanisms-based models to accurately interpolate and to extrapolate the data. Once this has been carried out, and assuming that the quality of the laboratory testing meets appropriate standards then one can expect the predictions of the supercomputer simulation to be sufficiently meaningful and accurate to be able to make design decisions.

Clearly this is a process which can only be used following rigorous quality assurance, checking of the accuracy of the simulations, and assessment of the level of risk associated with making decisions based upon the results thereof.

2.5. The impact of supercomputer simulation on design

Shown in Table 2.1 are the estimates, made in 1994, of effects of computer central processor unit (CPU) speed on run times of a particular software package known as DAMAGE XX which has been used by the author over a number of years in the analysis for the design of high-temperature components [4,5]. Although this data relates to developments over the last decade, they have proved to be reliable and provide a snapshop of likely similar developments over the next ten years. In the first column of the table is listed a range of computers which either have existed, are existing or are projected. In the second column the speed of the central processor unit is shown in Mega, Giga and Teraflops, i.e. floating point operations per second. In the final column is shown the computer run times for the same DAMAGE XX job. What can be seen is the way in which a particular supercomputer simulation will be accelerated over a period of approximately 10 years. At the top of the third column a figure of 200 hours is given for a IBM 3090 computer. This can be seen to be progressively reduced to a projected figure given by IBM research [6] of 3.6 seconds. This reflects an increase of speed of 2×10^5. This clearly demonstrates the strength of the forward technological pull brought about by the increase in computer speed, and by the availability of fast access storage capability. Projections do not stop at the figures given in this diagram, and discussions with experts [6] indicate that this is likely to be an ongoing trend. If the same rate of speed-up takes place over the next decade, then the run time for the software will be approximately 20 μs! At this point, it is worth noting that whilst the CPU speed of workstations is in the same domain as that of Supercomputers, this speed can rarely be harnessed if data intensive simulations are to be carried out with any rapidity. Therefore CPU speed should only be judged in the context of the overall style of useage of the computer- based

Table 2.1. Effect of Computer CPU speed on run-times of the Creep Continuum Damage Mechanics Solver Damage

COMPUTER	SPEED	DAMAGE XX RUN TIME
IBM 3090 (1986) 6 PROCESSOR	40 MFLOPS	200 h
CRAY X-MP (1988) 4 PROCESSORS	235 MFLOPS	
unvectorised		34 h
vectorised		255 min
multi-tasked		106 min
CRAY Y-MP (1992)	350 MFLOPS	71 min
CRAY C90 (1993)	1 GFLOP	25 min
JAPANESE NEC SX-3 (1994) vectorised	5.5 GFLOPS	2.6 min
IBM Research Projection (1991) by 2000	1 TFLOPS	3.6 s

simulation facility. The weak link in the chain which connects laboratory test data to the results of supercomputer simulation shown in Fig. 2 is therefore the quality of the materials constitutive equations or models, the accessibility to, and the quality of the associated materials databases. These features will be exemplified in the examples considered in later sections of the paper.

3. HIGH-TEMPERATURE DESIGN

3.1. Background

High-temperature design in the creep range of metallic components is largely carried out using design by code/methods, e.g. BS 5500, and ASME Code Case 1592. However, flexibility is provided within the codes for the designer to use design by analysis methods. The latter are essential in situations where the designer operates on or beyond the boundaries of applicability. It is in these cases where new and more accurate methods of analysis have appeal.

High-temperature design has always suffered from uncertainties particularly in three areas: (a) paucity and scatter in creep data, here conservatism has always enforced lower bound data to be used with a superimposed factor of ignorance; (b) the uncertainty of operational conditions e.g. temperature, stress level and their time histories, here conservatism enforces the use of pessimistic data; and (c) the accuracy of the analysis, this arises from the geometrical and material non-linearities, the complexity of which often necessitate the use of approximate or bounding computations. Scatter in creep data is now well understood and can be controlled if adequate temperature and stress-state control are maintained; however, cost is a limiting factor. The uncertainty of operating conditions is a factor which is always likely to be present to some degree. Lastly, the accuracy of the analysis technique, although strongly coupled with the need for accurate test data, is an area where dramatic change is taking place. This has been enabled by the increasing availability of low cost, high-performance computing and data storage, and the principal theme of this section is centred around these developments.

The theme of this section is the use of computational Continuum Damage Mechanics (CDM) with the finite element method to analyse a broad range of structural components using simple uni-axial creep data. The method has the advantage of providing traceability from the constitutive equations used through the physics of the deformation, damage and fracture processes involved to the fundamental microstructural behaviour. In this way starting from the physics of the microstructural processes it is not necessary to specify the type of computation required for each specific category of behaviour, instead all that need be adhered to is the rigour of the generic approach. Emphasis is therefore placed upon the identification of a physical mechanism and upon access to experimental data which both characterises and quantifies the strengths of these mechanisms. The advantage of the approach is that extrapolation from short term to long term behaviour is more reliable, since it is founded upon an accurate description of the physics of all the mechanisms involved. The section commences with an outline of CDM. This is followed by an overview of the generic approach to design provided by computational CDM using the finite element technique. A review is then given of how beneficial stress redistribution can be achieved by permitting widespread continuum damage to evolve; and, of how the multi-axial stress rupture criteria of the material influences component behaviour. It is then shown how the CDM approach can be used, without modification, to predict the behaviour of creep crack growth; and it is also shown how the characteristics of the material ductility may be changed to produce failure of engineering structures by either a widespread growth of damage, or by the growth of a highly localised region of damage to yield creep crack growth. The final section addresses the creep rupture of components which are comprised of several materials. The example chosen is a butt welded ferritic steel steam pipe.

3.2. What is CDM?

Fig. 3 shows a micrograph of a copper uni-axial testpiece tested at 250°C after failure. The failure surface is shown on the left-hand side of the figure and the widespread nature of the grain boundary damage may be observed over the entire

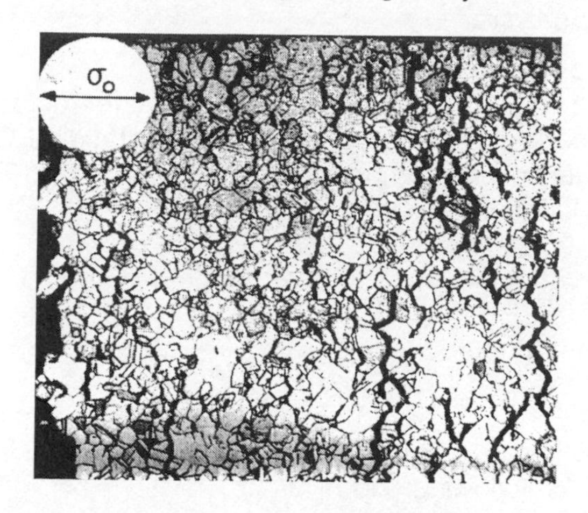

Figure 3 Micrograph of a copper uni-axial testpiece tested at 250°C after failure. The left-hand side shows failure section

region of the micrograph. What is evident is that a single crack has not predominated and propagated across the section. Hayhurst [7] has shown from this and other studies that provided the stress field is homogeneous then a field of damage nucleates and grows in a uniform way over the same region. The strain rate behaviour and the damage evolution rate behaviour may be described by the following equations:

$$\dot{\varepsilon}_{ij} = f(\sigma_{ij},\omega_1,\omega_2), \qquad \dot{\omega}_1 = g(\sigma_{ij},\omega_1,\omega_2), \qquad \dot{\omega}_2 = h(\sigma_{ij},\omega_1\,\omega_2) \qquad (3.1)$$

where ω_1 and ω_2 are rates of change of the damage state variables. Each mechanism of damage operates over a given domain of temperature, stress level and stress state. Cocks and Ashby [8] have shown that the micromechanisms can be modelled in this way and that the equations which describe the basic mechanisms can be rewritten in the above form to provide an accurate global description. It is this important step that enables one to achieve traceability from the global, or macro scale behaviour of the material back to the behaviour of the material at the micro scale. Guidance on the domains of temperature and stress level over which the mechanisms operate may be obtained from the mechanisms maps of Ashby [9], and on stress state dependence from the work of Cocks and Ashby [8].

Although in (3.1) only two damage state variables have been included, in practice there may be more. These could include the following: cavity nucleation and growth; ductile void growth; multiplication of dislocation substructures; and precipitate coarsening. In the work discussed herein constitutive equations will be used to model most of these mechanisms.

It will be shown below how this background can be used to form a systematic approach to design analysis.

3.3. A CDM approach to design analysis

Over the last two and a half decades a number of papers [10-15] have reported contributions to the establishment of computational Continuum Damage Mechanics as a route to high-temperature design analysis. The design/analysis procedure is set out in the list below.

Identify for the structure or component to be designed:

(a) Domains of temperature.
(b) Ranges of stress.
(c) Stress states.

Then carry out the following:

(d) Make a preliminary material selection using either Ashby's [16] material selection procedures for conceptual design; or available materials creep data bases.
(e) Access available and relevant databanks of creep curves at the appropriate temperatures, stresses and stress states.
(f) Check Ashby's mechanisms maps [9] to provide a steer on the mechanisms of deformation and rupture, in order to identify the appropriate mathematical models.
(g) Use numerical techniques [17] to fit the appropriate models to available creep data.
(h) Use the models in conjunction with computational CDM finite element based techniques.
(i) Determine limiting design factors such as global and local strain histories, damage field evolution, macro crack initiation and failure lifetimes.

In what follows a compendium of design cases and studies is presented to illustrate the power of the above design analysis approach.

3.4. CDM: The prediction of structural behaviour from simple stress raisers through complex stress states to creep crack growth

The establishment of the design approach set out above will now be traced and some of the important results will be illustrated.

A. Stress redistribution due to CDM

Shown in Fig. 4 is a mid-thickness micrograph of a part of quarter section of a copper tension panel tested at 250°C immediately prior to failure. The plate

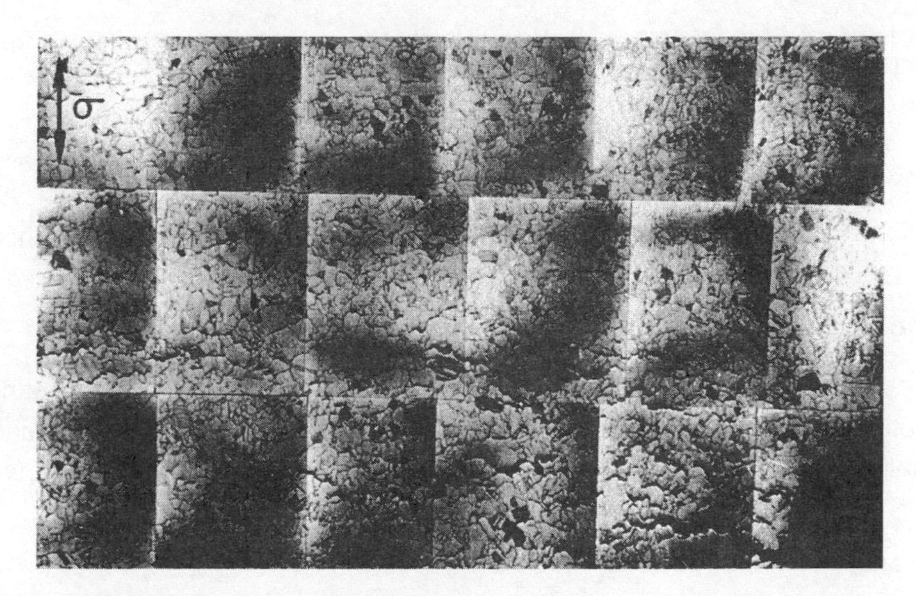

Figure 4 Micrograph of a part quarter-section of a copper tension panel, tested at 250°C, containing a central circular hole. The test was stopped just before failure.

contained a central circular hole, part of which is shown in the bottom right-hand corner of the figure. The left-hand boundary of the figure is of a region close to the edge of the plate. The top boundary of the figure is subjected to a remote steady uniform tension. It is evident from Fig. 4 that widespread continuum damage has taken place and the CDM calculations reported by Hayhurst et al [12] show that the damage formation is accompanied by considerable stress redistribution which, shortly after load up, produces an almost uniform stress across the minimum load bearing section of the plate. This leads to the result that the lifetime of the plate may be computed using the net section stress and uni-axial creep rupture data. The average experimental lifetimes are greater than the computed lifetimes by amounts equivalent to 3% and 7% on stress for the aluminium alloy and copper plates respectively. This result clearly shows the highly beneficial effects of stress redistribution due to the growth of continuum damage. To further test the effectiveness of stress redistribution, a similar plate was tested [18] in which the hole was replaced by a narrow slit of the same characteristic dimension. Again it was shown that stress redistribution rapidly nullified the initial high stress and strain gradients at the tip of the slit, and that the lifetime of the plate could again be predicted using the net section stress and uni-axial creep rupture data. The average experimental lifetimes are less than the computed lifetimes by amounts equivalent to 1% and 2% on stress respectively for the aluminium and copper plates containing slits. This result holds provided that the material behaves such that a major part of the strain is accumulated during tertiary creep [19].

B. Role of multi-axial stress rupture criteria

The creep strain rate equation (3.1) is given by

$$d\varepsilon_{ij} / dt = 3K\sigma_e^{n-1} S_{ij} t^m / 2(1-\omega)^n \tag{3.2}$$

where S_{ij} is the deviatoric stress $\sigma_{ij} - \delta_{ij}\sigma_{kk}/3$, σ_e is the effective stress $(3S_{ij}S_{ij}/2)^{1/2}$, ω is the creep damage, and K, m and n are material constants.

The damage rate equation is given by

$$d\omega / dt = M\Delta^{\chi}(\sigma_{ij}) t^m / (1+\phi)(1-\omega)^{\phi} \tag{3.3}$$

where $\Delta(\sigma_{ij})$ is the stress function $\alpha\sigma_1 + (1-\alpha)\sigma_e$, σ_1 is the maximum principal tension stress, and α, ϕ and M are material constants. Equation (3.3) can be normalised with respect to the uni-axial stress σ_0 to yield normalised stresses, $\Sigma_{ij} = \sigma_{ij}/\sigma_0$, $\omega = 0$ at $t = 0$, and $\omega = 1$ at failure to give

$$\Delta(\Sigma_{ij}) = 1 = \alpha\Sigma_1 + (1-\alpha)\Sigma_e \tag{3.4}$$

which defines the isochronous rupture surface in normalised stress space in terms of the parameter α [11].

In this section two materials will be discussed having different multi-axial stress rupture criteria. They are copper, for which α = 0.83 [4] which is approximately described by a maximum principal tension stress rupture criteria; and, an aluminium alloy for which α = 0, which is approximately described by an effective stress rupture criterion [11]. Both of these materials will be considered in this section when tested in axi-symmetrically notched tension bars. The latter have been established as a practical laboratory technique for subjecting materials to high values of the first stress invariant J_1 (= σ_{ii}) and to low values of the effective stress σ_e. Copper and aluminium notched bars when creep tested in this way show slight notch weakening and notch strengthening respectively. That is, for notch weakening the specimen has a rupture lifetime less than the lifetime of a plain bar specimen tested at the average stress acting at the notch throat; and conversely for notch strengthening the specimen has a rupture lifetime in excess of the lifetime of a plain bar tested at the net section stress. Copper shows typically 2% notch weakening and the aluminium alloy shows approximately 28% notch strengthening. As shown by Hayhurst et al [4] CDM is capable of predicting these results using (3.2) and (3.3). The success of the approach is entirely dependent upon a knowledge of accurate values of the multi-axial stress rupture criterion α for both materials.

In design terms this is an important result, since in components which involve changes in section, and at root radii, complex stress states are always generated. For these situations design calculation methods often assume that the rupture criterion is effective stress controlled (α = 0) when in fact many practical materials have α values approaching unity.

Hayhurst et al [4] have shown that when the circular notched bar geometry is replaced by the British Standard V-notch geometry then damage accumulation behaviour becomes highly localised to the notch root. It is then more akin to damage evolution behaviour observed in creep crack growth. Despite this, CDM was shown [4] to be capable of predicting this localised behaviour and creep rupture lifetimes. This result gave the hint that CDM should be capable to predicting creep crack growth, and led to a study of the behaviour of plane strain double-edged notched tension specimens [20], the results of which are outlined in the next section.

C. Prediction of creep crack growth using CDM

A double-edged cracked tension specimen was used to test creep rupture behaviour both in copper and in an aluminium alloy. The CDM prediction showed two dramatically different predictions. The copper testpiece showed a widespread redistribution of stress due to damage growth as shown in Fig. 5a, the result being that the singularities in stress and strain at the cracked tip were quickly relaxed, and behaviour was quickly established close to that of net section stress control. Specimen lifetimes were very accurately predicted. In the case of the aluminium alloy testpiece the predicted behaviour closely agreed with that observed, which showed highly localised damage which had grown on a plane inclined at 67° to the notch plane as shown in Fig. 5b. Damage growth by this mechanism took place at a decreasing rate until net section behaviour finally took over.

Both of these studies together with others on stainless steel double-edged cracked specimens, and on compact tension specimens [21] unambiguously demonstrated that CDM computations using the finite element solver DAMAGE XX [4,12, 20] is capable of predicting creep crack growth from uni-axial creep data and a knowledge of the multi-axial stress rupture criterion of the material.

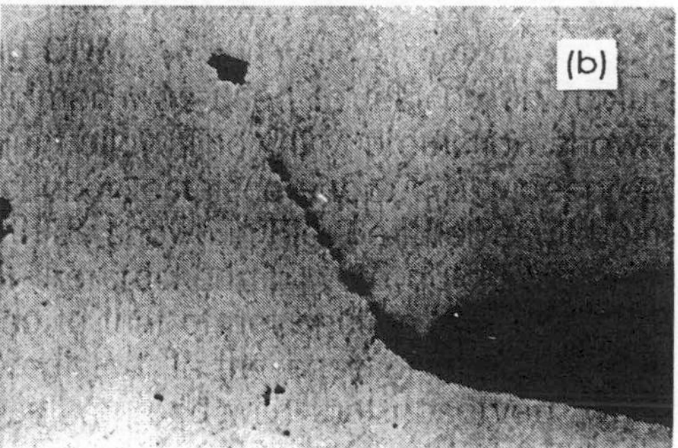

Figure 5 Mid-thickness micrographs of plane strain double-edged cracked tension specimens just before failure, (a) copper testpiece tested at 250°C and, (b) aluminium alloy testpiece tested at 210°C.

For internally and externally cracked aluminium alloy and copper specimens, predicted and experimental lifetimes agreed to within an amount equivalent to 1% on stress. For the same specimens in austenitic stainless steel experimental lifetimes are less than computed values by an amount equivalent to 4% on stress, and greater than computed values by an amount equivalent to 2% on stress for internally and externally cracked specimens respectively.

D. Overview

It has been clearly demonstrated that finite element based computations made using CDM, with DAMAGE XX, are capable of predicting a wide range of behaviours from net-section stress dominated behaviour to creep crack growth from a knowledge of:

(a) uni-axial creep curves
(b) multi-axial stress tests
(c) physically-based constitutive equations which incorporate (a) and (b).

This has been shown to work exceedingly well provided that no mechanism changes take place. Should this occur then the models require recalibration from new test data. Operation over two or more mechanisms requires the assemblage of the complete set of equations within the Continuum Damage Mechanics calculations. The use or rejection of a particular set of equations is governed by the temperature, stress, and stress state levels operating at a particular point within the structure as determined by the physics of the processes [8,9].

Provided the mechanisms are modelled correctly then the approach forms a rational approach for remnant life predictions, which are so dependent on the extrapolation of short term data to the long term often encountered in component operation.

In the next section of the paper the prediction of the creep behaviour of multi-material structures is considered. The particular problem chosen is that of a ferritic steel butt welded steam pipe subjected to: steady internal pressure. An overview of these studies is given in the next section.

3.5. Failure of a butt welded steam pipes

A. Internal pressure only

Work reported by Hall and Hayhurst [5] has investigated the possibility of characterising: (a) the creep deformation and rupture properties of the parent pipe material; (b) the heat affected zone region of the weldment; and (c) the aggregated properties of the weld material by formulating creep continuum damage mechanics constitutive equations of the type given by (3.2). They also investigated the possibility of using a finite element based creep CDM solver, DAMAGE XX [4,12], to predict the weld behaviour. The vessel to be studied is that tested by Coleman et al [22] under constant internal pressure of 45.5 MPa at a constant temperature of 565°C. A part diametral section of a butt welded pipe is shown in Fig. 6. The weld material and the pipe material are 2.25 Cr, 1 Mo, and 0.5 Cr, 0.5 Mo, 0.25 V ferritic steels respectively. The ends of the vessel were closed using unrestrained

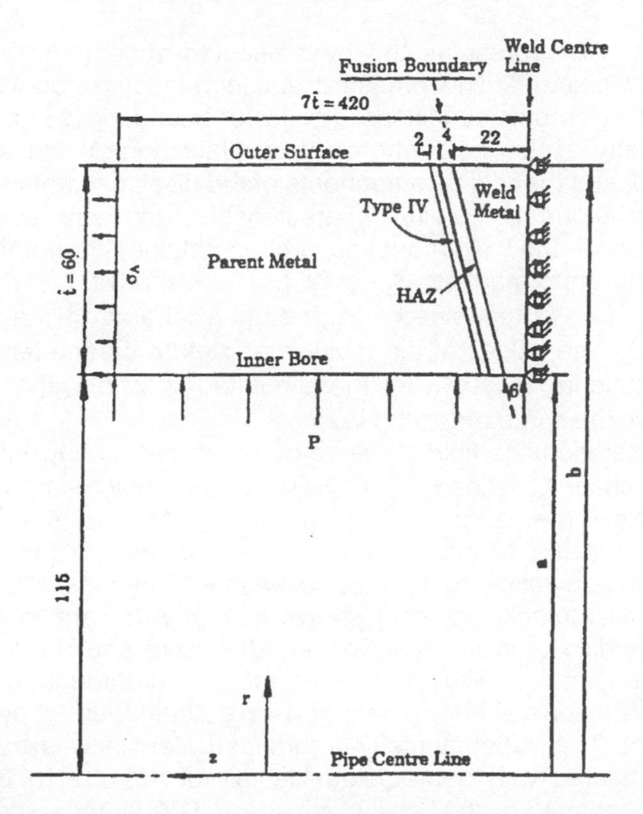

Figure 6 Part diametral section of a butt welded steam pipe showing axis of weld symmetry, geometry and boundary conditions. Dimensions are given in mm.

domed heads. The finite element idealisation chosen for the vessel is shown in Fig. 7 (a) where the three different material zones can be clearly identified by reference to the micrograph of the same section shown in Fig. 7 (b). The double

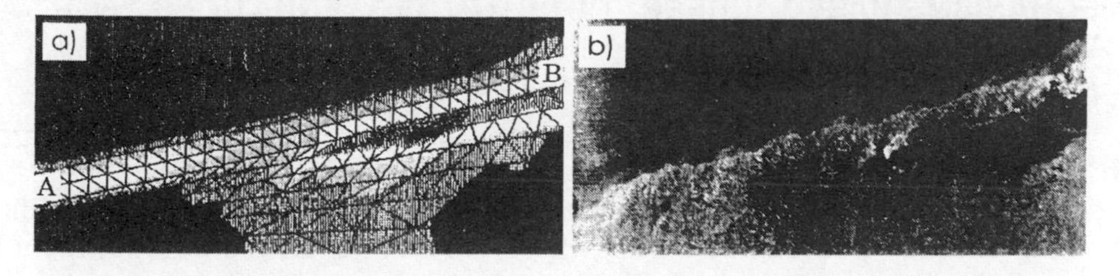

Figure 7. Predicted damage distribution (a) for the weld model shown in Fig. 6, indicating fusion boundary failure below the HAZ, denoted by AB, which may be compared with (b) the observed cracking in a diametral section at pipe failure.

row of elements AB represents the heat affected zone (HAZ), and the region below it is the weldment. The pressure and end-loading conditions have been arranged for axi-symmetric conditions. The lower boundary of Fig. 7 (a) shows the plane of symmetry of the weld where the displacement in the axial direction is maintained zero, and the radial components of the displacements are unspecified.

The creep deformation, failure characteristics, and constitutive equations of each material zone in the weld model have been obtained from uni-axial and multi-axial creep tests performed by Cane [23]. The value of the multi-axial stress rupture parameter α is determined from the uni-axial and multi-axial tests carried out by Cane [23]. The values of the multi-axial rupture parameter α are: $\alpha = 0.60$ for the parent material; $\alpha = 0.43$ for the simulated HAZ; and the weld material is assumed to have the same α as the HAZ.

The predicted damage field at failure of the vessel is shown in Fig. 7 (a). The high levels of change defined by $\omega > 0.80$ and indicated by the red zone immediately below the letter B may be interpreted as the formation of a macrocrack. The same situation may be seen in Fig. 7 (b) which shows a diametral micrograph taken from a failed vessel. The experimentally observed location of the microcrack, and that shown in Fig. 7 (a) are in extremely close agreement. In addition, on the test vessels, the axial and hoop components of strain and their variation with time were measured throughout the test. The predicted time variations of hoop strain and axial strain were shown to be in very close agreement. The experimentally-determined lifetime was approximately 46 000 h; this compares with a predicted lifetime of 43 882 h. Extremely close agreement, demonstrated the ability of the CDM approach to predict the microstructural response of the weld.

In this section a simple damage state variable has been used in the constitutive equations, and shown to provide excellent predictions of component behaviour. However, this is known to provide a satisfactory approach if one mechanism dominates. When several mechanisms are present it is necessary to use a multi-state variable approach, and this is now presented.

4. MECHANISMS-BASED CREEP CONSTITUTIVE EQUATIONS FOR AN ALUMINIUM ALLOY

4.1. Background

The amount of literature on the behaviour of aluminium alloys is very large both for uni-axial [24,25] and for multi-axial [26] stress states. Using these data, two approaches have been used to develop creep constitutive equations. The first is the phenomenological approach [27] which has been shown to be equivalent to a single state damage variable theory; and, the second is the approach of the materials scientist in which the physics of the microstructural processes are described either by experimental measurement [28] or by the use of mechanisms-based models [8]. In both approaches the descriptions used have to reflect the range of internal softening mechanisms and their relative strengths, for example:

the nucleation and growth of grain boundary cavities [29], the evolution of the dislocation sub-structure [30], and the evolution of the particulate microstructure (ageing) [31,32,33]. The effectiveness of each resulting constitutive equation has to be judged by its ability to describe the strain rates and lifetimes measured in long-term tests under both uni-axial and multi-axial stress.

Microstructural observations of damage in aluminium alloys has been made by Hayhurst [11]. Narrow grain boundary cracks were observed along grain boundary facets perpendicular to the direction of the maximum principal tension stress in both uni-axial and bi-axial stress creep tests. Similar observations were made by Johnson, Henderson and Khan [26] except that the grain boundary defects only became apparent in the latter stages of life. Both sets of evidence suggest that the growth of damage is dependent on the maximum principal tension stress σ_1.

However the bi-axial tests carried out by both Hayhurst [11] and Johnson, Henderson and Khan [26] unambiguously showed that the aluminium alloys studied did obey a maximum effective stress, σ_e, or von Mises stress rupture criterion. This is further corroborated by Hayhurst, Morrison and Leckie [34] who carried out tests in both uni-axial tension and compression and produced identical effective creep strain-time curves for the same stress level which exhibited primary, secondary and tertiary behaviour. The dichotomy between the observation of σ_1 controlled damage growth and the observed maximum effective stress rupture criterion is still in debate although a suggestion has been made that it is a consequence of tertiary creep being controlled by two damage state variables [35]. It is one of the issues that is addressed in this section.

In mechanics studies of components, ranging from stress raisers in plane stress, through notched bars, to plane strain cracked specimens, Hayhurst et al [4,10, 20, 36] have used (3.2) and (3.3).

Because of the success achieved in using these equations to predict the behaviour of aluminium components, and because it uses a single state variable theory which only strictly applies in situations where one physical mechanism dominates the rupture process, it will be used as a starting point for discussion. In the study reported here the results of three creep tests carefully conducted on an Aluminium Alloy (aluminium Cu, Fe, Ni, Mg, and Si alloy manufactured to British Standard Specification BS 1472) at 150°C are used together with a knowledge that the material satisfies a maximum effective stress rupture criterion [11]. In addition, it is well known that the stress dependence included in (3.2) and (3.3) does not describe behaviour over the entire stress range encountered in practice. To overcome this deficiency the sinh function has been used to provide a more accurate representation.

In the study, experimental data will be represented by the predictions of different theories which represent combinations of different mechanisms, and their appropriateness will be assessed with the objective of identifying the correct mechanisms. Having identified the controlling physical mechanisms for the aluminium alloy, the implications for determining the material constants in those equations will be explored.

4.2. Investigation of material behaviour predicted by different models

In the following section different models will be proposed and assessed for their ability to predict the observed uni-axial behaviour of the aluminium alloy.

A. Single state variable damage representation

Typical predictions obtained using (3.2) and (3.3) are compared with experimental results in Fig. 8 for three selected stress levels. Values of the material constants used are:

$$K = 3.511\times10^{-29}, \qquad M = 1.960\times10^{-23}, \quad \chi = 8.220,$$
$$n = 11.034, \qquad \phi = 12.107, \quad m = -0.3099;$$

and, Young's Modulus is E = 71.1 x 10^3; the units of all constants are given in MPa, h, and absolute strain. The unbroken lines represent the theoretical predictions made using (3.2) and (3.3) and the broken lines correspond to the experimental data. It may be seen from the test carried out at the stress level of 241.30 MPa that the stress sensitivity, described by the power law function in (3.2) and (3.3), does not correctly predict the primary-secondary creep rates. But most importantly, for all three tests the shape of the tertiary part of the creep curve is too abrupt, i.e. strain rates are too large. This clearly suggests that the single state variable damage description is not adequate, and that the stress sensitivity term is incorrect. The next set of equations to be considered aim to rectify these deficiencies.

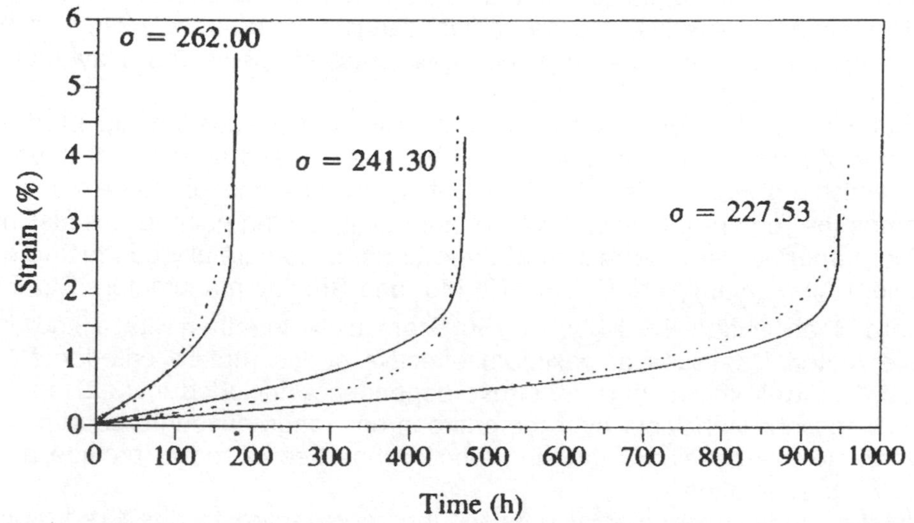

Figure 8. Comparison between experimental creep curves and theoretical predictions made using a single damage state variable theory. The solid lines denote theoretical predictions: the broken lines denote experimental results. Stress level σ is given in MPa

B. Two state variable representation

Othman, Hayhurst and Dyson [30] have developed and successfully used a multi-axial constitutive equation with two damage state variables to describe the behaviour of nickel-based superalloys. The two damage state variables represent two physical mechanisms which operate together, namely: softening which takes place in tertiary creep due to grain boundary cavity nucleation and growth, and to the multiplication of mobile dislocations. An additional feature of this model is the use of the sinh function to describe the stress sensitivity of creep rates over a wide stress range.

Most nickel-based superalloys undergoing creep spend the majority of their lifetimes in the tertiary stage of creep. Similar behaviour is observed for aluminium alloys, and so it seems appropriate to check the suitability of these equations for predicting the creep behaviour of the aluminium alloy discussed here.

The multi-axial equation set discussed by Othman et al [30] are given here with primary creep included:

$$\frac{d\varepsilon_{ij}}{dt} = \frac{3}{2}\frac{A}{(1-\omega_2)^n}\left\{\frac{\overline{S}_{ij}}{\sigma_e}\right\}\frac{1}{(1-\omega_1)}\sinh\{B\sigma_e(1-H)\}$$

$$\frac{dH}{dt} = \frac{h}{\sigma_e}\frac{A}{(1-\omega_2)^n}\frac{1}{(1-\omega_1)}\sinh\{B\sigma_e(1-H)\}\left(1-\frac{H}{H^*}\right)$$

$$\frac{d\omega_1}{dt} = CA\frac{(1-\omega_1)}{(1-\omega_2)^n}\sinh\{B\sigma_e(1-H)\} \tag{4.1}$$

$$\frac{d\omega_2}{dt} = DA\left(\frac{\sigma_1}{\sigma_e}\right)^{\nu} N\frac{\sinh\{B\sigma_e(1-H)\}}{(1-\omega_1)(1-\omega_2)^n}$$

where $n = B\sigma_e(1-H)\coth(B\sigma_e(1-H))$ A, B, H*, h, C, and D are material constants. The first damage state variable, ω_1, is defined from the physics of dislocation softening [30] to lie within the range 0 to 1 for mathematical convenience, and the second, ω_2 is defined from the physics of nucleation-controlled creep-constrained cavitation to vary from zero at t=0 to $\omega_2 = 0.3$ at failure.

Equation set (4.1) has been used to describe the behaviour of the same uni-axial test results for aluminium alloy presented in Fig. 8 using the following material constants:

$A = 2.96 \times 10^{-11}$, $B = 7.17 \times 10^{-2}$, $h = 1.37 \times 10^{5}$,
$H^* = 0.2032$, $C = 35$, $D = 6.63$,

the units of all constants are given in MPa, h and absolute strain. The parameters have been determined using the techniques discussed earlier in this section. Theoretically predicted values are compared with experimental data in Fig. 9.

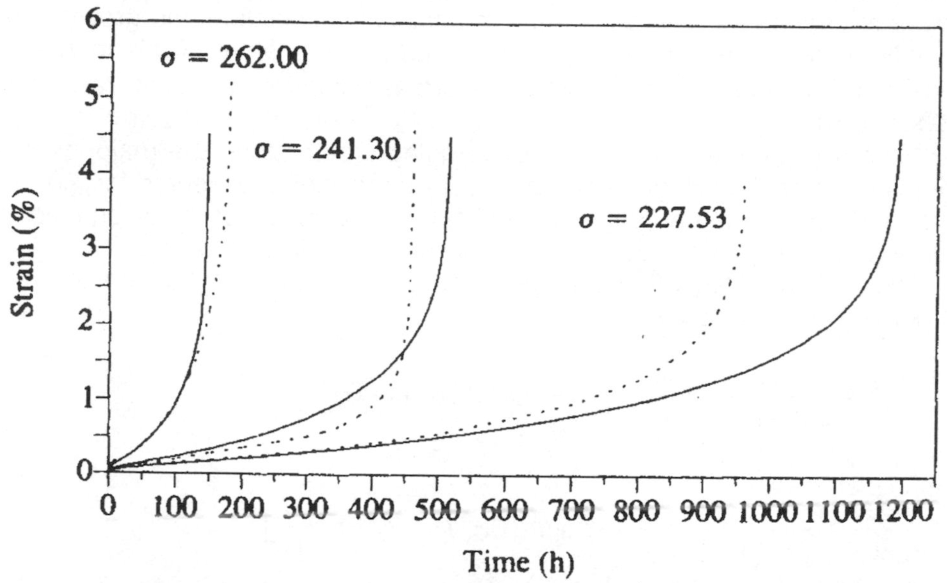

Figure 9. Comparison between experimental creep curves and theoretical predictions using a two damage state variable theory. The theory models damage evolution due to creep constrained cavitation and the multiplication of mobile dislocations. The solid lines denote theoretical predictions: the broken lines denote experimental results. Stress level σ is given in MPa

The primary and secondary creep strain levels and rates are well described due to the presence of the sinh term, and of the primary creep state variable, H, in equation set (4.1). The rupture ductility and the characteristic shape of the tertiary region are both well predicted, however the lifetimes are considerably in error. Following sensitivity studies carried out on the effect of the parameter C, in the third equation of set (4.1), on the quality of the theoretical predictions, it has been found that:

(a) The presence of the two state variables ω_1, and ω_2 provides an additional degree of freedom over that provided by (3.2) and (3.3), which enables the characteristic shape of the tertiary region to be predicted.
(b) That if a value of C is selected which gives a closer, but unsatisfactory, prediction of lifetimes, then the quality of the secondary creep predictions is highly unsatisfactory.

It is therefore reasonable to conclude that a two damage state variable theory is required to predict the shape of the tertiary region, but that the model given by equation set (4.1) which describes the multiplication of mobile dislocations is inappropriate. The governing second state-variable is thought to be that which models ageing, and equation set (4.1) will now be modified accordingly.

C. Two state variable model for damage due to creep cavitation and ageing

In the following sub-section a new constitutive equation set, proposed by Dyson [37] and based upon a new mechanism of creep in particle-hardened alloys, is applied to the aluminium alloy data. The stress level dependence of creep rate is described by a sinh function, and two damage state parameters are used to model tertiary creep softening caused by (a) grain boundary nucleation and growth, and (b) ageing of the particulate microstructure. Primary creep is also described by the model.

i) Uni-axial behaviour

The form of the constitutive equations proposed is given by:

$$\frac{d\varepsilon}{dt} = \frac{A}{(1-\omega_2)^n}\sinh\left[\frac{B\sigma(1-H)}{1-\Phi}\right]$$

$$\frac{dH}{dt} = \frac{h\,d\varepsilon}{\sigma\,dt}\left(1-\frac{H}{H^*}\right) \tag{4.2}$$

$$\frac{d\Phi}{dt} = \frac{K_c}{3}(1-\Phi)^4$$

$$\frac{d\omega_2}{dt} = \frac{DA}{(1-\omega_2)^n}\sinh\left(\frac{B\sigma(1-H)}{1-\Phi}\right)$$

where A, B, H^*, h, K_C and D are material constant and where n is given by

$$n = \frac{B\sigma(1-H)}{(1-\Phi)}\coth\left(\frac{B\sigma(1-H)}{1-\Phi}\right)$$

which approximates to $\{(B\sigma(1-H)/(1-\Phi)\}$ for most cases of interest. Material parameters, which appear in this model may be divided into three groups, namely (a) the constants h and H^* which describe primary creep; (b) the parameters A and B which characterise secondary creep; and (c) the parameters K_C and D responsible for damage evolution.

The equation set contains two damage state variables used to model tertiary softening mechanisms. The first damage state variable, Φ, is defined from the physics of ageing [38] to lie within the range 0 to 1 for mathematical convenience.

The second damage variable, ω_2, describes grain boundary creep constrained cavitation the magnitude of which is strongly sensitive to alloy composition and to processing route. The second equation in set (4.2) describes primary creep using the variable H, which varies from 0 at the beginning of the creep process to H^*, where H^* is the saturation value of H.

The equation set (4.2) has been used to predict the aluminium alloy behaviour for the data presented in Fig. 8. Comparison of the results predicted using the materials constants given in Table 4.1 with experimental data is given in Fig. 10. The figure shows that primary creep strains, minimum creep rates, the shapes of

Table 4.1. Optimised material constants

A $(h)^{-1}$	B $(MPa)^{-1}$	h (MPa)	H^* (-)	K_C $(h)^{-1}$	D (-)
2.960×10^{-11}	7.167×10^{-2}	1.370×10^{5}	0.2032	19.310×10^{-5}	6.630

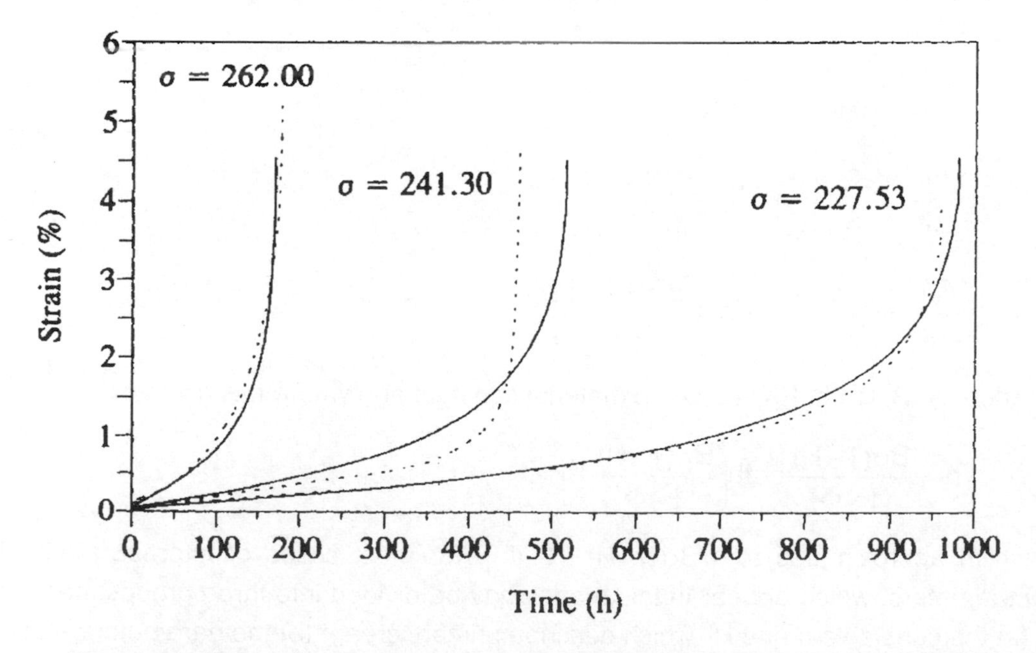

Figure 10. Comparison between experimental creep curves and theoretical predictions made using a two damage state variable theory. The theory models damage evolution due to creep constrained cavitation and ageing. The solid lines denote theoretical predictions and the broken lines denote experimental results. Stress level σ is given in MPa.

tertiary creep and lifetimes are all reasonably well predicted for the stress range studied. Hence, for uni-axial creep, the damage mechanisms of creep constrained cavitation and ageing appear appropriate.

ii) Multi-axial behaviour

The multi-axial behaviour of the aluminium alloy is now discussed under the headings of deformation and rupture and their normalisation.

a) Deformation

The strain rate equation in set (4.2) is now considered without the damage parameters Φ, ω_2, and primary creep variable H. It then leads to a relation of the following form:

$$d\varepsilon / dt = A \sinh (B\sigma) \tag{4.3}$$

This equation can be generalised for multi-axial conditions by assuming an energy dissipation rate potential:

$$\Psi = \frac{A}{B}\cosh(B_e) \tag{4.4}$$

Assuming normality and the associated flow rule, the multi-axial relation is given by:

$$\frac{d\varepsilon_{ij}}{dt} = \frac{\partial \Psi}{\partial \overline{S}_{ij}} = \frac{3}{2} A \left(\frac{\overline{S}_{ij}}{\sigma_e} \right) \sinh(B\sigma_e) \tag{4.5}$$

and by the introduction of the following terms:

$$\Sigma_{ij} = \sigma_{ij} / \sigma_o; \qquad \lambda_{ij} = \varepsilon_{ij} / \varepsilon_o; \qquad S_{ij} = \overline{S}_{ij} / \sigma_o$$

where $\varepsilon_o = \sigma_o / E$. The normalised time may be defined as:

$$\tau = \int_0^t (EA / \sigma_o) dt = (EA/\sigma_o) t = (A/\varepsilon_o) t$$

The constitutive equations set (4.2) may be rewritten using these parameters as:

$$\begin{aligned}
\frac{dv_{ij}}{d\tau} &= \frac{3}{2(1-\omega_2)^n} \left(\frac{S_{ij}}{\Sigma_e} \right) \sinh\left[\frac{\alpha \Sigma_e (1-H)}{1-\Phi} \right] \\
\frac{dH}{d\tau} &= \frac{h}{E\Sigma_e} \frac{1}{(1-\omega_2)^n} \sinh\left[\frac{\alpha \Sigma_e (1-H)}{1-\Phi} \right] \left(1 - \frac{H}{H^*} \right) \\
\frac{d\Phi}{d\tau} &= \frac{\varepsilon_o}{A} \frac{K_c}{3} (1-\Phi)^4
\end{aligned} \tag{4.6}$$

$$\frac{d\omega_2}{d\tau} = \frac{D\varepsilon_o}{(1-\omega_2)^n}\left(\frac{\Sigma_1}{\Sigma_e}\right)^{\nu} N \sinh\left[\frac{\alpha\Sigma_e(1-H)}{1-\Phi}\right]$$

where $\alpha = B\sigma_o$ and $n = \frac{\alpha\Sigma_e(1-H)}{1-\Phi}\coth\left[\frac{\alpha\Sigma_e(1-H)}{1-\Phi}\right]$.and N is a parameter used to indicate the state of loading; e.g. for σ_1 tensile, N = 1; and for σ_1 compressive, N = 0.

b) Rupture

Multi-axial rupture results are conveniently plotted in terms of isochronous surfaces (i.e. the shape of the locus of points having the same rupture time). Hayhurst [11] has investigated the bi-axial creep rupture behaviour of metals and alloys, and concluded that there are two extreme types of stress sensitive rupture behaviour, namely those described by the maximum principal stress sensitive rupture criterion on the one hand, and the effective stress sensitive rupture criterion on the other.

To achieve correspondence with the multi-axial behaviour obtained by Hayhurst [11] for aluminium alloys the isochronous rupture locus must be effectively stress controlled. In equation set [10] the last three expressions can be integrated for levels of σ_1 and σ_e to give lifetime t_o and then normalised with respect to the uni-axial stress σ_o required to give the same lifetime. The results

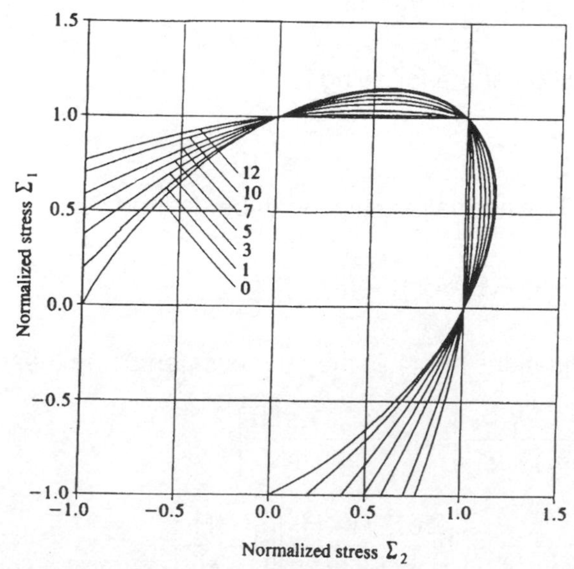

Figure 11. Isochronous rupture loci for bi-axial plain stress determined for damage evolution due to creep constrained cavitation with ageing using equation set (4.6) for σ_o = 262 MPa. The loci are given for the range of values of the stress state sensitivity index ν marked on the figure, ν = 12, 10, 7, ..., 0.

may be plotted on the normalised axes Σ_1 and Σ_2 to give isochronous rupture loci, as shown in Fig. 11 for $\nu = 0, 1, 3, 5, 7, 10, 12$ at the stress $\sigma_o = 262$ MPa. The loci have been found to be independent of ω_2 over the range of values $0.3 \geq \omega_2 \geq 0$. However, when the stress σ_o is changed the shape of isochronous rupture loci is also changed. It was found that in calculations of the loci for constant ω_2, their shape remained constant for a given stress.

In the next section, these equations are used to study the effect of the presence of extensometer ridges on cylindrical creep testpieces which are frequently used to establish materials databases.

5. EXPERIMENTAL AND THEORETICAL EVALUATION OF HIGH ACCURACY UNI-AXIAL CREEP TESTPIECES WITH SLIT EXTENSOMETER RIDGES

5.1. Background

The measurement of uni-axial creep strain in laboratory tests is frequently carried out using cylindrical bar testpieces on which ridges have been machined to identify the gauge length over which strain is to be measured. Mechanical extensometers are fitted to these ridges which are used to transfer the displacements occurring during creep to a location outside of the high-temperature furnace where low cost transducers can accurately measure displacements at ambient temperatures. The measured displacements are then used to compute the variation of strain with time. This type of testpiece and extensometer will be studied within this section.

A previous theoretical study carried out by Lin, Hayhurst and Dyson [39] has shown that the uni-axial strain measured in tensile testpieces using ridged specimens and extensometers do not agree with the true strains in the parallel section of the testpiece. The error levels have been shown by Lin, Hayhurst and Dyson [40] to be dependent upon the applied stress level, but principally upon the size of the gauge length. For Nickel superalloy specimens, of diameter 7.65 mm, with gauge lengths of 51 mm errors have been predicted in excess of 10%. For gauge lengths of 10 mm the error increases to typically 30%. These figures are for the low stress level of 118 MPa, which can be expected to double for the stress level of 250 MPa. Whilst gauge lengths of 10 mm are not frequently used in creep testing, they are however used in combined cyclic plasticity and creep testing.

The reason for these high errors has been shown by Lin, Hayhurst and Dyson [40] to be due to the circumferential reinforcement of the testpiece provided by the extensometer ridges. The ridge perturbs the stress, strain, and damage fields above and below the ridge, with the perturbations extending a distance of typically 1.5 times the diameter of the parallel sided region of the testpiece. The circumferential stress generated in the extensometer ridge is predominantly compressive and Lin, Hayhurst and Dyson [40] have shown how these stresses may be relieved by the introduction of slits into the ridges; and how the errors in measured creep strains can, as a consequence, be reduced by a factor of typically two. The same circumferential reinforcement effect has been demonstrated experimentally by Ohashi et al [41] in tubular specimens subjected to internal pressure. They tested a range of different specimens in which the geometry of the extensometer ridge was varied and shown to strongly affect the measured strains.

Since the results due to Lin, Hayhurst and Dyson [39, 40] are theoretical, there is a need to investigate them further, and to verify them experimentally; this is the major aim of the research reported here. The results of experiments are reported which have been carried out on unslitted and slitted extensometer ridged testpieces with different gauge lengths. The experimental results are compared with those determined theoretically from a knowledge of the uni-axial constitutive equations, and the multi-axial rupture criterion of the material. A precipitation hardened aluminium-magnesium-silicon alloy[1] has been selected for the experimental investigation and tests are reported which have been carried out at 150 ± 0.5°C. The experimental results obtained on testpieces with unslitted extensometer ridges are used to determine the true constitutive equations for the material. The constitutive equation set (4.6) model primary creep, ageing, and creep constrained cavitation as reported by Kowalewski, Hayhurst and Dyson [17]; and, they are used here in a Continuum Damage Mechanics Finite Element analysis, performed by the solver DAMAGE XX [5, 20], to predict the measured creep curves of the slitted extensometer ridged testpieces. These predictions are then compared with the experimental results. Conclusions are made on the effectiveness of the slit extensometer ridged testpiece design at achieving high-accuracy measurements of uni-axial creep strains.

5.2. Experimental programme

A. Material selection

The aluminium alloy was selected because of its availability, and for the existence of results previously obtained on this material by Kowalewski, Hayhurst and Dyson [17] for a different sample of the material, but prepared from the same material cast. It was convenient to test the material at a lower temperature of 150 ± 0.5°C; and, to select the material for its ease of machinability of the slitted extensometer ridges.

B. Testpiece manufacture

The basic testpieces, Fig. 12, were machined using a DNC Lathe to create the cylindrical specimen form, and a vertical machining centre was used to machine the end flats and loading pin holes. The extensometer ridges were slitted using a dressed circular slitting saw mounted with its arbor located vertically in the machining centre whilst the axis of the testpiece was mounted horizontally in a dividing head. A finished machined extensometer ridge is shown in Fig. 13. Twenty four slits were machined into each extensometer ridge to reduce the constraint to deformation in the hoop direction, in this way the slitted extensometer ridges were subjected mainly to conditions of plane stress. In addition to relieving the hoop stresses, the constraint to axial deformation provided by the

[1] Aluminium, Cu, Fe, Ni, Mg and Si alloy manufactured to British Standard Specification B.S. 1472.

extensometer ridges was also weakened by the introduction of the slits, since about 2/3 of the extensometer ridge material was removed.

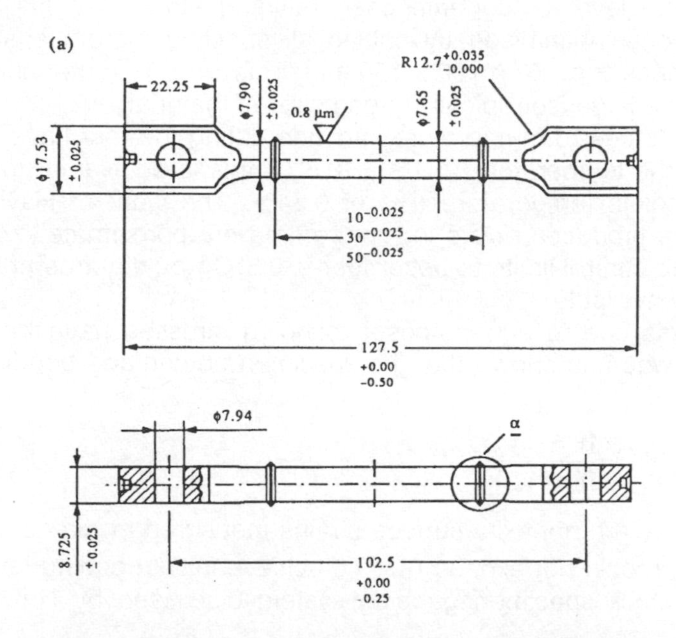

Figure 12. Engineering drawing of the testpiece showing plan and side elevation

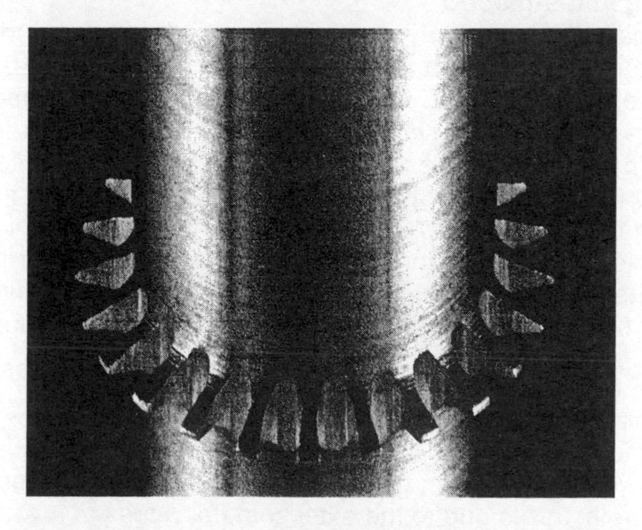

Figure 13. Photograph of slitted extensometer ridge

C. Experimental technique

There are three factors which are known to significantly influence the accuracy and repeatability of creep tests. They are: material variation; temperature control; and the level of superimposed bending stress. In this work material variation has been minimised by cutting all specimens from a single block of material of dimensions 120 x 120 x 150 mm. This block of material has been itself machined from a large isotropically forged billet of material.

The effects due to temperature variation during the test have been reduced by controlling the temperature to 150 ± 0.5°C; this is equivalent to a percentage change in absolute temperature (°K) of 0.24%. The work of Hayhurst [42] has shown that this produces errors in rupture time of approximately 7%. To reduce the temperature control limits to better than ± 0.5°C would not be practical with the test machinery available.

The effects due to superimposed bending stresses have been studied by Hayhurst [42] who has shown that the specimen percentage bending determined from

$$\text{Bending } (\%) = \frac{(\varepsilon_1 - \varepsilon_2)}{(\varepsilon_1 + \varepsilon_2)} \times 100$$

where ε_1 and ε_2 are uni-axial surface strains measured at diametrically opposed surface points should not exceed 6%. To achieve initial bending levels below 6% the universal block specimen gripping system discussed by Hayhurst [42] has been used.

In these ways the effect of material variability, temperature variation, and superimposed specimen bending stress have all been reduced.

D. Unslitted ridged testpiece

To investigate the effect of gauge length on the measured creep curves, three tests were carried out with gauge lengths of 50, 30 and 10 mm, all at the same stress level of 250.0 MPa. The measured creep curves are shown by the solid lines in Fig. 14. It may be seen that the creep curve for the testpiece gauge length of 50 mm has creep rates which are typically twice those for the creep curve measured for a gauge length of 10 mm. Also the lifetime for the testpiece with a gauge length of 50 mm is 76 hours, and that for the gauge length of 10 mm is 187 hours. This result shows the dramatic strengthening effect provided by the testpiece with the shorter gauge length. The strengthening effect is due to the elevation of the first stress invariant $(=\sigma_{ii})$ and the resultant suppression of the effective stress $\sigma_e (= (3S_{ij}S_{ij}/2)^{1/2})$, where S_{ij} is the deviatoric stress tensor) in the region of the extensometer ridges. When the extensometer ridges are sufficiently close an interaction takes place between the perturbed fields at each ridge. The experimental curve for the 30 mm gauge length is very little different to that for the 50 mm gauge length; this is due to the extensometer ridges

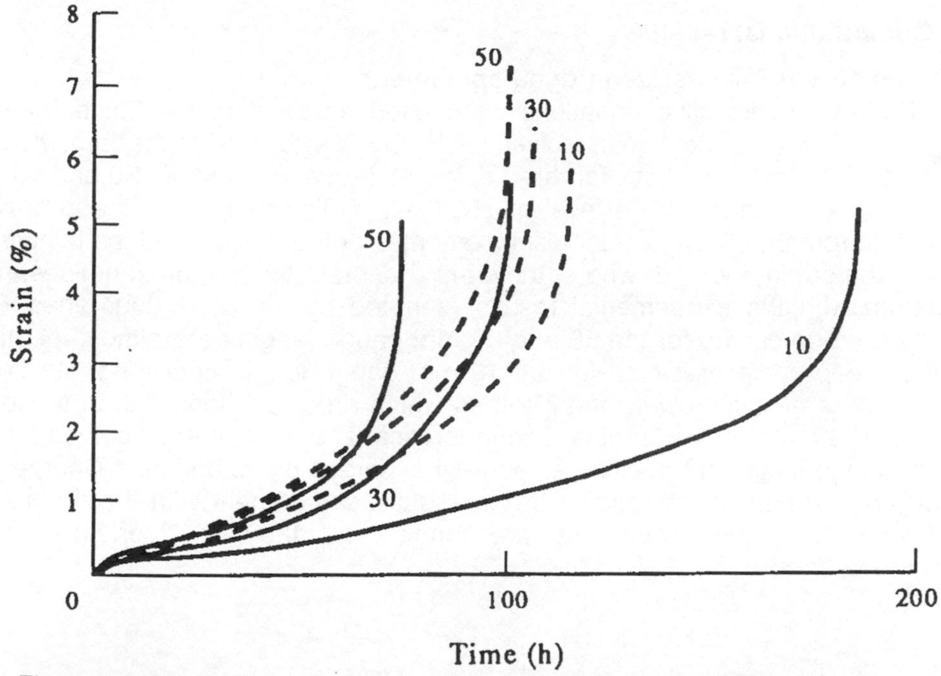

Figure 14. Comparison of experimental creep curves for unslitted and slitted ridged testpieces denoted by continuous and broken lines, respectively, for tests carried out at the stress level of 250.0 MPa. The numbers given on the figure denote the gauge lengths of 50, 30 and 10 mm.

E. Slitted ridged testpieces

To investigate the effect of gauge length on the creep curves measured by the slitted testpieces, tests have been carried out at a stress level of 250.0 MPa for the three gauge lengths 50, 30, 10 mm used with the unslitted testpieces. The resulting creep curves are shown in Fig. 14 by the broken lines. It may be seen that the measured creep rates in all three tests are much closer than those for the unslitted testpieces. The lifetimes are much closer than those for the unslitted testpieces; for example in the case of the 50 mm gauge length the lifetime is 100 hours and that for the 10 mm gauge length is 117 hours for the slitted ridges, compared with 76 hours and 187 hours, respectively, for the unslitted ridges. This clearly shows that the effect of the slitted extensometer ridges is to reduce the degree of interaction between the stress and strain fields at the ridges, particularly in the case of the 10 mm gauge length. Again, for the 50 and the 30 mm gauge length testpieces the creep curves are quite close to each other for the same reasons discussed in the case of the unslitted testpieces. This clearly demonstrates the improved accuracy of strain and lifetime measurements made by the slitted ridged testpiece.

5.3. Computational results

A. Creep curves for unslitted ridged specimens

The true constitutive equations were used as input to the Finite Element Continuum Damage Mechanics solver DAMAGE XX to predict the behaviour of unslitted creep test testpieces for three different gauge lengths: 50, 30 and 10 mm subjected to the same stress level of 250.0 MPa. The theoretical creep curves, computed from the extensometer displacements, for the unslitted ridged testpieces are presented in Fig. 15 where they are denoted by the broken lines, and compared with the experimental results, denoted by the continuous lines. The predicted creep curves for the 50 and 30 mm gauge lengths are almost identical, but the creep curve predicted for the 10 mm gauge length specimen has creep rates which are much lower, and lifetimes which are much longer than those for the 50 and 30 mm gauge lengths. The theoretical and experimental curves for the 10 mm gauge length are in close agreement, whereas the experimental curves for the 50 and 30 mm gauge lengths are in slight disagreement with the theoretical predictions; but, the differences are small, and they are certainly within experimental error.

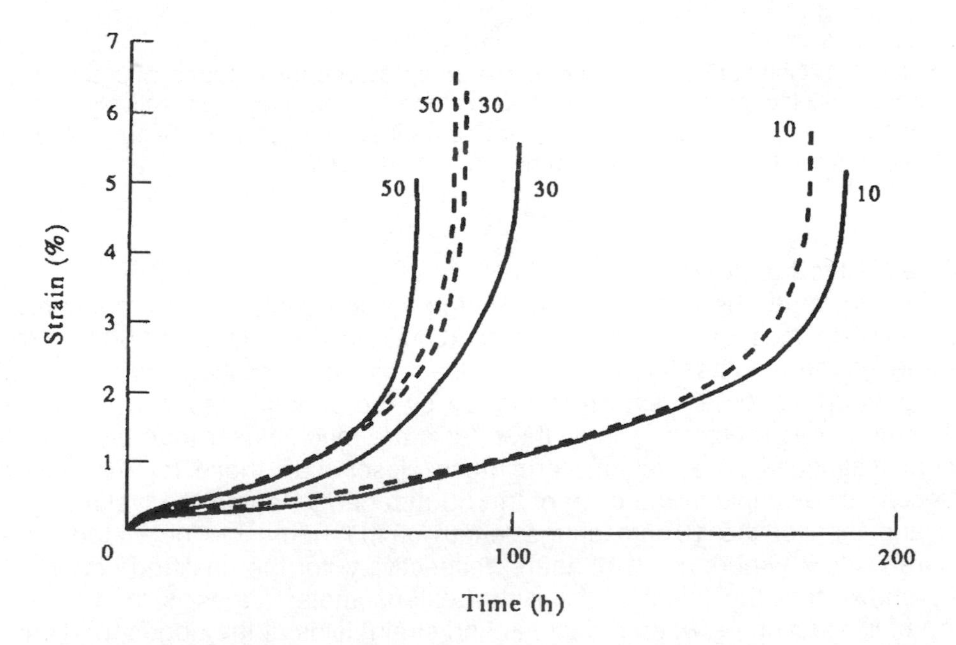

Figure 15. Comparison of creep curves for unslitted ridged testpieces of 50, 30 and 10 mm gauge lengths under the stress level of 250 MPa, computed from the extensometer displacement using the solver DAMAGE XX (broken lines), with the experimental curves presented previously in Fig. 14.

B. Creep curves for slitted ridged specimens

The true constitutive equations have been used in the solver DAMAGE XX to predict the creep curves at the stress level of 250.0 MPa for the slitted ridged testpiece; the curves are shown in Fig. 16 using the broken lines for the three gauge lengths: 50, 30, and 10 mm, where they are compared with the results of experiments denoted by the continuous lines. The predicted creep curves for the 50 and 30 mm gauge lengths are very close. The predicted creep curve for the 10 mm gauge length specimen, as in the case of the unslitted ridged specimen, shows creep rates which are lower, and lifetimes which are longer than for the 50 and 30 mm gauge lengths. However, the predicted degree of strengthening, expressed as a ratio of lifetime of the 10 mm gauge length testpiece to that for the 50 mm gauge length testpiece is 1.37; this compares with a value of the same ratio of 2.06 for the unslitted ridged testpiece. This theoretical result clearly confirms the experimental observations, and that the presence of the slits in the extensometer ridges relieves the constraint to deformation provided by the unslitted extensometer ridges.

By comparison of the predicted creep curves (broken lines) of Figs 15 and 16, in the primary and secondary stages, it may be seen that the creep strains for the 10 mm gauge length slitted ridged testpiece have increased relative to those for the unslitted ridged testpiece. Hence producing results closer to the expected curves, approximated to by those given for the 50 and 30 mm gauge length testpieces.

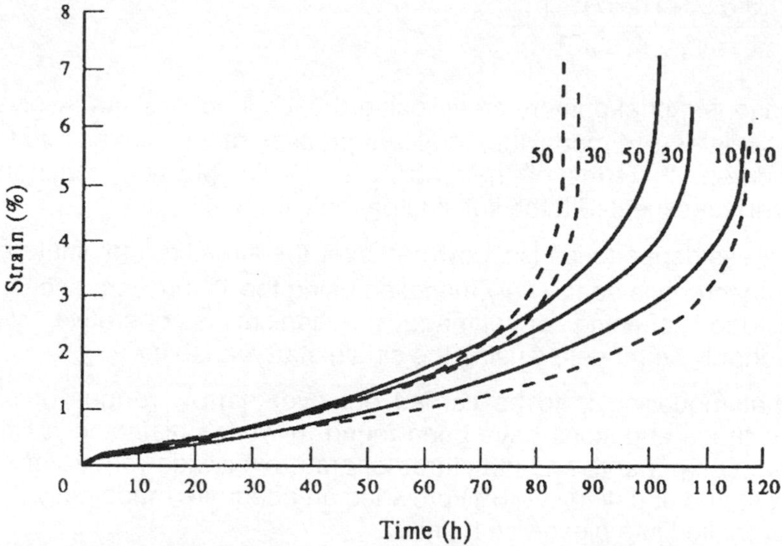

Figure 16. Comparison of creep curves for slitted ridged specimens of different gauge lengths computed from the extensometer displacements using the solver DAMAGE XX (broken lines), with the experimental creep curves (continuous lines). The numbers given on the figure denote the gauge lengths of 50, 30 and 10 mm. All creep curves are for the stress level of 250 MPa.

Comparison of strain levels in Fig. 16 for the theoretical curves with those for the experimental curves illustrates how primary and secondary creep are closely predicted; but, how the tertiary creep behaviour is somewhat in error for both the 50 and 30 mm gauge lengths. The errors between the theoretical and experimental curves for the 10 mm gauge length are small.

This clearly shows that, within experimental error, the true constitutive equations may be used to predict the creep strain behaviour of the slitted ridged testpieces. This section also highlights the importance of using the correct experimental procedures to establish materials databases for use in design. In the next section generalised procedures are discussed for the identification of mechanisms-based constitutive equations.

6. PHYSICAL MECHANISMS CONSTITUTIVE EQUATIONS AND THEIR CALIBRATION

6.1. Single damage state parameter model

This model has been introduced through (3.2) and (3.3) which can be written in the normalised form:

$$\begin{aligned} \frac{dv_{ij}}{dt} &= K\frac{3\sum_{e}^{n-1} t^{m}}{2(1-\omega)} \\ \frac{d\omega}{dt} &= \frac{M\left[\alpha\Sigma_{1} + (1-\alpha)\Sigma_{e}\right]^{\chi} t^{m}}{(1+\phi)(1-\omega)^{\phi}} \end{aligned} \qquad (6.1)$$

These equations model primary creep using the t^m function, and involve a single damage parameter ω to model the dominant process responsible for softening due to tertiary creep. In terms of their ability to model physical mechanisms the equations can be expected to be satisfactory only if:

(a) the stress dependence is n powered over the stress range of interest;
(b) primary creep is adequately modelled using the t^m function; and,
(c) only one softening or damage mechanism is operative, which can reasonably be modelled using the single state variable.

Provided that modelling is to be carried out over narrow ranges of stress and temperature these equations have been found to work satisfactorily. However, in practice, real components and structures operate over wide ranges of stress and temperatures, and equation set (6.1) provides an adequate model only if the stress domain is modelled in a piecewise manner.

The major deficiency is that the approach is only satisfactory over the range of data used to calibrate the model, and hence it is not possible to use it to extrapolate.

6.2. Multi-state variable modelling of aluminium alloys

The equations (4.6) determined in the previous section for an aluminium alloy are summarised here, for convenience, in normalised form for multi-axial conditions:

$$\frac{dv_{ij}}{dt} = \frac{3}{2}\frac{A}{(1-\omega)^n}\left(\frac{S_{ij}}{\sigma_e}\right)\sinh\left[\frac{B\sigma_e(1-H)}{(1-\Phi)}\right] \tag{6.2a}$$

$$\frac{dH}{dt} = \frac{h}{\sigma_e}\frac{A}{(1-\omega_2)^n}\sinh\left[\frac{B\sigma_e(1-H)}{(1-\Phi)}\right]\left(1-\frac{H}{H^*}\right) \tag{6.2b}$$

$$\frac{d\Phi}{dt} = \frac{K_c}{3}(1-\Phi)^4 \tag{6.2c}$$

$$\frac{d\omega_2}{dt} = \frac{DA}{(1-\omega_2)^n}\left\{\frac{\sigma_1}{\sigma_e}\right\}^{\nu}\sinh\left\{\frac{B\sigma_e(1-H)}{(1-\Phi)}\right\} \tag{6.2d}$$

In this equation set primary creep is represented by the parameter H; for constant stress, and with all other variables negligible, its evolution with time is shown in Fig. 17. At zero time its value is zero and it then increases to its saturation value H* at the end of primary creep. The parameter H models the change in dislocation structures during primary creep. Equation 12c models stress independent ageing due to particle coarsening which is shown schematically at the grain level, in Fig. 18. The parameter Φ monotonically increases from zero to unity. Equation 12d describes grain boundary creep constrained cavitation which is shown schematically in Fig. 19. Creep constrained cavitation can be either nucleation or growth controlled, for the aluminium alloy considered here only nucleation control will be considered.

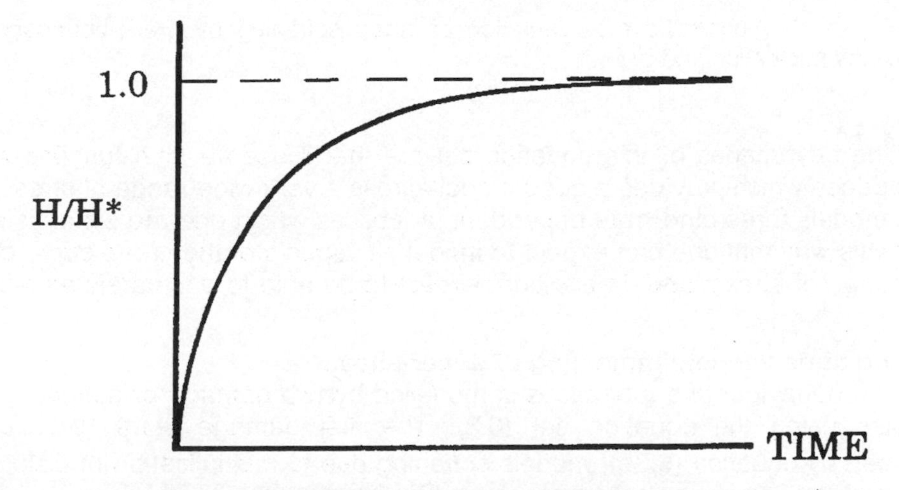

Figure 17. Change of Primary Creep State variable with time

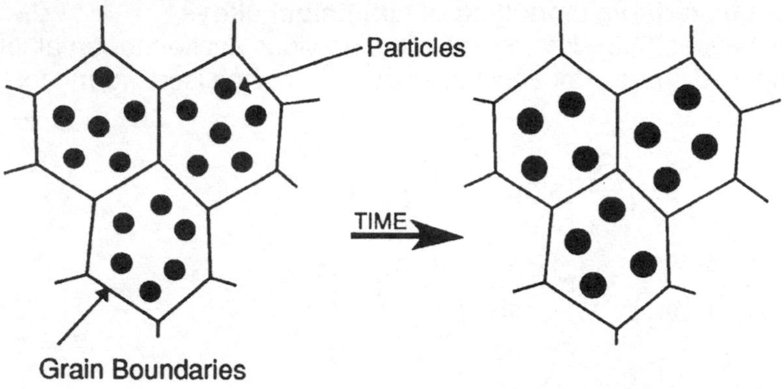

Figure 18. Stress independent ageing and particle coarsening

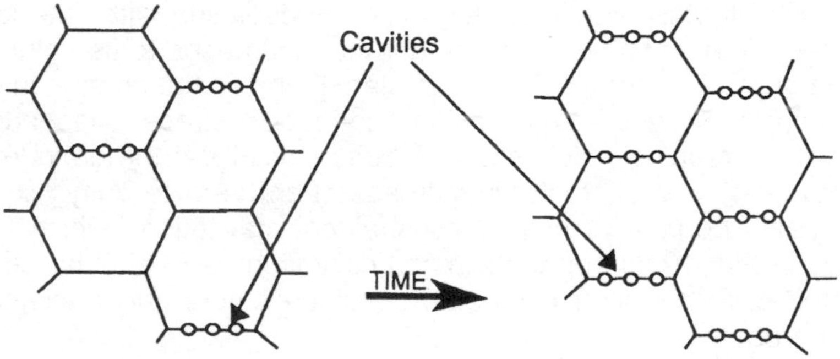

Figure 19. Schematic representation of creep softening by grain boundary cavity nucleation and growth

The advantages of this equation set are that it uses a sinh function stress dependence which provides a good model across a very wide range of stress, and that it models three other rate dependent processes which operate synergistically. It is in this way that one can expect to model all aspects of the creep curve over a wide range of stress; and, in addition, expect to be able to accurately extrapolate data.

6.3. Two state variable modelling of super-alloys

The behaviour of super-alloys is modelled by two damage or softening state variables using the equation set (6.3). The first damage state variable ω_1, described by equation (6.3b), models softening due to multiplication of dislocation sub-structures, as shown schematically in Fig. 20, and the second damage

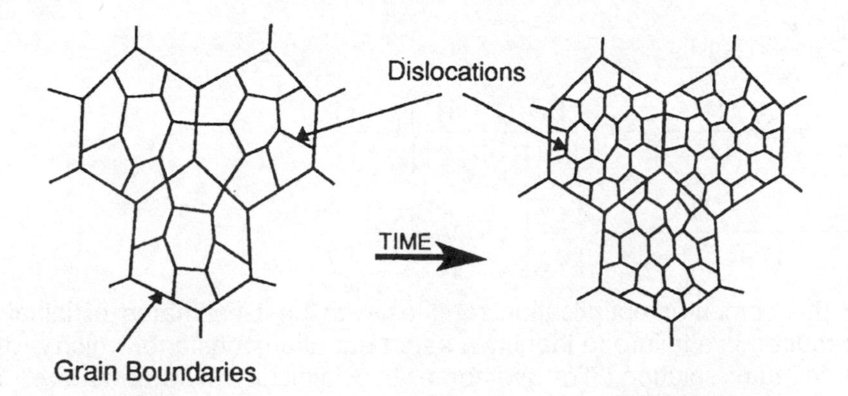

Figure 20. Schematic representation of creep softening by multiplication of dislocation sub-structures

variable ω_2, described by (6.3c), models softening due to grain boundary cavity nucleation and growth [17]

$$\frac{d\varepsilon_{ij}}{dt} = \frac{3}{2}\frac{A}{(1-\omega_1)(1-\omega_2)^2}\left(\frac{S_{ij}}{\sigma_e}\right)\sinh(B\sigma_e) \quad (6.3a)$$

$$\frac{d\omega_1}{dt} = CA\frac{(1-\omega_1)}{(1-\omega_2)^2}\sinh(B\sigma_e) \quad (6.3b)$$

$$\frac{d\omega_2}{dt} = \frac{DA}{(1-\omega_1)(1-\omega_2)^n}\left(\frac{\sigma_1}{\sigma_e}\right)^{\nu}\sinh(B\sigma_e) \quad (6.3c)$$

6.4. Synergy between many internal physical processes

In a previous section it was described how the appropriate combination of mechanisms required to describe the behaviour of an Aluminium alloy was identified. The formalism used was to take the general set of equations, which incorporates all internal state variables and their respective couplings, as outlined in equation set (6.4). However this equation set contains nine independent constants, each of which has to be determined to provide an adequate fit to the experimental data. Attempts to use available numerical, schemes such as those employed by Dunne et al [43] for simpler equation sets have not proved successful.

$$\frac{d\varepsilon_{ij}}{dt} = \frac{3}{2}\frac{A}{(1-\omega_1)(1-\omega_2)^2}\left(\frac{S_{ij}}{\sigma_e}\right)\sinh\left[\frac{B\sigma_e(1-H)}{(1-\Phi)}\right] \quad (6.4a)$$

$$\frac{dh}{dt} = \frac{h}{\sigma_e}\frac{A}{(1-\omega_1)(1-\omega_2)^n}\sinh\left\{\frac{B\sigma_e(1-H)}{(1-\Phi)}\right\}\left\{1-\frac{H}{H^*}\right\} \quad (6.4b)$$

$$\frac{d\Phi}{dt} = \frac{K_c}{3}(1-\Phi)^4 \quad (6.4c)$$

$$\frac{d\omega_1}{dt} = CA\frac{(1-\omega_1)}{(1-\omega_2)^n}\sinh\left\{\frac{B\sigma_e(1-H)}{(1-\Phi)}\right\}\left\{1-\frac{H}{H^*}\right\} \quad (6.4d)$$

$$\frac{d\omega_2}{dt} = \frac{DA}{(1-\omega_1)(1-\omega_2)^n}\left\{\frac{\sigma_1}{\sigma_e}\right\}^{\nu}\sinh\left\{\frac{B\sigma_e(1-H)}{(1-\Phi)}\right\} \quad (6.4e)$$

Entry to the computer optimisation routines with best estimates of initial values always resulted in a failure to identify a set of equation constants which provided a global minimum solution. To overcome this difficulty Kowalewski et al [17] developed a new approach. It involves taking each mechanism or its allied equation and isolating it from the others in such a way that a good estimate of the constants associated with the mechanism can be determined. When good estimates have been determined for as many parameters as possible, the values are used as starting values in the numerical optimisation of the Kernel function:

$$\text{Minimum} = \sum_{I=1}^{a}\left\{\sum_{j=1}^{b}\left(\varepsilon_j^p - \varepsilon_j^{exp}\right)^2\right\}_i + Z_i\left(t_i^p - t_i^{exp}\right)/t_i^{exp} \quad (6.5)$$

where a is the number of creep curves, b is the number of points per curve, ε^p and ε^{exp} are the predicted and experimental values of strain respectively, Z_i is an amplification constant, t^p and t^{exp} denote predicted and experimental lifetimes, respectively. The second term in the expression (6.5) is invoked only when $t^p > t^{exp}$.

Proceeding in this way and by exclusion or inclusion of each of the equations 14b-14e, the quality of fit to the experimental data can be judged by the magnitude of the Kernel function (6.5). Hence it is possible to identify those profound or dominant physical mechanisms. It is in this way that the appropriate mechanisms-based constitutive equations can be identified from the available data base.

7. DISCUSSION AND CONCLUSIONS

The role of super computer simulation in modelling the behaviour of materials and components has been highlighted as an important part of the wealth creation process. To perform simulations/analyses involving the wide range of materials used in industrial applications, it has been shown necessary to be able to access materials databases, to identify the relevant constitutive equations, and to use appropriate finite element analysis techniques for the computer simulation process.

The selection of constitutive equations which relate to the physics of the deformation, damage, and failure processes is necessary if meaningful predictions from extrapolations are to be made using such techniques. High-temperature creep Continuum Damage Mechanics has been used as an example. The power of Continuum Damage Mechanics has been highlighted as a means of predicting the high-temperature behaviour of structures ranging from modest stress raisers, to components in which creep crack growth takes place.

The example of a ridged testpiece has also been used both to demonstrate the power of the approach, and to highlight the importance of carrying out laboratory testing, or materials metrology, under appropriate conditions. Failure to do so can result in significant errors being introduced.

Presented in the paper is a multi-damage state variable approach to materials modelling. Each state variable is linked to a physical mechanism of damage, and the evolution of the variable is determined through a rate equation for each variable. All variables interact synergistically through coupling with the strain-rate constitutive equation.

Numerical techniques are discussed for the identification of the dominant mechanisms/constitutive equations from laboratory test data. Numerical optimisation schemes are outlined for the determination of the parameters in the constitutive equations. The importance is explained of obtaining accurate starting values for the generalised optimisation schemes; and, techniques are outlined for obtaining best estimates of starting values. The approach outlined is general and accurate, but does rely on the availability of good quality laboratory data.

As supercomputers become cheaper and faster the need to use them for more challenging simulations will become even greater. However, the principle barrier to progress will be the availability of high-quality data over a sufficiently broad range of conditions.

As a principle conclusion the urgent need must be stressed for widespread testing programs to be initiated which are directed at producing good quality long-term materials data. Failure to do this will result in an inability to harness supercomputer simulation techniques as a wealth creator.

REFERENCES

1. Preiss, K., Integration of Materials Data into Concurrent Engineering, Proceedings CODATA Workshop on Materials Data for Computer Aided Engineering, February 1993, Frankfurt, Germany.

2. Hayhurst, D. R., Computer Aided Engineering: integrating themes, In Materials and Engineering Design: The Next Decade, Ed. B. F. Dyson and D. R. Hayhurst, The Institute of Metals, London, 1989,113-117.

3. Ashby, M. F., Materials Selection in conceptual design, In Materials and Engineering Design: The Next Decade, Ed. B. F. Dyson and D. R. Hayhurst, The Institute of Metals, London, 1989, 13-25.

4. Hayhurst, D. R., Dimmer, P. R. and Morrison, C. J., Development of Continuum Damage in the Creep Rupture of Notched Bars, Phil. Trans. R. Soc. Lond., A, 311, (1984), 103-129.

5. Hall, F. R. and Hayhurst, D. R., Continuum Damage Mechanics Modelling of High Temperature Deformation and Failure in a Pipe Weldment, Proc. Roy. Soc. Lond. A, 433, (1991), 383-403.

6. EPSRC, Strategic Users, IBM/SERC Supercomputing Joint Study, Science and Engineering Research Council (SERC), Rutherford and Appleton Laboratory, Proceedings, April, 1991, 1-15.

7. Hayhurst, D. R., Engineering approaches to high-temperature design, Pineridge Press, Swansea, Chapter 3, 1983, 85-176.

8. Cocks, A. C. F. and Ashby, M. F., On creep fracture by void growth, Progress in Materials Science, 27, (1982), 189-244.

9. Ashby, M. F., A first report on deformation-mechanisms maps, Acta Metall., 20, (1972), 887-892.

10. Hayhurst, D. R., Stress redistribution and rupture due to creep in a uniformly stretched thin plate containing a circular hole, J. Appl. Mech., 1, 40, (1973), 244-256.

11. Hayhurst, D. R., Creep rupture under multi-axial states of stress, J. Mech. Phys. Solids, 20, (1972), 381-390.

12. Hayhurst, D. R., Dimmer, P. R. and Chernuka, M. W., Estimates of the creep rupture lifetime of structures using the finite element method, J. Mech. Phys. Solids, 23, (1975), 335-355.

13. Hayhurst, D. R. and Storakers, B., Creep rupture of the Andrade shear disc, Proc. Roy. Soc. Lond. A, 349, (1976), 369-382.

14. Hayhurst, D. R., Creep Continuum Damage Mechanics: A unifying theme in high-temperature design, High-temperature Structural Design, ESIS 12, Edited by L. H. Larsson, 1992, Mechanical Engineering Publications, London, 317-334.

15. Hayhurst, D. R., Thermodynamic Modeling and Materials Data Engineering, Spinger-Verlag, Berlin Heidelberg, Chapter 4, 1998, 189-224.

16. Ashby, M. F. and Abel, C. A., Material Selection to Resist Creep, Phil. Trans. R. Soc. Lond. A., 351,(1995), 451-466.

17. Kowalewski, Z. L., Hayhurst, D. R. and Dyson, B. F., Mechanisms-based creep constitutive equations for an aluminium alloy, J. Strain Analysis, 29, (1994), 309-316.

18. Hayhurst, D. R., Morrison, C. J. and Leckie, F. A., The effect of stress concentrations on the creep rupture of tension panels, J. Appl. Mech., 42, (1975), 613-618.

19. Goodall, I. W. and Ainsworth, R. A., Failure of structures by creep, Proc. 3rd Int. Conf. Press. Vess. Tech., Tokyo, Vol. II, (1977), ASME, 871-882.

20. Hayhurst, D. R., Brown, P. R. and Morrison, C. J., The role of continuum damage in creep crack growth, Phil. Trans. R. Soc., Lond., A, 311, (1984), 131-158.

21. Hall, F. R., Hayhurst, D. R. and Brown, P. R., Prediction of plane-strain creep-crack growth using Continuum Damage Mechanics, Int. J. Damage Mech., 5, (1996), 353-393.

22. Coleman, M. C., Parker, J. D., Walters, D. J. and Williams, J. A., The deformation behaviour of thick walled pipes at elevated temperatures, Proceedings of the International Conference on Materials (ICM 3), 2, (1979), (Pergamon Press, Cambridge), 193-202.

23. Cane, B. J., Collaborative programme on the correlation of test data for high-temperature design of welded steam pipes: presentation and analysis of materials data, CEGB Laboratory Note No. RD/L/21; 01N81, 1981.

24. Garofalo, F., Fundamentals of Creep and Creep-Rupture in Metals, MacMillan, New York, 1965.

25. Dorn, J. E., Creep and Fracture of Metals at High Temperatures, NPL Symposium, London, H.M.S.O., (1956), 89.

26. Johnson, A. E., Henderson, J. and Khan, B., Complex-Stress Creep Relaxation and Fracture of Metallic Alloys, H.M.S.O., Edinburgh, 1962.

27. Kachanov, L. M., Time of the Fracture Process under Creep Conditions, Izv. Akad. Nauk. SSSR, Otd. Tech. Nauk. 8, (1958), 26,

28. Greenwood, G. W., Creep Life and Ductility, Int. Congress on Metals, Cambridge, 1973. Microstructure and the Design of Alloys, 2, (1973), 91.

29. Dyson, B. F., and McLean, D., Creep of Nimonic 80A in Torsion and Tension, Met. Sci., 11, (1977), 37-45.

30. Othman, A. M., Hayhurst, D. R., and Dyson, B. F., Skeletal Point Stresses in Circumferentially Notched Tension Bars undergoing Tertiary Creep Modelled with Physically-based Constitutive Equations, Proc. Roy. Soc.,Lond., A, 441, (1993), 343-358.

31. Ashby, M. F., and Dyson, B. F., Creep Damage Mechanics and Mechanisms, Advances in Fracture Research, Ed. by S. R. Valluri et al., Pergamon Press, 3-36, 1984.

32. Dyson, B. F. and McLean, M., Particle Coarsening σ_0 and Tertiary Creep, Acta Met, 31, (1983), 17-27.

33. Church, J. M., Creep and Creep Crack Growth Behaviour and Rupture of Tubular Components under Multi-Axial Stress States, PhD. Thesis, University of Sheffield, November 1992.

34. Hayhurst, D. R., Morrison, C. J., and Leckie, F. A., Constitutive Relations for Creep Damage under Multi-Axial Non-Proportional Loading, Proc IUTAM Symposium on Constitutive Relations for Finite Deformation of Polycrystalline Metals, 54-76, Beijing 1991., Pub. Peking University Press, Beijing and Springer Verlag, Heidelberg.

35. Dyson, B. F. and Gibbons, T. B., Tertiary Creep in Nickel-Base Superalloys: Analysis of Experimental Data and Theoretical Synergies, Acta Metall. 35, (1987), 2355-2369.

36. Hayhurst, D. R., Leckie, F. A. and Morrison, C. J., Creep Rupture of Notched Bars, Proc. R. Soc. Lond, A, 360, (1978), 243-264.

37. Dyson, B. F., A New Mechanism and Constitutive Law for Creep of Precipitation Hardened Engineering Alloys, NPL Report DMM A102, 1993.

38. Lifshitz, I.M. and Slyozon, V.V., The Kinetics of Precipitation from Supersaturated Solid Solutions, J. Phys. Chem. Solids, 19, 1/2, (1961), 35-50.

39. Lin, J., Hayhurst, D.R. and Dyson, B.F., The Standard Ridged Uniaxial Testpiece: Computed Accuracy of Creep Strain. J. of Strain Analysis, 28, 2, (1993), 101-115.

40. Lin, J., Hayhurst, D.R., and Dyson, B.F., A New Design of Uniaxial Creep Testpiece with Slit Extensometer Ridges for Improved Accuracy of Strain Measurement. J. Mech. Sci., 35, 1, (1993), 63-78.

41. Ohashi, Y., Tokuda, M., and Yamashita, H., Effect of Third Invariant of Stress Deviator on Plastic Deformation of Mild Steel, J. Mech. Phys. Solids, 23, (1975), 295-323.

42. Hayhurst, D.R., The Effect of Test Variables on Scatter in High-Temperature Tensile Creep-Rupture Data, Int. J. Mech. Sci., 16, (1974), 829-841.

43. Dunne, F.P.E., Othman, A.M., Hall, F.R. and Hayhurst, D.R., Representations of uni-axial creep curves, using continuum damage mechanics, Int. J. Mech. Sci., 32, 11, (1990), 945-957.

THERMODYNAMICALLY FOUNDED CDM MODELS FOR CREEP AND OTHER CONDITIONS

J.-L. Chaboche
Office National d'Etudes et de Recherches Aérospatiales, Châtillon, France

ABSTRACT

The fundamental concepts of Continuum Damage Mechanics are reviewed, with the objective to conform to a sufficiently general thermodynamic framework. The theories are developed at a macroscopic level, with the capability to describe various types of materials, ductile or brittle, metallic, concrete, composites,... etc., in an unique framework. For that, we consider both elasticity coupled with damage, plasticity and viscoplasticity coupled with damage and the damage growth equations themselves. Moreover, we discuss the important but difficult problems associated with the damage deactivation effects that can take place under compressive loadings.

A special attention is focused on the nature of damage state variables, in correspondence with the choices made in the various thermodynamic potentials, and on the various coupling possibilities. Some applications are presented on metallic materials, concerning ductile damage, creep damage and creep-fatigue interaction. We also discuss briefly the different levels for the inelastic/damage structural analysis and the applications of CDM to the Local Approaches to Fracture. Some illustrations are also given for Metal Matrix Composites and Ceramic Matrix Composites.

1. INTRODUCTION

For some twenty years, much work has been devoted to Continuum Damage Mechanics (CDM), based on the concepts initially introduced by Kachanov [1] and Murzewski [2], then Rabotnov [3]. Damage is considered as a material degrading process subsequent but not identical to irreversible deformation processes. The defects that occur correspond to cumulative localized dislocations (in metals) with a much more pronounced irreversible character than plastic strain. In addition, many brittle materials are deformed by damage.

After the work of the Russian School, Continuum Damage Mechanics was first developed in Europe, essentially for applications to creep in metallic materials. The English School made a remarkable contribution at the beginning of the 1970s, with the work of Leckie and Hayhurst [4, 5] in particular. There were Polish [6], Swedish [7] and Japanese [8] contributions as well. The basic concepts of Damage Mechanics were described theoretically in France, in particular through thermodynamic formalism [9, 10, 11]. It was also in France that applications were systematically sought for other types of damage (fatigue, ductile fracture) and other types of materials [9, 12].

It was not until the beginning of the 1980s that Damage Mechanics was recognized (actually rediscovered) in the United States, with the work of Krajcinovic [13, 14], Ortiz, [15], Ju [16], then Chow [17], Voyiadjis [18] and many others. This scientific area has since been developing rapidly throughout the world, as regards both basic research (where the problems are far from solved) and applications.

Fracture Mechanics considers cracks *per se*, modifying structural boundary conditions. By contrast, Continuum Damage Mechanics accounts for defects through a homogenization concept and describes their growth macroscopically, while remaining in the framework of Continuum Mechanics (CM). References [7, 19, 20, 14, 10] give relatively complete and general descriptions of CDM. In this course, it is discussed from the viewpoints outlined below.

In the first part, we introduce general elementary concepts, distinguishing between irreversible strain, damage and macroscopic propagation. The definitions being applied to damage variables are described. Finally, a review of the Thermodynamics of Irreversible Processes (TIP) with internal variables examines the problem from the energy standpoint, both for the state variables and for the dissipation processes associated with damage.

In the second one, the theoretical framework of CDM is developed, considering with some attention the state equations, i.e. the damaged elastic behavior, and the dissipative aspects, coupled plasticity and damage, with several optional choices. The third and fourth parts examine the applications to metallic materials and structures, considering successively the ductile damage, the creep damage, creep–fatigue interaction problems, and the various levels for structural damage analyses.

Finally, we specify some CDM applications for brittle materials, especially composite materials. This is the occasion to discuss the difficult problems associated with the damage deactivation processes that take place under compressive–like loading conditions. The capabilities of some specific models are discussed and illustrated for two classes of composite materials.

2. CONTINUUM DAMAGE MECHANICS AND THERMODYNAMICS

2.1. General notions and concepts

A. Principal characters of damaging effects

As it was recalled above, material damage corresponds to the creation of defects that are already relatively large, in any case irreversible, much more than plastic deformations. It is not possible here to describe a detailed list of the various damages that can develop in various kinds of materials. Let us simply enumerate creep damage, fatigue damage, ductile damage in metals, including nucleation of cavities or microcracks, their growth and coalescence, and brittle damage in rocks, concrete or composites, ..., etc. Some of these processes are described in details in other chapters of these Lecture Notes [21, 22, 23].

B. Deformation, damage and crack propagation

A distinction should be made between deformation, damage and crack propagation, which may correspond to three consecutive states of the material:

- Irreversible deformations are described in the framework of continuum mechanics. Their irreversibility (on a macroscopic scale) is not complete, since the material can very well be redeformed to restore the initial shape. This is illustrated in Fig. 2.1a for tension–compression. On the contrary, we assume that damage corresponds to a complete degradation (barring complete reprocessing of the material). Fig. 2.1b shows the typical (ideal) case of a quasi–elastic brittle material whose nonlinear behavior is basically due to damage, observed by the drop in the modulus of elasticity (during unloading), whereas irreversible deformation is negligible on the contrary.

- The concept of continuum damage concerns the possibility of using continuum mechanics for calculations of damaged structures. In order to work on a representative

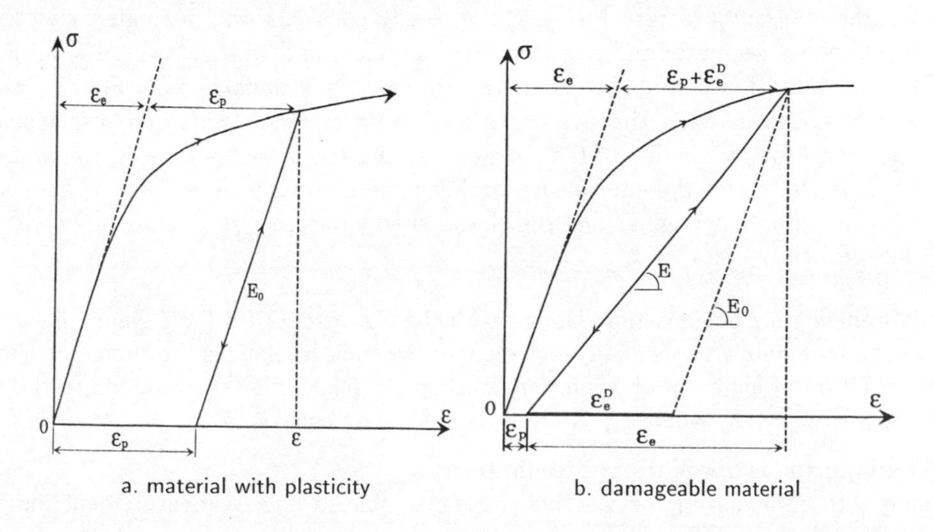

Figure 2.1: Irreversible deformation in plasticity and brittle damage

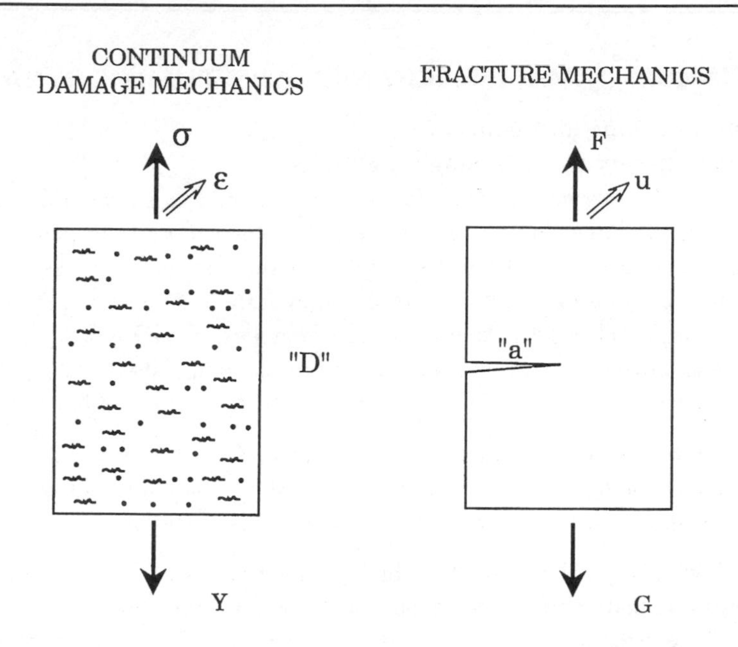

Figure 2.2: The RVE of Damage Mechanics and Fracture Mechanics

volume element (RVE) with stress and strain variables defined as averages, it is necessary for the defects to be sufficiently numerous and small. Fig. 2.2a schematically shows this RVE concept associated with CDM. The damage variable denoted D is initially assumed equal to zero, $D = 0$, when there is no decohesion of the material.

- On the contrary, Fracture Mechanics generally considers only a single macroscopic crack whose geometry and size is clearly identified, and which crosses the structure which material is considered as continuum (generally undamaged). Fig. 2.2 shows a comparison between the two situations, using external loads and displacements applied to the part in lieu of the average stresses and strains applied to the material RVE. It also shows the equivalence between the thermodynamic force Y associated with the diffuse damage D and the elastic energy release rate G associated with the unit growth of crack a.

It is immediately obvious that Damage Mechanics reaches its limit when the diffuse damage combines into a single main crack on a macroscopic level. This situation corresponds to the definition of initiation of a main macroscopic crack. It also corresponds to the final situation, with $D = D_c$, where D_c is often considered as equal to 1.

C. Damage measurements and definitions

Along with definition of the variables, it is desirable to have a measurement method, even if it is sometimes difficult to use in practice. Four types of measurements can be used

[24]: (i) Measurement by remaining life. This is the measurement of interest to the engineer for calculating the lifetime of parts. (ii) Microstructural measurements (volume fraction of defects, cavities, microcracks). The concept of net stress mentioned below refers to this type of measurement. (iii) Measurements of physical parameters. It is possible to measure the variation due to damage of parameters such as the density [25], resistivity [26] or acoustic emission. A model is obviously necessary to establish mechanical damage. (iv) Measurements of the variation in mechanical behavior. Such measurements are the best suited to mechanical modeling. A distinction is made between two methods, one based on the concept of net section (and net stress) and the other on the concept of effective stress, both introduced here in the particular uniaxial case:

- *The net stress* is the average stress applied to the resistant section (or net section) of the damaged specimen. As shown in Fig. 2.3, this is carried out on the actual geometry of the specimen, after taking macroscopic deformations into account. The net stress σ^* is deduced from the Cauchy stress σ by a reduction in section factor taking into account the average reduction ω in section due to voids and cracks defined by ω. This is expressed:

$$\sigma^* = \frac{S}{S^*}\sigma = \frac{\sigma}{1-\omega}. \tag{2.1}$$

  Such an approach was used for instance by Murakami [8, 20] to generalize the concept of creep damage in polycrystals directly to the anisotropic case.

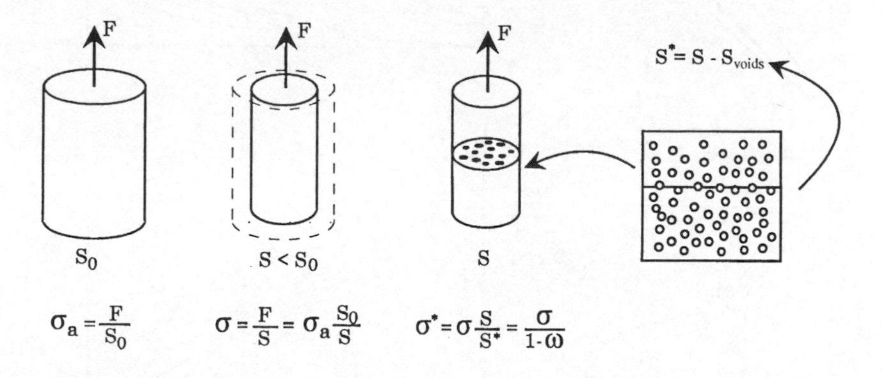

Figure 2.3: Apparent stress, true stress and net stress

- On the contrary, *effective stress* [9] takes into account stress concentrations in the neighborhood of the defects. It is based on the measured macroscopic behavior of the damaged medium. By analogy with the previous case, we define an effective section $\tilde{S}$ and the effective uniaxial stress is then:

$$\tilde{\sigma} = \frac{S}{\tilde{S}}\sigma = \frac{\sigma}{1-D}, \tag{2.2}$$

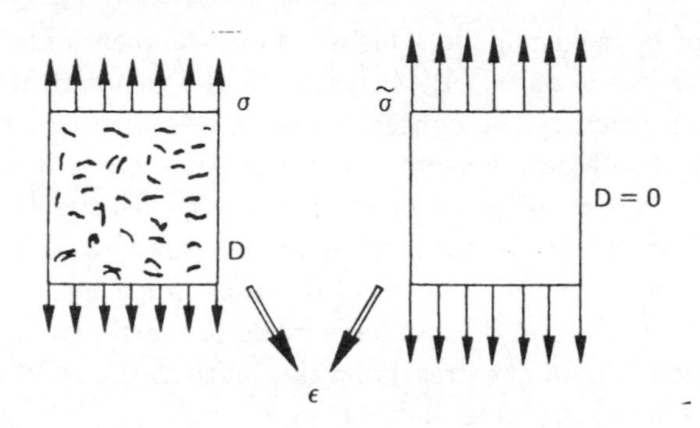

Figure 2.4: Definition of effective stress

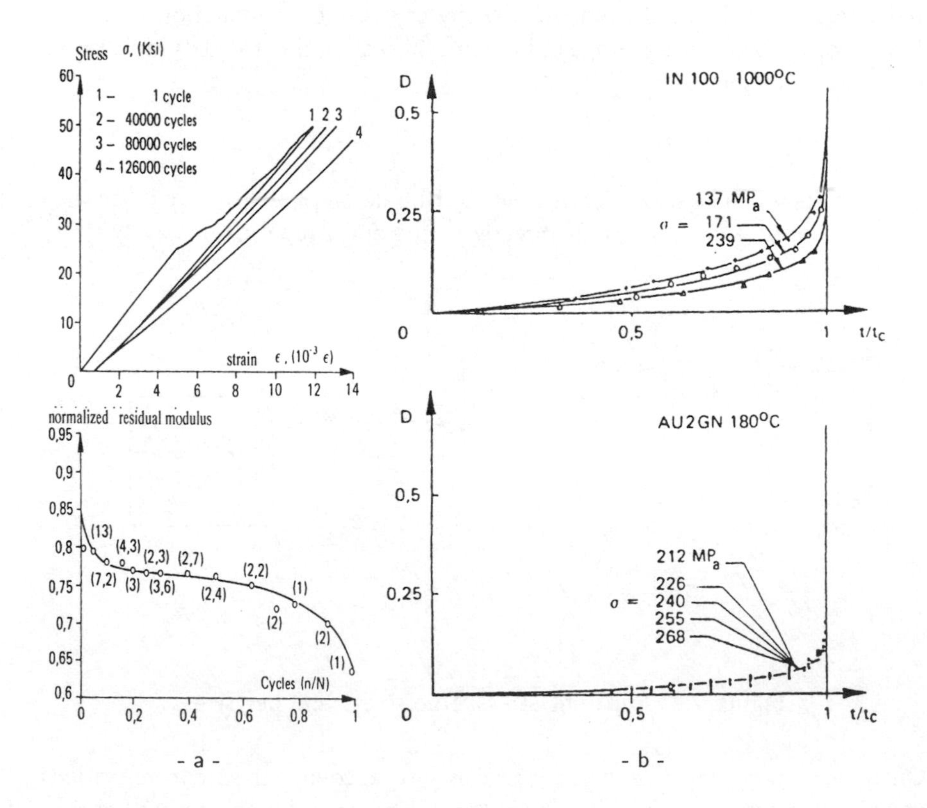

Figure 2.5. Damage measurements by effective stress a. Fatigue of a graphite/epoxy composite [0/90] — top: stress/strain curves; — bottom: residual modulus of elasticity, b. Creep damage measurement by tertiary creep. — top: IN100 superalloy; — bottom: AU2GN aluminum alloy

where D represents the macroscopic effect of the mechanical behavior degradation. This concept is generalized to the multiaxial case either by the strain equivalence hypothesis [9] or by the energy equivalence hypothesis [27]. The first hypothesis is stated as follows:

The effective stress $\tilde{\sigma}$ is the one which would have to be applied to an element of undamaged material (all other things being equal) so that it is deformed in the same way (same strain) as a damaged element subjected to the current stress σ (Fig. 2.4).

The variation in mechanical behavior can be measured through the variation of various parameters (the parameters' choice depend on the type of application): (i) Variation in the modulus of elasticity used for ductile plastic damage [11] as well as for fatigue [28], in particular composite fatigue, as is shown by the example of Fig. 2.5a [29]. (ii) Plastic or viscoplastic behavior [10, 30]. Fig. 2.5b shows the example of two typical metallic alloys. (iii) Cyclic behavior for Low Cycle Fatigue of metallic materials [31]. Fig. 2.6, taken from reference [28], shows that this method is applicable to materials of an unstable cyclic behavior. (iv) It is also possible to use ultrasonic wave propagation velocity instead of directly measuring the moduli of elasticity. This technique was successfully developed for measuring up to nine parameters in anisotropic composite materials [32].

As will be seen below, the effective stress is used in the constitutive laws in place of the Cauchy stress to describe the impact of the damage on the macroscopic behavior of

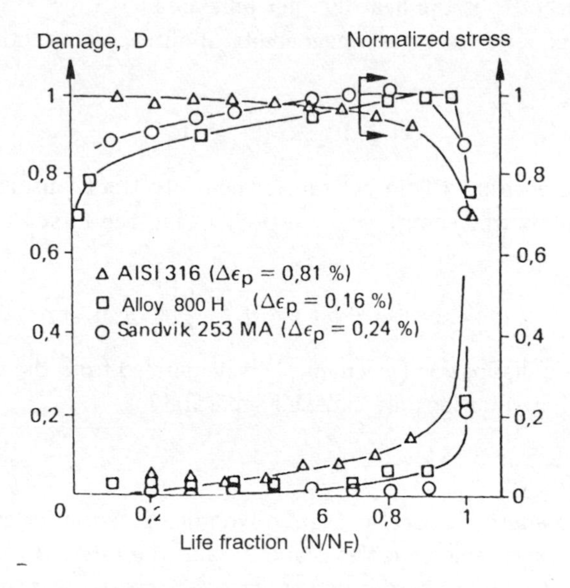

Figure 2.6. Fatigue damage growth curves measured on three materials and corresponding variation in normalized stress, from [28]

the material. As an initial approximation, it is the same quantity which is used regardless of the type of behavior analyzed (elastic, plastic, viscoplastic).

2.2. Basic thermodynamic concepts

A. State equations

Continuum Damage Mechanics can be developed in the framework of the general concepts of the Theory of Irreversible Processes (TIP) with internal variables [33, 12, 34]. This section recalls the general principles in the framework of small transformations. The total strain is divided into the sum of the elastic strain $\underset{\sim}{\varepsilon}^{\mathrm{e}}$, plastic strain $\underset{\sim}{\varepsilon}^{\mathrm{p}}$ and thermal expansion $\underset{\sim}{\varepsilon}^{\theta}$. It is possible to combine $\underset{\sim}{\varepsilon}^{\mathrm{e}}$ and $\underset{\sim}{\varepsilon}^{\theta}$ into thermoelastic strain $\underset{\sim}{\varepsilon}^{\mathrm{e}\theta}$, which is a state variable corresponding to reversible effects:

$$\underset{\sim}{\varepsilon} = \underset{\sim}{\varepsilon}^{\mathrm{e}} + \underset{\sim}{\varepsilon}^{\mathrm{p}} + \underset{\sim}{\varepsilon}^{\theta} = \underset{\sim}{\varepsilon}^{\mathrm{e}\theta} + \underset{\sim}{\varepsilon}^{\mathrm{p}} \tag{2.3}$$

The observable variables are the strain $\underset{\sim}{\varepsilon}$ and the temperature T. We use the Local State method in the framework of conventional Continuum Mechanics: the thermodynamic state of a material point (and the RVE surrounding it) is assumed independent of the thermodynamic state of the neighboring elements (local nature of the constitutive and damage laws).

In this framework, the First Principle of Thermodynamics expresses energy conservation:

$$\rho \dot{e} = \underset{\sim}{\sigma} : \dot{\underset{\sim}{\varepsilon}} + r - \mathrm{div}\underline{\boldsymbol{q}} \tag{2.4}$$

where e is the specific internal energy, ρ is the density, r is the density of the internal heat production (radiation) and $\underline{\boldsymbol{q}}$ is the heat flux per unit area.

The Second Principle expresses the irreversibility of entropy production by the following inequality:

$$\rho \dot{\mathcal{S}} + \mathrm{div}\frac{\underline{\boldsymbol{q}}}{T} - \frac{r}{T} \geq 0 \tag{2.5}$$

where $\mathcal{S}$ is the specific entropy. It can be transformed into the Clausius–Duhem inequality by means of the Helmholz free energy $\psi = e - T\mathcal{S}$, using the First Principle to eliminate radiation:

$$\underset{\sim}{\sigma} : \dot{\underset{\sim}{\varepsilon}} - \rho(\dot{\psi} + \mathcal{S}\dot{T}) - \underline{\boldsymbol{q}}.\frac{\underline{\underline{\mathbf{grad}T}}}{T} \geq 0 \tag{2.6}$$

We assume that intrinsic dissipation (mechanical) is decoupled from the thermal dissipation, so that the two following equations are satisfied separately:

$$\Phi_{\mathrm{int}} = \underset{\sim}{\sigma} : \dot{\underset{\sim}{\varepsilon}} - \rho(\dot{\psi} + \mathcal{S}\dot{T}) \geq 0 \qquad \text{and} \qquad \Phi_{th} = -\underline{\boldsymbol{q}}.\frac{\underline{\underline{\mathbf{grad}T}}}{T} \geq 0 \tag{2.7}$$

The Helmholtz free energy is used as thermodynamic potential or state potential. The state variables are the observable variables $\underset{\sim}{\varepsilon}$ and T, and the internal variables. Among the latter, we distinguish between the hardening variables denoted α_j and the damage variables denoted d_k. The elasticity and plasticity processes are assumed separate, such that:

$$\psi = \psi(\underset{\sim}{\varepsilon}, \underset{\sim}{\varepsilon}^{\mathrm{p}}, \alpha_j, d_k, T) = \psi_e(\underset{\sim}{\varepsilon} - \underset{\sim}{\varepsilon}^{\mathrm{p}}, d_k, T) + \psi_p(\underset{\sim}{\varepsilon}^{\mathrm{p}}, \alpha_j, d_k, T) \tag{2.8}$$

A classical demonstration, not reproduced here, allows us to eliminate all the reversible processes from the Clausius–Duhem inequality. This yields the state equations:

$$\underset{\sim}{\sigma} = \rho \frac{\partial \psi}{\partial \underset{\sim}{\varepsilon}} = \rho \frac{\partial \psi}{\partial \underset{\sim}{\varepsilon}^{e\theta}} = \rho \frac{\partial \psi}{\partial \underset{\sim}{\varepsilon}^{e}} \qquad \mathcal{S} = -\frac{\partial \psi}{\partial T} \tag{2.9}$$

which express the thermodynamic forces associated with the observable state variables. In addition, we can postulate the forces associated with the internal variables (or affinities):

$$\underset{\sim}{\sigma}^{p} = -\rho \frac{\partial \psi}{\partial \underset{\sim}{\varepsilon}^{p}} = \rho \frac{\partial \psi_e}{\partial \underset{\sim}{\varepsilon}^{e}} - \rho \frac{\partial \psi_p}{\partial \underset{\sim}{\varepsilon}^{p}} = \underset{\sim}{\sigma} - \rho \frac{\partial \psi_p}{\partial \underset{\sim}{\varepsilon}^{p}} = \underset{\sim}{\sigma} - \underset{\sim}{\boldsymbol{A}}_p \tag{2.10}$$

$$A_j = \rho \frac{\partial \psi}{\partial \alpha_j} \quad \text{and} \quad \underset{\sim}{\boldsymbol{A}}_p = \rho \frac{\partial \psi_p}{\partial \underset{\sim}{\varepsilon}^{p}} \quad \text{and} \quad y_k = -\rho \frac{\partial \psi}{\partial d_k} \tag{2.11}$$

The + and − signs in (2.11) are arbitrary. They correspond to the intuitive idea that hardening produces an increase in mechanical strength whereas damage causes a decrease in strength.

B. Complementary dissipative laws

The intrinsic dissipation is used to develop plasticity and viscoplasticity laws that satisfy automatically the Second Principle. It expresses now:

$$\Phi_{\text{int}} = \underset{\sim}{\sigma} : \dot{\underset{\sim}{\varepsilon}}^{p} - \underset{\sim}{\boldsymbol{A}}_p : \dot{\underset{\sim}{\varepsilon}}^{p} - \sum_j A_j \dot{\alpha}_j + \sum_k y_k \dot{d}_k = \underset{\sim}{\sigma}^{p} : \dot{\underset{\sim}{\varepsilon}}^{p} - \sum_j A_j \dot{\alpha}_j + \sum_k y_k \dot{d}_k \geq 0 \tag{2.12}$$

It is recalled that Φ_{int} expresses the heat power dissipated by the material element during the (visco)plastic and damage processes. It should also be noted that $\underset{\sim}{\varepsilon}^{p}$ can be considered equivalently as a hardening variable, with $\underset{\sim}{\boldsymbol{A}}_p$ as associated thermodynamic force, or as a plastic strain which dissipates with $\underset{\sim}{\sigma}^{p} = \underset{\sim}{\sigma} - \underset{\sim}{\boldsymbol{A}}_p$. Below, we use the second interpretation. It was seen that the state equations are derived from a thermodynamic potential. Following the Generalized Standard Medium (GSM) approach, we can also derive the complementary equations, i.e. the evolution equations for the internal variables associated with the irreversible processes, from a dissipation "pseudo–potential" [35]. The pseudo–potential must depend on all the rates of change of the internal variables:

$$\phi = \phi(\dot{\underset{\sim}{\varepsilon}}^{p}, \dot{\alpha}_j, \dot{d}_k\, ;\, \underset{\sim}{\varepsilon}^{e}, \underset{\sim}{\varepsilon}^{p}, \alpha_j, d_k, T) \tag{2.13}$$

It may also depend on the internal variables themselves, considered as parameters. The following normality rule is assumed:

$$\sigma^{p} = \frac{\partial \phi}{\partial \dot{\underset{\sim}{\varepsilon}}^{p}} \qquad A_j = -\frac{\partial \phi}{\partial \dot{\alpha}_j} \qquad y_k = \frac{\partial \phi}{\partial \dot{d}_k} \tag{2.14}$$

It is actually more practical to express the pseudo–potential(s) in terms of the thermodynamic forces. The Legendre–Fenschel transformation yields a pseudo–potential:

$$\phi^* = \phi^*(\underset{\sim}{\sigma}^{p}, A_j, y_k\, ;\, \underset{\sim}{\varepsilon}^{e}, \underset{\sim}{\varepsilon}^{p}, \alpha_j, d_k, T) \tag{2.15}$$

with the normality rule:

$$\dot{\underset{\sim}{\varepsilon}}^{\mathrm{p}} = \frac{\partial \phi^*}{\partial \underset{\sim}{\sigma}^{\mathrm{p}}} \qquad \dot{\alpha}_j = -\frac{\partial \phi^*}{\partial A_j} \qquad \dot{d}_k = \frac{\partial \phi^*}{\partial y_k} \tag{2.16}$$

Therefore, assuming the pseudo–potential is convex and positive and that it contains the origin ($\phi^*(0) = 0$), satisfaction of the Second Principle is demonstrated a priori for any evolution:

$$\Phi_{\mathrm{int}} = \underset{\sim}{\sigma}^{\mathrm{p}} : \frac{\partial \phi^*}{\partial \underset{\sim}{\sigma}^{\mathrm{p}}} + \sum_j A_j \frac{\partial \phi^*}{\partial A_j} + \sum_k y_k \frac{\partial \phi^*}{\partial y_k} \geq \phi^* \geq 0 \tag{2.17}$$

Above, we neglected thermal dissipation, assumed positive. The conventional heat conduction law is the Fourier law:

$$\underline{\boldsymbol{q}} = -k \ \underline{\mathbf{grad}T} \tag{2.18}$$

where k is the thermal conductivity of the medium, assumed isotropic. Fourier's law could easily be generalized to an anisotropic continuum, using a second rank tensor for conductivity. The heat equation is deduced from the First Principle (2.4) by replacing $\rho \dot{e}$ by its expression taken from $\psi = e - T\mathcal{S}$, then $\dot{\psi}$ by its expression using the state variables. In addition, we take the Fourier law into account, which then yields $\mathrm{div}\underline{\boldsymbol{q}} = -k\Delta T$. This yields:

$$k\Delta T = \rho T \dot{\mathcal{S}} - \Phi_{\mathrm{int}} - r \tag{2.19}$$

Taking $\mathcal{S} = -\partial\psi/\partial T$ and the other state equations into account and introducing the specific heat $c = T\,\partial\mathcal{S}/\partial T$ finally yields the heat equation:

$$k\Delta T = \rho c \dot{T} - \Phi_{\mathrm{int}} - r : -T \left(\frac{\partial \underset{\sim}{\sigma}}{\partial T} : \dot{\underset{\sim}{\varepsilon}}^{\mathrm{e}} + \frac{\partial \underset{\sim}{A}_{\mathrm{p}}}{\partial T} : \dot{\underset{\sim}{\varepsilon}}^{\mathrm{p}} + \sum_j \frac{\partial A_j}{\partial T} \dot{\alpha}_j + \sum_k \frac{\partial y_k}{\partial T} \dot{d}_k \right) \tag{2.20}$$

The conventional heat equation corresponds to the special case neglecting intrinsic dissipation (negligible plastic deformation, hardening and damage) and radiation r. In addition, we use the specific heat at fixed deformation

$$c_\varepsilon = T \left. \frac{\partial \mathcal{S}}{\partial T} \right|_\varepsilon \tag{2.21}$$

which eliminates the isentropic coupling term $-T\,\partial\underset{\sim}{\sigma}/\partial T : \underset{\sim}{\varepsilon}^{\mathrm{e}}$, finally yielding

$$k\Delta T = \rho\, c_\varepsilon \dot{T} \tag{2.22}$$

In the case of applications with large deformations at a high rate, it is no longer necessarily possible to neglect the corresponding dissipation. The complete heat equation (2.20) must then be used.

C. Energy dissipated as heat and stored energy

The thermodynamic framework gives us an interpretation of the energies dissipated as heat or stored in the material. In particular, the intrinsic dissipation representing the power dissipated by the production of heat during the irreversible processes includes three types of terms [36]:

$$\Phi_{\text{int}} = \underset{\sim}{\sigma} : \dot{\underset{\sim}{\varepsilon}}^{\text{p}} - \underset{\sim}{A}_{\text{p}} : \dot{\underset{\sim}{\varepsilon}}^{\text{p}} - \sum_j A_j \dot{\alpha}_j + \sum_k y_k \dot{d}_k \tag{2.23}$$

The first term on the right–hand side is the irreversible power supplied. During monotonic tension, for instance, its integral corresponds to the area under the curve $\sigma - \epsilon^{\text{p}}$:

$$\int_0^{\epsilon^p} \sigma \, d\varepsilon^{\text{p}}$$

The integral of the sum of the next two terms is the energy stored in the material, W_s, by the hardening mechanisms associated with plasticity. The last term denotes the energy dissipated as heat by the damage processes. Schematically, we can separate these energies as shown in Fig. 2.7.

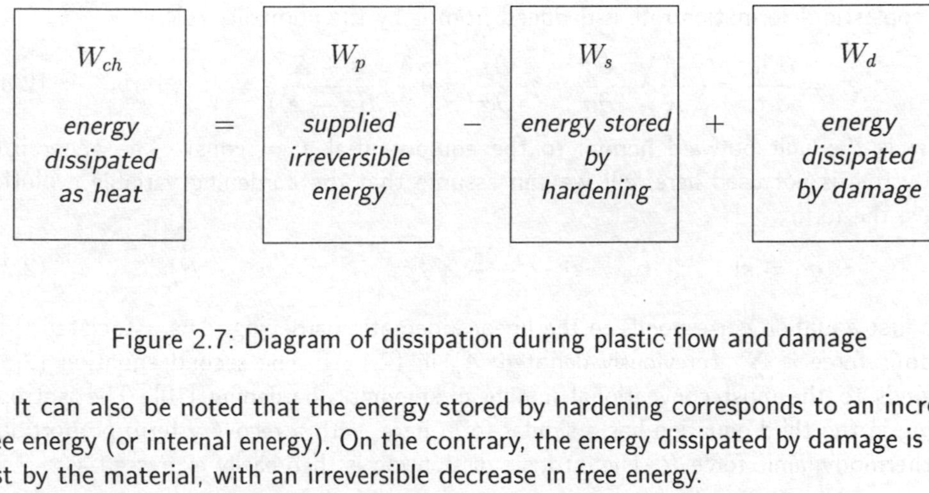

W_{ch}		W_p		W_s		W_d
energy dissipated as heat	=	*supplied irreversible energy*	−	*energy stored by hardening*	+	*energy dissipated by damage*

Figure 2.7: Diagram of dissipation during plastic flow and damage

It can also be noted that the energy stored by hardening corresponds to an increase in free energy (or internal energy). On the contrary, the energy dissipated by damage is energy lost by the material, with an irreversible decrease in free energy.

D. Example: application to elasto–visco–plasticity

In the example below, we use two kinematic hardening variables denoted $\underset{\sim}{\alpha}_1$ and $\underset{\sim}{\alpha}_2$ (second order tensors) and an isotropic hardening variable denoted r (not to be confused with radiation). The free energy is assumed to have the form:

$$\rho\psi = \frac{1}{2} \underset{\sim}{\varepsilon}^{\text{e}} : \underset{\approx}{\Lambda} : \underset{\sim}{\varepsilon}^{\text{e}} + \frac{1}{3} C_1 \underset{\sim}{\alpha}_1 : \underset{\sim}{\alpha}_1 + \frac{1}{3} C_2 \underset{\sim}{\alpha}_2 : \underset{\sim}{\alpha}_2 + \psi_r(r) \tag{2.24}$$

where $\underset{\approx}{\Lambda}$ represents the tensor of elastic stiffness (fourth order) and C_1, C_2 are hardening coefficients. With (2.9), the first term gives the elasticity law:

$$\underset{\sim}{\sigma} = \rho \frac{\partial \psi}{\partial \underset{\sim}{\varepsilon}^{\text{e}}} = \underset{\approx}{\Lambda} : \underset{\sim}{\varepsilon}^{\text{e}} \tag{2.25}$$

The other terms supply the thermodynamic forces associated with the hardening variables:

$$\underset{\sim}{X}_1 = \rho \frac{\partial \psi}{\partial \underset{\sim}{\alpha}_1} = \frac{2}{3} C_1 \underset{\sim}{\alpha}_1 \qquad \underset{\sim}{X}_2 = \rho \frac{\partial \psi}{\partial \underset{\sim}{\alpha}_2} = \frac{2}{3} C_2 \underset{\sim}{\alpha}_2 \tag{2.26}$$

$$\underset{\sim}{X} = \underset{\sim}{X}_1 + \underset{\sim}{X}_2 \qquad R = \rho \frac{\partial \psi}{\partial r} = \frac{\partial \psi_r}{\partial r} \tag{2.27}$$

We assume that the center of the elasticity domain is expressed by the sum $\underset{\sim}{X}$ and that the evolution of its radius is the thermodynamic force R. This domain is assumed to satisfy the von Mises criterion:

$$f(\underset{\sim}{\sigma}, \underset{\sim}{X}, R) = J(\underset{\sim}{\sigma} - \underset{\sim}{X}) - R - \sigma_y = \left(\frac{3}{2}(\underset{\sim}{s} - \underset{\sim}{X}) : (\underset{\sim}{s} - \underset{\sim}{X})\right)^{1/2} - R - \sigma_y \tag{2.28}$$

In *"associated viscoplasticity"*, we assume the existence of the viscoplastic potential of the form:

$$\Omega_p = \frac{K}{n+1} \left\langle \frac{f}{K} \right\rangle^{n+1} \tag{2.29}$$

The viscoplastic deformation rate is deduced from it by the normality rule:

$$\dot{\underset{\sim}{\varepsilon}}^{\mathrm{p}} = \frac{\partial \Omega_p}{\partial \underset{\sim}{\sigma}} = \left\langle \frac{f}{K} \right\rangle^n \frac{\partial f}{\partial \underset{\sim}{\sigma}} = \dot{p} \frac{\partial f}{\partial \underset{\sim}{\sigma}} = \dot{p} \frac{3}{2} \frac{\underset{\sim}{s} - \underset{\sim}{X}}{J(\underset{\sim}{s} - \underset{\sim}{X})} = \sqrt{\frac{3}{2}} \, \dot{p} \, \underset{\sim}{n} \tag{2.30}$$

where $\underset{\sim}{n}$ is the unit outward normal to the equipotential $f = \mathrm{const}$. The generalized normality rule is not used here, but we can assume that the hardening variable evolution laws have the form:

$$\dot{\underset{\sim}{\alpha}}_1 = \dot{\underset{\sim}{\varepsilon}}^{\mathrm{p}} \qquad \dot{\underset{\sim}{\alpha}}_2 = \dot{\underset{\sim}{\varepsilon}}^{\mathrm{p}} - \frac{3}{2} \frac{\gamma_2}{C_2} \underset{\sim}{X}_2 \dot{p} \qquad \dot{r} = \dot{p} - \frac{g}{c} R \dot{p} \tag{2.31}$$

The first equation corresponds to the linear kinematic hardening. The associated thermodynamic force is $\underset{\sim}{X}_1$ (previously denoted $\underset{\sim}{A}_{\mathrm{p}}$ in (2.11)). The second equation (2.31) corresponds to the most conventional nonlinear kinematic hardening [10]. The isotropic hardening in the third equation has a similar form here, with a recovery term proportional to the thermodynamic force R. The intrinsic dissipation is then easily expressed as:

$$\begin{aligned} \Phi_{\mathrm{int}} &= \underset{\sim}{\sigma} : \dot{\underset{\sim}{\varepsilon}}^{\mathrm{p}} - \underset{\sim}{X}_1 : \dot{\underset{\sim}{\alpha}}_1 - \underset{\sim}{X}_2 : \dot{\underset{\sim}{\alpha}}_2 - R \dot{r} \\ &= (\underset{\sim}{\sigma} - \underset{\sim}{X}_1 - \underset{\sim}{X}_2) : \dot{\underset{\sim}{\varepsilon}}^{\mathrm{p}} - R\dot{p} + \frac{3}{2} \frac{\gamma_2}{C_2} \underset{\sim}{X}_2 : \underset{\sim}{X}_2 \dot{p} + \frac{g}{c} R^2 \dot{p} \\ &= (J(\underset{\sim}{\sigma} - \underset{\sim}{X}) - R) \dot{p} + \frac{3}{2} \frac{\gamma_2}{C_2} \underset{\sim}{X}_2 : \underset{\sim}{X}_2 \dot{p} + \frac{g}{c} R^2 \dot{p} \\ &= (f + \sigma_y) \dot{p} + \frac{3}{2} \frac{\gamma_2}{C_2} \underset{\sim}{X}_2 : \underset{\sim}{X}_2 \dot{p} + \frac{g}{c} R^2 \dot{p} \geq 0 \end{aligned} \tag{2.32}$$

The last term of (2.32) includes three terms and it is obvious that all of them are positive. The Second Principle is therefore verified a priori. The first term corresponds to the sum of the viscous stress $\sigma_v = f$ (outside the elastic domain) and the initial threshold σ_y. The corresponding energy is entirely dissipated as heat. The second and third terms correspond to the recovery terms in the nonlinear hardening equations. They are also dissipated as heat.

From this, it immediately results that the energy stored by the linear kinematic hardening mechanism is effectively equal to $\frac{1}{3}\, C_1\, \underset{\sim}{\alpha}_1 : \underset{\sim}{\alpha}_1$. On the contrary, the energy stored by nonlinear kinematic hardening is obtained by the integral under curve X_2 minus the integral of term $\frac{3}{2}\frac{\gamma_2}{C_2}\, \underset{\sim}{\boldsymbol{X}}_2 : \underset{\sim}{\boldsymbol{X}}_2\, \mathrm{d}p$. It is easily shown that we also find $\frac{1}{3}\, C_2\, \underset{\sim}{\alpha}_2 : \underset{\sim}{\alpha}_2$.

It is important to note that only the presence of the recovery terms, in particular the nonlinear kinematic hardening term, can express the decrease often observed in ratio $W_{\mathrm{s}}/W_{\mathrm{p}}$ of the stored energy to the plastically dissipated energy (area under curve $\sigma - \varepsilon_p$), which occurs during the increase in plastic deformation for monotonic loading. According to the calorimetric measurements made by Chrysochoos [37, 38] during hardening tests on three materials, it was demonstrated [36] that the combination of two kinematic variables and isotropic hardening enabled to express both the mechanical behavior (stress/plastic strain curve) and stored energy (ratio $W_{\mathrm{s}}/W_{\mathrm{p}}$).

E. Energy dissipated by damage

The state potential, assumed to be the Helmholtz free energy (2.8), includes the damage in the elastic term ψ_{e} and the hardening term ψ_{p}. Here, we start by assuming that the damage is not included in ψ_{p}. In addition, we consider only one scalar variable D for the time being to simplify the notations and do not include the temperature or thermal expansion. We therefore state:

$$\rho\,\psi = \frac{1}{2}\,\underset{\sim}{\varepsilon}^{\mathrm{e}} : \underset{\approx}{\tilde{\Lambda}}(D) : \underset{\sim}{\varepsilon}^{\mathrm{e}} + \psi_{\mathrm{p}}(\alpha_j) \tag{2.33}$$

where $\underset{\approx}{\tilde{\Lambda}}(D)$ represents the elastic stiffness tensor of the damaged material. Hooke's law and the thermodynamic force associated with damage D are given by:

$$\underset{\sim}{\sigma} = \rho\,\frac{\partial\psi}{\partial\underset{\sim}{\varepsilon}^{\mathrm{e}}} = \underset{\approx}{\tilde{\Lambda}} : \underset{\sim}{\varepsilon}^{\mathrm{e}} \qquad\qquad y = -\frac{\partial\psi}{\partial D} = -\frac{1}{2} : \underset{\sim}{\varepsilon}^{\mathrm{e}} : \frac{\partial\underset{\approx}{\tilde{\Lambda}}}{\partial D} : \underset{\sim}{\varepsilon}^{\mathrm{e}} \tag{2.34}$$

It can be demonstrated that the latter is equal to half the elastic energy released by damage evolution at fixed stress and temperature:

$$y = \frac{1}{2}\,\frac{\mathrm{d}W_{\mathrm{e}}}{\mathrm{d}D}\bigg|_{\sigma,T} \tag{2.35}$$

For this, it is sufficient to define the elastic energy variation by $dW_{\mathrm{e}} = \underset{\sim}{\sigma} : \mathrm{d}\underset{\sim}{\varepsilon}^{\mathrm{e}}$, calculate $\underset{\sim}{\sigma}$ by the state equation (2.9), differentiate and multiply by $\underset{\sim}{\varepsilon}^{\mathrm{e}}$:

$$\underset{\sim}{\varepsilon}^{\mathrm{e}} : \mathrm{d}\underset{\sim}{\sigma} = \underset{\sim}{\varepsilon}^{\mathrm{e}} : \underset{\approx}{\tilde{\Lambda}} : \mathrm{d}\underset{\sim}{\varepsilon}^{\mathrm{e}} + \underset{\sim}{\varepsilon}^{\mathrm{e}} : \frac{\partial\underset{\approx}{\tilde{\Lambda}}}{\partial D} : \underset{\sim}{\varepsilon}^{\mathrm{e}}\,\mathrm{d}D = \underset{\sim}{\sigma} : \mathrm{d}\underset{\sim}{\varepsilon}^{\mathrm{e}} - 2\,y\,\mathrm{d}D \tag{2.36}$$

For a constant stress, this yields the announced result. Variable y therefore corresponds exactly to the elastic energy release rate for a crack, a general parameter of Fracture Mechanics usually denoted G and which has been used in basic fracture theories since Griffith.

There is actually a total analogy between Fracture Mechanics and Continuum Damage Mechanics with the following correspondence:

Fracture Mechanics :		**Damage Mechanics :**
Structure	⇔	*R V E*
External forces	⇔	*Stress applied to the R V E*
Displacement of the load application points	⇔	*General deformation Crack*
Crack (length, area)	⇔	*Damage variable (diffuse)*
Elastic energy release rate G	⇔	*Thermodynamic force y (elastic energy release rate)*

Fig. 2.8 illustrates the partition of the energies dissipated during a tensile test. Curve OA'B' represents the hardening evolution during plastic flow OAB. Parts AB and BC correspond to plastic flow and the increase in elastic strain during the damage process (represented here for a constant stress). The total energy dissipated is divided into (1) energy stored in the material (hardening), (2) energy dissipated as heat and (3) energy released during the damage process and converted into heat.

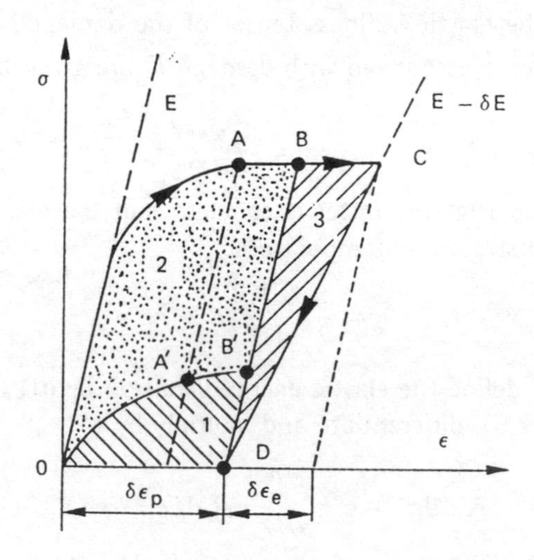

Figure 2.8: Diagram of dissipation during plastic flow and damage

3. THEORETICAL FOUNDATIONS OF DAMAGE MECHANICS

In this part, we discuss the general formulation of Continuum Damage Mechanics, beginning with a more detailed definition of the damage variables and how they affect mechanical behavior. This will be done in the framework of the elasticity laws or state equations and by using the concept of effective stress to formulate the plasticity and viscoplasticity laws coupled with damage. The difficult problem of the effects of damage deactivation will be discussed separately in the particular case of composites, Section 6.

3.1. Damage variables and state equations

A. Tensorial nature of the damage variables

The simplest damage variables to be introduced are obviously the scalar variables. Unfortunately, the defects forming the damage, i.e. cavities, interface decohesions and microcracks, are very often oriented by the loading which created them. It is therefore necessary to use tensorial variables to describe the directional nature of the damage.

A first way of describing the directional character is to use a scalar function defined on a vector space (on $\mathbb{R}^3$) for each RVE around a material point, which associates any direction $\underline{n}$ of space with a defect probability density $p(\frac{a^3}{V}, \underline{n})$ which direction is perpendicular to $\underline{n}$. More specifically, $p(\frac{a^3}{V}, \underline{n})$ is associated with the relative crack density $\frac{a^3}{V}$ where a is the radius of a crack assumed circular (disk) as an initial approximation. Reasoning on material symmetries, due in particular to [39], shows that a series expansion of this probability density yields only even moments, i.e. limiting ourselves to the fourth order:

$$\omega = \frac{1}{V} \int_V \frac{a^3}{V}\; p(\frac{a^3}{V}, \underline{n})\; \mathrm{d}V \tag{3.1}$$

$$\underset{\sim}{\omega} = \frac{1}{V} \int_V \frac{a^3}{V}\; p(\frac{a^3}{V}, \underline{n})\; \underline{n} \otimes \underline{n}\; \mathrm{d}V \tag{3.2}$$

$$\underset{\approx}{\omega} = \frac{1}{V} \int_V \frac{a^3}{V}\; p(\frac{a^3}{V}, \underline{n})\; \underline{n} \otimes \underline{n} \otimes \underline{n} \otimes \underline{n}\; \mathrm{d}V \tag{3.3}$$

We can therefore use the following types of variables to express the elastic behavior of the damaged material, depending on the required accuracy and type of application:

1. *One scalar variable:* The damage is considered isotropic, with no privileged orientation. For initially isotropic material, we can therefore write the elastic potential with a single scalar damage variable D:

   $$\rho\psi = \frac{1}{2}(1-D)\; \underset{\sim}{\varepsilon}^{\mathrm{e}} : \underset{\approx}{\Lambda} : \underset{\sim}{\varepsilon}^{\mathrm{e}} \tag{3.4}$$

   where $\underset{\approx}{\Lambda}$ is the initial elastic stiffness tensor of the undamaged material. This formulation was used from the outset by Continuum Damage Mechanics. It is the simplest approach, very practical and is still very widely used, in particular because of the work of Lemaître [40]. It led to many developments for damage processes of all types and many types of materials. One of its features is to define the thermodynamic force associated with the damage as:

$$Y = -\rho \frac{\partial \psi}{\partial D} = \frac{1}{2} \underset{\sim}{\varepsilon}^{\mathrm{e}} : \underset{\approx}{\Lambda} : \underset{\sim}{\varepsilon}^{\mathrm{e}} \tag{3.5}$$

i.e. the elastic energy of the effective undamaged material.

2. *Two scalar variables:* The isotropic property of the damage (i.e. conservation of the initial isotropy) is more general as expressed by using two scalars, D and Δ, rather than only one [41, 16]. The first is associated with the hydrostatic part of the energy and the other with the remainder:

$$\rho \psi = \frac{1}{2} \lambda(1 - \Delta) \, (\varepsilon^{\mathrm{e}}_{kk})^2 + \mu(1 - D) \, \varepsilon^{\mathrm{e}}_{ij} \varepsilon^{\mathrm{e}}_{ij} \tag{3.6}$$

3. *Several scalar variables* associated with predefined material directions. This is mainly the case for composites in which microcracks are often oriented by the structure and direction of the constituents (plies, reinforcements, yarns, fibers). In this case, there are as many scalar variables as there are privileged directions. Some applications will be considered in Section 6.

4. *A second–rank tensor:* This is the type of variable most frequently used to express the anisotropy introduced by damage in initially isotropic materials. It corresponds to the minimum complexity of an anisotropic theory. The validity of such a second–rank tensor can be demonstrated from the geometric standpoint of the reduction in resistant section or net section [42, 8, 43]. It should be noted that a second–rank tensor alone is not sufficient to describe completely the anisotropy induced in the elastic behavior by damage. It is also necessary to state how this second–rank tensor acts on the fourth–rank tensor characterizing the behavior (elastic stiffness or compliance). This is the role of the damage effect tensors which will be defined below.

5. *A fourth–rank tensor.* This is the lowest–order damage variable capable of describing the damage–induced anisotropy (regardless of whether the material was initially isotropic or anisotropic). Used for the first time by Chaboche [44], this fourth–rank damage variable is naturally introduced through the concept of effective stress based on the strain equivalence principle (Paragraph i. of Section 3.1.B.) In this case, the damage state variable, a fourth–rank tensor, directly acts as damage effect tensor mentioned above for a theory based on a second–rank tensor. Certain theories have been developed since using the elasticity tensor directly as state variable [15, 45].

B. Effective stress concept

We do not indicate here the three–dimensional and anisotropic generalization of the *net stress concept*, that was introduced in Section 2.1.C. This generalized concept has been used for example by Murakami and Ohno [8] for application to the case of creep in

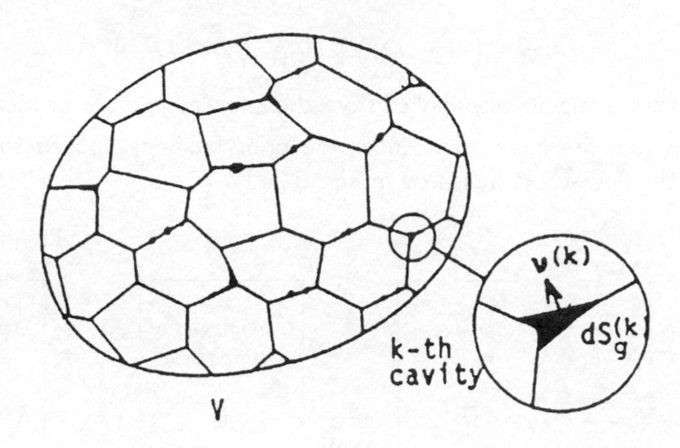

Figure 3.1: Net section in the case of cavitated grain boundaries

polycrystalline metals (Fig. 3.1). We concentrate on the definitions made in terms of the mechanical behavior of the damaged material, based either on the *strain equivalence* or on the *energy equivalence* principles.

i. Effective stress tensor based on strain equivalence

The definition of the effective stress tensor $\tilde{\underset{\sim}{\sigma}}$ given in [44] is that it is the stress tensor that would have to be applied to the RVE of undamaged material to obtain the same strain tensor as the one observed on the damaged RVE subjected to the current stress tensor $\underset{\sim}{\sigma}$.

In elasticity, the damaged material law is expressed by equation (2.34), whereas that of the undamaged material is given by (2.25). Eliminating $\underset{\sim}{\varepsilon}^{e}$ between the two equations yields:

$$\tilde{\underset{\sim}{\sigma}} = \underset{\approx}{M}^{-1} : \underset{\sim}{\sigma} \tag{3.7}$$

where $\underset{\approx}{M}$ is the damage effect operator, a fourth–rank tensor, which can be expressed in terms of $\underset{\approx}{\Lambda}$ and $\tilde{\underset{\approx}{\Lambda}}$, the elasticity tensors of the undamaged material and damaged material respectively:

$$\underset{\approx}{M} = \tilde{\underset{\approx}{\Lambda}} : \underset{\approx}{\Lambda}^{-1} \tag{3.8}$$

To reduce this equation to an expression similar to the one used in the uniaxial case (2.2), we set $\underset{\approx}{M} = \underset{\approx}{I} - \underset{\approx}{D}^{*}$, where $\underset{\approx}{I}$ is the fourth-rank unit tensor $\underset{\approx}{I} = \frac{1}{2}\left(\underset{\sim}{1} \underline{\otimes} \underset{\sim}{1} + \underset{\sim}{1} \overline{\otimes} \underset{\sim}{1}\right)$. Here $\underset{\approx}{D}^{*} = \underset{\approx}{I} - \tilde{\underset{\approx}{\Lambda}} : \underset{\approx}{\Lambda}^{-1}$ denotes the fourth–rank damage tensor which can be measured through the measure of $\tilde{\underset{\approx}{\Lambda}}$. These choices amount to assuming that the elastic stiffness tensor of the damaged RVE is expressed $\tilde{\underset{\approx}{\Lambda}} = (\underset{\approx}{I} - \underset{\approx}{D}^{*}) : \underset{\approx}{\Lambda}$. Actually, this tensor is generally not symmetrical. Therefore, considering $\underset{\approx}{D}$ as state variable, it is preferable to use:

$$\tilde{\Lambda} = \frac{1}{2}\left((I - D) : \Lambda + \Lambda : (I - D^T)\right) \tag{3.9}$$

In practice, the theory is generally applied from the knowledge of D by means of a damage growth equation (and its integration under the prior loadings applied to the RVE). We therefore apply the flow–chart indicated in Fig. 3.2.

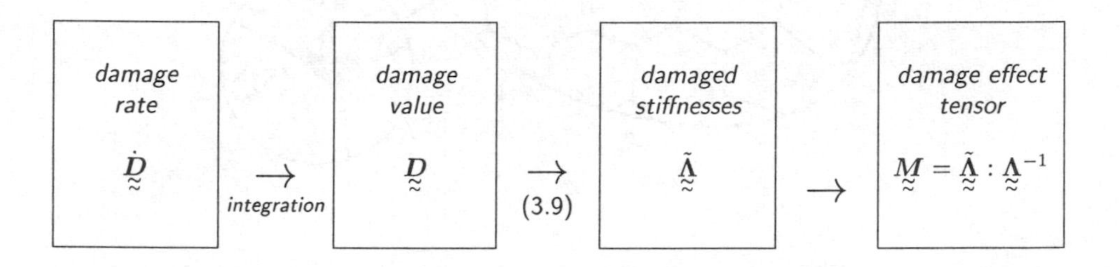

Figure 3.2: Flow–chart of the strain equivalence

ii. Effective stress tensor based on energy equivalence

In this case, the definition is given by [27]: the elastic energy of the material under stress σ and strain ε^e is the same as that of the effective (undamaged) material subjected to the effective stress $\tilde{\sigma}$ and effective elastic strain $\tilde{\varepsilon}^e$. We then have:

$$\begin{aligned} W_e &= \frac{1}{2}\tilde{\sigma} : \tilde{\varepsilon}^e = \frac{1}{2}\tilde{\varepsilon}^e : \Lambda : \tilde{\varepsilon}^e = \frac{1}{2}\tilde{\sigma} : S : \tilde{\sigma} \\ &= \frac{1}{2}\sigma : \varepsilon^e = \frac{1}{2}\varepsilon^e : \tilde{\Lambda} : \varepsilon^e = \frac{1}{2}\sigma : \tilde{S} : \sigma = \tilde{W}_e \end{aligned} \tag{3.10}$$

where $\tilde{\Lambda}$ and $\tilde{S}$ are the stiffness and compliance respectively of the damaged material. The effective stress and the effective strain (different here from the actual strain) are necessarily:

$$\tilde{\sigma} = M^{-1} : \sigma \qquad \text{and} \qquad \tilde{\varepsilon}^e = M^T : \varepsilon^e \tag{3.11}$$

where M is again the damage effect tensor. We also obviously have:

$$\tilde{\Lambda} = M : \Lambda : M^T \qquad \text{and} \qquad \tilde{S} = M^{-T} : S : M^{-1} \tag{3.12}$$

It should be noted that this somewhat changes the meaning of the damage parameter by comparison with the above case. If we wanted to preserve a fourth–rank tensor D, we would write:

$$\tilde{\Lambda} = (I - D) : \Lambda : (I - D^T) \tag{3.13}$$

automatically symmetrical, but where D now acts quadratically. The quadratic form of equation (3.13) means that this approach cannot be identified with any measured change in $\tilde{\Lambda}$ (as was the case with the above approach).

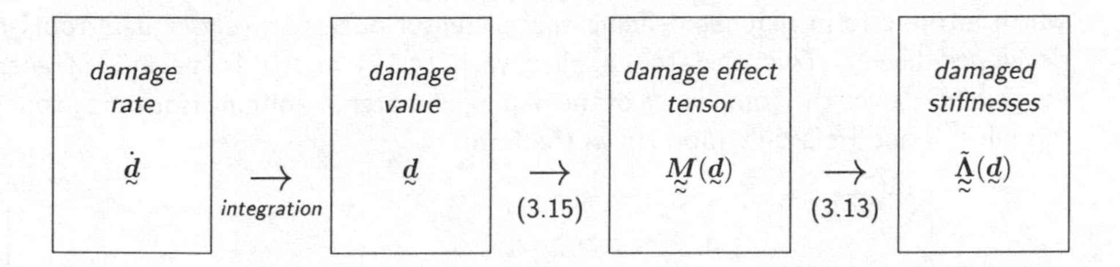

Figure 3.3: Flow–chart of the energy–equivalence

Actually, it is preferably used with a second–rank damage tensor $\underset{\sim}{d}$ to directly express the damage effect tensor $\underset{\approx}{M}(\underset{\sim}{d})$. Various possibilities are indicated below. It should be noted that the order of definition of the operators is necessarily different from that used for the strain equivalence. In effect, we now follow the flow–chart shown on Fig. 3.3.

iii. A few possible forms for the damaged elasticity law

The matrix forms are given in the principal damage system, using the Voigt convention which orders the second–rank tensor $\underset{\sim}{\varepsilon}$ as a column vector $(\varepsilon_{11},\ \varepsilon_{22},\ \varepsilon_{33},\ 2\,\varepsilon_{23},\ 2\,\varepsilon_{31},\ 2\,\varepsilon_{12})$. In certain cases, the intrinsic form is given using the tensor products $\otimes\ \underline{\otimes}\ \overline{\otimes}$.

- For the *scalar damage variables* used for composites, essentially the additive form is used, such that:

$$\tilde{\underset{\approx}{\Lambda}} = \underset{\approx}{\Lambda} - \sum_{i=1}^{m} \delta_i \underset{\approx}{\boldsymbol{K}}_i \tag{3.14}$$

  where δ_i, $(i = 1, ...m)$ are the scalar damage variables oriented by the components. If the composite is globally orthotropic, $m = 3$, and the fourth–rank tensors $\underset{\approx}{\boldsymbol{K}}_i$ characteristic of the material are easily expressed in the principal system of the material. They contain a very limited number of material parameters.

- When using *second–rank tensors* as damage variables, the multiplicative form used in energy equivalence [27] can be expressed in the principal damage system by the "diagonal" matrix:

$$\underset{\approx}{M}(\underset{\sim}{d}) = \begin{bmatrix} 1-d_1 & 0 & 0 & & & \\ 0 & 1-d_2 & 0 & & & \\ 0 & 0 & 1-d_3 & & & \\ & & & \sqrt{(1-d_2)\,(1-d_3)} & & \\ & & & & \sqrt{(1-d_3)\,(1-d_1)} & \\ & & & & & \sqrt{(1-d_1)\,(1-d_2)} \end{bmatrix} \tag{3.15}$$

which intrinsic form requires defining special tensor operators (e.g. square root) not developed herein. This operator, applied with (3.12) in the framework of energy equivalence, gives the compliance of the damaged material, initially isotropic (Young's modulus E and Poisson's ratio ν), in the form:

$$\tilde{\underset{\approx}{S}} = \frac{1}{E}\begin{bmatrix} \frac{1}{(1-d_1)^2} & -\frac{\nu}{(1-d_1)(1-d_2)} & -\frac{\nu}{(1-d_1)(1-d_3)} & & & \\ sym & \frac{1}{(1-d_2)^2} & -\frac{\nu}{(1-d_2)(1-d_3)} & & & \\ sym & sym & \frac{1}{(1-d_3)^2} & & & \\ & & & \frac{1+\nu}{(1-d_2)(1-d_3)} & & \\ & & & & \frac{1+\nu}{(1-d_3)(1-d_1)} & \\ & & & & & \frac{1+\nu}{(1-d_1)(1-d_2)} \end{bmatrix}$$

- Again *with a second–rank tensor, but with strain equivalence*, we can use a damaged stiffness tensor with the form:

$$\tilde{\underset{\approx}{\Lambda}}(\underset{\sim}{d}) = \underset{\approx}{\Lambda} - [\underset{\approx}{D}(\underset{\sim}{d}) : \underset{\approx}{K}]_s \tag{3.16}$$

where $\underset{\approx}{K}$ is a fourth–rank tensor characteristic of the material (which depends in particular on its initial symmetries) and $\underset{\approx}{D}(\underset{\sim}{d})$ is a damage effect operator, expressed, for instance:

$$\underset{\approx}{D}(\underset{\sim}{d}) = \xi\left(\underset{\sim}{1} \otimes \underset{\sim}{d}\right)_s + \frac{1-\xi}{2}\left(\underline{\underset{\sim}{1} \otimes \underset{\sim}{d}} + \underset{\sim}{1}\overline{\otimes}\underset{\sim}{d}\right)_s \tag{3.17}$$

We can very well use the initial elastic stiffness $\underset{\approx}{K}$ as characteristic tensor $\underset{\approx}{K} = \underset{\approx}{\Lambda}$, which yields equation (3.9) used classically. We can also select ξ small and $\underset{\approx}{K} = E\underset{\approx}{I}$ to obtain a solution close to that given by micromechanics.

- Finally, with *a fourth–rank tensor*, we can either use an equation such as (3.16) above (multiplicative form), with $\underset{\approx}{D}$ now as a state variable, or even consider the elasticity stiffness as a state variable.

3.2. State couplings — dissipative couplings

This section discusses construction of the damage laws themselves, at least in their general formulation. We are of course obliged to ensure a certain consistency between the elastic constitutive law, the damage growth and the couplings between plasticity and damage.

A. Different forms of state couplings

Above, we mainly considered coupling between the damage and the material elasticity law. The concepts of effective stress and the expressions given in Section 3.1.B. for introducing the damage effect apply to the elastic behavior. A first question arises concerning the state equations which, in the context of mechanical behavior, involve both elasticity laws and hardening laws or rather the relations between the state variables of the hardening processes and the associated thermodynamic forces, generally used in the plasticity criteria: should the damage be included or not in the free energy term ψ_{p} associated with hardening? Three types of theories can be identified:

- (i) *The initial theory does not consider the presence of this coupling* [9]. It therefore supplies an easy interpretation of the thermodynamic force associated with the damage. As was seen above in Section 2.2.E., it is an elastic energy release similar to that used in Linear Fracture Mechanics, where the local plasticity in the crack tip is neglected (small scale yielding). This theory can be used even in presence of plasticity and hardening. It simply means that there is no coupling between hardening and damage.

- (ii) *The theory suggested by Cordebois and Sidoroff* [27] introduces state variable β, which is a scalar measure of the cumulative damage playing a similar role for damage to that played by the isotropic hardening variable r for plasticity. The free energy is then written:

$$\psi = \psi_\mathrm{e}(\underset{\sim}{\varepsilon}^\mathrm{e},\, D) + \psi_\mathrm{p}(\alpha_j) + \psi_\mathrm{d}(\beta) \tag{3.18}$$

where α_j are the set of hardening variables and D is used here as scalar variable (to simplify, the temperature is omitted). We denote the thermodynamic forces associated with D and β as Y and B:

$$Y = -\rho \frac{\partial \psi_\mathrm{e}}{\partial D} \qquad\qquad B = \rho \frac{\partial \psi_\mathrm{d}}{\partial \beta} \tag{3.19}$$

It will be seen below how these variables B and β are used to construct the damage growth law. As above, there is no coupling between hardening and damage, but it should be immediately mentioned that this approach modifies the interpretation of the energy dissipated by damage, which was given by the first approach, since instead of $Y\dot{D}$, the intrinsic dissipation contains two terms:

$$\Phi_\mathsf{int} = Y\dot{D} - B\dot{\beta} \tag{3.20}$$

- (iii) *The coupling between damage and hardening can be considered independently of variable β above*, with:

$$\psi = \psi_\mathrm{e}(\underset{\sim}{\varepsilon}^\mathrm{e},\, D) + \psi_\mathrm{p}(\alpha_j, D) \tag{3.21}$$

In this case, we change the meaning of the thermodynamic force associated with D, which now contains two terms:

$$Y = -\rho \frac{\partial \psi}{\partial D} = Y_\mathrm{e} + Y_\mathrm{p} = -\rho \frac{\partial \psi_\mathrm{e}}{\partial D} - \rho \frac{\partial \psi_\mathrm{p}}{\partial D} \tag{3.22}$$

Considering the damage to be isotropic, with the same factor for the two terms ψ_e and ψ_p, we can write:

$$\psi = (1-D)\left[\psi_\mathrm{e}^\mathrm{o}(\underset{\sim}{\varepsilon}^\mathrm{e}) + \psi_\mathrm{p}^\mathrm{o}(\alpha_j)\right] \quad \text{where} \quad \rho\,\psi_\mathrm{e}^\mathrm{o}(\underset{\sim}{\varepsilon}_\mathrm{e}) = \frac{1}{2}\,\underset{\sim}{\varepsilon}_\mathrm{e} : \underset{\approx}{\mathbf{\Lambda}} : \underset{\sim}{\varepsilon}_\mathrm{e} \tag{3.23}$$

We then obtain the state equations and the corresponding effective stresses, again in the isotropic case:

$$\underset{\sim}{\sigma} = \rho \frac{\partial \psi_\mathrm{e}}{\partial \underset{\sim}{\varepsilon}^\mathrm{e}} = (1-D)\underset{\approx}{\mathbf{\Lambda}} : \underset{\sim}{\varepsilon}^\mathrm{e} \qquad\qquad A_j = \rho \frac{\partial \psi}{\partial \alpha_j} = (1-D)\rho \frac{\partial \psi_\mathrm{p}^\mathrm{o}}{\partial \alpha_j} \tag{3.24}$$

$$\tilde{\underset{\sim}{\sigma}} = \frac{\underset{\sim}{\sigma}}{1-D} = \rho \frac{\partial \psi_{\mathrm{e}}^{\mathrm{o}}}{\partial \underset{\sim}{\varepsilon}^{\mathrm{e}}} \qquad \tilde{A}_j = \frac{A_j}{1-D} = \rho \frac{\partial \psi_{\mathrm{p}}^{\mathrm{o}}}{\partial \alpha_j} \tag{3.25}$$

This approach was followed by Saanouni [46] and by Ju [16]. It is the approach most widely used today, since it offers a more "symmetric" construction (see Paragraph 3.4.A.iii.).

B. Coupling of dissipation

In the case of the variation laws as well, there are various possibilities according to whether it is considered that the plasticity and damage mechanisms are the same or not:

- (i) *The first approach*, used by Lemaître [11, 40], consists of considering only a single mechanism governed by plasticity with a single dissipation potential, with a normality rule involving only one plastic multiplier:

$$\phi^* = F(\underset{\sim}{\sigma}, A_j, Y\,;\, D) = F_{\mathrm{p}}(\underset{\sim}{\sigma}, A_j\,;\, D) + F_{\mathrm{d}}(Y\,;\, D) \tag{3.26}$$

$$\dot{\underset{\sim}{\varepsilon}}^p = \dot{\lambda} \frac{\partial F}{\partial \underset{\sim}{\sigma}} = \dot{\lambda} \frac{\partial F_{\mathrm{p}}}{\partial \underset{\sim}{\sigma}} \qquad \dot{\alpha}_j = -\dot{\lambda} \frac{\partial F}{\partial A_j} = -\dot{\lambda} \frac{\partial F_{\mathrm{p}}}{\partial A_j} \tag{3.27}$$

$$\dot{D} = \dot{\lambda} \frac{\partial F}{\partial Y} = \dot{\lambda} \frac{\partial F_{\mathrm{d}}}{\partial Y} \tag{3.28}$$

In the case of time–independent mechanisms, multiplier $\dot{\lambda}$ is determined by the consistency condition on the plasticity. In viscoplasticity, $\dot{\lambda}$ is given as a known function of $\underset{\sim}{\sigma}$, A_j [47].

However, with this approach, the distinction between plasticity and damage appears only as a separation between $F_{\mathrm{p}} \equiv f_{\mathrm{p}}$ and $F_{\mathrm{d}} \equiv f_{\mathrm{d}}$, the first corresponding to the plastic flow criterion and the second to the damage criterion equipped with a threshold. It can be seen that damage progresses only when there is plastic flow. Similarly, beyond the damage initiation threshold, there cannot be any plasticity without a corresponding increase in damage. This therefore appears as a relatively strong limitation.

- (ii) *The second approach* consists of considering the two mechanisms and the two associated criteria separately, using a quasi Generalized Standard Media theory. We therefore have two independent dissipation potentials and two independent multipliers:

$$F_{\mathrm{p}}(\underset{\sim}{\sigma}, A_j;\, D) \qquad f_{\mathrm{p}}(\underset{\sim}{\sigma}, A_j;\, D) \leq 0 \tag{3.29}$$

$$F_{\mathrm{d}}(Y, B\,;\, D, \beta) \qquad f_{\mathrm{d}}(Y, B\,;\, D, \beta) \leq 0 \tag{3.30}$$

$$\dot{\underset{\sim}{\varepsilon}}^p = \dot{\lambda}_{\mathrm{p}} \frac{\partial F_{\mathrm{p}}}{\partial \underset{\sim}{\sigma}} \qquad \dot{\alpha}_j = -\dot{\lambda}_{\mathrm{p}} \frac{\partial F_{\mathrm{p}}}{\partial A_j} \tag{3.31}$$

$$\dot{D} = \dot{\lambda}_{\rm d} \frac{\partial F_{\rm d}}{\partial Y} \qquad \dot{\beta} = -\dot{\lambda}_{\rm d} \frac{\partial F_{\rm d}}{\partial B} \tag{3.32}$$

It can be seen that D and β are included as parameters in both potentials. For instance, D is included in $f_{\rm p}$ through the effective stress concept, as will be seen in Section 3.2.A.–C. The two independent multipliers, $\dot{\lambda}_{\rm p}$ for plasticity and $\dot{\lambda}_{\rm d}$ for damage, are determined by the two consistency conditions associated with the two criteria $\dot{f}_{\rm p} = f_{\rm p} = 0$, $\dot{f}_{\rm d} = f_{\rm d} = 0$, at least in the rate–independent case. In the rate–dependent case, the two multipliers can be given as two known functions of the corresponding variables.

This dissociation of the two types of mechanisms is illustrated in Fig. 3.4 (equivalent von Mises stress $\sigma_{\rm eq}$ versus hydrostatic pressure $\sigma_{\rm H}$), clearly showing the possibility of accessing the two surfaces independently ($f_{\rm p} = 0$ and $f_{\rm d} = 0$) and causing them to progress, again in the framework of a rate–independent theory (with isotropic variations). A similar approach is used by most authors in this area of CDM [16, 27, 48, 17, 49, 50, 30, 36].

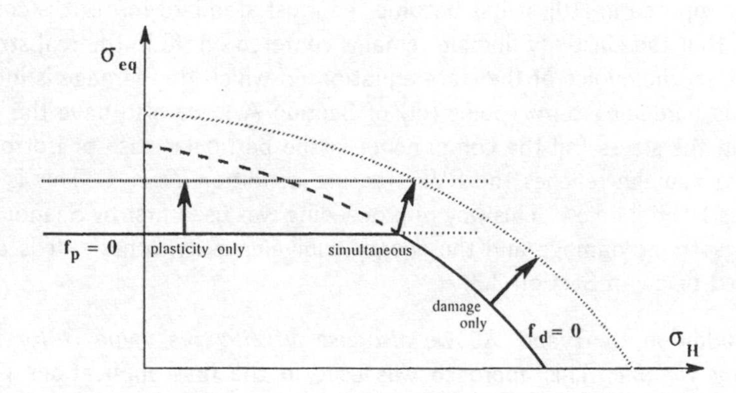

Figure 3.4: Coupling of a plasticity criterion and a damage criterion

C. A few possibilities for the elastic limit criterion

Damage influences the plastic flow by affecting either the elastic domain or the hardening law. Although not necessary, the effective stress concept supplies a restrictive choice which decreases the number of possibilities and parameters depending on the material, while giving satisfactory qualitative and quantitative results. This is the approach we shall use, but there are still a significant number of possible variants:

- (i) *It is assumed that only the effective stress is replaced, but that the hardening variables are not amplified by damage.* This is the choice made by Benallal [47] and Lemaître [40]. The elastic limit criterion is then expressed as:

$$f = f(\tilde{\underset{\sim}{\sigma}}, \underset{\sim}{X}, R) = J\left(\tilde{\underset{\sim}{\sigma}} - \underset{\sim}{X}\right) - R - \sigma_y \leq 0 \tag{3.33}$$

where $\tilde{\underset{\sim}{\sigma}}$ is defined by one of the choices mentioned in Section 3.1.B., for instance equation (3.7). σ_y represents the initial elastic limit of the material. In Lemaître's approach [40], limited to isotropic damage, we have:

$$f = J\left(\frac{\underset{\sim}{\sigma}}{1-D} - \underset{\sim}{X}\right) - R - \sigma_y \leq 0 \tag{3.34}$$

With this approach, the elastic domain does not remain centered on $\underset{\sim}{X}$. However, when approaching fracture ($D = 1$), $\underset{\sim}{\sigma}$ effectively approaches 0, as is shown by Exercise 3 of Section 3.3.C.

- (ii) *In addition to the effective stress, we use the effective kinematic hardening variable* $\tilde{\underset{\sim}{X}}$, determined from $\underset{\sim}{X}$ by the same operation:

$$\tilde{\underset{\sim}{X}} = \underset{\approx}{M}^{-1} : \underset{\sim}{X} \tag{3.35}$$

$$f = f(\tilde{\underset{\sim}{\sigma}}, \tilde{\underset{\sim}{X}}, R) = J\left(\tilde{\underset{\sim}{\sigma}} - \tilde{\underset{\sim}{X}}\right) - R - \sigma_y \leq 0 \tag{3.36}$$

Initially proposed in [10], it has become the most standard form. It is consistent with the fact that the elasticity domain remains centered on $\underset{\sim}{X}$ in the real stress space. In addition, by the choice of the state equation, in which the damage is included in the kinematic hardening term (point (iii) of Section A.), we also have the possibility of canceling the stress (all the components in the particular case of isotropic damage) when the damage reaches the RVE fracture condition $D = 1$. This is examined in Exercises 1 and 2 below. This way of proceeding was used first by Saanouni [46] in the case of isotropic damage and the energy equivalence hypothesis. It is the approach developed below in Section 3.3.

- (iii) In addition to $\tilde{\underset{\sim}{\sigma}}$ and $\tilde{\underset{\sim}{X}}$, *we also use an effective value* $\tilde{R}$ *for the isotropic hardening variable.* This approach was used in the first applications [9], [30], and is also used by Saanouni for the case of isotropic damage. We thus set: $\tilde{R} = \frac{R}{1-D}$ However, since this equation is difficult to generalize in the case of anisotropic damage, we do not consider this hypothesis.

3.3. General approach for plasticity/damage coupling

We consider plasticity (or viscoplasticity) with kinematic and isotropic hardening. To simplify the equations, we do not include hardening recovery phenomena, although they could also be included in the same theoretical framework. The expressions are given for the isothermal case, but their generalization to the anisothermal case is straightforward.

From the standpoint of thermodynamics, we assume the existence of a state potential, free energy, in form (iii) of Section 3.2.A. For dissipation, we use the *quasi–standard GSM theory* and assume the existence of several independent dissipation potentials and several multipliers (as already mentioned for choice (ii) of Section 3.2.B.)

The initial medium is assumed to be anisotropic, described by Hill's criterion for plasticity. The damage is also considered anisotropic, but it is unnecessary here to detail the tensorial nature of the damage variables, whose evolution is not described in detail until Section 3.4. They are simply denoted d without any further precision.

A. State equations

The state equations are given by the free energy thermodynamic potential:

$$\rho\psi = \rho(\psi_{\mathrm{e}} + \psi_{\mathrm{p}}) = \frac{1}{2}\,\underset{\sim}{\varepsilon}^{\mathrm{e}} : \underset{\approx}{\tilde{\Lambda}}(d) : \underset{\sim}{\varepsilon}^{\mathrm{e}} + \sum_i \frac{1}{2}\,\underset{\sim}{\alpha}_i : \underset{\approx}{\tilde{C}}_i(d) : \underset{\sim}{\alpha}_i + \rho\psi_{\mathrm{r}}(r,\, d) \tag{3.37}$$

where $\underset{\sim}{\alpha}_i$ are the kinematic hardening state variables (second–rank tensors), indefinite in number. We use only one isotropic hardening variable r. We then have:

$$\underset{\sim}{\sigma} = \rho\frac{\partial\psi}{\partial\underset{\sim}{\varepsilon}^{\mathrm{e}}} = \underset{\approx}{\tilde{\Lambda}}(d) : \underset{\sim}{\varepsilon}^{\mathrm{e}} \tag{3.38}$$

$$\underset{\sim}{X}_i = \rho\frac{\partial\psi}{\partial\underset{\sim}{\alpha}_i} = \underset{\approx}{\tilde{C}}_i(d) : \underset{\sim}{\alpha}_i \qquad R = \rho\frac{\partial\psi}{\partial r} = \rho\frac{\partial\psi_{\mathrm{r}}}{\partial r} \tag{3.39}$$

The thermodynamic forces associated with the damage are specified later. To simplify the expression, we assume that all the $\underset{\approx}{C}_i$ are proportional with one another, i.e. that: $\underset{\approx}{C}_i = c_i\,\underset{\approx}{C}$. We can therefore define the damage effect operator $\underset{\approx}{M}$ for the elasticity law and the damage effect operator $\underset{\approx}{N}$ for the kinematic hardening law:

$$\underset{\sim}{\tilde{\sigma}} = \underset{\approx}{M}^{-1} : \underset{\sim}{\sigma} \qquad \underset{\sim}{\tilde{X}}_i = \underset{\approx}{N}^{-1} : \underset{\sim}{X}_i \tag{3.40}$$

Using strain equivalence yields:

$$\underset{\approx}{M} = \underset{\approx}{\tilde{\Lambda}} : \underset{\approx}{\Lambda}^{-1} \qquad \underset{\approx}{N} = \underset{\approx}{\tilde{C}} : \underset{\approx}{C}^{-1} \tag{3.41}$$

On the contrary, using energy equivalence, we can choose to take $\underset{\approx}{M} = \underset{\approx}{N}$, adding similar relations for the effective "strains":

$$\underset{\sim}{\tilde{\varepsilon}} = \underset{\approx}{M}^{\mathrm{T}} : \underset{\sim}{\varepsilon} \qquad \underset{\sim}{\tilde{\varepsilon}}^{\mathrm{e}} = \underset{\approx}{M}^{\mathrm{T}} : \underset{\sim}{\varepsilon}^{\mathrm{e}}$$

$$\underset{\sim}{\tilde{\varepsilon}}^{p} = \underset{\approx}{M}^{\mathrm{T}} : \underset{\sim}{\varepsilon}^{p} \qquad \underset{\sim}{\tilde{\alpha}}_i = \underset{\approx}{M}^{\mathrm{T}} : \underset{\sim}{\alpha}_i \tag{3.42}$$

B. Plasticity evolution laws

The plasticity/viscoplasticity laws include an elasticity domain $f \leq 0$ (which can be reduced to a point in viscoplasticity). We assume Hill's criterion, defined by the fourth–rank tensor $\underset{\approx}{H}$, characteristic of the damaged material and its symmetries:

$$f = \left\|\underset{\sim}{\tilde{\sigma}} - \underset{\sim}{\tilde{X}}\right\|_{\mathrm{H}} - R - k = \left((\underset{\sim}{\tilde{\sigma}} - \underset{\sim}{\tilde{X}}) : \underset{\approx}{H} : (\underset{\sim}{\tilde{\sigma}} - \underset{\sim}{\tilde{X}})\right)^{\frac{1}{2}} - R - k \tag{3.43}$$

where $\underset{\sim}{\tilde{X}} = \sum_i \underset{\sim}{\tilde{X}}_i$. As potential associated with plasticity, we choose the following additive expression in which the second and third terms are introduced to produce the kinematic and isotropic dynamic recovery effects:

$$F_{\mathrm{p}} = f + \frac{1}{2}\sum_i \gamma_i\,\underset{\sim}{\tilde{X}}_i : \underset{\approx}{Q} : \underset{\sim}{\tilde{X}}_i + \frac{1}{2}\,\frac{g}{c}\,R^2 \tag{3.44}$$

The generalized normality rule then gives, using (3.40):

$$\dot{\underset{\sim}{\varepsilon}}^p = \dot{\lambda}\,\frac{\partial F_{\rm p}}{\partial \underset{\sim}{\sigma}} = \dot{\lambda}\,\frac{\underset{\approx}{\boldsymbol{M}}^{-\rm T} : \underset{\approx}{\boldsymbol{H}} : (\tilde{\underset{\sim}{\sigma}} - \tilde{\underset{\sim}{\boldsymbol{X}}})}{\left\|\tilde{\underset{\sim}{\sigma}} - \tilde{\underset{\sim}{\boldsymbol{X}}}\right\|_{\rm H}} = \dot{\lambda}\,\underset{\approx}{\boldsymbol{M}}^{-\rm T} : \underset{\sim}{\boldsymbol{n}} \tag{3.45}$$

$$\begin{aligned} \dot{\underset{\sim}{\alpha}}_i = -\dot{\lambda}\,\frac{\partial F_{\rm p}}{\partial \underset{\sim}{\boldsymbol{X}}_i} &= \dot{\lambda}\,\frac{\underset{\approx}{\boldsymbol{N}}^{-\rm T} : \underset{\approx}{\boldsymbol{H}} : (\tilde{\underset{\sim}{\sigma}} - \tilde{\underset{\sim}{\boldsymbol{X}}})}{\left\|\tilde{\underset{\sim}{\sigma}} - \tilde{\underset{\sim}{\boldsymbol{X}}}\right\|_{\rm H}} - \gamma_i\,\underset{\approx}{\boldsymbol{N}}^{-\rm T} : \underset{\approx}{\boldsymbol{Q}} : \tilde{\underset{\sim}{\boldsymbol{X}}}_i\,\dot{\lambda} \\ &= \underset{\approx}{\boldsymbol{N}}^{-\rm T} : \left(\underset{\approx}{\boldsymbol{M}}^{\rm T} : \dot{\underset{\sim}{\varepsilon}}^p - \gamma_i\,\underset{\approx}{\boldsymbol{Q}} : \tilde{\underset{\sim}{\boldsymbol{X}}}_i\,\dot{\lambda}\right) \end{aligned} \tag{3.46}$$

$$\dot{r} = -\dot{\lambda}\,\frac{\partial F_{\rm p}}{\partial R} = \dot{\lambda} - \frac{g}{c}\,R\,\dot{\lambda} \tag{3.47}$$

where $\underset{\sim}{\boldsymbol{n}}$ denotes the "unit" direction (in a space transformed by Hill's criterion):

$$\underset{\sim}{\boldsymbol{n}} = \frac{\underset{\approx}{\boldsymbol{H}} : (\tilde{\underset{\sim}{\sigma}} - \tilde{\underset{\sim}{\boldsymbol{X}}})}{\left\|\tilde{\underset{\sim}{\sigma}} - \tilde{\underset{\sim}{\boldsymbol{X}}}\right\|_{\rm H}} \qquad \left\|\underset{\sim}{\boldsymbol{n}}\right\|_{H^{-1}} = \left(\underset{\sim}{\boldsymbol{n}} : \underset{\approx}{\boldsymbol{H}}^{-1} : \underset{\sim}{\boldsymbol{n}}\right)^{\frac{1}{2}} = 1 \tag{3.48}$$

In rate–independent plasticity, multiplier $\dot{\lambda}$ is determined by the consistency condition $f = \dot{f} = 0$, possibly coupled with a similar condition for the damage criterion (this determination will be examined in Paragraph 3.4.A.iii.) In viscoplasticity, $\dot{\lambda}$ is expressed, for instance, as a power function:

$$\dot{\lambda} = \left(\dot{\underset{\sim}{\varepsilon}}^p : \underset{\approx}{\boldsymbol{M}} : \underset{\approx}{\boldsymbol{H}}^{-1} : \underset{\approx}{\boldsymbol{M}}^{\rm T} : \dot{\underset{\sim}{\varepsilon}}^p\right)^{\frac{1}{2}} = \left\langle \frac{f}{K} \right\rangle^n \tag{3.49}$$

C. Asymptotic behavior on approaching fracture

In this section, we examine the evolution of the stress and the hardening variables on approaching fracture of the RVE, as D approaches 1. As illustration, we show that with the approaches used, the stress necessarily approaches 0. This point is important for numerical reasons, in particular in the framework of CDM applications with *local approaches to fracture*, Section 5.3.

We consider rate–independent plasticity and use the version with isotropic damage to simplify the explanation. In an anisotropic case, only certain components of the stress would cancel out, depending on the fracture criterion selected. Of course, the demonstration is not valid unless we neglect possible bifurcation and localization phenomena which occur in structural analysis. Herein, we only examine the "fundamental" solution of the isolated RVE.

We conduct three analyses in parallel, by strain equivalence, energy equivalence and a law not coupled on the hardening terms (approach followed by Lemaître [40]). To simplify, it is assumed that the elasticity operators $\underset{\approx}{\boldsymbol{\Lambda}}$ and $\underset{\approx}{\boldsymbol{C}}$ are co–linear and we use only one nonlinear kinematic hardening variable. This case is considered generic for all other cases. There is no isotropic hardening.

In the framework of rate–independent plasticity, we use an elasticity domain and a dissipation potential in the form:

$$f = \left|\left|\tilde{\underset{\sim}{\sigma}} - \tilde{\underset{\sim}{X}}\right|\right|_{\mathrm{H}} - k = \left((\tilde{\underset{\sim}{\sigma}} - \tilde{\underset{\sim}{X}}) : \underset{\approx}{H} : (\tilde{\underset{\sim}{\sigma}} - \tilde{\underset{\sim}{X}})\right)^{\frac{1}{2}} - k \tag{3.50}$$

$$F_{\mathrm{p}} = f + \frac{1}{2}\gamma\,\tilde{\underset{\sim}{X}} : \underset{\approx}{C}^{-1} : \tilde{\underset{\sim}{X}} \tag{3.51}$$

in which we chose $\underset{\approx}{Q} = \underset{\approx}{C}^{-1}$ as dynamic recovery term (to be compared with (3.44) of Section 3.3.)

- *Exercise 1: Strain equivalence:*

  We assume $\underset{\approx}{M} = \underset{\approx}{N}$ and the effective stresses are given by:

$$\tilde{\underset{\sim}{\sigma}} = \underset{\approx}{\Lambda} : (\underset{\sim}{\varepsilon} - \underset{\sim}{\varepsilon}^p) = \underset{\approx}{M}^{-1} : \underset{\sim}{\sigma} \qquad \tilde{\underset{\sim}{X}} = \underset{\approx}{C} : \underset{\sim}{\alpha} = \underset{\approx}{M}^{-1} : \underset{\sim}{X} \tag{3.52}$$

  where the damage effect operator is given by (3.41). We determine the stress and strain rates from potential F_{p}:

$$\dot{\underset{\sim}{\varepsilon}}^p = \dot{\lambda}\,\frac{\partial F_{\mathrm{p}}}{\partial \underset{\sim}{\sigma}} = \dot{\lambda}\,\frac{\underset{\approx}{M}^{-\mathrm{T}} : \underset{\approx}{H} : (\tilde{\underset{\sim}{\sigma}} - \tilde{\underset{\sim}{X}})}{\left|\left|\tilde{\underset{\sim}{\sigma}} - \tilde{\underset{\sim}{X}}\right|\right|_{\mathrm{H}}} = \dot{\lambda}\,\underset{\approx}{M}^{-\mathrm{T}} : \underset{\sim}{n} \tag{3.53}$$

$$\begin{aligned} \dot{\underset{\sim}{\alpha}} = -\dot{\lambda}\,\frac{\partial F_{\mathrm{p}}}{\partial \underset{\sim}{X}} &= \dot{\underset{\sim}{\varepsilon}}^p - \gamma\,\underset{\approx}{M}^{-\mathrm{T}} : \underset{\approx}{C}^{-1} : \tilde{\underset{\sim}{X}}\,\dot{\lambda} \\ &= \dot{\underset{\sim}{\varepsilon}}^p - \gamma\,\underset{\approx}{M}^{-\mathrm{T}} : \underset{\sim}{\alpha}\,\dot{\lambda} = \underset{\approx}{M}^{-\mathrm{T}} : (\underset{\sim}{n} - \gamma\,\underset{\sim}{\alpha})\,\dot{\lambda} \end{aligned} \tag{3.54}$$

  We assume loading to be monotonic. When the plastic strain and damage increase indefinitely, stabilization occurs for $||\underset{\sim}{\alpha}|| \to 1/\gamma$ (regardless of the damage growth included in operator $\underset{\approx}{M}$). Therefore, since $\tilde{\underset{\sim}{X}} = \underset{\approx}{C} : \underset{\sim}{\alpha}$, the norm of effective stress tensor $\tilde{\underset{\sim}{X}}$ approaches a scalar value such as C/γ, as for the case without damage (Fig. 3.5). During plastic flow, we necessarily have:

$$\left|\left|\tilde{\underset{\sim}{\sigma}} - \tilde{\underset{\sim}{X}}\right|\right|_{\mathrm{H}} = \underset{\sim}{n} : (\tilde{\underset{\sim}{\sigma}} - \tilde{\underset{\sim}{X}}) = k \tag{3.55}$$

$$\text{i.e.} \quad \tilde{\underset{\sim}{\sigma}} = \tilde{\underset{\sim}{X}} + k\,\underset{\approx}{H}^{-1} : \underset{\sim}{n} \quad \text{or} \quad \underset{\sim}{\sigma} = \underset{\sim}{X} + k\,\underset{\approx}{M} : \underset{\approx}{H}^{-1} : \underset{\sim}{n} \tag{3.56}$$

  Now we apply the hypothesis of damage isotropy, replacing $\underset{\approx}{M}$ by $(1 - D)\,\underset{\approx}{I}$. This yields:

$$\underset{\sim}{\sigma} = \underset{\sim}{X} + k\,(1 - D)\,\underset{\approx}{H}^{-1} : \underset{\sim}{n} \tag{3.57}$$

  As we approach fracture, i.e. as D approaches 1, we have the announced result: the norm of $\underset{\sim}{X}$ approaches 0, since $\underset{\sim}{X} = (1 - D)\tilde{\underset{\sim}{X}}$. Accordingly, the norm of $\underset{\sim}{\sigma}$ also approaches 0. This evolution is schematically illustrated in Fig. 3.5 for uniaxial tension, for both effective stresses and real stresses.

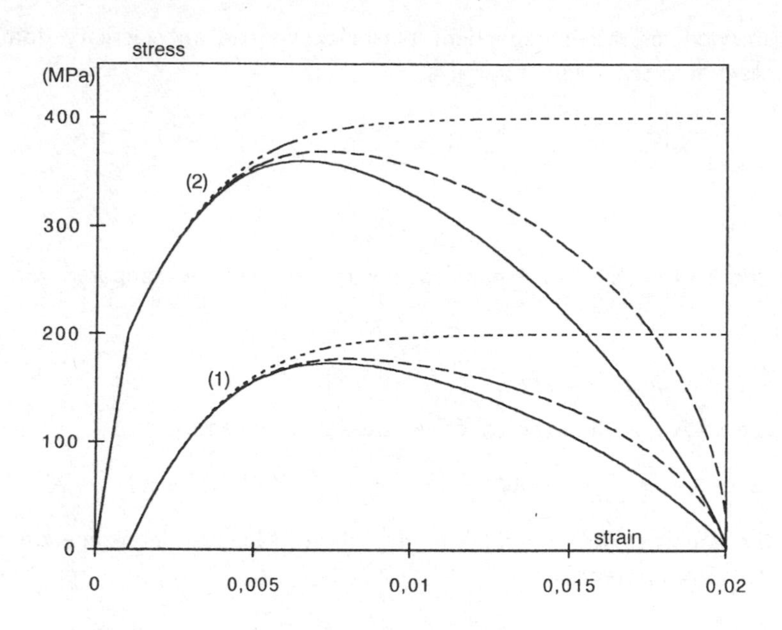

Figure 3.5. Plastic/damage coupling under tension ((1) X and (2) σ): ——— strain equivalence; – – –– energy equivalence; ...... undamaged curves. $E = 200\,000\,\mathrm{MPa}$, $C = 100\,000\,\mathrm{MPa}$, $k = 200\,\mathrm{MPa}$, $\gamma = 500$, $\varepsilon_{p_r} = .02$.

- *Exercise 2: Energy equivalence*

Equations (3.52) are supplemented by (3.42) but we no longer have (3.41) as above and the state equations are:

$$\underset{\sim}{\sigma} = \tilde{\underset{\approx}{\Lambda}} : \underset{\sim}{\varepsilon}^{\mathrm{e}} \qquad \underset{\sim}{X} = \tilde{\underset{\approx}{C}} : \underset{\sim}{\alpha} \tag{3.58}$$

$$\tilde{\underset{\sim}{\sigma}} = \underset{\approx}{\Lambda} : \tilde{\underset{\sim}{\varepsilon}}^{\mathrm{e}} \qquad \tilde{\underset{\sim}{X}} = \underset{\approx}{C} : \tilde{\underset{\sim}{\alpha}} \tag{3.59}$$

It should be noted that the damage effect operator, again a fourth–rank tensor, should be considered as homogeneous to the "square root" of the previous tensor, used in strain equivalence. The rate equations are then written:

$$\dot{\underset{\sim}{\varepsilon}}^{p} = \dot{\lambda}\, \underset{\approx}{M}^{-\mathrm{T}} : \underset{\sim}{n} \tag{3.60}$$

$$\dot{\underset{\sim}{\alpha}} = \dot{\underset{\sim}{\varepsilon}}^{p} \; - \gamma \underset{\approx}{M}^{-\mathrm{T}} : \underset{\approx}{C}^{-1} : \tilde{\underset{\sim}{X}}\; \dot{\lambda} = \dot{\underset{\sim}{\varepsilon}}^{p} - \gamma\, \underset{\sim}{\alpha}\, \dot{\lambda} \; = \; \underset{\approx}{M}^{-\mathrm{T}} : (\underset{\sim}{n} - \gamma\, \tilde{\underset{\sim}{\alpha}})\; \dot{\lambda} \tag{3.61}$$

It is now $\tilde{\underset{\sim}{\alpha}}$ which norm approaches $1/\gamma$ when the plastic strain and damage increase. In addition, the norm of $\tilde{\underset{\sim}{X}}$ approaches C/γ. Obviously, (3.56) remain valid and, in the case of isotropic damage, we find that $\underset{\sim}{X}$ and $\underset{\sim}{\sigma}$ cancel out when $D \to 1$.

The evolution during a tensile test are similar to the above case. The dashed–line curves of Fig. 3.5 were plotted assuming that damage growth versus plastic strain was given by $D = (\varepsilon_{\rm p}/\varepsilon_{pR})^2$.

- *Exercise 3: Theory without hardening/damage coupling*

In the case of the theory used by Lemaître [40], the effective stress concept is not used for the kinematic hardening variable (see 3.2.C.-(i)). Here we show that it gives globally similar results, as illustrated by Fig. 3.6. We limit ourselves to the isotropic case with (3.34) for the elasticity domain. The damage growth law is then given by:

$$\dot{\underset{\sim}{\varepsilon}}^p = \frac{\dot{\lambda}}{1-D}\,\underset{\sim}{n} \qquad \dot{\underset{\sim}{\alpha}} = (1-D)\,\dot{\underset{\sim}{\varepsilon}}^p - \gamma\,\underset{\sim}{\alpha}\,\dot{\lambda} = (\underset{\sim}{n} - \gamma\,\underset{\sim}{\alpha})\,\dot{\lambda} \tag{3.62}$$

When $D \to 1$ and the plastic deformation increases indefinitely, the norm of variable $\underset{\sim}{\alpha}$ approaches $1/\gamma$, but in this case the norm of $\underset{\sim}{X} = \frac{2}{3}\,C\underset{\sim}{\alpha}$ approaches C/γ. However, the flow criterion applies with:

$$\tilde{\underset{\sim}{\sigma}} = \underset{\sim}{X} + \frac{2}{3}\,k\,\underset{\sim}{n} \qquad \text{and} \qquad \underset{\sim}{\sigma} = (1-D)\left(\underset{\sim}{X} + \frac{2}{3}\,k\,\underset{\sim}{n}\right) \tag{3.63}$$

Therefore, $\underset{\sim}{\sigma}$ still approaches $\underset{\sim}{0}$ when D approaches 1, as is shown by Fig. 3.6. It should be noted that the results of this latter approach are very similar to those given for the first method with hardening/damage coupling.

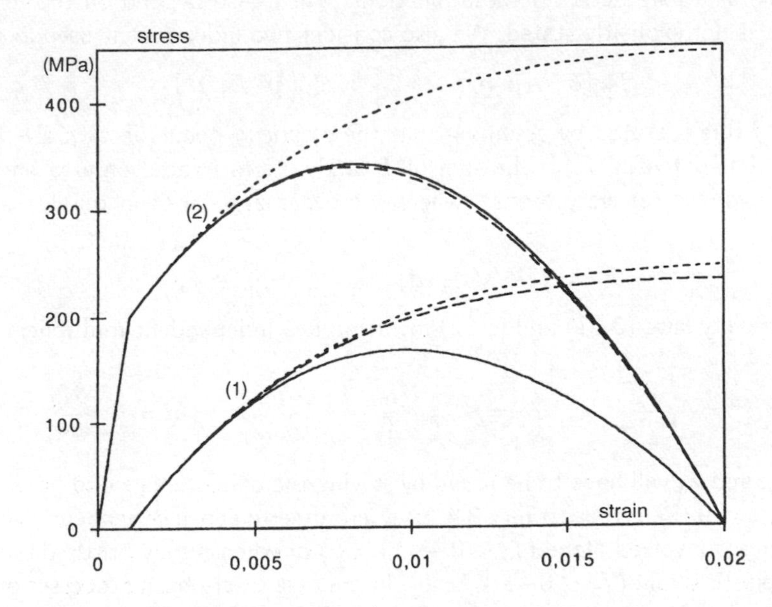

Figure 3.6. Plasticity–damage coupling in tension ((1) X and (2) σ): ______ strain equivalence; — — — theory uncoupled over X; ...... curves without damage. $E = 200\,000\,\mathrm{MPa}$, $C = 50\,000\,\mathrm{MPa}$, $k = 200\,\mathrm{MPa}$, $\gamma = 200$, $\varepsilon_{p_r} = .02$.

3.4. Damage evolution laws

Below we discuss the general forms of the criteria and evolution equations commonly used to describe damage growth. They can be classified in three categories: (i) those related to a purely standard thermodynamical approach (GSM: Generalized Standard Media), such as that used by Lemaître [40]; (ii) those resulting from a *"quasi–standard"* formulation in which a distinction is made, for dissipation, between the plasticity potential, the damage potential and the corresponding multipliers; (iii) finally, those which do not refer to a dissipation potential.

We distinguish between the rate–independent models used for ductile fracture of metallic materials and for brittle materials, concretes and composites, and the rate–dependent models, essentially related to creep in metals.

A. Rate–independent formalism

i. Normality laws

Unless specified, we use a second–rank tensor $\underset{\sim}{\boldsymbol{d}}$ as damage variable. In certain applications, we will limit ourselves to a scalar variable D. We use the quasi–standard formalism already mentioned in Paragraph 3.2.B.(ii), with two independent criteria:

$$f_{\mathrm{p}}\left(\underset{\sim}{\boldsymbol{\sigma}},\, A_j;\; \underset{\sim}{\boldsymbol{d}}\right) \leq 0 \qquad\qquad f_{\mathrm{d}}\left(\underset{\sim}{\boldsymbol{y}},\, B;\; \underset{\sim}{\boldsymbol{d}}\right) \leq 0 \tag{3.64}$$

which delimit the elasticity (non–plasticity) and non–damage domains in the stress space and in the space of the thermodynamic forces associated with damage. The state variable $\underset{\sim}{\boldsymbol{d}}$ can be used as a parameter for these functions, which also depend on the temperature, although this is not explicitly stated. We also consider two independent pseudo–potentials:

$$F_{\mathrm{p}}\left(\underset{\sim}{\boldsymbol{\sigma}},\, A_j;\; \underset{\sim}{\boldsymbol{d}}\right) \qquad\qquad F_{\mathrm{d}}\left(\underset{\sim}{\boldsymbol{y}},\, B;\; \underset{\sim}{\boldsymbol{d}}\right) \tag{3.65}$$

The normality rule is stated by assuming that the plasticity occurs in a *quasi–associated"* manner, i.e. that potential F_{p} is the sum of f_{p} and a term related only to hardening (to produce the dynamic recovery terms), whereas the damage varies in purely *"associated"* mode:

$$F_{\mathrm{p}} = f_{\mathrm{p}} + q(A_j;\, \underset{\sim}{\boldsymbol{d}}) \qquad\qquad F_{\mathrm{p}} \equiv f_{\mathrm{p}} \tag{3.66}$$

With the normality laws (3.31) and (3.32) involving two independent multipliers $\dot{\lambda}_{\mathrm{p}}$ and $\dot{\lambda}_{\mathrm{d}}$ we write:

$$\dot{\underset{\sim}{\varepsilon}}^{p} = \dot{\lambda}_{\mathrm{p}} \frac{\partial f_{\mathrm{p}}}{\partial \underset{\sim}{\boldsymbol{\sigma}}} \qquad \dot{\alpha}_j = -\dot{\lambda}_{\mathrm{p}} \frac{\partial f_{\mathrm{p}}}{\partial A_j} - \dot{\lambda}_{\mathrm{p}} \frac{\partial q}{\partial A_j} \qquad \dot{\underset{\sim}{\boldsymbol{d}}} = \dot{\lambda}_{\mathrm{d}} \frac{\partial f_{\mathrm{d}}}{\partial \underset{\sim}{\boldsymbol{y}}} \tag{3.67}$$

Multipliers $\dot{\lambda}_{\mathrm{p}}$ and $\dot{\lambda}_{\mathrm{d}}$ will have to be found by solving the consistency conditions $f_{\mathrm{p}} = \dot{f}_{\mathrm{p}} = 0$ and $f_{\mathrm{d}} = \dot{f}_{\mathrm{d}} = 0$. Referring to Fig. 3.4, they are determined independently in the cases where plasticity is involved alone ($f_{\mathrm{d}} < 0 \rightarrow \dot{\lambda}_{\mathrm{d}} = 0$) or when purely brittle damage occurs without any plastic strain ($f_{\mathrm{p}} < 0 \rightarrow \dot{\lambda}_{\mathrm{p}} = 0$). In the case where both processes are involved simultaneously, it is necessary to consider the possibility of coupling between the two. It should also be noted that multipliers $\dot{\lambda}_{\mathrm{p}}$ and $\dot{\lambda}_{\mathrm{d}}$ must necessarily be positive. Otherwise, it is considered that there is unloading with respect to either the plasticity criterion or the damage criterion (and then we write $\dot{\lambda}_{\mathrm{p}} = 0$ or $\dot{\lambda}_{\mathrm{d}} = 0$).

ii. Form of the non–damage criterion

For the rate independent case the damage evolution is governed by the way the non–damage criterion is modified. It can be parameterized by its size in the $\underset{\sim}{y}$ space. This is the method used by Ladevèze [41], Allix [51], Simo and Ju [45] and many other authors. The criterion is expressed as:

$$f_{\rm d} = \Omega(\underset{\sim}{y}) - \omega \leq 0 \tag{3.68}$$

and the evolution is subjected to the conditions:

$$\dot{\underset{\sim}{d}} = \dot{\lambda}_{\rm d} \frac{\partial f_{\rm d}}{\partial \underset{\sim}{y}} \qquad \dot{\lambda}_{\rm d} = \dot{\omega} \tag{3.69}$$

which amounts to considering that ω is the "memory" of the maximum size reached by the criterion instead of being associated with the cumulative damage. In effect, if $f_{\rm d} < 0$, there is no change. On the contrary, if $\dot{\lambda}_{\rm d} > 0$, i.e. if there is an increase in the damage, the consistency condition requires that:

$$\omega(t) = \max \left[\omega_o, \ \max_{\tau \leq t} \Omega(\underset{\sim}{y}(\tau)) \right] \tag{3.70}$$

More generally, the size of the non–damage criterion can be made to depend on the damage variables themselves, considered as parameters:

$$f_{\rm d} = \Omega(\underset{\sim}{y}) - \omega(\underset{\sim}{d}) \leq 0 \tag{3.71}$$

Function $\Omega(\underset{\sim}{y})$ can be selected as a quadratic norm of $\underset{\sim}{y}$.

iii. Consistency condition

We can now completely analyze the consistency condition of rate–independent models involving both plasticity and damage. We consider only the situation in which the free energy corresponding to hardening also depends on the damage (as described in Paragraph 3.2.A.-(iii)). And we consider only the hypothesis of strain equivalence to define the effective stress. It is shown that in this case, determination of the plasticity and damage multipliers is decoupled, at least for controlled loadings under total strain which are the important ones for numerical applications ("displacement based" finite element codes). Otherwise, if only the elastic part of the free energy is affected by the damage, the same controlled total strain situation leads to coupling when determining the two multipliers $\dot{\lambda}_{\rm p}$ and $\dot{\lambda}_{\rm d}$. We then have a linear system to solve and more complex expressions.

To simplify the demonstration to the utmost, we make the following simplifying assumptions: (i) Only kinematic hardening is considered, with a single variable. In addition, we consider only linear kinematic hardening; (ii) The damage is assumed to be entirely active. The conclusions would be same if it were not, but the computations would be more complicated; (iii) The damage itself is assumed to be isotropic. The thermodynamic force is denoted Y. The damage effect tensor is then $\underset{\approx}{M} = (1-D)\,\underset{\approx}{I}$.

In the framework of the above simplifying assumptions, the free energy and the state equations are expressed:

$$\psi = (1-D)\left[\psi_{\rm e}^{\rm o}(\underset{\approx}{\varepsilon} - \underset{\approx}{\varepsilon}^{p}) + \psi_{\rm p}^{\rm o}(\underset{\approx}{\alpha})\right] \tag{3.72}$$

$$\underset{\sim}{\sigma} = \rho \frac{\partial \psi}{\partial \underset{\sim}{\varepsilon}} = \tilde{\underset{\approx}{\Lambda}} : (\underset{\sim}{\varepsilon} - \underset{\sim}{\varepsilon}^p) = (1 - D)\, \underset{\approx}{\Lambda} : (\underset{\sim}{\varepsilon} - \underset{\sim}{\varepsilon}^p) \tag{3.73}$$

$$\underset{\sim}{X} = \rho \frac{\partial \psi}{\partial \underset{\sim}{\alpha}} = \tilde{\underset{\approx}{C}} : \underset{\sim}{\alpha} = (1 - D)\, \underset{\approx}{C} : \underset{\sim}{\alpha} \tag{3.74}$$

such that:

$$\tilde{\underset{\sim}{\sigma}} = \underset{\approx}{M}^{-1} : \underset{\sim}{\sigma} = \frac{\underset{\sim}{\sigma}}{1 - D} = \underset{\approx}{\Lambda} : (\underset{\sim}{\varepsilon} - \underset{\sim}{\varepsilon}^p) \tag{3.75}$$

$$\tilde{\underset{\sim}{X}} = \underset{\approx}{M}^{-1} : \underset{\sim}{X} = \frac{\underset{\sim}{X}}{1 - D} = \underset{\approx}{C} : \underset{\sim}{\alpha} \tag{3.76}$$

The thermodynamic force associated with the damage includes two terms:

$$Y = Y_{\mathrm{e}} + Y_{\mathrm{p}} = \rho \left(\psi_{\mathrm{e}}^{\mathrm{o}}(\underset{\sim}{\varepsilon} - \underset{\sim}{\varepsilon}^p) + \psi_{\mathrm{p}}^{\mathrm{o}}(\underset{\sim}{\alpha}) \right) = \frac{1}{2}\ (\underset{\sim}{\varepsilon} - \underset{\sim}{\varepsilon}^p) : \underset{\approx}{\Lambda} : (\underset{\sim}{\varepsilon} - \underset{\sim}{\varepsilon}^p) \tag{3.77}$$

The elastic limit criterion and the non–damage surface have the following form:

$$f_{\mathrm{p}} = \parallel \tilde{\underset{\sim}{\sigma}} - \tilde{\underset{\sim}{X}} \parallel_{\mathrm{H}} - k \leq 0 \qquad\qquad f_{\mathrm{d}} = Y - \omega(D) \leq 0 \tag{3.78}$$

and the normality rules yield:

$$\dot{\underset{\sim}{\varepsilon}}^p = \dot{\lambda}_{\mathrm{p}}\, \frac{\partial f_{\mathrm{p}}}{\partial \underset{\sim}{\sigma}} = \dot{\lambda}_{\mathrm{p}}\, \underset{\approx}{M}^{-\mathrm{T}} : \underset{\sim}{n} = \frac{\dot{\lambda}_{\mathrm{p}}}{1 - D}\, \underset{\sim}{n} \qquad\qquad \dot{D} = \dot{\lambda}_{\mathrm{d}}\, \frac{\partial f_{\mathrm{p}}}{\partial Y} = \dot{\lambda}_{\mathrm{d}} \tag{3.79}$$

Assuming controlled loading in total strain space, the consistency condition on plasticity is expressed simply without using $\dot{D}$ or $\dot{\lambda}_{\mathrm{d}}$:

$$\begin{aligned} \dot{f}_{\mathrm{p}} &= \frac{\partial f_{\mathrm{p}}}{\partial \tilde{\underset{\sim}{\sigma}}} : \left(\dot{\tilde{\underset{\sim}{\sigma}}} - \dot{\tilde{\underset{\sim}{X}}} \right) = \underset{\sim}{n} : \left(\dot{\tilde{\underset{\sim}{\sigma}}} - \dot{\tilde{\underset{\sim}{X}}} \right) = \underset{\sim}{n} : \underset{\approx}{\Lambda} : (\dot{\underset{\sim}{\varepsilon}} - \dot{\underset{\sim}{\varepsilon}}^p) \\ &= \underset{\sim}{n} : \underset{\approx}{\Lambda} : \dot{\underset{\sim}{\varepsilon}} - \frac{\underset{\sim}{n} : \underset{\approx}{\Lambda} : \underset{\sim}{n}}{1 - D}\, \dot{\lambda}_{\mathrm{p}} - \frac{\underset{\sim}{n} : \underset{\approx}{C} : \underset{\sim}{n}}{1 - D}\, \dot{\lambda}_{\mathrm{p}} - \gamma\, \frac{\underset{\sim}{n} : \underset{\sim}{X}}{(1 - D)^2}\, \dot{\lambda}_{\mathrm{p}} = 0 \end{aligned} \tag{3.80}$$

which directly yields:

$$\dot{\lambda}_{\mathrm{p}} = \frac{1}{h} \underset{\sim}{n} : \underset{\approx}{\Lambda} : \dot{\underset{\sim}{\varepsilon}} \qquad\qquad h = \frac{\underset{\sim}{n} : \left(\underset{\approx}{\Lambda} + \underset{\approx}{C} \right) : \underset{\sim}{n}}{1 - D} - \gamma\, \frac{\underset{\sim}{n} : \underset{\sim}{X}}{(1 - D)^2} \tag{3.81}$$

Multiplier $\dot{\lambda}_{\mathrm{d}}$ is then determined by the consistency condition on the damage criterion using:

$$\dot{Y} = \dot{Y}_{\mathrm{e}} + \dot{Y}_{\mathrm{p}} = (\underset{\sim}{\varepsilon} - \underset{\sim}{\varepsilon}^p) : \underset{\approx}{\Lambda} : (\dot{\underset{\sim}{\varepsilon}} - \dot{\underset{\sim}{\varepsilon}}^p) + \underset{\sim}{\alpha} : \underset{\approx}{C} : \dot{\underset{\sim}{\alpha}} = \tilde{\underset{\sim}{\sigma}} : (\dot{\underset{\sim}{\varepsilon}} - \dot{\underset{\sim}{\varepsilon}}^p) + \tilde{\underset{\sim}{X}} : \dot{\underset{\sim}{\alpha}} \tag{3.82}$$

Taking the plasticity criterion into account (3.78) one find successively:

$$\tilde{\underset{\sim}{\sigma}} : \dot{\underset{\sim}{\varepsilon}} - \frac{k_{\mathrm{p}}}{1 - D}\, \dot{\lambda}_{\mathrm{p}} - \omega'(D)\, \dot{\lambda}_{\mathrm{d}} = 0 \tag{3.83}$$

$$\dot{\lambda}_{\mathrm{d}} = \frac{\tilde{\underset{\sim}{\sigma}} : \dot{\underset{\sim}{\varepsilon}} - \frac{k_{\mathrm{p}}}{1-D}\, \dot{\lambda}_{\mathrm{p}}}{\omega'(D)} = \frac{1}{(1 - D)\, \omega'(D)} \left(\underset{\sim}{\sigma} - \frac{k_{\mathrm{p}}}{h}\, \underset{\sim}{n} : \underset{\approx}{\Lambda} \right) : \dot{\underset{\sim}{\varepsilon}} \tag{3.84}$$

Just above, ω' denotes the derivative of ω. The simplicity of (3.81) and (3.84) is obvious. This is not the case when hardening/damage couplings are constructed differently, since the damage rate, and therefore $\dot{\lambda}_{\mathrm{d}}$, is involved in the consistency condition on the plasticity (3.81).

iv. Tangent operator

In structural computations, for the case of rate–independent laws, we often require the *"tangent operator"* to the stress/strain behavior, as defined by:

$$\dot{\underset{\sim}{\sigma}} = \underset{\approx}{\boldsymbol{K}} : \dot{\underset{\sim}{\varepsilon}} \tag{3.85}$$

The expression of this tangent operator is sometimes relatively complicated. Below, only three exercises are given, under the same conditions as above, with scalar damage. It is first shown that in the elastic case, the formulation used, based on the thermodynamic framework, leads to a symmetric tangent operator (principal symmetry of a fourth–rank tensor). However, this is not always the case, as is shown by Exercise 2. Finally, in the coupled elastoplastic case (Exercise 3), it is shown that the tangent operator is never symmetric, even for the simplest linear kinematic hardening law. The introduction of more complex hardening can in no way change this result.

- *Exercise 1: Elasticity coupled with the damage — standard approach.* The state equations for stress and the thermodynamic force associated with damage (with $\underset{\sim}{\varepsilon} = \underset{\sim}{\varepsilon}^{\mathrm{e}}$) are written:

$$\underset{\sim}{\sigma} = \underset{\approx}{\tilde{\boldsymbol{\Lambda}}} : \underset{\sim}{\varepsilon} \qquad\qquad Y = \frac{1}{2}\, \underset{\sim}{\varepsilon} : \underset{\approx}{\boldsymbol{\Lambda}} : \underset{\sim}{\varepsilon} \tag{3.86}$$

The consistency condition for the non–damage surface allows us to write the evolution equation for D expressed as:

$$\dot{D} = \dot{\lambda}_{\mathrm{d}} = \frac{\dot{Y}}{\omega'(D)} \tag{3.87}$$

Deriving the state equations (3.86) and taking the isotropy of D into account, $\underset{\approx}{\tilde{\boldsymbol{\Lambda}}} = (1-D)\,\underset{\approx}{\boldsymbol{\Lambda}}$ yields:

$$\dot{\underset{\sim}{\sigma}} = (1-D)\,\underset{\approx}{\boldsymbol{\Lambda}} : \dot{\underset{\sim}{\varepsilon}} - \underset{\approx}{\boldsymbol{\Lambda}} : \underset{\sim}{\varepsilon}\,\dot{D} \qquad\qquad \dot{Y} = \underset{\sim}{\varepsilon} : \underset{\approx}{\boldsymbol{\Lambda}} : \dot{\underset{\sim}{\varepsilon}} \tag{3.88}$$

Combining (3.87) and (3.88) leads to:

$$\dot{\underset{\sim}{\sigma}} = (1-D)\,\underset{\approx}{\boldsymbol{\Lambda}} : \dot{\underset{\sim}{\varepsilon}} - \frac{(\underset{\approx}{\boldsymbol{\Lambda}} : \underset{\sim}{\varepsilon})\,(\underset{\sim}{\varepsilon} : \underset{\approx}{\boldsymbol{\Lambda}} : \dot{\underset{\sim}{\varepsilon}})}{\omega'(D)} \tag{3.89}$$

which leads to the tangent operator of (3.85):

$$\underset{\approx}{\boldsymbol{K}} = (1-D)\,\underset{\approx}{\boldsymbol{\Lambda}} - \frac{1}{\omega'(D)}\,\underset{\approx}{\boldsymbol{\Lambda}} : (\underset{\sim}{\varepsilon} \otimes \underset{\sim}{\varepsilon}) : \underset{\approx}{\boldsymbol{\Lambda}} = \underset{\approx}{\tilde{\boldsymbol{\Lambda}}} - \frac{1}{\omega'(D)}\,\tilde{\underset{\sim}{\sigma}} \otimes \tilde{\underset{\sim}{\sigma}} \tag{3.90}$$

This operator obviously exhibits the principal symmetry.

- *Exercise 2: Elasticity coupled with the damage — nonstandard approach.* The non-standard character is due to the use of a damage criterion based on strain instead of being based on the thermodynamic force associated with the damage. For instance, using Mazars' criterion [52] in the particular case where all the principal strains are positive yields:

$$f_{\rm d} = \| \underset{\sim}{\varepsilon} \| - \omega(D) - k_{\rm d} \leq 0 \tag{3.91}$$

The damage law is then expressed:

$$\dot{D} = \dot{\lambda}_{\rm d} = \frac{1}{\omega'(D)} \frac{\underset{\sim}{\varepsilon} : \dot{\underset{\sim}{\varepsilon}}}{\| \underset{\sim}{\varepsilon} \|} \tag{3.92}$$

The constitutive law derived from (3.88) yields:

$$\dot{\underset{\sim}{\sigma}} = (1 - D)\, \underset{\approx}{\Lambda} : \dot{\underset{\sim}{\varepsilon}} - \frac{(\underset{\approx}{\Lambda} : \underset{\sim}{\varepsilon})\; (\underset{\sim}{\varepsilon} : \dot{\underset{\sim}{\varepsilon}})}{\omega'(D) \;\; \| \underset{\sim}{\varepsilon} \|} \tag{3.93}$$

giving the tangent operator:

$$\underset{\approx}{K} = (1 - D)\, \underset{\approx}{\Lambda} - \frac{1}{\omega'(D) \;\; \| \underset{\sim}{\varepsilon} \|}\, \underset{\approx}{\Lambda} : (\underset{\sim}{\varepsilon} \otimes \underset{\sim}{\varepsilon}) = \tilde{\underset{\approx}{\Lambda}} - \frac{1}{\omega'(D) \;\; \| \underset{\sim}{\varepsilon} \|}\, \tilde{\underset{\sim}{\sigma}} \otimes \underset{\sim}{\varepsilon} \tag{3.94}$$

which is no longer symmetric.

- *Exercise 3: Elastoplasticity coupled with damage.* We use the approach with state coupling between hardening and damage. The state equations are given by (3.73) for the behavior and by (3.77) for Y. It was seen that the plasticity multiplier was expressed simply by (3.81), whereas the damage multiplier was given by (3.84). The derived constitutive law thus yields:

$$\begin{aligned}
\dot{\underset{\sim}{\sigma}} &= (1 - D)\, \underset{\approx}{\Lambda} : (\dot{\underset{\sim}{\varepsilon}} - \dot{\underset{\sim}{\varepsilon}}^p) - \underset{\approx}{\Lambda} : (\underset{\sim}{\varepsilon} - \underset{\sim}{\varepsilon}^p)\, \dot{D} = \tilde{\underset{\approx}{\Lambda}} : \dot{\underset{\sim}{\varepsilon}} - \underset{\approx}{\Lambda} : \underset{\sim}{n}\, \dot{\lambda}_{\rm p} - \tilde{\underset{\sim}{\sigma}}\, \dot{\lambda}_{\rm d} \\
&= \tilde{\underset{\approx}{\Lambda}} : \dot{\underset{\sim}{\varepsilon}} - \frac{\tilde{\underset{\sim}{\sigma}} \otimes \tilde{\underset{\sim}{\sigma}}}{\omega'(D)} : \dot{\underset{\sim}{\varepsilon}} - \underset{\approx}{\Lambda} : \underset{\sim}{n}\, \dot{\lambda}_{\rm p} - \frac{k_{\rm p}}{(1 - D)\, \omega'(D)}\, \tilde{\underset{\sim}{\sigma}}\, \dot{\lambda}_{\rm p} \\
&= \tilde{\underset{\approx}{\Lambda}} : \dot{\underset{\sim}{\varepsilon}} - \frac{\tilde{\underset{\sim}{\sigma}} \otimes \tilde{\underset{\sim}{\sigma}}}{\omega'(D)} : \dot{\underset{\sim}{\varepsilon}} - \frac{1}{h} \left[\underset{\approx}{\Lambda} : \underset{\sim}{n} - \frac{k_{\rm p}\, \tilde{\underset{\sim}{\sigma}}}{(1 - D)\, \omega'(D)} \right] \left(\underset{\sim}{n} : \underset{\approx}{\Lambda} : \dot{\underset{\sim}{\varepsilon}} \right)
\end{aligned} \tag{3.95}$$

such that the tangent operator is expressed:

$$\underset{\approx}{K} = \tilde{\underset{\approx}{\Lambda}} - \frac{\tilde{\underset{\sim}{\sigma}} \otimes \tilde{\underset{\sim}{\sigma}}}{\omega'(D)} - \frac{1}{h}\, \underset{\approx}{\Lambda} : (\underset{\sim}{n} \otimes \underset{\sim}{n}) : \underset{\approx}{\Lambda} + \frac{k_{\rm p}}{h\, (1 - D)\, \omega'(D)}\, (\tilde{\underset{\sim}{\sigma}} \otimes \underset{\sim}{n}) : \underset{\approx}{\Lambda} \tag{3.96}$$

With the first two terms, we effectively find the above elastic case. The third term is symmetric and corresponds to pure plasticity without damage. It is the last term associated with plasticity/damage coupling *which is necessarily nonsymmetric.*

B. Rate–dependent laws

Actually, the rate–dependent laws are easier both to formulate and to use, provided the *"quasi–standard"* formalism mentioned above is accepted. Below, we describe three ways of defining the damage law, giving the example of metal creep damage in each case.

It is first necessary to describe the difficulty of the problem to be solved, illustrated by the case of creep. It requires being able to distinguish sufficiently between the viscoplasticity laws (rate–dependent version of the plasticity) and the damage laws themselves, at least from two aspects:

- The form of the multiaxial criterion itself. In effect, it is known that there are major differences in the stress space between the form of the metal viscoplastic equipotentials, which practically obey the von Mises criterion, with quasi–identity between tensile and compressive behavior (for instance), and the form of isochronous surfaces (surfaces with an equal time to creep fracture) which, on the contrary, may exhibit a very large difference between tension and compression.

- The nonlinearity of the viscosity functions associated with the viscoplasticity law and with the damage law. For the case of the power functions alone, it is known that the exponents differ significantly.

Below, we use the appropriate notations for a second–rank damage tensor $\underset{\sim}{\boldsymbol{d}}$ and the associated thermodynamic force $\underset{\sim}{\boldsymbol{y}}$, although it is not always necessary (since isotropic scalar damage is used in certain applications). In addition, it is recalled that all the functions and criteria may depend strongly on the temperature (thermally active phenomena) even if T is not explicitly specified.

i. Standard approach

This approach consists of considering only one dissipation potential for all the viscoplasticity and damage mechanisms. As for the rate–independent case, we can use an elasticity domain $f_{\mathrm{p}} \leq 0$ in the space of generalized stresses $\underset{\sim}{\boldsymbol{\sigma}}, A_j$ and a non–damage domain $f_{\mathrm{d}} \leq 0$ in the space of thermodynamic forces associated with the damage:

$$f_{\mathrm{p}} = f_{\mathrm{p}}(\underset{\sim}{\boldsymbol{\sigma}},\, A_j\,;\; \alpha_j,\, \underset{\sim}{\boldsymbol{d}}) \qquad\qquad f_{\mathrm{d}} = f_{\mathrm{d}}(\underset{\sim}{\boldsymbol{y}}\,;\; \underset{\sim}{\boldsymbol{\varepsilon}}^{\mathrm{e}},\, \underset{\sim}{\boldsymbol{d}}) \tag{3.97}$$

However, contrary to the rate–independent case, these domains may be exceeded ($f_{\mathrm{p}} > 0$ or $f_{\mathrm{d}} > 0$), causing viscoplasticity and damage. $\underset{\sim}{\boldsymbol{\varepsilon}}^{\mathrm{e}}, \alpha_j$ and $\underset{\sim}{\boldsymbol{d}}$ may be involved in these criteria as parameters. We can construct the potential in two different ways:

- *Laws without multiplicative function:* The dissipation potential is the sum of two nonlinear functions, F_{p} and F_{d}, constructed from f_{p} and f_{d}, for instance of the power functions:

$$\phi^* = F_{\mathrm{p}}(\underset{\sim}{\boldsymbol{\sigma}},\, A_j\,;\; \alpha_j,\, \underset{\sim}{\boldsymbol{d}}) + F_{\mathrm{d}}(\underset{\sim}{\boldsymbol{y}}\,;\; \underset{\sim}{\boldsymbol{\varepsilon}}^{\mathrm{e}},\, \underset{\sim}{\boldsymbol{d}}) \tag{3.98}$$

$$= \frac{K}{n+1}\left\langle \frac{f_{\mathrm{p}}}{K} \right\rangle^{n+1} + \frac{A}{r+1}\left\langle \frac{f_{\mathrm{d}}}{A} \right\rangle^{r+1} \tag{3.99}$$

The rates of the state variables are then derived from this potential by the generalized normality rule:

$$\dot{\underset{\sim}{\boldsymbol{\varepsilon}}}^{p} = \frac{\partial \phi^*}{\partial \underset{\sim}{\boldsymbol{\sigma}}} = \frac{\partial F_{\mathrm{p}}}{\partial \underset{\sim}{\boldsymbol{\sigma}}} = \left\langle \frac{f_{\mathrm{p}}}{K} \right\rangle^{n} \frac{\partial f_{\mathrm{p}}}{\partial \underset{\sim}{\boldsymbol{\sigma}}} \tag{3.100}$$

$$\dot{\alpha}_j = -\frac{\partial \phi^*}{\partial A_j} = -\frac{\partial F_{\mathrm{p}}}{\partial A_j} = -\left\langle \frac{f_{\mathrm{p}}}{K} \right\rangle^{n} \frac{\partial f_{\mathrm{p}}}{\partial A_j} \tag{3.101}$$

$$\dot{\underset{\sim}{d}} = \frac{\partial \phi^*}{\partial \underset{\sim}{y}} = \frac{\partial F_{\rm d}}{\partial \underset{\sim}{y}} = -\left\langle \frac{f_{\rm d}}{A} \right\rangle^r \frac{\partial f_{\rm d}}{\partial \underset{\sim}{y}} \tag{3.102}$$

By proceeding in this way, we distinguish between the nonlinear effects contained in the power functions (with, for instance, different exponents r and n). However, this approach is restrictive as regards the form of the multiaxial damage criterion. In practice, it is relatively difficult to reconstruct a criterion in the space of forces $\underset{\sim}{y}$ which gives the isochronous surface forms observed in creep. The only way of proceeding is then to consider that $f_{\rm d}$ also depends on the elastic strains, acting as parameters, and allowing the stress to be introduced indirectly by the elastic constitutive law. In this case, the choice of the expression of criterion $f_{\rm d}$ in the stress space is again free. This approach is purely standard. However, it uses the state variables as parameters, which is relatively artificial. The use of hardening variables α_j is necessary to introduce the nonlinear kinematic hardening laws by adding a quadratic term in A_j to $f_{\rm p}$ and subtracting the corresponding quadratic term in α_j, which is identical to it by the state equations.

- *Laws with a single multiplicative function.* This is the approach used by Benallal [47] and Lemaître [40]. We again write the dissipation potential as the sum of two terms:

$$\phi^* = F_{\rm p}(\underset{\sim}{\sigma},\, A_j\, ;\, \underset{\sim}{d}) + F_{\rm d}(\underset{\sim}{y}\, ;\, \underset{\sim}{d}) \tag{3.103}$$

but the generalized normality rule now involves a single multiplier function for both the viscoplasticity and the damage mechanisms:

$$\dot{\lambda} = \dot{\lambda}(\underset{\sim}{\sigma},\, A_j,\, \underset{\sim}{y},\, \underset{\sim}{d}) \qquad \dot{\underset{\sim}{\varepsilon}}^p = \dot{\lambda}\, \frac{\partial \phi^*}{\partial \underset{\sim}{\sigma}} = \dot{\lambda}\, \frac{\partial F_{\rm p}}{\partial \underset{\sim}{\sigma}} \tag{3.104}$$

$$\dot{\alpha}_j = -\dot{\lambda}\, \frac{\partial \phi^*}{\partial A_j} = -\dot{\lambda}\, \frac{\partial F_{\rm p}}{\partial A_j} \qquad \dot{\underset{\sim}{d}} = \dot{\lambda}\, \frac{\partial \phi^*}{\partial \underset{\sim}{y}} = \dot{\lambda}\, \frac{\partial F_{\rm d}}{\partial \underset{\sim}{y}} \tag{3.105}$$

Now, exactly as for rate–independent plasticity, we can consider $F_{\rm p}$ as the sum of $f_{\rm p}$ and a quadratic term in A_j, as in equation (3.44), and $F_{\rm d}$ as identical to $f_{\rm d}$. This approach is therefore less artificial than the first one. It is however much more restrictive, since it is no longer possible to distinguish between the viscosity and damage nonlinearities, since they have the same nonlinear function as a factor. In particular, it is no longer possible to have two different exponents.

ii. Quasi–standard approach

In this case, the existence of two independent dissipation potentials and two viscosity functions, also independent, is assumed simultaneously for the two mechanisms, viscoplasticity and damage:

$$F_{\rm p} = F_{\rm p}(\underset{\sim}{\sigma},\, A_j\, ;\, \underset{\sim}{d}) \qquad F_{\rm d} = F_{\rm d}(\underset{\sim}{y}\, ;\, \underset{\sim}{d}) \tag{3.106}$$

$$\dot{\lambda}_{\rm p} = \dot{\lambda}_{\rm p}(\underset{\sim}{\sigma},\, A_j\, ;\, \underset{\sim}{d}) \qquad \dot{\lambda}_{\rm d} = \dot{\lambda}_{\rm d}(\underset{\sim}{y}\, ;\, \underset{\sim}{d}) \tag{3.107}$$

such that the generalized normality rule is now written:

$$\dot{\underset{\sim}{\varepsilon}}^p = \dot{\lambda}_{\rm p}\,\frac{\partial F_{\rm p}}{\partial \underset{\sim}{\sigma}} \qquad \dot{\alpha}_j = -\dot{\lambda}_{\rm p}\,\frac{\partial F_{\rm p}}{\partial A_j} \qquad \dot{\underset{\sim}{d}} = \dot{\lambda}_{\rm d}\,\frac{\partial F_{\rm d}}{\partial \underset{\sim}{y}} \tag{3.108}$$

Here again, as for the laws described just above, we can choose $F_{\rm p}$ as the sum of $f_{\rm p}$ and a quadratic term in A_j (to introduce the nonlinear kinematic hardening recovery terms) and $F_{\rm d}$ identical to $f_{\rm d}$. Otherwise, to return to the case of the first laws without multiplier functions, we can choose the viscoplasticity and damage multipliers as two power functions:

$$\dot{\lambda}_{\rm p} = \left\langle \frac{f_{\rm p}}{K} \right\rangle^n \qquad \dot{\lambda}_{\rm d} = \left\langle \frac{f_{\rm d}}{A} \right\rangle^r \tag{3.109}$$

This "quasi–standard" approach therefore avoids the rather artificial introduction of state variables $\underset{\sim}{\varepsilon}^{\rm e}$ and α_j, while combining the advantages of the two types of equations mentioned in the previous section. If we want, we can even express $\dot{\lambda}_{\rm d}$ directly as a function of the stresses (instead of forces $\underset{\sim}{y}$), which amounts exactly to Hayhurst's criterion [4] for creep. It is however clear that by construction, dissipation remains always positive:

$$\begin{aligned} \Phi_{\rm int} &= \underset{\sim}{\sigma} : \dot{\underset{\sim}{\varepsilon}} - \sum_j A_j\,\dot{\alpha}_j + \underset{\sim}{y} : \dot{\underset{\sim}{d}} \\ &= \dot{\lambda}_{\rm p}\left(\underset{\sim}{\sigma} : \frac{\partial F_{\rm p}}{\partial \underset{\sim}{\sigma}} + \sum_j A_j\,\frac{\partial F_{\rm p}}{\partial A_j} \right) + \dot{\lambda}_{\rm d}\,\underset{\sim}{y} : \frac{\partial F_{\rm d}}{\partial \underset{\sim}{y}} \geq 0 \end{aligned} \tag{3.110}$$

By definition, if the two potential functions $F_{\rm p}$ and $F_{\rm d}$ are positive and convex, and vanish at the origin, the two terms which are factors of $\dot{\lambda}_{\rm p} \geq 0$ and $\dot{\lambda}_{\rm d} \geq 0$ are independently positive. Obviously, the dissipation was also positive in each of the two approaches of the previous section.

iii. Damage as a fourth–rank tensor

The above quasi–standard approach is practically the only one which allows a damage criterion based on stresses and anisotropic damage described by a fourth–rank tensor to be included, while ensuring that the Second Principle is verified a priori. For this purpose, it is sufficient to choose:

$$F_{\rm d} = f_{\rm d}(\underset{\approx}{Y}) = \parallel \underset{\approx}{Y} \parallel_{\rm Q} - k_{\rm d} = \left(\underset{\approx}{Y} : (\underset{\approx}{Q} :: \underset{\approx}{Y}) \right)^{\frac{1}{2}} - k_{\rm d} \tag{3.111}$$

where $\underset{\approx}{Q}$ is a fourth–rank tensor dependent on the material and $k_{\rm d}$ is a threshold, possibly zero (as in conventional creep laws). Symbol :: denotes tensorial summation over four indices. In this case, we have:

$$\dot{\lambda}_{\rm d} = \dot{\lambda}_{\rm d}(\underset{\sim}{\sigma}\,;\ \underset{\approx}{D}) \qquad \dot{\underset{\approx}{D}} = \dot{\lambda}_{\rm d}\,\frac{\partial F_{\rm d}}{\partial \underset{\approx}{Y}} = \dot{\lambda}_{\rm d}\,\frac{\underset{\approx}{Q} :: \underset{\approx}{Y}}{\parallel \underset{\approx}{Y} \parallel {\rm Q}} \tag{3.112}$$

We can even choose the linear version [30], in which tensor $\underset{\approx}{Q}$ gives the direction of the damage rate, in the principal system of effective stresses. We then set:

$$F_{\mathrm{d}} = \left\langle \mathrm{tr}\underset{\approx}{Q} : \underset{\approx}{Y} \right\rangle - k_{\mathrm{d}} = \left\langle \underset{\approx}{Q} :: \underset{\approx}{Y} \right\rangle - k_{\mathrm{d}} = \langle Q_{ijkl} Y_{ijkl} \rangle - k_{\mathrm{d}} \tag{3.113}$$

such that:

$$\dot{\underset{\approx}{D}} = \dot{\lambda}_{\mathrm{d}}\, H(\underset{\approx}{Q} :: \underset{\approx}{Y})\, \underset{\approx}{Q} \tag{3.114}$$

where H is the Heaviside function. The Second Principle is again verified of course since:

$$\Phi_{\mathrm{int}} = \underset{\approx}{Y} :: \dot{\underset{\approx}{D}} = \dot{\lambda}_{\mathrm{d}}\, H(\underset{\approx}{Q} :: \underset{\approx}{Y})\, \underset{\approx}{Q} :: \underset{\approx}{Y} = \dot{\lambda}_{\mathrm{d}} \left\langle \underset{\approx}{Q} :: \underset{\approx}{Y} \right\rangle \geq 0 \tag{3.115}$$

4. APPLICATION OF CDM TO METALLIC MATERIALS

In this part, we treat some examples of macroscopic damage models that are built up in the framework of Continuum Damage Mechanics, considering successively ductile damage, creep damage and creep–fatigue interaction. Illustrations are given for some specific metallic materials, in the situations where coupling effects between plasticity and damage are not considered. In the next Section we will examine the corresponding coupling effects.

4.1. Ductile plastic damage

A. General

A few examples of plasticity/damage coupling equations were given in Section 3.3.. They are well suited to the case of metallic materials and ductile damage. In particular, we were able to study all the aspects involved in modeling:

- Selection of the effective stress in elasticity in the elastic limit criterion and plastic flow law,
- Selection of the hardening law and its coupling mode with the damage,
- The general principles for writing the criteria and damage laws were recalled in Section 3.4., in particular for the rate–independent formulations which are the ones to be used for the case of ductile damage,
- It was seen how the consistency conditions associated with plasticity and damage apply in the case of a rate–independent theory (Paragraph 3.4.A.iii.),
- Finally, we examined determination of the tangent operator associated with the combination of the elasticity law and plasticity law, both coupled with the damage law.

Below, we give only one example of such a model, the one developed by Lemaître [11], entirely based on thermodynamic concepts. Damage and hardening are assumed isotropic, with a single mechanism governed by the increase in cumulative plastic strain. We use a single dissipation potential and a single multiplier, common to plasticity and damage. In addition, the thermodynamic force Y associated with the damage D plays an essential role for fully expressing the multiaxial nature of damage growth. In spite of the restrictions related to the standard thermodynamic framework and the extremely simplistic and uncomplicated character of the growth model, this approach gives reasonable results that are consistent with many experimental data. Numerous attempts to generalize the same model to other concepts (kinematic hardening, viscoplasticity, creep, fatigue, etc.) are given in [40].

B. Formulation based on thermodynamics

The state equations are defined by the free energy thermodynamic potential, i.e. in the isotropic case:

$$2\,\rho\,\psi = (1-D)\;\left[\lambda\;(\varepsilon^{\mathrm{e}}_{kk})^2 + 2\mu\;\varepsilon^{\mathrm{e}}_{ij}\varepsilon^{\mathrm{e}}_{ij}\right] \tag{4.1}$$

We then find:

$$\sigma_{ij} = \rho\frac{\partial\psi}{\partial\varepsilon^{\mathrm{e}}_{ij}} = (1-D)\;\left(\lambda\;\varepsilon^{\mathrm{e}}_{kk}\;\delta_{ij} + 2\mu\;\varepsilon^{\mathrm{e}}_{ij}\right) \tag{4.2}$$

$$Y = -\rho\frac{\partial\psi}{\partial D} = \lambda\;(\varepsilon^{\mathrm{e}}_{kk})^2 + 2\mu\;\varepsilon^{\mathrm{e}}_{ij}\;\varepsilon^{\mathrm{e}}_{ij} \tag{4.3}$$

The thermodynamic force Y associated with damage can also be expressed as follows, taking into account the elasticity law (4.2) and the usual relations between λ, μ and E, ν:

$$Y = \frac{\sigma^2_{\mathrm{eq}}}{2\,E\,(1-D)^2}\left[\frac{2}{3}\;(1+\nu) + 3\,(1-2\,\nu)\;\left(\frac{\sigma_{\mathrm{H}}}{\sigma_{\mathrm{eq}}}\right)^2\right] \tag{4.4}$$

where σ_{eq} is the equivalent von Mises stress, σ_{H} is the hydrostatic stress, E is Young's modulus and ν is Poisson's ratio.

The damage growth law introduced is fully consistent with the standard thermodynamic framewoιк discussed in Paragraph 3.2.B.(i). The dissipation potential is chosen in additive form (3.26). More specifically, the damage part of the potential is a quadratic function of Y:

$$F_d = \frac{S}{2\,(1-D)}\;\left(\frac{Y}{S}\right)^2 \tag{4.5}$$

and in the particular case of rate–independent plasticity, we can write:

$$\dot{D} = \dot{\lambda}\;\frac{\partial F_{\mathrm{p}}}{\partial Y} = \frac{Y}{S}\;\frac{\dot{\lambda}}{1-D} = \frac{Y}{S}\;\dot{p} \tag{4.6}$$

Taking into account a simple hardening law, defined by a power function of the accumulated plastic strain, the selected yield surface has the form:

$$f_{\mathrm{p}} = \frac{\sigma_{\mathrm{eq}}}{1-D} - R - \sigma_y = \frac{\sigma_{\mathrm{eq}}}{1-D} - K\;p^{1/m} - \sigma_y \le 0 \tag{4.7}$$

where K, m and S are coefficients depending on the material. By combining (4.6) and (4.7) with (4.4), when plastic flow occurs ($f_{\mathrm{p}} = 0$), we obtain a differential equation for growth of the ductile plastic damage. Since m is very large compared with 2, we can neglect the variation of p to obtain a linear growth law between D and p:

$$\dot{D} = \frac{K^2}{2\,ES}\,\frac{\sigma_{\rm eq}^2}{2\,E\,(1-D)^2}\left[\frac{2}{3}\,(1+\nu)+3\,(1-2\,\nu)\,\left(\frac{\sigma_{\rm H}}{\sigma_{\rm eq}}\right)^2\right]\,\dot{p} \qquad (4.8)$$

This equation can be integrated in the case of radial loading for which $\sigma_{\rm eq}/\sigma_{\rm H}$ is constant, by selecting the following limits:

$p < p_{\rm D}$ (damage threshold) $\rightarrow D = 0$

$p = p_{\rm R}$ (RVE fracture) $\rightarrow D = D_{\rm c}$

We then find [11]:

$$D = \frac{D_{\rm c}}{p_{\rm R}-p_{\rm D}}\,\left\langle\left[\frac{2}{3}\,(1+\nu)+3\,(1-2\,\nu)\,\left(\frac{\sigma_{\rm H}}{\sigma_{\rm eq}}\right)^2\right]\,p - p_{\rm D}\right\rangle \qquad (4.9)$$

After identification of constants $p_{\rm D}, p_{\rm R}, D_{\rm c}$ and m from tensile tests, this model is capable of correctly reproducing the influence of triaxial nature of the stresses on the ultimate strength [11] (Fig. 4.1). It gives results comparable to McClintock's model [53] or Rice and Tracey's model [54] and can be used for the metal forming limit curves [55]. The thermodynamic framework used by Rousselier [56] for the case of large strains leads to similar results. Fig. 4.2, taken from [57], illustrates the effects of strain/ductile fracture damage coupling for a steel.

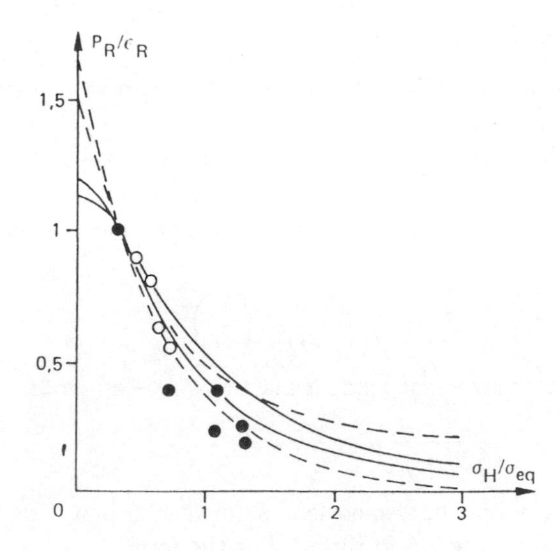

Figure 4.1. Influence of the triaxial nature of stresses on the ultimate strength of two steels. ——— domain covered by McClintock and Rice–Tracey's models; – – – – – domain covered by Lemaître's model

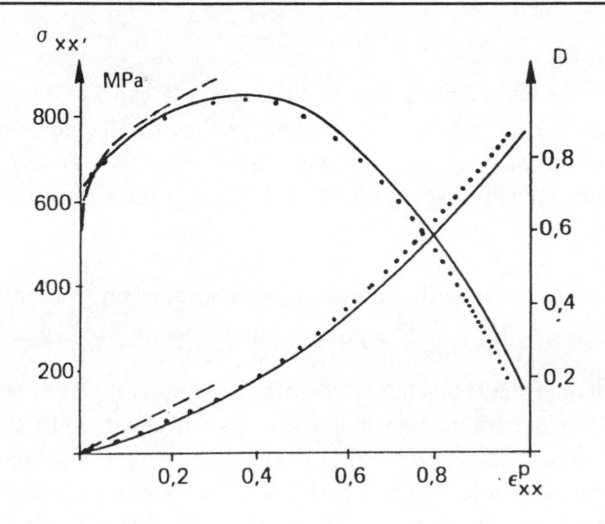

Figure 4.2. Strain/ductile damage coupling. Application to tension on A151-1010 low carbon rolled steel. — — —— experiment; ———— plasticity; .... viscoplasticity

4.2. Creep damage

A. Uniaxial creep equations

The oldest CDM model is Kachanov's [1] which, again in the uniaxial case, is expressed:

$$\dot{\omega} = \left[\frac{\sigma}{A(1-\omega)}\right]^r \tag{4.10}$$

The concept of net stress is obvious here. The larger the section of cavitated grain boundaries, the larger the increase in cavity growth, with an unstable mechanism (and $\omega \to 1$ at fracture, to simplify). Rabotnov's model [3] is slightly more sophisticated:

$$\dot{\omega} = \left(\frac{\sigma}{A}\right)^r (1-\omega)^{-k} \tag{4.11}$$

It gives a better description of the tertiary creep curves (by coupling with viscoplastic behavior). It is used below in Paragraph B. and in Section 5.1.

For a creep test (at constant stress), damage law (4.11) is integrated for ω varying between 0 and 1 and t varying between 0 and t_c. The solution is found easily, yielding:

$$\omega = 1 - \left(1 - \frac{t}{t_c}\right)^{\frac{1}{k+1}} \qquad t_c = \frac{1}{k+1}\left(\frac{\sigma}{A}\right)^{-r} \tag{4.12}$$

The power function (4.12b) between σ and the time to fracture t_c is relatively well verified experimentally. It should be noted that exponent r decreases with the temperature, as is the case of most materials for thermally activated phenomena. A relatively general property for most metallic materials can also be noted: exponent r is slightly lower than exponent N of Norton's law, the secondary creep law (power function $\dot{\varepsilon}_{ps} = (\sigma/K)^N$). Generally, we observe $0.6N < r < N$.

B. Multiaxial creep

Creep damage depends on the stress mulitiaxiality. It can be seen that, depending on the material, the multiaxial criterion corresponds either to the maximum principal stress (copper), or to a criterion more like the von Mises criterion (aluminum alloys) or other combinations (various steels). Two types of multiaxial criteria were proposed for isotropic materials:

- *A product type criterion* with parameter α depending on the material:

$$\chi(\underset{\sim}{\sigma}) = (\Sigma_1)^{\alpha}\,(\sigma_{\mathrm{eq}})^{1-\alpha} \tag{4.13}$$

  resulting from micromechanical considerations. However, this criterion puts too much emphasis on the maximum principal stress Σ_1. In effect, with this criterion, a pure uniaxial compressive stress ($\Sigma_1 = 0$!) does not cause any damage. It is however well known that damage occurs (except perhaps in the case of copper) due to intergranular shear phenomena and local transverse stresses caused by inhomogeneities in the grain orientation and Poisson's effects.

- *Hayhurst's criterion* [4] is preferable, since it combines three invariants additively: the maximum principal stress Σ_1, the hydrostatic pressure σ_{H} and the equivalent von Mises invariant σ_{eq}:

$$\chi(\underset{\sim}{\sigma}) = \alpha\,\Sigma_1 + 3\beta\,\sigma_{\mathrm{H}} + (1-\alpha-\beta)\,\sigma_{\mathrm{eq}} \tag{4.14}$$

  This criterion includes two parameters, α and β, that depend on the material. It allows damage under pure uniaxial compression ($\Sigma_1 = 0,\ \sigma_{\mathrm{H}} = -|\sigma|/3,\ \sigma_{\mathrm{eq}} = |\sigma|$) when $\alpha + 2\beta < 1$.

  The creep damage law under multiaxial conditions (but with scalar, isotropic damage) is written easily by analogy with the uniaxial equation (4.11):

$$\dot{\omega} = \left\langle \frac{\chi(\underset{\sim}{\sigma})}{A} \right\rangle^{r} (1-\omega)^{-k} \tag{4.15}$$

C. Anisotropic creep damage laws

The anisotropic models of creep damage, of which there are actually very few, generally obey the same hypotheses: the damage growth law is a scalar equation qualitatively relating the damage growth rate to the loading variables and cumulative damage:

$$\dot{\omega} = f(\underset{\sim}{\sigma}, \omega) \tag{4.16}$$

The two theories described briefly below for tensorial damage (second– or fourth–rank) both include the use of several multiplicative factors, shown schematically on Fig. 4.3.

i. Murakami–Ohno's formulation [8]

The damage is described by a second–rank tensor. The net stress tensor is used (Section 2.1.C.) to characterize multiaxiality and the trace of the damage tensor is used to introduce nonlinearity of the damage growth. The direction of the rate of $\underset{\sim}{d}$ is expressed as

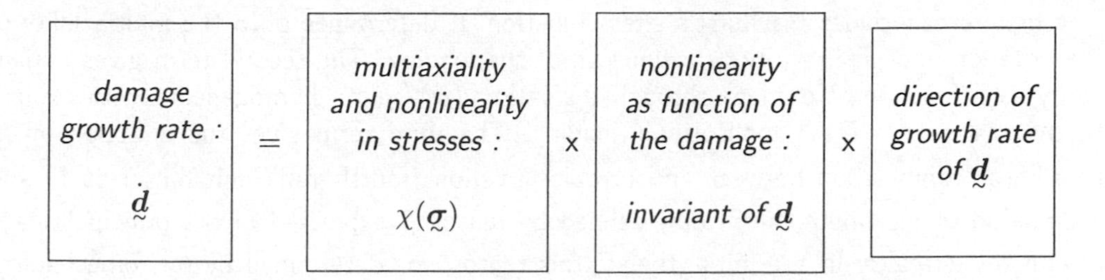

Figure 4.3: Structure of the creep damage rate equation

a linear combination between an isotropic criterion ($\gamma = 0$) and a direction determined by the direction of the maximum principal value of net stress ($\gamma = 1$):

$$\dot{\underset{\sim}{d}} = B\ \langle\chi(\underset{\sim}{\sigma}^*)\rangle^r\ \left[\mathrm{Tr}\,(\underset{\sim}{1} - \underset{\sim}{d})\right]^{-k}\ \left[(1-\gamma)\underset{\sim}{1} + \gamma\, \underline{\nu}^{(1)} \otimes \underline{\nu}^{(1)}\right] \tag{4.17}$$

In the first term, the stress function $\chi(\underset{\sim}{\sigma}^*)$ can be defined by Hayhurst's criterion (4.14). The term involving the damage depends on invariant $\mathrm{Tr}\,(\underset{\sim}{d})$. In the third term, parameter γ sets the direction of the growth rate of $\underset{\sim}{d}$ and $\underline{\nu}^{(1)}$ denotes the maximum principal direction of the net stress tensor $\underset{\sim}{\sigma}^*$. Obviously, (4.17) degenerates to Rabotnov's equation (4.15) for the case of isotropic damage. The only material parameters to be identified are:

- Exponent r and constant B, determined directly from the relation between stress and time to fracture in pure creep under uniaxial loading. These parameters are strongly dependent on the temperature,

- Exponent k obtained from the data on damage measured during tertiary creep (e.g. through the concept of effective stress, as indicated in Section 2.1.C.),

- Coefficients α, β of Hayhurst's criterion; a few multiaxial creep elements are then necessary,

- Parameter γ which can be accessed through metallographic data (decohesion directions), by coupling with the elastic or viscoplastic behavior or finally through non–proportional loading tests (e.g. tension then torsion).

Recent applications of this anisotropic creep theory are summarized in Skrzypek and Ganczarski [58].

ii. Chaboche's formulation [44]

In this theory, the variable used to describe the current damage state is a fourth–rank tensor $\underset{\approx}{D}$. The damage law has the following form, similar to (4.17) of the previous theory:

$$\dot{\underset{\approx}{D}} = \left\langle \frac{\chi(\tilde{\underset{\sim}{\sigma}})}{A} \right\rangle^r \left[\frac{\chi(\tilde{\underset{\sim}{\sigma}})}{\chi(\underset{\sim}{\sigma})}\right]^k \left[(1-\gamma)\underset{\approx}{I} + \gamma\, \underset{\approx}{\Gamma}\right] \tag{4.18}$$

The first term includes Hayhurst's stress function. It determines both the multiaxiality of the criterion and nonlinearity as a function of the loading. The second term gives nonlinearity as a function of damage through the ratio $\chi(\tilde{\underset{\sim}{\sigma}})/\chi(\underset{\sim}{\sigma})$, homogeneous with respect to the scalar $(1-D)^{-1}$ for isotropic loading. The third term gives the direction of $\dot{\underset{\approx}{D}}$ as a linear combination between an isotropic direction (fourth–rank unit tensor as $\underset{\approx}{I}$) and a direction of maximum anisotropy, defined by tensor $\underset{\approx}{\Gamma}$ expressed in the principal space of effective stresses. In the initial theory, this tensor was determined by micromechanical analysis for a network of parallel microcracks, by analogy with the effective elastic behavior. The anisotropy of the evolution of $\underset{\approx}{D}$ allows prediction of a different total time to fracture depending on the non–proportionality of loading. Fig. 4.4 illustrates the case of tensile creep for a time t^* continued by torsional creep for the same equivalent stress $\chi(\underset{\sim}{\sigma})$ (t_σ is the time to tensile fracture and t_R is the time to total "tension + torsion" fracture, whereas ξ is a material parameter involved in tensor $\underset{\approx}{\Gamma}$).

4.3. Fatigue and creep/fatigue interaction

A. A fatigue damage accumulation model

A simple model proposed by Chaboche [31, 59] allows nonlinear fatigue damage accumulation to be described, while correctly expressing the Wöhler curves for periodic conditions:

$$\frac{\mathrm{d}D}{\mathrm{d}N}=\frac{D^{\alpha(\sigma_\mathrm{a},\sigma_\mathrm{mean})}}{\langle\sigma_\mathrm{u}-\sigma_\mathrm{max}\rangle}\left[\frac{\sigma_\mathrm{a}}{M(\sigma_\mathrm{mean})}\right]^\beta \tag{4.19}$$

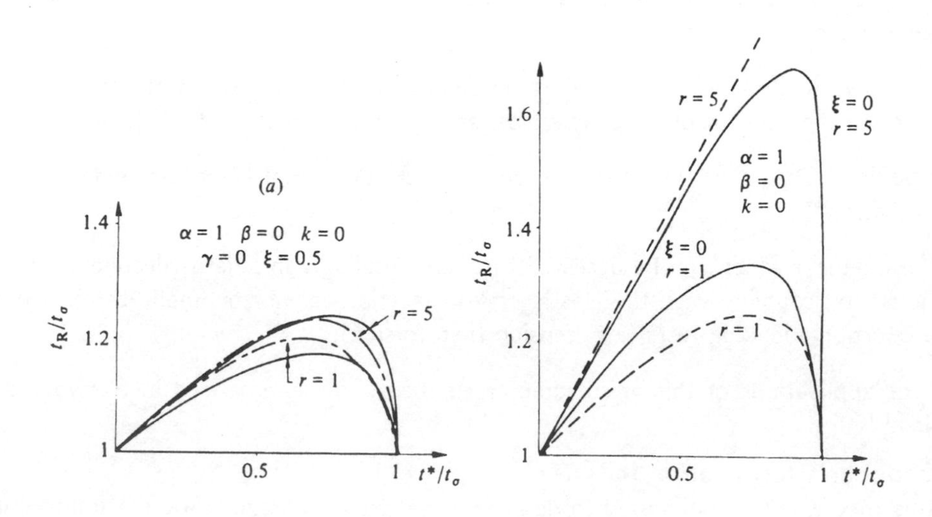

Figure 4.4. Calculation of time to fracture for creep under tension then torsion. (a) $\xi=.5$, (b) $\xi=0$, ———Chaboche's theory; –. – . – . – .– Murakami-Ohno's theory; ——- Kachanov's theory

$$\alpha = 1 - a\,\langle \sigma_{\mathrm{a}} - \sigma_{\mathrm{l}}(\sigma_{\mathsf{mean}}) \rangle \qquad\qquad M = M_o\left(1 - b\,\frac{\sigma_{\mathsf{mean}}}{\sigma_{\mathrm{u}}}\right) \tag{4.20}$$

where σ_{a} is the applied stress amplitude, σ_{u} the UTS (Ultimate Tensile Strength), σ_{l} the fatigue limit for the loading considered, given by a linear dependency on the mean–stress σ_{mean} and M is a linear function in σ_{mean} (similar to σ_{l}). A and M_o are coefficients characteristic of the material. The key to this model is exponent α, which depends on the stresses.

B. Principle of creep/fatigue interaction model

One advantage of Continuum Damage Mechanics is to provide a natural way of predicting the interactions between damages of different physical types. Direct summing, justified by the effective stress concept, i.e. summing not the physical defects themselves but their mechanical effects, leads to a damage growth law of the form [10]:

$$\mathrm{d}D = f_{\mathrm{c}}(\sigma,\, D)\,\mathrm{d}t + f_{\mathrm{F}}(\sigma_{\mathsf{max}},\, \sigma_{\mathsf{mean}},\, D)\,\mathrm{d}N \tag{4.21}$$

where functions f_{c} and f_{F} are deduced from the selected equations and determined for pure creep and pure fatigue. First, we shall examine the justification of (4.21). Actually, fatigue damage is of a different nature than creep damage (transcrystalline microcracks near the surface of the part in the first case, cavities on the grain boundaries throughout the volume in the second case). The interaction at structural microscale involves complex mechanisms when both types of damage are present simultaneously, i.e. when the loading involves both cycling and hold times at high temperature. In the general CDM approach, creep damage, called D_{c}, and fatigue damage D_{F} correspond to the mechanical effects of the physical defects, through the concept of effective stress. We then see how the effective stress due to the creep damage, then the effective stress due to fatigue damage (the opposite would give the same end result) can be defined successively for a given state D_{c}, D_{F}. This is illustrated for tension by:

$$\sigma \xrightarrow[D_{\mathrm{c}}]{} \sigma_{\mathrm{c}} = \frac{\sigma}{1 - D_{\mathrm{c}}} \xrightarrow[D_{\mathrm{F}}]{} \tilde{\sigma} = \frac{\sigma_{\mathrm{c}}}{1 - D_{\mathrm{F}}} = \frac{\sigma}{(1 - D_{\mathrm{c}})\,(1 - D_{\mathrm{F}})} \cong \frac{\sigma}{1 - D} \tag{4.22}$$

It can be seen that the approximation $D = D_{\mathrm{c}} + D_{\mathrm{F}}$ amounts to neglecting the second–order product $D_{\mathrm{c}}\, D_{\mathrm{F}}$.

The fatigue damage model indicated above, (4.19), expresses the effects of nonlinear damage accumulation. Before combining it with the creep equations of Section 4.2., it must be made capable of expressing the actual mechanical behavior of the material damaged by fatigue as well. However, it is known that such damage does not become mechanically perceptible until very late, and therefore corresponds to very nonlinear growth. We therefore use a one–to–one change of variable in (4.19)

$$D \longleftrightarrow 1 - (1 - D)^{\beta+1}, \tag{4.23}$$

which respects the limits ($D = 0$ and $D = 1$) and does not change anything in the fatigue lives predicted by (4.19) (regardless of the loadings considered). This yields:

$$\frac{\mathrm{d}D}{\mathrm{d}N} = \frac{\left[1 - (1 - D)^{\beta+1}\right]^{\alpha(\sigma_a, \sigma_{\mathrm{mean}})}}{\langle \sigma_u - \sigma_{\mathrm{max}} \rangle} \left[\frac{\sigma_a}{M(\sigma_{\mathrm{mean}})\,(1 - D)}\right]^{\beta} \tag{4.24}$$

and it is this equation which is combined with the creep law to simulate fatigue/creep interaction.

This model was successfully used on many alloys. It has the advantage of being a predictive model, since the coefficients characterizing the material are nearly all determined by pure fatigue tests (at high temperature; relatively high frequencies from 10 to 100 Hz) and pure creep tests (only coefficients a of the fatigue model, (4.20) and k of the creep model, (4.18), can be fitted to experimental data involving interaction effects).

C. Possibilities for improvement

Such a formulation can be improved for the physical processes, but at the cost of greater complexity. For instance, environmental effects can be involved in two different ways:

- *In the fatigue term*, by including a frequency effect (depending on the environment). This effect can be deduced from the oxidation kinetics [60] and taken into account in a damage growth law leading to microinitiation.

  To preserve the consistency of the model, the fatigue/creep interactions must not be taken into account until after initiation, when the fatigue microcracks propagate into the regions damaged by creep. In such a model, the time affects two levels, as shown in the diagram of Fig. 4.5: before microinitiation by the oxidation effect on the fatigue term, then by interaction between creep and micropropagation. Work has been conducted on this subject since the end of the 1980s, in particular in references [61, 62, 63].

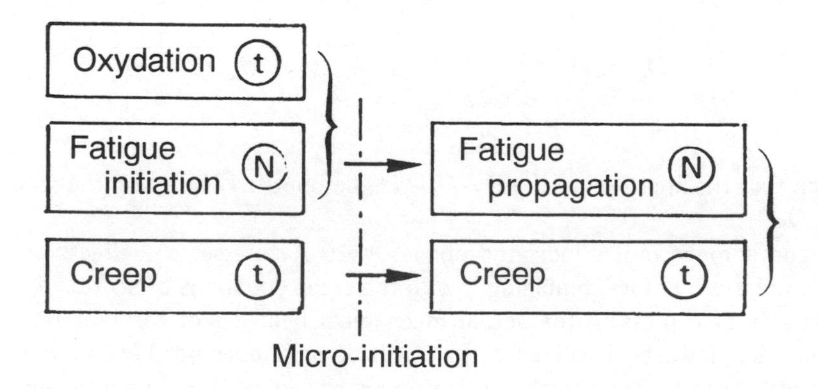

Figure 4.5. Diagram of time/cycle interactions in a model with microinitiation/micropropagation separation

- *In the creep term*, but this is not explicitly included in the current formulations. Actually, it is reasonable to consider that oxidation implicitly affects the creep terms (and the corresponding tests). Let us assume that the real process is described by:

$$\mathrm{d}D = f(\sigma,\, D,\, \phi)\,\mathrm{d}t \qquad\qquad \mathrm{d}\phi = g(\sigma,\, D,\, \phi)\,\mathrm{d}t$$

$$\mathrm{d}\varepsilon_{\mathrm{p}} = h\left(\frac{\sigma}{1-D},\, \phi\right)\,\mathrm{d}t \tag{4.25}$$

  These equations give the kinetics of the creep damage, the variable ϕ describing oxidation and plastic strain (with coupling by the effective stress hypothesis). If we now assume that oxidation does not directly affect the deformation process, ϕ vanishes from the last equation. Then, for a given stress, in the case of a creep test for instance, we can integrate the first two equations (4.25) and obtain a relation between ϕ and D. Substituting it in the first equation formally yields a system such as:

$$\mathrm{d}D = F(\sigma,\, D)\,\mathrm{d}t \qquad\qquad \mathrm{d}\varepsilon_{\mathrm{p}} = H\left(\frac{\sigma}{1-D}\right)\,\mathrm{d}t \tag{4.26}$$

  which can be identified with a pure creep equation. This can be done analytically using the power functions of σ, D, ϕ.

D. Fatigue/creep/oxidation interaction model

To illustrate the first type of improvement mentioned above, we give the results obtained by Gallerneau's model [60] for single crystal superalloys (protected against oxidation). The model is not described in details. It is consistent with the diagram shown above in Fig. 4.6, with:

- An initiation law of a type similar to (4.19), but coupled with oxidation,
- An oxidation law,
- A micropropagation law of a type similar to (4.24), but corrected for the initiation part, already taken into account,
- A pure creep law identified by tensile creep tests.

The two time–dependent kinetics (creep and oxidation) are expressed using temperature dependencies satisfying the Arrhénius equation (thermally activated phenomena), whereas the fatigue law is identified as independent of the temperature by normalizing the stress by the static fracture stress $\sigma_{\mathrm{u}}(T)$ (two different expressions of $\sigma_{\mathrm{u}}(T)$ are used, one for the base metal in the propagation part and the other for the protection, which has a large brittle region, in the initiation part). The model contains very few material coefficients and operates over a very large temperature range. Fig. 4.7 and 4.8 give the predictions of the model at 950 and 1100 °C for reversed loadings .

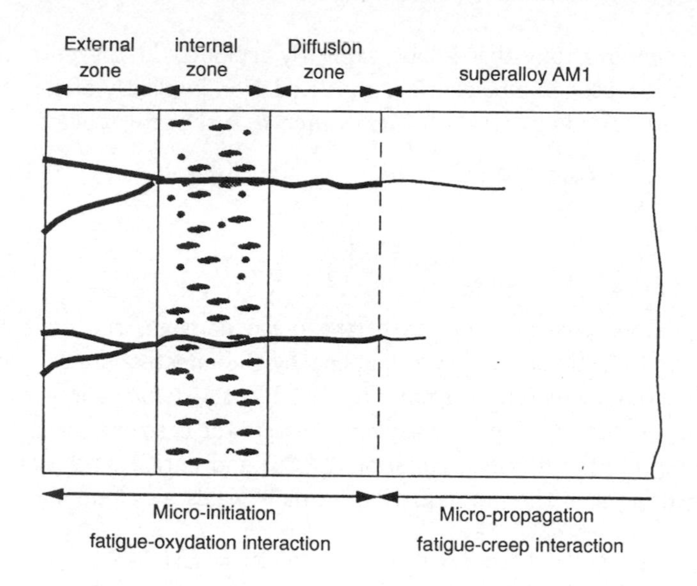

Figure 4.6: Microinitiation/micropropagation on a coated superalloy

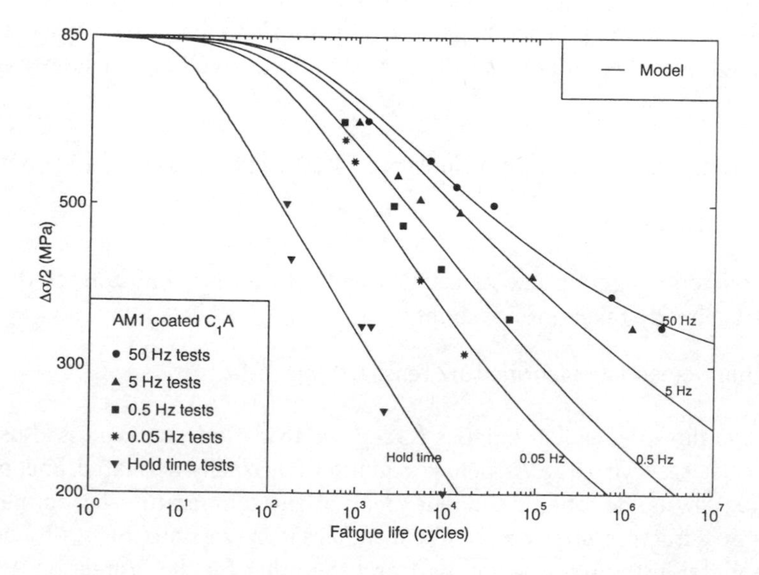

Figure 4.7. Experimental and computed Wöhler curves on coated AM1 at 950° C reversed loading

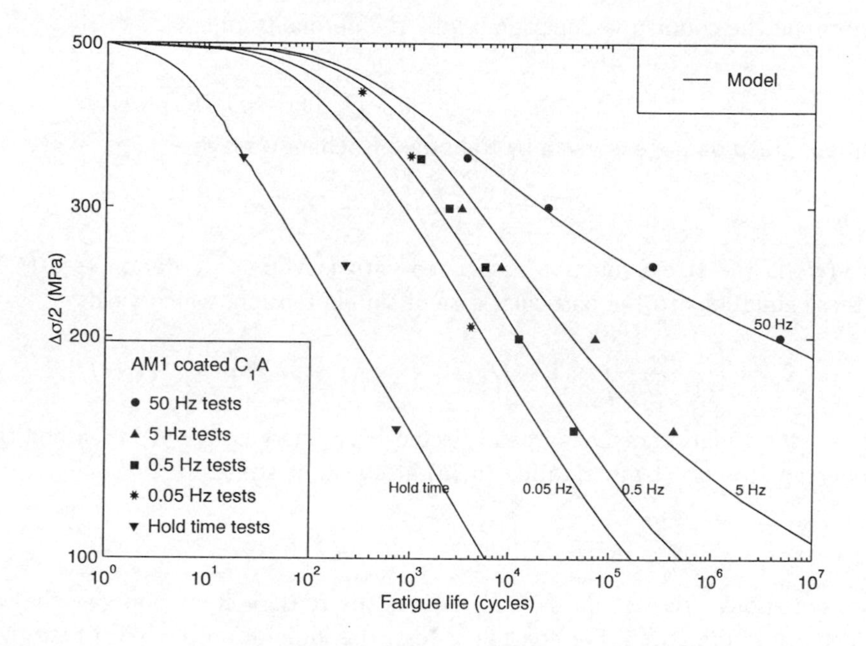

Figure 4.8. Experimental and computed Wöhler curves on coated AM1 at 1100° C — reversed loading

5. STRUCTURAL COMPUTATIONS AND LOCAL APPROACHES TO FRACTURE

5.1. Viscoplasticity coupled with damage

The general formalism was developed in detail in Sections 3.3. and 3.4. for both time-independent plasticity and for viscoplasticity. Here we give a few additional elements in the context of the applications, considering only the case of viscoplasticity coupled with damage. More specifically, to arrive at an exercise which can be solved analytically, we limit ourselves to the case of isotropic elasto–visco–plasticity with isotropic hardening and isotropic damage, i.e. a scalar variable D. We assume a viscoplastic potential in the form:

$$\Omega = \frac{K}{n+1}\left[\frac{\tilde{\sigma}_{\mathrm{eq}}}{K\ p^{1/m}}\right]^n \tag{5.1}$$

where $\tilde{\sigma}_{\mathrm{eq}}$ is the second invariant, i.e. the equivalent von Mises stress, applied to the effective stress tensor, and p is the accumulated plastic strain defined by its rate:

$$\dot{p} = \sqrt{\frac{2}{3}\,\underset{\sim}{\dot{\varepsilon}}^{\mathrm{p}} : \underset{\sim}{\dot{\varepsilon}}^{\mathrm{p}}} \tag{5.2}$$

We determine the coupled viscoplastic law by the normality rule:

$$\dot{\underset{\sim}{\varepsilon}}^{\mathrm{p}} = \frac{3}{2}\,\dot{p}\,\frac{\underset{\sim}{\sigma}'}{\sigma_{\mathrm{eq}}} \qquad \dot{p} = \frac{1}{1-D}\left[\frac{\tilde{\sigma}_{\mathrm{eq}}}{(1-D)\,K\,p^{1/m}}\right]^n \tag{5.3}$$

In addition, creep damage is given by Rabotnov-Kachanov's law:

$$\dot{D} = \left\langle \frac{\chi(\underset{\sim}{\sigma})}{A} \right\rangle^r (1-D)^{-k}, \tag{5.4}$$

where $\chi(\underset{\sim}{\sigma})$ is the stress function associated with Hayhurst's criterion (see (4.14)). We apply these equations to the particular case of simple tension, which yields:

$$\dot{\varepsilon}_{\mathrm{p}} = \left(\frac{\sigma}{K\,\varepsilon_{\mathrm{p}}^{1/m}}\right)^n (1-D)^{-n-1} \qquad \dot{D} = \left(\frac{\sigma}{A}\right)^r (1-D)^{-k} \tag{5.5}$$

First of all, the variation of D can be neglected for primary creep. We then find the end of primary creep strain $\varepsilon_{\mathrm{p}}^*$ by integrating (5.5a) at constant stress:

$$\varepsilon_{\mathrm{p}}^* = \left[\frac{n+m}{m}\left(\frac{\sigma}{K}\right)^n t\right]^{\frac{m}{n+m}} \tag{5.6}$$

The two equations (5.5), coupled by D, express the tertiary creep and give the expression of the creep fracture strain. For constant stress, the integration of (5.5b) first gives:

$$D = 1 - \left(1 - \frac{t}{t_{\mathrm{c}}}\right)^{\frac{1}{k+1}} \qquad t_c = \frac{1}{k+1}\left(\frac{\sigma}{A}\right)^{-r} \tag{5.7}$$

then the constitutive equation is written:

$$\mathrm{d}\varepsilon_{\mathrm{p}} = \left(1 - \frac{t}{t_{\mathrm{c}}}\right)^{-\frac{n+1}{k+1}} \left(\frac{\sigma}{K}\right)^n \mathrm{d}t \tag{5.8}$$

Integrating for a constant σ yields:

$$\varepsilon_{\mathrm{p}} = \varepsilon_{\mathrm{p}}^* + (\varepsilon_{\mathrm{pR}} - \varepsilon_{\mathrm{p}}^*)\left[1 - \left(\frac{1 - t/t_{\mathrm{c}}}{1 - t^*/t_{\mathrm{c}}}\right)^{\frac{k-n}{k+1}}\right] \tag{5.9}$$

$$\varepsilon_{\mathrm{pR}} = \varepsilon_{\mathrm{p}}^* + \frac{k-1}{k+n}\left(\frac{\sigma}{K(\varepsilon_{\mathrm{p}}^*)^{1/m}}\right)^n t_{\mathrm{c}}\left(1 - \frac{t^*}{t_{\mathrm{c}}}\right)^{\frac{k-n}{k+1}} \qquad t^* = \frac{m}{n+m}\left(\frac{\sigma}{K}\right)^{-n} (\varepsilon_{\mathrm{p}}^*)^{\frac{n+m}{m}} \tag{5.10}$$

The exponents are generally in the following order: $r \leq n \leq k$. We deduce that the ultimate strain $\varepsilon_{\mathrm{p}R}$ is a decreasing function of the applied stress under creep. Fig. 5.1 shows that the agreement is good for the IN100 alloy, both for tertiary creep and for ultimate strain. The following construction is simplified by neglecting primary creep (i.e. when $m \to \infty$). In this case:

$$\varepsilon_{\mathrm{p}} = \varepsilon_{\mathrm{pR}}\left[1 - \left(1 - \frac{t}{t_{\mathrm{c}}}\right)^{\frac{k-n}{k+1}}\right] \qquad \varepsilon_{\mathrm{pR}} = \frac{k+1}{k-n}\,\dot{\varepsilon}_{\mathrm{p_s}}\,t_{\mathrm{c}} \tag{5.11}$$

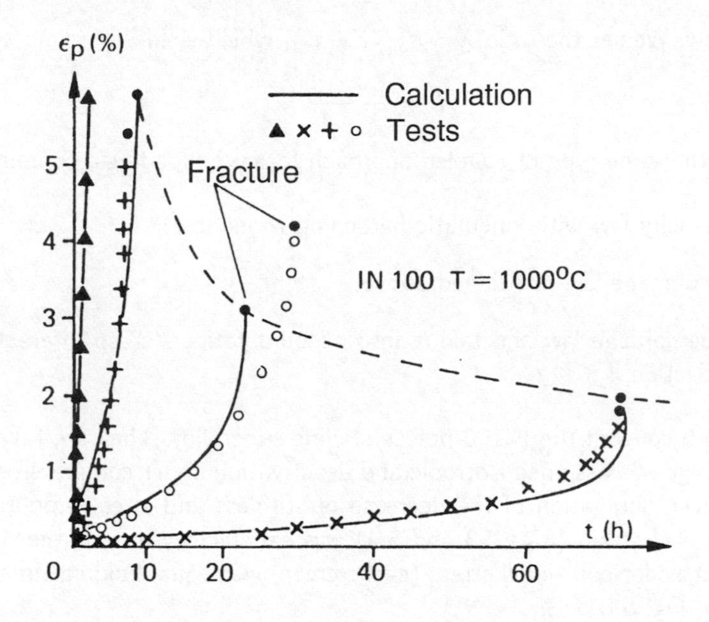

Figure 5.1: Tertiary creep model and prediction of ductility on the IN100 superalloy

where $\dot{\varepsilon}_{p_s}$ is given by the secondary creep law $\dot{\varepsilon}_{p_s} = (\sigma/K)^n$. Fig. 5.2, normalized, shows how the tertiary creep part can be interpreted to determine the value of coefficient k of the

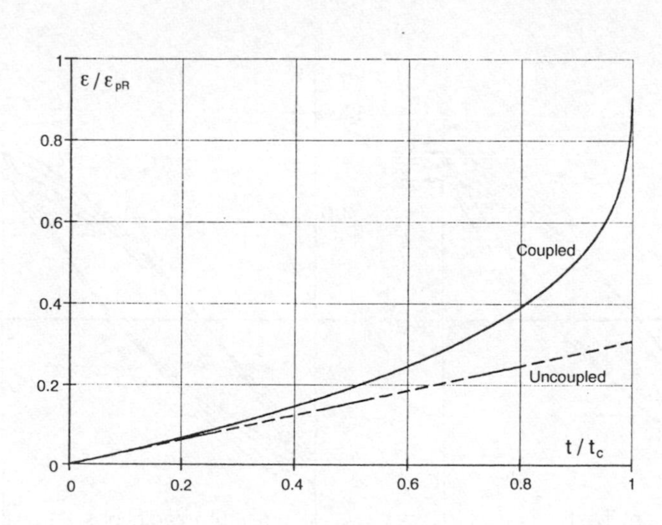

Figure 5.2. Schematic description of tertiary creep and the ratio between ultimate strains calculated with and without coupling (time vs strain)

creep damage law. We set the ratio $\lambda = \varepsilon_{\mathrm{PR}}/(\dot{\varepsilon}_{\mathrm{P_s}}\, t_{\mathrm{c}})$, which yields:

$$k = \frac{1 + \lambda n}{\lambda - 1} \tag{5.12}$$

We can use the same type of coupled approach for cyclic loadings, combining:

- A viscoplasticity law with kinematic hardening (nonlinear),
- The creep damage law mentioned above,
- The fatigue damage law and taking into account fatigue/creep interaction as mentioned in Section 4.3.

Figs 5.3 to 5.5 concern the IN100 polycrystalline superalloy. They are taken from [30]. Here again, damage was assumed isotropic and deactivation under compressive loading was neglected. However, simulation of the decrease of stiffness and stress amplitude during a test under controlled strain (Figs 5.3 and 5.4) was satisfactory as was the description of the tertiary effect under controlled stress (cyclic creep with equal holding times for tension and compression, Fig. 5.5).

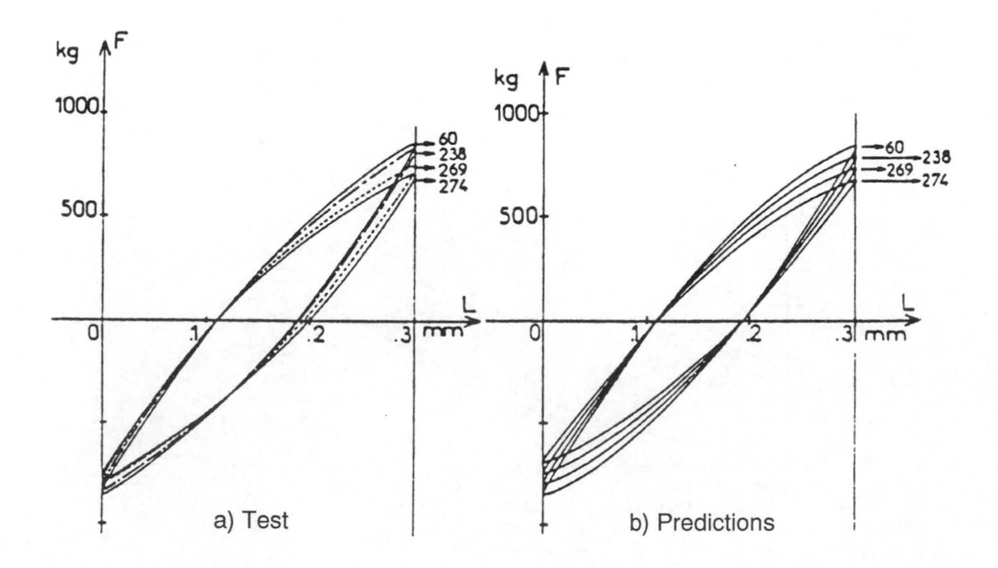

Figure 5.3. Cyclic elasto–visco–plastic behavior coupled with damage. IN100 superalloy at 1000° C. Test with controlled elongation

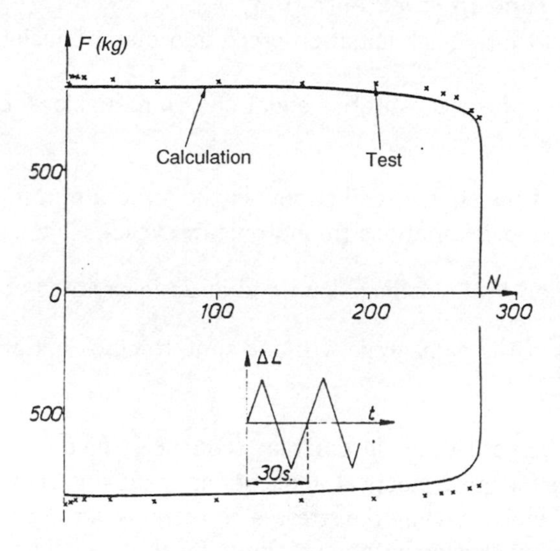

Figure 5.4: Simulation of the test of Fig. 5.3 based on the number of cycles

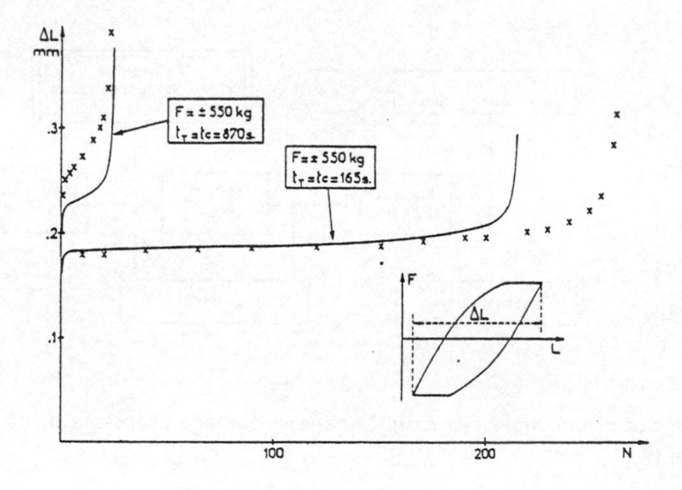

Figure 5.5: Simulation of the two cyclic creep tests of IN100 at 1000° C

5.2. Prediction of time to crack initiation

The mechanical part of a crack initiation prediction method includes four basic steps:

- (i) Determination of the constitutive equations, in particular for describing the cyclic inelastic behavior
- (ii) Calculation of the stresses and strains in the structure from the applied loadings, temperatures and load variations (monotonic or cyclic)
- (iii) Determination of the damage laws and initiation criteria
- (iv) Calculation of the damage growth and time to crack initiation from the stresses calculated in (ii).

These steps may be carried out in different sequences. In the conventional approach, which is the one used for most practical applications, coupling between the behavior and damage is assumed when calculating the stresses and strains in the part (step (ii)). Fig. 5.6, taken from [40], illustrates the computation diagram for this case. This amounts to assuming that the calculated stress states without damage (during the stabilized cycle, for instance) continue to prevail up to the crack initiation. This neglects the additional redistribution induced by coupling with damage and the prediction is conservative, as was shown for the comparisons made in the case of creep [64]. The differences are reasonable, at least for the fatigue/creep life context, which generally justifies this decoupled approach.

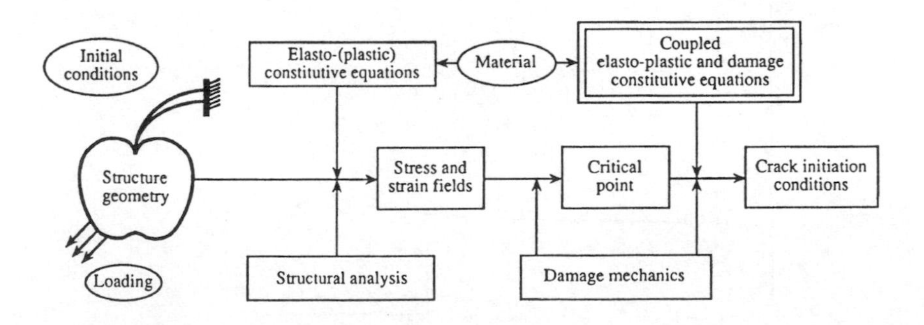

Figure 5.6. Schematics of uncoupled structural analysis, damage processes being calculated in a local post-treatment

The modern approach, so far used for structures of a reasonable size, consists of taking coupling into account when calculating the structure behaviour. Steps (ii) and (iv) are combined in a single computation, obviously much more costly since it requires calculating the entire life of the part. Fig. 5.7, taken from [40], shows the diagram corresponding to this calculation. In the case of low-cycle fatigue, accounting simultaneously for the

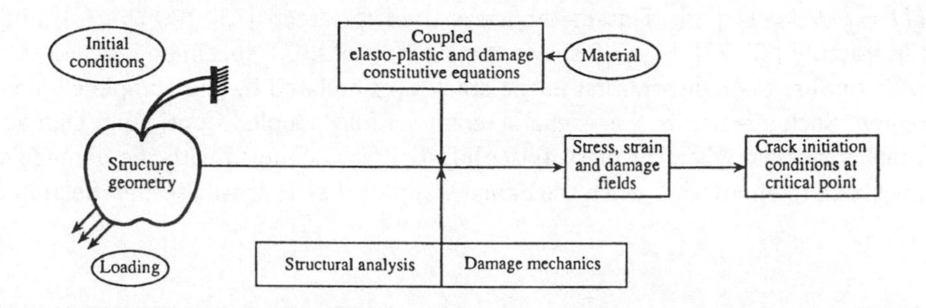

Figure 5.7: Structural computation diagram with coupling by Damage Mechanics

stress redistribution caused by cyclic plasticity and by damage requires calculating *"all"* the consecutive cycles, which is currently unfeasible, even on the largest computers. However, *cycle jump* methods and algorithms which can facilitate this type of analysis and lead to real industrial applications in the near future, have been developed [65, 66].

5.3. Local approaches to fracture

A. The principles of local approaches

Fracture Mechanics methods based on phenomenological relations between cracking rate and global parameters such as the stress intensity factor, are extremely practical and relatively easy to use. Such methods are very valuable tools with a wide range of applications. However, in certain cases, for instance when plasticity is not confined, or in the case of creep, such propagation laws may prove defective or be more complicated to use.

The purpose of local approaches to fracture is to provide a substitution tool, which, although possibly more costly, is better founded physically and has a more predictive character. These methods consist of calculating the crack tip stress and strain fields as accurately as possible (taking all the history effects into account) and applying a local fracture criterion (to the most stressed material element in the crack tip). Two kinds of techniques can be used in the framework of finite element methods:

- Application of the criterion at a critical distance from the crack tip (generally on a finite element) followed by release of the crack tip node when the criterion is satisfied (taking the stress redistribution induced by this "discrete" crack propagation into account). This method was successfully used for fatigue [67, 68, 69], ductile fracture [70], and creep [71]. It has the drawback of using discrete increments and depending on the fineness of the mesh.

- Taking total coupling into account by CDM. Continuous crack growth is then described by a gradual decrease in the local strength of the damaged material. The crack is the locus of the material points for which critical damage is reached

($D = D_c = 1$) [72]. This method was used for creep [73, 74, 75, 76], and ductile fracture [77, 78]. Fig. 5.8 shows the example of a CT specimen cracked by creep, with total stress redistribution in the crack tip simulated by the completely damaged region. Such a redistribution is characteristic of fully coupled Continuum Damage Mechanics, at least when the elasto(visco)plasticity law coupled with the damage causes the stress to approach 0 when the damage approaches 1, as was seen in Section 3.3.C.

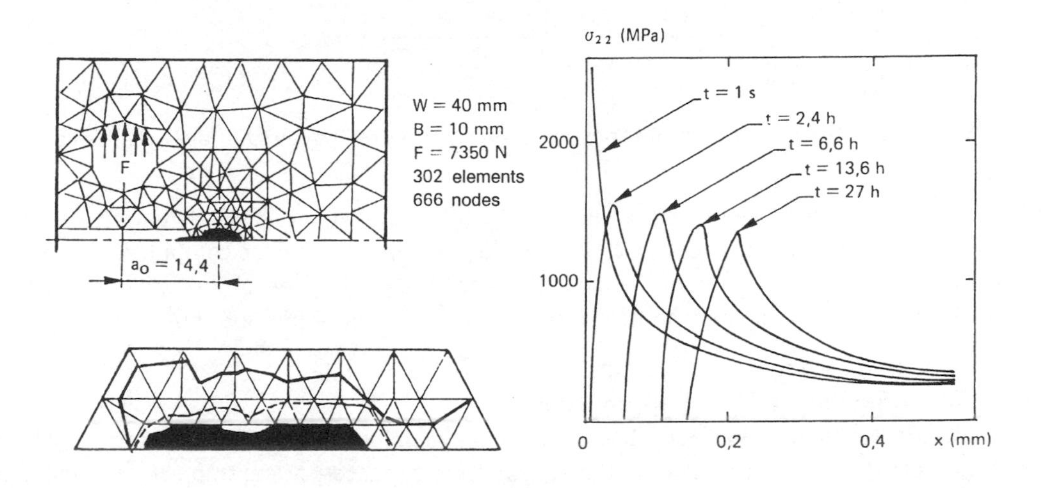

Figure 5.8. Simulation of crack propagation by creep using a local approach. CT specimen, INCO 718 alloy at 600° C

B. Numerical problems associated with total coupling

The second type of local approach is attractive, because it corresponds to a complete prediction conducted using constitutive models identified on simple experiments. Unfortunately, the numerical computations show that convergence is not obtained with the fineness of the 2D mesh. This is due to a localization effect when the volume element reaches instability conditions [78, 79, 80]. For instance, in the case of creep of a CT specimen, only the first row of elements is completely fractured and the global result (calculated crack growth rate) depends strongly on the height of the first row. Several techniques can be used to overcome this difficulty [79, 81]:

- (i) Place a lower limit on the size of the finite elements using a concept of characteristic volume associated with defect statistics [77, 78],
- (ii) Introduce higher-order gradients in the damage constitutive equation for strain and stress,
- (iii) Use localization limiters [79], in particular a strain defined different than locally,

- (iv) Use a non–local damage law for definition of the damage growth law, while preserving a local definition for strain [81, 74, 80].

It is the latter approach (iv) which is developed briefly below. The approach used in [80] concerns the damage rate, for instance for creep. The non–local definition of this rate is given by:

$$\dot{D}(x) = \frac{\int_\Omega \phi(x, \xi)\, \dot{\overline{D}}(\xi)\, \mathrm{d}\xi}{\int_\Omega \phi(x, \xi)\, \mathrm{d}\xi} \tag{5.13}$$

where x and ξ denote the point considered and any point in its vicinity Ω. In the integral, $\dot{\overline{D}}$ is the growth law such as that defined locally as in (4.15) for creep. For the neighborhood function ϕ, we can choose a Gaussian function:

$$\phi = \exp - \left(\frac{\| x - \xi \|}{d^*} \right)^2 \tag{5.14}$$

where d^* represents a characteristic distance on the scale of the material microstructure (for instance, grain size for creep in polycrystalline alloys).

The use of this non–local definition of damage has several advantages:

- Convergence with the fineness of the mesh is ensured, while keeping the characteristic distance d^* constant: the completely damaged region, i.e. the crack, which preserves a finite distance, extending through the thickness of several elements;

- Consecutive reloadings of the Gauss points during fracture of the neighboring points (because of stress redistribution) are completely eliminated. The variations are thus more continuous and lead to improving the efficiency of the integration algorithm;

- The concept of Representative Volume Element is taken into account objectively, which is necessary for damage phenomena, to integrate a sufficient number of defects in the volume element. The size of the finite elements corresponding to discretization of a problem of Continuum Mechanics is no longer related to the RVE size, which depends essentially on the material.

6. DAMAGE MODELS IN BRITTLE MATERIALS

Brittle materials are those in which no major strain occurs before damage and ultimate fracture of the RVE (or part). Excluding glass and solid ceramics, not covered herein, this type of behavior is encountered in concretes and composites. Very often, brittleness is observed macroscopically even when the matrix on the local scale is capable of deforming (considerably in the case of certain organic matrices, but locally, which means that the composite preserves a generally brittle behavior). One of the main features is therefore the fact that the macroscopic behavior can be expressed by an elasticity law coupled with the damage. Below, we very briefly review a few uses of Continuum Damage Mechanics for this type of situation. We consider applications only for some composite materials.

6.1. Application to various composites

A. Scales

Before discussing composite damage, the scales used and the main mechanisms involved should be recalled. They include:

- *The scale of the structure* or the composite part (wing box structure, panels, stiffened panels, etc.), often made of a laminate. Certain very macroscopic approaches then consider the laminate as a macroscopically homogeneous material. However, such approaches have become obsolete. The finite element method is now used for designing parts in the framework of a "plate" or "shell" or "3D" representation, dissociating the kinematic aspect (discretization of displacements) from the behavioral aspect which must be analyzed for each ply.

- *The scale of the elementary ply* (Fig. 6.1) which is the scale we use for developing the macroscopic constitutive equations (scale sometimes called "mesoscale"). It concerns the unidirectional ply in Organic Matrix Composites or sometimes fabrics (plain weaves for SiC/SiC composites, satin weaves for C/PMR15 composites, etc.). The composite material (yarn + matrix or fiber + matrix) is then considered macroscopically as a continuum, even in presence of damage (transverse cracks for instance). In this framework, the interface phenomena between plies can be modeled using a special medium (interface layer, as in the work of LMT-Cachan [41], [82], [83], [51]).

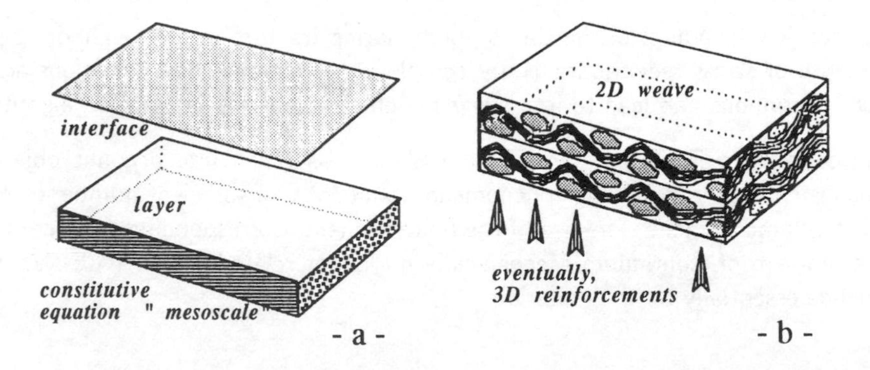

Figure 6.1: Various components of laminated composites

- *The microscale* is the scale taking into account the elementary behavior of the constituents: matrices, fibers, fiber/matrix interfaces, etc. In certain cases, the behavior of the woven composite is analyzed by a dual micro/macro transition (fiber + matrix → yarn, then yarn + matrix → composite fabric).

- *An even finer scale* would be the one involved in the fiber/matrix interfaces (diffusion regions, different interphases, etc.) which is outside the framework of this discussion.

B. A hierarchical approach

In this context, the work conducted over the passed few years at ONERA, in particular the research of P.M Lesne and J.F Maire [84], [85], [86] led to developing a hierarchical approach to the constitutive equations for the various composite systems (Fig. 6.2):

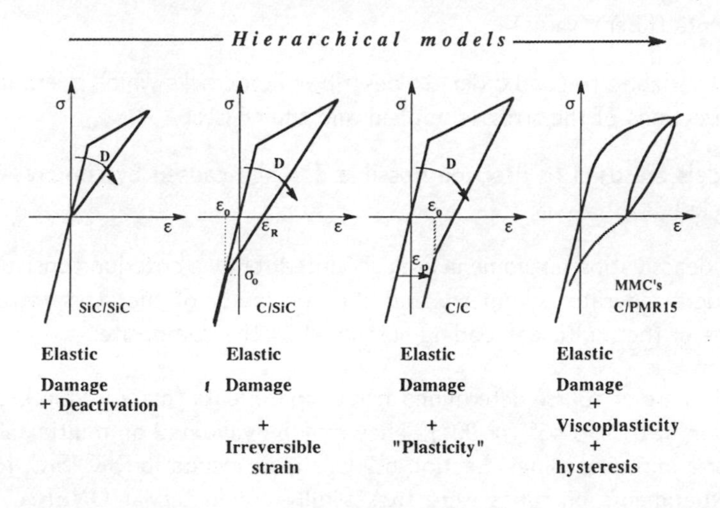

Figure 6.2: A hierarchical approach to the behavior of composites

- The basic configuration is woven ceramic matrix composites (CMC's), such as SiC/SiC, whose behavior is essentially elastic with nonlinearity caused by the damage (decrease in the elastic properties) as well as very clear damage deactivation effects under compression. The models obtained, already very sophisticated, are implemented in industrial computation codes. They are detailed below.
- Irreversible strains due to damage and plasticity type effects were introduced for C/SiC and C/C composites.
- For C/PMR15 composites used in SNECMA engines, damageable viscoplasticity and/or viscoelasticity models were developed to account for the anisotropies, the hysteresis effects during the loading and unloading cycles, creep and relaxation phenomena, and partial or total recovery effects.
- These macroscopic models are now being applied to organic matrix composites (OMC's) used as unidirectional laminates. Different but similar versions have been developed for SiC/Ti metallic matrix composites (MMC's) using approaches with micro/macro transitions [87], [88], [89].

The general approach is in the context of Continuum Thermodynamics with internal variables, and more specifically in the framework of CDM. For plasticity and viscoplasticity problems, kinematic hardening concepts are used in a relatively conventional scheme (for metals). For damage, the more sophisticated versions take the following aspects into account:

- Scalar damage variables associated with microcracks oriented by the highest strength components (fibers, yarns)
- Tensorial variables (second order) to describe microcracks which orientation is related to the directions of the stresses applied and their history
- The models are used to describe possible damage caused by compression (splitting type cracks)
- Damage deactivation phenomena are accounted for by a criterion consistent with both the elasticity operator symmetry and the continuity of the stress/strain responses regardless of the multiaxial loading sustained by the çomposite.

These models are of course determined from experiments (macroscopic), generally tension/compression at 0° and 45° (or 90°). They may be validated on multiaxial experiments under simple or complex loadings. Section 6.2.C. shows results for SiC/SiC, for which tension/torsion experiments on tubes were successfully conducted at ONERA, supplying an exceptionally rich database [90]. Tension/internal pressure tests were also conducted on the same specimens. A few results are given below. From all the work on the various types of composites, it appears that there are currently three types of difficulties or shortcomings in the macroscopic models:

- Damage deactivation is insufficient for certain shear stresses because of the restrictive "stress/strain response continuity" condition. Research is being done to "expand" this deactivation condition based on micromechanical analyses of the crack closure phenomena and the associated energy storages/dissipations;
- Loading/unloading hysteresis phenomena and the associated residual strains in systems without viscosity or plasticity of the matrix. This point is not discussed herein. It requires taking into account fiber/matrix/interface/matrix crack phenomena involving decohesion/interface friction mechanisms. This is clearly an area where research is required;
- Fracture criteria which can be of two sorts: either involving critical values as material parameters or preferably by systematically searching for the instability/bifurcation conditions associated with the structural analyses themselves.

C. Application to the SiC/Ti metallic matrix composite

To illustrate the possibilities of this hierarchical approach to the constitutive/damage equations for composites, below we give a few comparisons between experiments and simulations using models developed in the framework of this approach. The case of ceramic/ceramic composites that can be suitably simulated by damageable elastic models will be discussed later (Section 6.2.). Here we give only a brief review for composites with a significant inelastic behavior, without developing the models themselves.

The material consists of long fibers in a titanium matrix. It is used unidirectionally and is tested in the longitudinal direction and the 90° direction at temperatures of 450° C and 550° C. A macroscopic model of the elasto–visco–plastic behavior coupled with damage was developed [87], [88] based on a micromechanical analysis and micro/macro transition. This model takes into account the residual manufacturing stresses, the viscoplastic behavior of the matrix, combined damage in the fiber and matrix and damage deactivation effects.

Figs 6.3 and 6.4 show that the model gives a correct description for tension in the longitudinal direction, for which the behavior is essentially elasto–visco–plastic (with high stiffness) and for tension in the transverse direction, for which strong damage (fiber/matrix interface fractures) with a significant decrease in the modulus of elasticity visible during unloading is observed, followed by the development of plastic strain.

Fig. 6.5 gives an example of controlled cyclic transverse loading under stress (increasing levels) in which the same observations are made. In addition, for high amplitudes, the plasticity varies cyclically both under compression and under tension. On the contrary, at low amplitudes, damage deactivation is observed instead on the transition to compression around a zero strain.

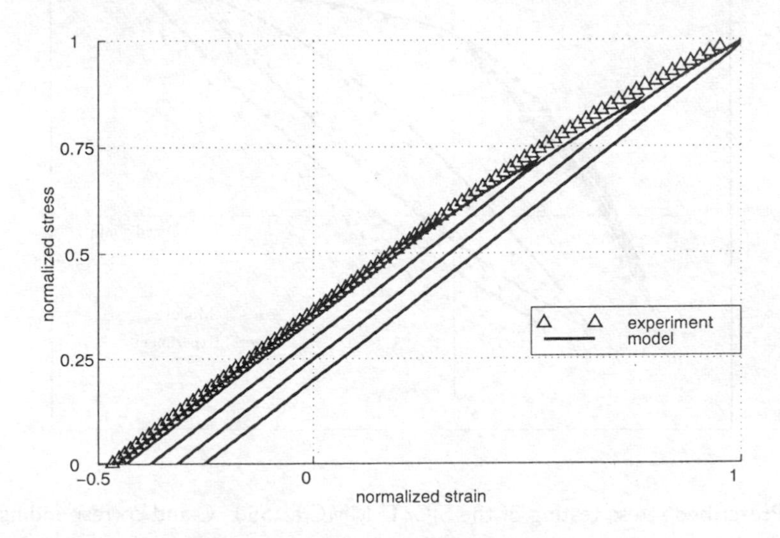

Figure 6.3: Longitudinal tension on a SiC/Ti composite at 450° C

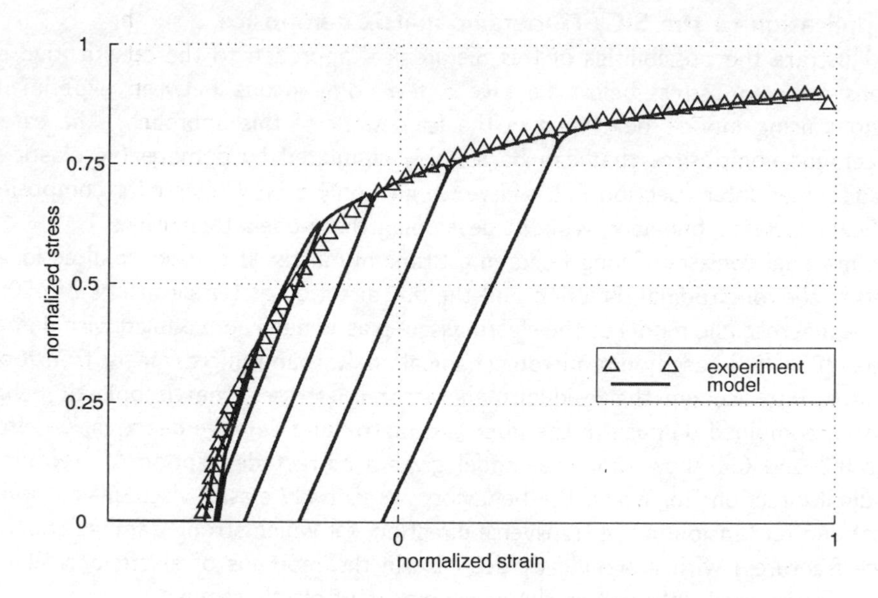

Figure 6.4: Transverse tension on a SiC/Ti composite at 450° C

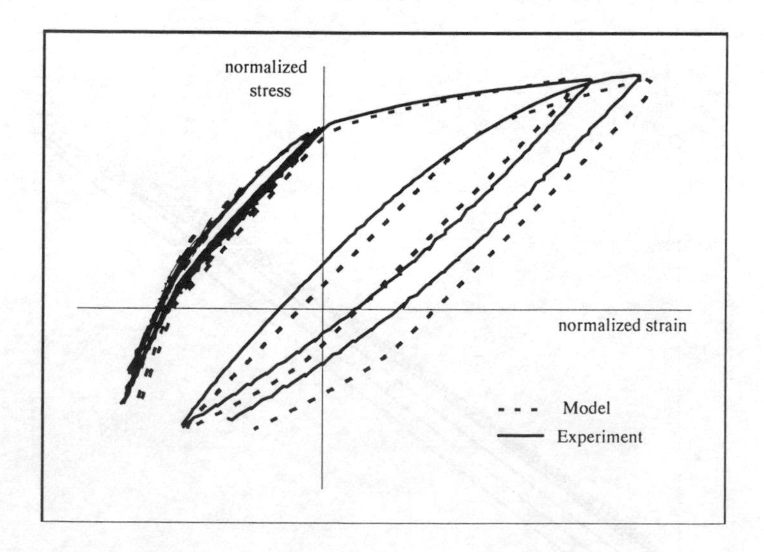

Figure 6.5. Prescribed stress testing of the SiC/Ti MMC at 550° C and corresponding simulation

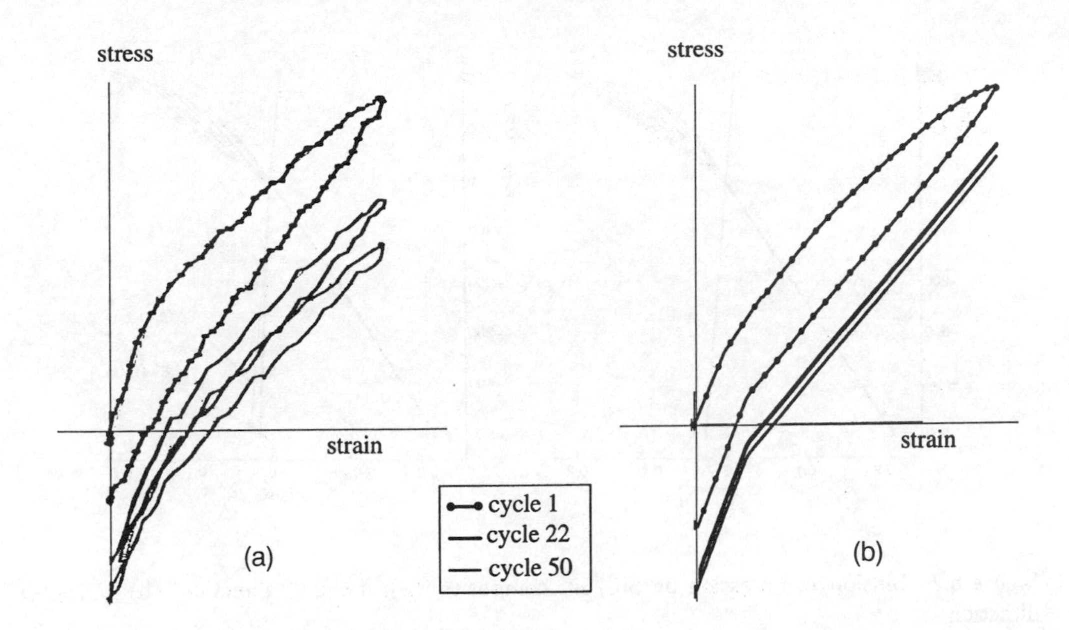

Figure 6.6. Cyclic loading with prescribed strain on an SiC/Ti MMC: (a) experiment from [91]; (b) simulation

Finally, Fig. 6.6 shows the simulation of an experiment with controlled (repeated) strain, which even more clearly shows the deactivation phenomenon that occurs for a zero or slightly positive total strain. The significant relaxation of the corresponding stress can be observed on deactivation (change of the modulus of elasticity).

6.2. Modeling of ceramic/ceramic composites

Here we specialize the theories given in Section 3. and adapt them to the case of brittle composites. The possibility of irreversible strain caused by the damage (not introduced in Section 3.) and some plastic strain is however assumed. The modeling of damage deactivation effects for compressive-like loadings is central in the present case (see Fig. 6.7).

A. State law

The internal variables are either strains or damage variables. In particular, we identify: $\underset{\sim}{\varepsilon}^{\rm c}$, the strain tensor at the damage deactivation point; $\underset{\sim}{\varepsilon}^{\rm p}$, the plastic strain tensor (due to matrix inelasticity mechanisms); δ_i, $i = 1, 2, 3$, three scalar damage variables representing the microcracks which directions are given by the initial principal anisotropy axes of the material, i.e. the directions of the reinforcements (fibers, yarns, etc.); and $\underset{\sim}{d}$, the tensorial damage variable (second order), representing the microcracks which directions are related to those of the loading which caused them. The free energy is postulated in the form:

$$\psi = \frac{1}{2}(\underset{\sim}{\varepsilon} - \underset{\sim}{\varepsilon}^{\rm p}) : \underset{\approx}{\Lambda}_o : (\underset{\sim}{\varepsilon} - \underset{\sim}{\varepsilon}^{\rm p}) + \frac{1}{2}(\underset{\sim}{\varepsilon} - \underset{\sim}{\varepsilon}^{\rm c}) : \Delta\underset{\approx}{\Lambda}^{\rm eff} : (\underset{\sim}{\varepsilon} - \underset{\sim}{\varepsilon}^{\rm c}) \tag{6.1}$$

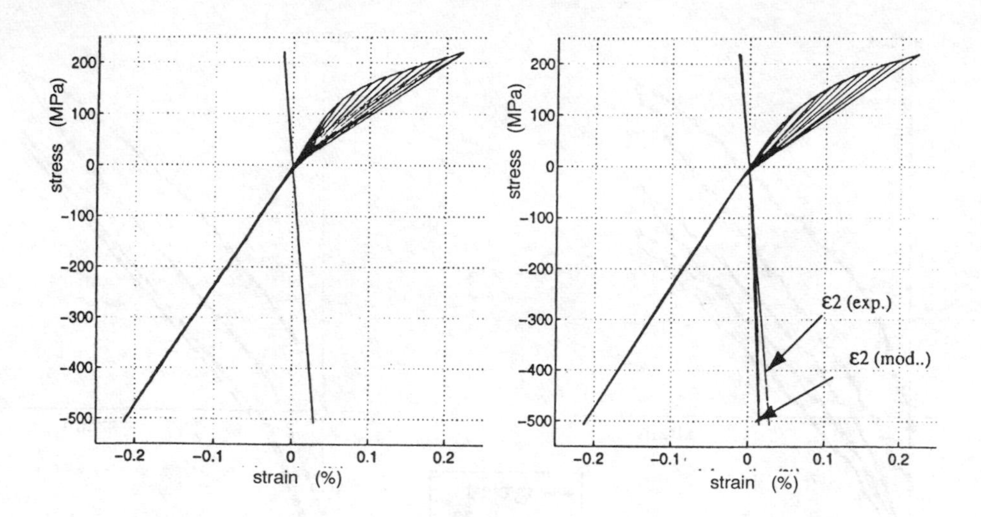

Figure 6.7. Tension/compression on SiC/SiC composite. (a) in the 0° direction; (b) in the 45° direction

From this we deduce the stress tensor:

$$\underset{\sim}{\sigma} = \frac{\partial \psi}{\partial \underset{\sim}{\varepsilon}} = \left(\underset{\approx}{\Lambda}_o + \Delta \underset{\approx}{\Lambda}^{\text{eff}}\right) : (\underset{\sim}{\varepsilon} - \underset{\sim}{\varepsilon}^{\text{c}}) + \underset{\approx}{\Lambda}_o : (\underset{\sim}{\varepsilon}^{\text{c}} - \underset{\sim}{\varepsilon}^{\text{p}}) \tag{6.2}$$

This equation contains the term corresponding to the undamaged material and the term affected by damage, which also contains the closure strain (with the unilateral condition).

The stress state corresponding to $\underset{\sim}{\varepsilon} = \underset{\sim}{\varepsilon}^{\text{c}}$ is given by $\underset{\sim}{\sigma}^{\text{c}} = \underset{\approx}{\Lambda}_o : (\underset{\sim}{\varepsilon} - \underset{\sim}{\varepsilon}^{\text{p}})$. Under pure tension, it corresponds either to a negative value (Fig. 6.8a) or to a positive value (Fig. 6.8b). The plastic strain $\underset{\sim}{\varepsilon}^{\text{p}}$ is defined with reference to the initial undamaged behavior (stiffness $\underset{\approx}{\Lambda}_o$) with a released configuration from the deactivation state $\underset{\sim}{\varepsilon}^{\text{c}}, \underset{\sim}{\sigma}^{\text{c}}$, as illustrated in Fig. 6.8a. The damage effect is introduced by fourth-rank operators with $\underset{\approx}{\Lambda}^{\text{eff}} = \underset{\approx}{\tilde{\Lambda}}$ given by the equation:

$$\Delta \underset{\approx}{\tilde{\Lambda}} = -\sum_{i=1}^{3} \delta_i \left[\underset{\approx}{\boldsymbol{A}}_i : \underset{\approx}{\boldsymbol{K}}_o\right]_s - \left[\underset{\approx}{\boldsymbol{D}}(\underset{\sim}{d}) : \underset{\approx}{\boldsymbol{K}}_o\right]_s \tag{6.3}$$

in which all the damage is considered active. The subscript s indicates the tensorial principal symmetrization. For scalar damage δ_i, operator $\underset{\approx}{\boldsymbol{A}}_i$ is expressed in the fixed directions $\underline{\boldsymbol{p}}_i, \underline{\boldsymbol{q}}_i, \underline{\boldsymbol{r}}_i$ associated with the reinforcements. For instance, for woven composite, $\underline{\boldsymbol{p}}_i$ is the vector orthogonal to the "i" yarns in the plane of the fabric, $\underline{\boldsymbol{q}}_i$ is the vector in the direction

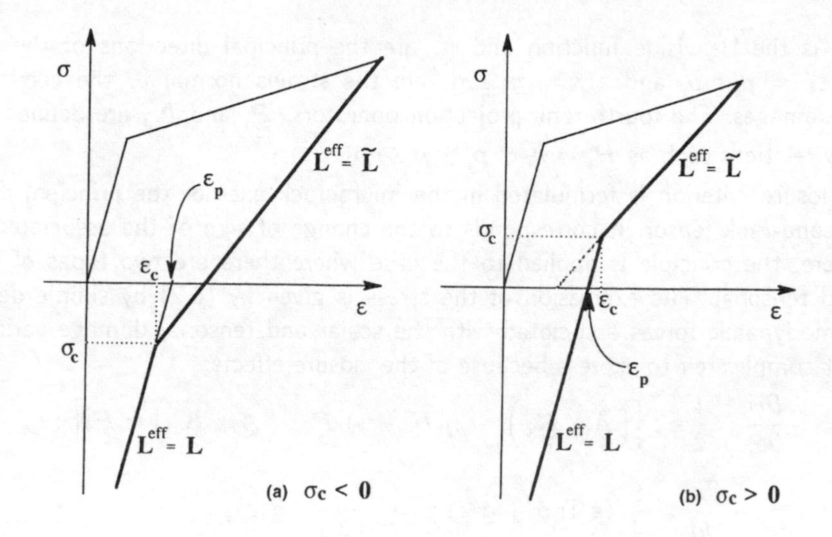

Figure 6.8. Tension/compression deactivation diagram for two closure positions and corresponding definition of plastic strain

of the yarn, and $\underline{\boldsymbol{r}}_i$ is the vector orthogonal to the plane. $\underset{\approx}{\boldsymbol{A}}_i$ has the form:

$$\begin{aligned} \underset{\approx}{\boldsymbol{A}}_i \;=\; & \alpha_{11}\, \underline{\boldsymbol{p}}_i \otimes \underline{\boldsymbol{p}}_i \otimes \underline{\boldsymbol{p}}_i \otimes \underline{\boldsymbol{p}}_i + \alpha_{12} \left[(\underline{\boldsymbol{p}}_i \otimes \underline{\boldsymbol{p}}_i) \otimes (\underline{\boldsymbol{q}}_i \otimes \underline{\boldsymbol{q}}_i)\right]_s \\ & +\alpha_{13} \left[(\underline{\boldsymbol{p}}_i \otimes \underline{\boldsymbol{p}}_i) \otimes (\underline{\boldsymbol{r}}_i \otimes \underline{\boldsymbol{r}}_i)\right]_s + \beta_{12} \left[(\underline{\boldsymbol{p}}_i \otimes \underline{\boldsymbol{q}}_i)_s \otimes (\underline{\boldsymbol{p}}_i \otimes \underline{\boldsymbol{q}}_i)_s\right] \\ & +\beta_{13} \left[(\underline{\boldsymbol{p}}_i \otimes \underline{\boldsymbol{r}}_i)_s \otimes (\underline{\boldsymbol{p}}_i \otimes \underline{\boldsymbol{r}}_i)_s\right] \end{aligned} \tag{6.4}$$

where $\alpha_{11}, \alpha_{12}, \alpha_{13}, \beta_{12}, \beta_{13}$ are coefficients depending on the material. In the particular case of orthotropic fabric, i.e. with orthogonal yarns, the formulation is simplified since there are only three vectors, $\underline{\boldsymbol{p}}_1, \underline{\boldsymbol{p}}_2, \underline{\boldsymbol{p}}_3$, orthogonal in pairs. For the tensorial variable, the damage effect tensor is given by (3.17) of Section 3.1.B..

B. Damage deactivation

The modelling of the damage deactivation effect, corresponding to the closure of some microcracks under compressive-like loadings, is a difficult problem. It has been discussed in [92]. Several theories from the literature are not acceptable, due to the presence of discontinuous stress-strain responses for general multiaxial non-proportional loading conditions. The proposed approach follows the proposal by Chaboche [93] of a deactivation that forces response continuity. A similar formulation has been reused in other works, like [94], [95]. When one of the damages is deactivated, we use $\underset{\approx}{\boldsymbol{\Lambda}}^{\text{eff}}$ given by:

$$\begin{aligned} \Delta \underset{\approx}{\boldsymbol{\Lambda}}^{\text{eff}} \;=\; & \Delta \underset{\approx}{\tilde{\boldsymbol{\Lambda}}} + \eta_1 \sum_{i=1}^{3} \delta_i H(-\varepsilon_i)\; \underset{\approx}{\boldsymbol{P}}_i : \left[\underset{\approx}{\boldsymbol{A}}_i : \underset{\approx}{\boldsymbol{K}}_o\right]_s : \underset{\approx}{\boldsymbol{P}}_i \\ & +\eta_2 \sum_{i=1}^{3} H(-\varepsilon_{n_i}) \underset{\approx}{\boldsymbol{N}}_i : \left[\underset{\approx}{\boldsymbol{D}}(\underset{\sim}{\boldsymbol{d}}) : \underset{\approx}{\boldsymbol{K}}_o\right]_s : \underset{\approx}{\boldsymbol{N}}_i \end{aligned} \tag{6.5}$$

where H is the Heaviside function and $\underline{n}_i$ are the principal directions of the tensorial damage. $\varepsilon_i = \underline{p}_i.\underset{\sim}{\varepsilon}.\underline{p}_i$ and $\varepsilon_{n_i} = \underline{n}_i.\underset{\sim}{\varepsilon}.\underline{n}_i$ are the strains normal to the corresponding principal damages. The fourth-rank projection operators, $\underset{\approx}{P}_i$ and $\underset{\approx}{N}_i$ are defined from $\underline{p}_i$ and $\underline{n}_i$ by relations such as $\underset{\approx}{P}_i = \underline{p}_i \otimes \underline{p}_i \otimes \underline{p}_i \otimes \underline{p}_i$.

The closure criterion is formulated in the microcrack axes or the principal directions of the second-rank tensor. It corresponds to the change of sign of the associated normal strain. Here, the principle is applied to the case where there are two types of damage, scalar and tensorial. The expression of the stress is given by (6.2) by simple derivation. The thermodynamic forces associated with the scalar and tensorial damage variables are somewhat complicated to express because of the closure effects:

$$y_i = -\frac{\partial \psi}{\partial \delta_i} = \frac{1}{2}\underset{\sim}{\varepsilon} : \left[\left(\underset{\approx}{A}_i : \underset{\approx}{K}_{\mathrm{o}} \right)_{\mathrm{s}} - \eta_1 H(-\varepsilon_i)\, \underset{\approx}{P}_i : \left(\underset{\approx}{A}_i : \underset{\approx}{K}_{\mathrm{o}} \right)_{\mathrm{s}} : \underset{\approx}{P}_i \right] : \underset{\sim}{\varepsilon} \tag{6.6}$$

$$\begin{aligned} \underset{\sim}{y} \;=\; & -\frac{\partial \psi}{\partial \underset{\sim}{d}} = \frac{\xi}{4}\left(\underset{\sim}{\varepsilon}\, \mathrm{Tr}\, \underset{\sim}{\tilde{\sigma}} + \underset{\sim}{\tilde{\sigma}}\, \mathrm{Tr}\, \underset{\sim}{\varepsilon} \right) + \frac{1-\xi}{2}\left(\underset{\sim}{\tilde{\sigma}}.\underset{\sim}{\varepsilon} \right)_{\mathrm{s}} \\ & -\eta_2 \sum_{j=1}^{3} H(-\varepsilon^*_{n_j}) \left[\frac{\xi}{2}\left(\underset{\sim}{\varepsilon}^*_j\, \mathrm{Tr}\, \underset{\sim}{\tilde{\sigma}}^*_j + \underset{\sim}{\tilde{\sigma}}^*_j\, \mathrm{Tr}\, \underset{\sim}{\varepsilon}^*_j \right) + \frac{1-\xi}{2}\left(\underset{\sim}{\tilde{\sigma}}^*_j.\underset{\sim}{\varepsilon}^*_j \right)_{\mathrm{s}} \right] \end{aligned} \tag{6.7}$$

where $\underset{\sim}{\tilde{\sigma}} = \underset{\approx}{K}_{\mathrm{o}} : \underset{\sim}{\varepsilon}$, $\quad \underset{\sim}{\varepsilon}^*_j = \underset{\approx}{N}_j : \underset{\sim}{\varepsilon} = \varepsilon^*_{n_j}\, \underline{n}_j \otimes \underline{n}_j$, $\quad \underset{\sim}{\tilde{\sigma}}^*_j = \underset{\approx}{K}_{\mathrm{o}} : \underset{\sim}{\varepsilon}^*_j$.

C. Damage growth law

The damage laws result directly from the general formalism given in Paragraph 3.4.A.2. The criteria associated with the scalar damage are multiple interrelated criteria defined in the space of thermodynamic forces y_i and have the form:

$$f_i = g_i \left(\sum_{j=1}^{3} a_{ij} \langle y_j \rangle \right) - \left\langle r_{\mathrm{o}} + \sum_{j=1}^{3} b_{ij}\delta_j \right\rangle \leq 0, \qquad i = 1, 2, 3 \tag{6.8}$$

where $a_{11} = a_{22} = a_{33} = b_{11} = b_{22} = b_{33} = 1$. The coupling coefficients such as a_{12} are used to adjust the shape of the non–damage surface in the region $y_1 > 0$, $y_2 > 0$, whereas b_{12} is used to adjust the variation of criterion $f_2 = 0$ when damage d_1 varies ($\delta_2 = 0$) or vice versa. When forces y_i are negative, we exclude them from the criterion (McCauley's symbol). Functions g_i describe the damage growth kinetics and r_{o} defines the initial threshold.

We use a special form for the tensorial variable:

$$\begin{aligned} f \;=\; & g\left[\chi \left(\underset{\sim}{y}^+ : \underset{\approx}{Q} : \underset{\sim}{y}^+ \right)^{\frac{1}{2}} + (1-\chi)\, \mathrm{Tr}\, \underset{\sim}{y} \right] \\ & -\rho_{\mathrm{o}} - \zeta\, \mathrm{Tr}\, \underset{\sim}{d} - (1-\zeta)\, \mathrm{Tr}\, (\underset{\sim}{y}.\underset{\sim}{d}) \leq 0 \end{aligned} \tag{6.9}$$

which involves a combination of two invariants. In the first, for $\chi = 1$, the initial anisotropy of the composite is included by tensor $\underset{\approx}{Q}$. Moreover $\underset{\sim}{y}^+$ denotes the positive part of tensor $\underset{\sim}{y}$. The second ($\chi = 0$) is introduced to have as special case a material in which the damage varies isotropically. The role of ζ is explained below.

The damage evolution equations are then given by:

$$\dot{\delta}_i = \sum_{j=1}^{3} \dot{\mu}_j \frac{\partial f_j}{\partial y_i} \qquad \dot{\underset{\sim}{d}} = \dot{\mu} \frac{\partial f}{\partial \underset{\sim}{y}} \tag{6.10}$$

in which the scalar multipliers satisfy the consistency conditions such that:

$$\begin{aligned} \dot{\mu}_j &= 0 \quad \text{if} \quad f_j < 0 \text{ or } \dot{f}_j < 0 \\ \dot{\mu}_j &\neq 0 \quad \text{if} \quad f_j = \dot{f}_j = 0 \end{aligned} \tag{6.11}$$

It can be noted that because of the linearity of the free energy as a function of the damage variables, the thermodynamic forces are independent of it and the multipliers of the scalar and tensorial criteria are therefore determined in a decoupled manner.

A key point to be stressed concerns couplings between directions in the case of non–proportional multiaxial loadings. For instance, let us assume a uniaxial damaging load in direction 1, followed by unloading and a new loading in orthogonal direction 2. Fig. 6.9 schematically illustrates what is given by the different criteria:

- The few existing experiments ([96], [90], [86]) appear to indicate that the non–damage domain is less extended in direction 2 after damaging in direction 1 (which would not have been the case without prior damage);

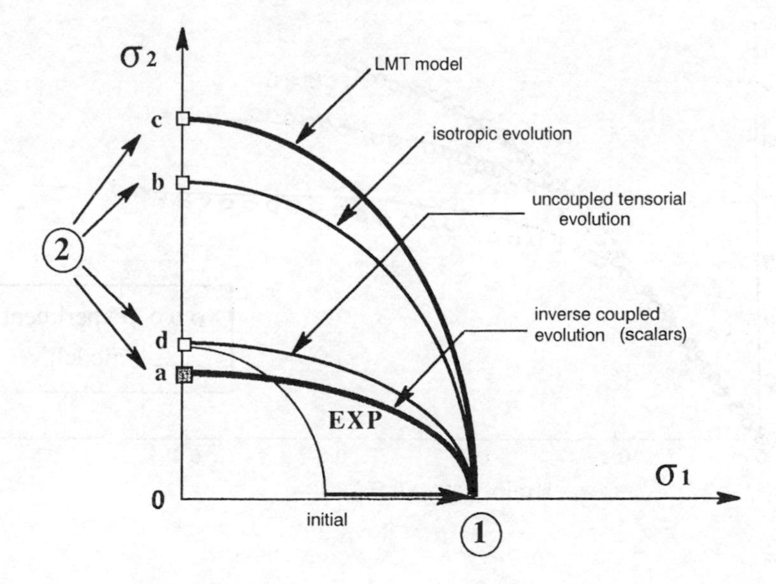

Figure 6.9. Diagram of damage under bi–axial loading: (a) experiment and evolution with inverse coupling (scalar variables); (b) isotropic evolution; (c) LMT model; (d) decoupled evolution (tensorial variable)

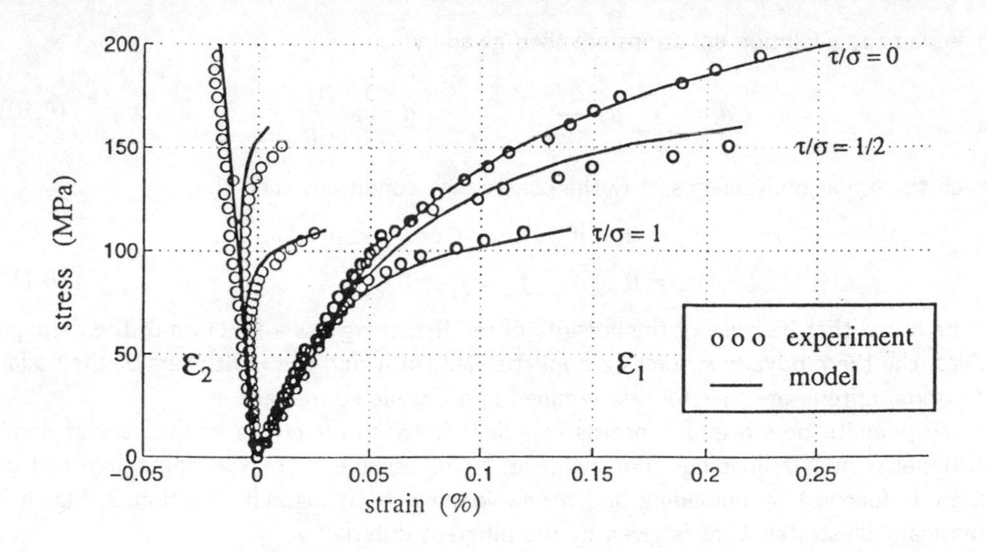

Figure 6.10: Tension/torsion test on SiC/SiC. Tension response

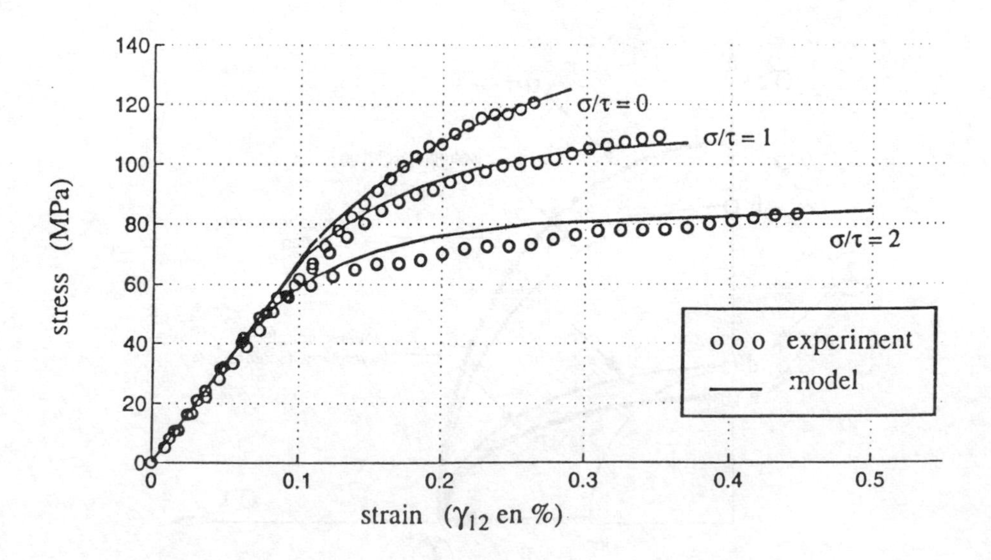

Figure 6.11: Tension/torsion test on SiC/SiC. Shear response

- A model such as the one used by Ladevèze [82] gives a new non–damage surface more extended in direction 2 than in direction 1 (which is contrary to experimental data);

- The tensorial model (6.9) with $\chi = 1$ and $\zeta = 1$ gives the same extension in both directions. In effect, the threshold varies isotropically in a purely scalar manner (criterion with isotropic evolution);

- Multiple coupled criteria give results closer to the experimental data, with a decrease in the non–damage threshold in direction 2. For this it is sufficient to choose negative coupling terms such as b_{12} in (6.6);

- Finally, the tensorial criterion of (6.9) with $\chi = 1$ and $\zeta = 0$ allows the threshold to be kept unchanged in direction 2 while the damage and corresponding threshold increase in direction 1. To summarize (and in the principal *ad hoc* space), the threshold function varies with $y_1 d_1 + y_2 d_2$, which means that after damage d_1 caused by $y_1 > 0$ with y_2 and d_2 equal to zero, the threshold in direction 2 for $y_2 = 0$ remains equal to the initial threshold. The evolution with this criterion are therefore decoupled between the directions. This criterion is used below to model the axial tension/internal pressure tests on SiC/SiC composites.

D. Examples of use

We give a few results from a very complete experimental analysis conducted on a SiC/SiC composite produced by SEP. This work includes tension/compression testing (monotonic or cyclic, increasing amplitude loads), tension/torsion testing on tubes and tension/compression/internal pressure testing on tubes.

The model is determined from experimental tension/compression data in the 0 and 45° directions. Fig. 6.7 shows a comparison for the 0° direction. The agreement is satisfactory, both for axial strain and transverse strain. It can be seen that the damage deactivation effect is extremely marked for this type of material, with almost complete recovery of the initial modulus of elasticity after compressive loading. In this case, the deactivation strain $\underset{\sim}{\varepsilon}^c$ was taken as equal to zero.

Figs 6.10 and 6.11 [90], [86] show the results obtained on a thin tube under tension/torsion for proportional monotonic loadings with various load ratios. The model very correctly predicts the multiaxiality effects, although it was not determined on such experiments. The transverse strains of Fig. 6.10 should be noted in particular.

The bi–axial tension/internal pressure test on the tube of Fig. 6.12 [86] corresponds to cyclic tensile loading (with increasing maxima) continued by internal pressure loading, also at increasing levels (the end effect is almost completely compensated). The lower part of the figure shows excellent agreement between the calculated and measured modulus of elasticity (secant modulae).

It can be seen that the stress level in direction 2 (circumferential) causing the start of the drop in modulus E_{22} is only slightly higher than the initial threshold under axial tension,

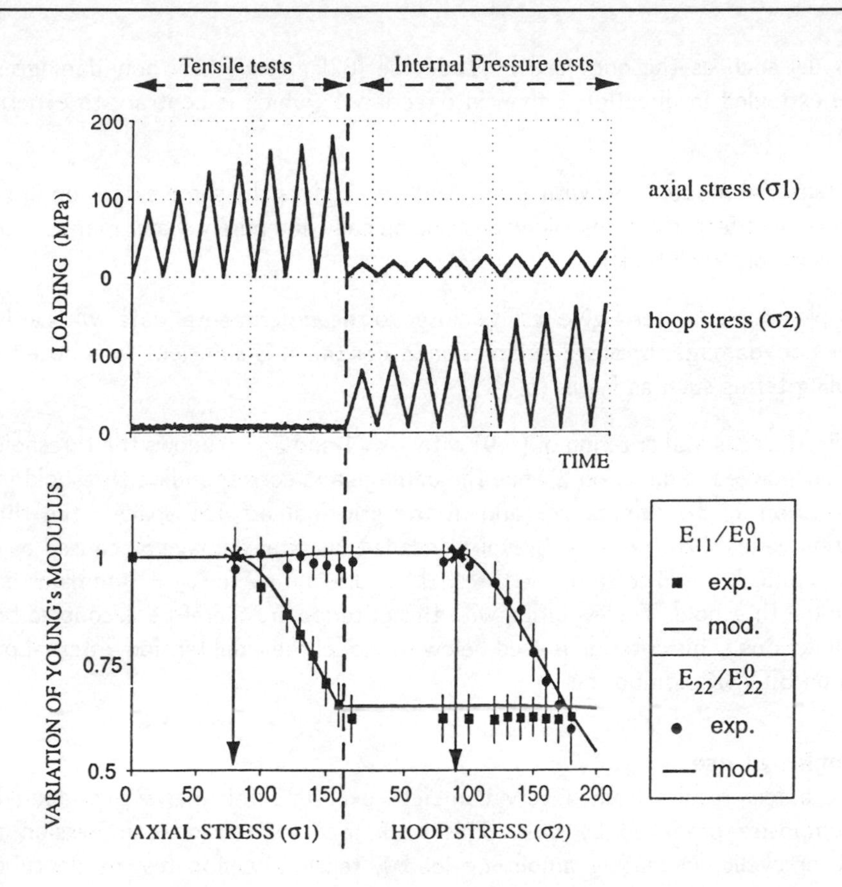

Figure 6.12. Prediction of the variation of Young's modulus (directions 1 and 2) on SiC/SiC for incremental tensile loading cycles followed by incremental internal pressure loadings.

leading to the drop in modulus E_{11}. This test clearly shows the requirement for a combined criterion such as the one given by (6.9) to keep the damage threshold from changing in direction 2 while the damage progresses by tension in direction 1.

REFERENCES

1. Kachanov L.M. : Time of the rupture process under creep conditions, Isv. Akad. Nauk. SSR. Otd Tekh. Nauk., 8(1958), pp. 26–31.
2. Murzewski J.W. : The tensor of failure and its application to determination of strength of welded joints, Bull. Acad. Polon. Sci., Série Sci. Tech., VI(1958), pp. 159–164.
3. Rabotnov Y.N. : Creep problems in structural members, North-Holland, 1969.
4. Hayhurst D.R. : Creep rupture under multiaxial state of stress, J. Mech. Phys. Sol., 20(1972), pp. 381–390.

5. Leckie F.A. and Hayhurst D.R. : Creep rupture of structures, Proc. Royal Soc. London, 340(1974), pp. 323–347.
6. Chrzanowski M. : Use of the damage concept in describing creep-fatigue interaction under prescribed stress, J. Mech. Sci., 18(1976), pp. 69–73.
7. Hult J. : Continuum Damage Mechanics. Capabilities limitations and promises, in: Mechanisms of Deformation and Fracture, Pergamon, Oxford, 1979 pp. 233–347.
8. Murakami S. and Ohno N. : A continuum theory of creep and creep damage, in: Creep in Structures (Edited by A. Ponter and D. Hayhurst), 3rd IUTAM Symp., Springer-Verlag, 1980 pp. 422–443, leicester.
9. Chaboche J.L. : Sur l'utilisation des variables d'état interne pour la description de la viscoplasticité cyclique avec endommagement, in: Problèmes Non Linéaires de Mécanique, 1977 pp. 137–159, symposium Franco-Polonais de Rhéologie et Mécanique, Cracovie.
10. Lemaître J. and Chaboche J.L. : Mécanique des Matériaux Solides, Dunod, Paris, 1985.
11. Lemaître J. : A Continuum Damage Mechanics model for ductile fracture, J. of Engng. Mat. Technol., 107(1985), pp. 83–89.
12. Germain P. , Nguyen Q.S. and Suquet P. : Continuum Thermodynamics, J. of Applied Mechanics, 50(1983), pp. 1010–1020.
13. Krajcinovic D. and Fonseka G.U. : The Continuous Damage theory of brittle materials, Parts 1 and 2, J. of Applied Mechanics, 48(1981), pp. 809–824.
14. Krajcinovic D. : Continuum Damage Mechanics, Applied Mechanics Reviews, 37(1984), pp. 1–6.
15. Ortiz M. : A constitutive theory for the inelastic behavior of concrete, Mech. Mater., 4(1985), pp. 67–93.
16. Ju J.W. : On energy-based coupled elastoplastic damage theories : constitutive modeling and computational aspects, Int. J. Solids Structures, 25(1989), pp. 803–833.
17. Chow C.L. and Wang J. : An anisotropic theory of Continuum Damage Mechanics for ductile fracture, Engineering Fracture Mechanics, 27(1987), pp. 547–558.
18. Voyiadjis G.Z. and Kattan P. : Damage of fiber reinforced composite materials with micromechanical characterization, Int. J. Solids Structures, 30(1993), pp. 2757–2778.
19. Chaboche J.L. : Continuous Damage Mechanics : a tool to describe phenomena before crack initiation, Nuclear Engineering and Design, 64(1981), pp. 233–247.
20. Murakami S. : Notion of Continuum Damage Mechanics and its application to anisotropic creep damage theory, J. of Engng. Mat. Technol., 105(1983), pp. 99.
21. Hayhurst D.R. : Materials data bases and mechanisms-based constitutive equations for use in design, in: Creep and Damage in Materials and Structures (Edited by H. Altenbach and J. Skrzypek), Springer, 1999 .
22. Krempl E. : Cyclic creep and creep fatigue interaction, in: Creep and Damage in Materials and Structures (Edited by H. Altenbach and J. Skrzypek), Springer, 1999 .
23. Skrzypek J. : Material damage models for creep failure analysis and design of structures, in: Creep and Damage in Materials and Structures (Edited by H. Altenbach and J. Skrzypek), Springer, 1999 .

24. Lemaître J. and Chaboche J.L. : Mechanics of Solid Materials, Cambridge University Press, Cambridge, U.K., 1990.
25. Jonas J.J. and Baudelet B. : Effect of crack and cavity generation on tensile stability, Acta Metallurgica, 25(1977), pp. 43–50.
26. Cailletaud G. , Policella H. and Baudin G. : Mesure de déformation et d'endommagement par mesure électrique, La Recherche Aérospatiale, (1980), pp. 69–75.
27. Cordebois J.P and Sidoroff F. : Anisotropie élastique induite par endommagement, Col. Euromech 115, Editions du CNRS, 1982, 1979 Grenoble.
28. Plumtree A. and Nilsson J.O. : Damage mechanics applied to high temperature fatigue, Journées Internationales de Printemps, 1986 Paris.
29. Charewicz A. and Daniel I.M. : Fatigue damage mechanisms and residual properties of graphite/epoxy laminates, Engineering Fracture Mechanics, 25(1986).
30. Chaboche J.L. : Description phénoménologique de la viscoplasticité cyclique avec endommagement, Doctorat d'Etat, Université Pierre et Marie Curie, Paris 6, Paris, 1978.
31. Chaboche J.L. : Une loi différentielle d'endommagement de fatigue avec cumulation non linéaire, Revue Française de Mécanique, (1974), pp. 71–82.
32. Baste S. , Guerjouma R. El and Gérard A. : Mesure de l'endommagement anisotrope d'un composite céramique-céramique par une méthode ultrasonore, Revue de Physique Appliquée, 24(1989), pp. 721–731.
33. Germain P. : Cours de Mécanique des Milieux Continus, volume I, Masson, Paris, 1973.
34. Sidoroff F. : On the formulation of plasticity and viscoplasticity with internal variables, Arch. Mech., Poland, 27(1975), pp. 807–819.
35. Halphen B. and Nguyen Q.S. : Sur les matériaux standards généralisés, J. de Mécanique, 14(1975), pp. 39–63.
36. Chaboche J.L. : Unified cyclic viscoplastic constitutive equations : development, capabilities and thermodynamic framework, Academic Press Inc., 1996 pp. 1–68.
37. Chrysochoos A. : Bilan énergétique en élastoplasticité, grandes déformations, J. Mécanique Théorique et Appliquée, 4(1985).
38. Chrysochoos A. : Dissipation et blocage d'énergie lors d'un écrouissage en traction simple, Doctorat d'Etat, Université de Montpellier, Montpellier, 1987.
39. Leckie F.A. and Onat E.T. : Tensorial nature of damage measuring internal variables, in: Physical Non-Linearities in Structural Analysis, IUTAM Symp., Springer, 1980 Senlis.
40. Lemaître J. : A course on Damage Mechanics, Springer, 1992.
41. Ladevèze P. : Sur une théorie de l'endommagement anisotrope, Rapport Interne 34, Laboratoire de Mécanique et Technologie, Cachan, 1983.
42. Vakulenko A.A. and Kachanov M.L. : Continuum theory of medium with cracks, Mech. of Solids, english transl. of Mekhanika Tverdogo Tela (in Russian), 6(1971), pp. 145–151.
43. Kachanov M. : Continuum model of medium with cracks, J. of the Engineering Mechanics Division, 106(1980), pp. 1039–1051.
44. Chaboche J.L. : Le concept de contrainte effective appliqué à l'élasticité et à la viscoplasticité en présence d'un endommagement anisotrope, Col. Euromech 115, Editions

du CNRS, 1982, 1979 pp. 737–760, grenoble.
45. Simo J.C. and Ju J.W. : Strain- and stress-based Continuum Damage models - I - Formulation, Int. J. Solids Structures, 23(1987), pp. 821–840.
46. Saanouni K. : Sur l'analyse de la fissuration des milieux élasto-viscoplastiques par la théorie de l'endommagement continu, Doctorat d'Etat, Université de Compiègne, 1988.
47. Benallal A. : Thermoviscoplasticité et endommagement des structures, Doctorat d'Etat, Université Pierre et Marie Curie, Paris 6, 1989.
48. Dragon A. : On phenomenological description ofrock-like materials with account for kinetics of brittle fracture, Archivium Mechaniki Stosowanej, 28(1976), pp. 13–30.
49. Hansen N.R. and Schreyer H.L. : Thermodynamically consistent theories for elastoplasticity coupled with damage, in: Damage Mechanics and Localization, volume 142/AMD, ASME, 1992 pp. 53–67.
50. Zhu Y.Y. and Cescotto S. : Fully coupled elasto-visco-plastic damage theory for anisotropic materials, Int. J. Solids Structures, 32(1995), pp. 1607–1641.
51. Allix O. , Ladevèze P. , Dantec E. Le and Vittecoq E. : Damage Mechanics for composite laminates under complex loading, in: Yielding, Damage and Failure of Anisotropic Solids (Edited by J. Boehler), EGF5, Mechanical Engineering Publications, London, 1990 pp. 551–569.
52. Mazars J. : Mechanical damage and fracture of concrete structures, in: Advances in Fracture Research, volume 4, Pergamon Press, Oxford, 1982 pp. 1499–1506.
53. McClintock F. : A criterion for ductile fracture by the growth of holes, J. of Applied Mechanics, (1968).
54. Rice J.R. and Tracey D.M. : On the ductile enlargement of voids in triaxial stress fields, J. Mech. Phys. Sol., 17(1969), pp. 201.
55. Lemaître J. : How to use Damage Mechanics, Nuclear Engng and Design, 80(1984), pp. 233–245.
56. Rousselier G. : Finite deformation constitutive relations including ductile fracture damage, in: Three-Dimensional Constitutive Relations and Ductile Fracture (Edited by Nemat-Nasser), North-Holland Publ. Comp.,1981, 1980 pp. 331–355, dourdan.
57. Benallal A. , Billardon R. , Doghri I. and Moret-Bailly L. : Crack initiation and propagation analyses taking into account initial strain hardening and damage fields, in: Numerical Methods in Fracture Mechanics (Edited by L. et al.), Pineridge Press, 1987 pp. 337–351, san Antonio, Texas.
58. Skrzypek J. : Application of the orthotropic damage growth rule to variable principal directions, Int. J. Damage Mechanics, 7(1998), pp. 180–206.
59. Chaboche J.L. and Lesne P.M. : A non-linear continuous fatigue damage model, Fatigue and Fracture of Engineering Materials and Structures, 11(1988), pp. 1–17.
60. Gallerneau F. : Etude et modélisation de l'endommagement d'un superalliage monocristallin revêtu pour aube de turbine, Doctorat d'Université, École Nationale Supérieure des Mines de Paris, 1995.
61. Lesne P.M. and Cailletaud G. : Creep-fatigue interaction under high frequency loading, ICM5, 1987 Beijing, Chine.

62. Savalle S. and Cailletaud G. : Microamorçage, micropropagation et endommagement, La Rech. Aérospatiale, (1982), pp. 395–411.
63. Lesne P.M. and Savalle S. : A differential damage rule with microinitiation and micropropagation, La Rech. Aérospatiale, (1987), pp. 33–47.
64. Benallal A. , Billardon R. and Marquis D. : Prévision de l'amorçage et de la propagation des fissures par la mécanique de l'endommagement, Int. Spring Meeting "Fatigue at High Temperature", 1986 Paris.
65. Savalle S. and Culié J.P. : Méthodes de calcul associées aux lois de comportement cyclique et d'endommagement, La Rech. Aérospatiale, (1978).
66. Lesne P.M. and Savalle S. : An efficient "cycles jump technique" for viscoplastic structure calculations involving large number of cycles, 2nd Int. Conf. on "Computational Plasticity", 1989 Barcelone.
67. Jr. J.C. Newman : Finite element analysis of fatigue crack propagation including the effecte of crack closure, Ph.d. thesis, VPI,Blacksburg, 1974.
68. Anquez L. : Elastoplastic crack propagation fatigue and failure, La Rech. Aérospatiale, (1983), pp. 15–39.
69. Kruch S. , Chaboche J.L. and Prigent P. : A fracture mechanics based fatigue-creep-environment crack growth model for high temperature, volume 59, 1994 pp. 141–148.
70. Devaux J.C. and Mottet G. : Déchirure ductile des aciers faiblement alliés : modèles numériques, Rapport 79-057, Framatome, 1984.
71. Walker K.P. and Wilson D.A. : Constitutive modeling of engine materials, Report FR17911 (AFWAL-TR-84-4073), PWA, 1984.
72. Lemaître J. : Local approach of fracture, Engng. Fracture Mechanics, 25(1986), pp. 523–537.
73. Hayhurst D.R. , Dimmer P.R. and Chernuka M.W. : Estimates of the creep rupture lifetime of structures using the finite element method, J. Mech. Phys. Sol., 23(1975), pp. 335.
74. Saanouni K. and Lesne P.M. : Non-local damage model for creep crack growth prediction, 2nd MECAMAT Int. Seminar, "High Temperature Fracture Mechanisms and Mechanics", Mech. Engineering Publications, 1987 Dourdan, France.
75. Liu Y. , Murakami S. and Kanagawa Y. : Mesh dependence and stress singularity in finite element analysis of creep crack growth by Continuum Damage Mechanics approach, Eur. J. Mech., A/Solids, 13(1994), pp. 395–417.
76. Hall F.R. , Hayhurst D.R. and Brown P.R. : Prediction of plane-strain creep crack growth using Continuum Damage Mechanics, Int. J. Damage Mechanics, 5(1996), pp. 353–402.
77. Rousselier G. , Devaux J.C. and Mottet G. : Ductile initiation and crack growth in tensile specimens. Application of Continuum Damage Mechanics, SMIRT 8, 1985 Brussels.
78. Billardon R. : Fully coupled strain analysis of ductile fracture, 1st MECAMAT Int. Seminar on Local Approaches of Fracture, 1986 Moret-sur-Loing, France.
79. Bazant Z.P. and Belytschko T. : Localization and size effect, 2nd Int. Conf. on Constitutive Laws for Engineering Materials : Theory and Applications, Elsevier, 1987 Tucson.

80. Saanouni K. , Chaboche J.L. and Lesne P.M. : On the creep crack growth prediction by a non local damage formulation, Eur. J. Mech., A/Solids, 8(1989), pp. 437–459.
81. Bazant Z.P. and Pijaudier-Cabot G. : Modeling of distributed damage by non local continuum with local straintrain hardening and damage fields, in: Numerical Methods in Fracture Mechanics (Edited by L. et al.), Pineridge Press, 1987 pp. 411–432, san Antonio, Texas.
82. Ladevèze P. , Gasser A. and Allix O. : Damage mechanics modelling for ceramic composites, J. of Engng. Mat. Technol., 116(1994).
83. Ladevèze P. : A damage computational approach for composites : Basic aspects and micromechanical relations, Computational Mechanics, 17(1995), pp. 142–150.
84. Saanouni K. and Lesne P.M. : Sur la description phénoménologique des déformations anélastiques dans les composites endommageables, C. R. Acad. Sci. Paris, t.315(1992), pp. 1165–1170.
85. Maire J.F. and Chaboche J.L. : A new formulation of continuum Damage Mechanics for composite materials, Aerospace Science and Technology, 4(1997), pp. 247–257.
86. Maire J.F. and Lesne P.M. : A damage model for ceramic matrix composites, Aerospace Science and Technology, 4(1997), pp. 259–266.
87. Kruch S. , Chaboche J.L. and Pottier T. : Two-scale viscoplastic and damage analysis of a metal matrix composite, in: Danmage and Interfacial Debonding in Composites (Edited by G. Voyiadjis and D. Allen), Elsevier, 1996 pp. 45–55.
88. Pottier T. , Kruch S. and Chaboche J.L. : Analyse de l'endommagement macroscopique d'un composite à matrice métallique, JNC10, AMAC, 1996 pp. 1361–1371, paris.
89. Chaboche J.L. , Lesné O. and Pottier T. : Continuum Damage Mechanics of composites : towards a unified approach, in: Damage Mechanics in Engineering Materials (Edited by G. Voyiadjis, J. Ju and J. Chaboche), Elsevier Science B.V., 1998 pp. 3–26.
90. Maire J.F. and Pacou D. : Essais de traction-compression-torsion sur tubes composites céramique-céramique, JNC10, AMAC, 1996 pp. 1225–1234, paris.
91. Legrand N. , Rémy L. , Dambrine B. and Molliex L. : Etude micromécanique de la fatigue oligocyclique d'un composite à matrice métallique base titane renforcé par des fibres unidirectionnelles de CiC, JNC10, AMAC, 1996 pp. 1349–1360, paris.
92. Chaboche J.L. : Damage induced anisotropy : on the difficulties associated with the active/passive unilateral condition, Int. J. Damage Mechanics, 1(1992), pp. 148–171.
93. Chaboche J.L. : Development of Continuum Damage Mechanics for elastic solids sustaining anisotropic and unilateral damage, Int. J. Damage Mechanics, 2(1993), pp. 311–329.
94. Curnier A. , He Q. and Zysset P. : Conewise linear elastic materials, J. Elasticity, 37(1995), pp. 1–38.
95. Halm D. and Dragon A. : A model of anisotropic damage by mesocrack growth - Unilateral effects, Int. J. Damage Mechanics, 5(1996), pp. 384–402.
96. Siquiera C. : Développement d'un essai biaxial sur plaques composites et utilisation pour la modélisation du comportement d'un matériau SMC, Doctorat d'Université, Université de Franche-Comté, 1993.

CREEP-PLASTICITY INTERACTION

E. Krempl
Rensselaer Polytechnic Institute, Troy, NY, USA

ABSTRACT

Inelastic analyses are needed for predicting the lifetime of one–of–a–kind components operating under severe conditions of loading and environment. Examples are power generation and propulsion machinery and chemical processing plants. The components are subjected to steady and cyclic loading which involves nonlinear monotonic and cyclic behavior. In addition, rate dependence, creep and relaxation are present. These phenomena must be captured in a constitutive equation suitable for inelastic analyses.

The "unified" viscoplasticity theory based on overstress (VBO) was formulated using the experiment–based approach. Servo–controlled testing machines and strain measurements on the gage length are employed to measure the response to a given input. The responses contain the influence of the changing microstructure. In VBO no yield surface is used and rate dependence is present at every temperature. The high temperature formulation has a static recovery term and provides for softening. Softening is the result of diffusion, which counteracts the hardening due to inelastic straining. Low and high homologous temperature qualitative properties such as the behavior under very fast, very slow monotonic, rate–dependent loading are investigated in addition to creep, relaxation and cyclic loading. The low homologous temperature formulation admits a long–time asymptotic solution, which is thought to apply to the "flow" stress region of the stress–strain diagram where special relaxation properties are observed in real materials and in the model. References to papers by the author and his students show that primary, secondary and tertiary creep can be modeled with VBO in addition to anomalous behavior. Inelastic strains are observed and modeled in recovery experiments at zero stress.

1. INTRODUCTION

Modern society expects predictability and reliability in the performance of engineering structures in addition to their safe operation at a competitive cost. Engineering practices to incorporate these demands in one–of–a–kind structures differ from those employed for mass produced items. The latter are usually subjected to simulated service testing to insure safety and reliability. For large structures and machinery as well as for large manufacturing processes this approach is not possible because of the cost and the time involved. It is not possible for example to test a prototype power plant for 30 years before producing it!

Consequently, for one–of–a–kind machinery or manufacturing processes, the performance, reliability and safety must be assessed early in the design stage, long before the machine is built or the process started. Examples are energy generating equipment (steam and gas turbines, refineries, pressure vessels) or propulsion machinery (jet engines and rocket motors), or the manufacturing of large components. In these cases, analyses of safety, reliability and performance must be made prior to actual production.

The increasing use of electricity in the 1930s led to the construction of new plants initially designed for more than twenty years of steady–state operation. Long–term creep rupture was a failure mode, which had to be prevented. Creep rupture was observed to occur in the quasi–elastic region of the stress–strain diagram. Conventional methods of using yield as a failure criterion were unsafe and therefore not applicable. Accordingly creep theory was developed Bailey [1], Odqvist and Hult [2], Rabotnov [3]). Creep failure was more limiting than plastic deformation and creep analysis provided an independent means of analysis of structures operating at elevated temperatures.

The situation changed when the power plants designed for steady–state operation were required to respond to the cyclical power requirements of modern society. The demand fluctuated with the time of the day and the seasons. Steam and gas turbines had to be started up and shut down. These relatively few start ups and shutdowns caused severe thermal stresses in the thick components which were designed for steady–state operation. Local plastic deformation occurred and led to unexpected low–cycle fatigue failures. Similar operation histories were found in jet engines, which had also frequent start–stop operations. The thermal stresses were so severe that plastic deformation was found to occur especially in stress raisers where low–cycle fatigue failure was observed. Worldwide investigations on low–cycle fatigue at room and elevated temperatures were undertaken and design rules to prevent low–cycle fatigue failure were developed. Low–cycle fatigue, high–cycle fatigue, creep, overload and buckling became failure modes that had to be considered in the design stage, long before the structure was ever built. For a review of the life prediction aspect the reader is referred to Sehitoglu [4]; Krempl [5]; Woodford [6]) and the references cited therein.

While these developments in the field stimulated the modern design needs, two other changes in technology became significant factors: Finite element programs and the improvement of mechanical testing capabilities through advancements in computer systems and other electronic equipment. The former made the nonlinear analysis economically possible and the latter helped to provide the needed knowledge of inelastic material behavior.

The analyses of field failures showed that cracking occurred at stress raisers where the region of inelastic behavior was confined to a small zone. This zone is generally considered to be under "strain control" by the surrounding elastic deformation. The new testing machines using servo–control and strain measurements on the gage length were able to simulate strain control. Such simulations could not have been reliably performed with the old testing technology. Starting in the 1970s, strain–controlled, cyclic deformation studies proliferated and provided information on the deformation behavior as well as on the lifetime of engineering materials. Aside from continuous cycling, hold–times under constant stress or strain were introduced to simulate the actual situation as close as possible. It became evident that the hold–times had a significant effect on lifetime and on the deformation behavior. The results of low–cycle fatigue studies strongly suggested that a cycle to cycle analysis is needed. There were, and still are, many voices claiming that detailed analyses are not needed since they unnecessarily increase the design cost. However, the advent of economical computing power counteracts the claim of excessive cost. It is now possible to perform nonlinear analyses economically in the design office, and indeed it is being done but with old material models that are not commensurate with the available computing power and the experiments obtained with modern testing equipment.

There is still another factor that supported the development of new material models, the revival of continuum mechanics. From the turn of the 19th century to the late 1940s mechanics of materials consisted essentially of solving boundary value problems in linear elasticity. After World War II the emphasis shifted to the basics of representing material behavior in stress analysis. The concepts of bodies, forces, kinematics, conservation laws valid for all continua and the constitutive equation, which represents a particular material, were clearly delineated. When new material behavior was found the on–going developments in continuum mechanics were available for use. Considerable progress was made in nonlinear fluids and finite elasticity, which is a widely used model for rubber.

The metallic materials of interest here provide a special challenge for continuum mechanics since they change their internal structure during inelastic deformation. Due to these changes in the internal make–up the deformed material is in essence a new material. It is known from materials science that mechanical properties are very sensitive to the internal structure of a metal or an alloy. As an example the yield strength of copper can easily change by a factor of two through cold work.

To account for these internal changes the author has proposed an experiment–based approach. This approach is a variant of a motto used in materials science. It states that the *current microstructure* **and the current loading conditions** *determine(s) the current response* (Bold added by the author). The addition of the bold face text is necessary since a metal or alloy that is in the same state will exhibit different behavior in a creep and in a relaxation test for which the test conditions are zero stress rate and zero total strain rate, respectively. If a model is based on experimentally observed behavior and the model reproduces the observed behavior, then the model obviously includes the effects of changing microstructure.

2. PHENOMENOLOGY IN MATERIALS MODELING

2.1. "Physical" vs. empirical approaches

The approach discussed here is frequently labeled empirical with the connotation that it is inferior to the physical, micromechanism–oriented approach. Before a conclusion can be reached the objectives of the research have to be defined.

Here the purpose is to have a mathematical model of the deformation behavior of a material in stress and lifetime analyses of engineering structures. Because of this requirement the model must be commensurate with modern computational methods. It must be formulated in three dimensions and must be a good model of the material.

In engineering applications long–lasting, steady motions are interrupted by cyclic loading. It is therefore necessary to consider monotonic loading at different rates, cyclic loading and creep and relaxation motions. It is not sufficient to consider only one motion or only steady–state conditions. Transients are important as well since the severity of loading is frequently highest in this type of loading.

The physical approaches have usually different objectives. Their aims are to elucidate the internal mechanisms that enable the macroscopic deformation and to learn from them how to improve the performance of to–be–made new materials. As a consequence a particular motion is selected, experiments are performed in a uniaxial setting and the mechanisms are delineated. Examples are the Ashby maps for creep and plasticity, see Frost and Ashby [7]. The mechanisms for steady–state operation are well documented. But even in this case the constants for the creep and plasticity equations must be determined from macroscopic experiments. "Theory gives the form of the equation; but experimental data are necessary to set the constants which enter it," page 6 of Frost and Ashby [7]. Moreover certain mechanisms are identified by macroscopic evidence. An example is the "Power–law Breakdown," that is observed at high stresses, see Frost and Ashby [7], page 13. It is stated that the power–law breakdown " is evidently a transition from climb–controlled to glide–controlled flow (Fig. 2.5)". The statement does not mention the mechanisms first, rather the failure of the power law to predict the creep rate gives rise to a physical explanation.

An excellent description of the creep and plasticity deformation mechanisms is provided. But this information is of limited usefulness for other motions. It is never mentioned that plasticity and creep ensue under completely different boundary conditions in testing. It is unlikely that the equations derived from creep or plasticity mechanisms can give useful results for other motions, say for cyclic loading.

2.2. Constitutive equation vs. response function

To understand the aforementioned assertion it is necessary to describe the difference between a response of a material to a given loading or input and a constitutive equation (CE) or material model.

In mechanical testing an input to a specimen in the interval (0 to t) causes an output in the interval (0 to t), where t is current time. The input and the changing properties (such as changes in dislocation density) of the specimen influence the output. If a different input is used then a different output has to be expected in general.

The constitutive equation of a material (CE) must be synthesized or invented from the input–output pairs. Finding a CE requires the solution to an inverse problem, which, even in the case of linear behavior, may not be unique. The response function is not a CE. It is only the answer (response) of the material to a specific input or question. Creep is the answer to a constant stress, relaxation is the answer to a constant strain input. The CE must be such that different outputs can be obtained for different inputs such as creep and relaxation. Only the response functions are manipulated in a physical approach. They do not obtain the CE.

In a widely quoted book Frost and Ashby [7] analyze the mechanisms of creep. The mathematical form of different response functions is related to different observed micromechanisms. Only response functions are manipulated and no CE is given. No guidance is provided on how to predict say for example relaxation using the data of the book.

From an engineering standpoint such guidance would be helpful. Frost and Ashby [7] do not consider cyclic loading and do not address the difference between a response function and a CE. So there is no guidance from the authors how to proceed.

A stress analyst will have to generalize the response functions used by Frost and Ashby [7] to a CE. In this book the power law is frequently discussed and this response is now used for application to cyclic loading by writing $\dot{\gamma} = \frac{\dot{\sigma}_s}{\mu} + B\left(\frac{|\sigma_s|}{\mu}\right)^n \frac{\sigma_s}{|\sigma_s|}$ where the notation by Frost and Ashby [7] is used. The shear modulus is μ, $\dot{\gamma}$ and σ_s is the shear strain rate and the shear stress, respectively. From dimensional considerations the constants B and μ are not independent, only their combination can be determined, no matter what the underlying mechanism may be, Tachibana and Krempl [8]. Were this equation now used to simulate a strain controlled cyclic test no hardening or softening and no Bauschinger effect would be predicted. This is not the prediction that one would expect from a physical model.

This result is not surprising. After all, the stress is constant and the deformation is monotone in the creep test. In contrast, the stress is continuously changing in a cyclic test.

A proper CE is a mathematical expression that can only be solved after the input has been used to specialize the CE. The equations describing the Standard Linear Solid (SLS) and VBO are CEs. Different differential equations result when the CEs of the SLS or VBO are specialized for creep (stress rate equal to zero) or relaxation (strain rate is zero). Then the modified CE can be integrated to yield the creep (relaxation) curve, which is the answer of the CE to the question "constant stress (strain)". If the conditions for cyclic loading are inserted into the CE then the solution of the modified equation should predict the cyclic behavior. See Appendix A for some examples.

It is much easier to find a mathematical expression to describe a response than to find the CE. A response function only needs a curve fit. The CE has to be synthesized or invented from all the known input–output pairs which is a difficult task. Consequently many response functions are used as a CE, see the remarks on pp. 105–106 and 115–116 of Krempl [9]. If response functions are employed, perhaps in modified form, an incomplete description of the material will be used.

In the physical, mechanism oriented approaches, CE is almost never established. Instead response functions are employed most of the time as material models.

Macroscopically, the state of stress and strain in the gage section is uniform and the specimen serves as a representative volume element provided certain conditions are met. The wall thickness of the specimen allows a sufficient number of grains through the thickness. We assume that these conditions are met and consider the specimen a representative volume element. Then the output to a given input contains information on the changing internal structure of the material. We will not be able to identify individual mechanisms since the volume element integrates over all mechanisms, but their combined effects influence the output of the volume element. Moreover, the specimen response is directly in a form that can be used in formulating continuum laws.

The question arises as to how many different inputs are needed to characterize a metal or alloy. There are no clear–cut answers to this question as some say that an infinite number of inputs are needed. To restrict the domain, we will consider a uniaxial state of stress only. A minimum number of different inputs include monotonic loading at different rates, in load and/or displacement control, creep (stress rate is zero), relaxation (strain rate is zero) and cyclic loading. From the outputs to these inputs a constitutive equation can be invented. This CE must meet the requirement to produce under identical inputs the same outputs as the real material. Continuum mechanics and mathematics enable a three dimensional CE for use in stress analysis. Bi–axial tests serve as a check on the predictions of the model.

From this procedure it is clear that the phenomenological CE for a given material is not necessarily a unique model. Synthesizing the CE from various input–output pairs is not a unique process. The CE contains, however, essential features of the real material. For the motions described above realistic answers will be obtained. This cannot be assured for response functions.

The physical approaches use almost exclusively uniaxial tests and no physically based tool seems to be available to generalize the information to three dimensions. The usual method is to introduce effective stresses and strains, but that is not sufficient. These quantities apply only in the isotropic case and are invariants of the stress or strain tensors that are based on continuum mechanics and not on physical mechanisms.

It appears that the physical approach and the experiment–based approach proposed by the writer have a common background. Both use the same uniaxial experiment to obtain material information. The experiment–based approach employs the results of the experiment directly to formulate a CE. The physical approach determines the micro mechanisms first but then uses the same data as the experiment–based approach to fit response functions rather than finding CEs. It is not clear how the micromechanism based approaches can be applied to motions other than for which they have been curve–fitted.

2.3. Curve–fitting and the absence of physical input. The experiment–based approach

Frequently, the phenomenological approach is downgraded as being not physical and consisting only of curve fitting. Even the physical approach has to engage in curve fitting if data useful for engineering analysis are to be provided. The physical theory does not provide the constants, see the above quotation from Frost and Ashby [7], page 6. In the experiment–based approach curve fitting is also necessary.

The invention of a CE does not involve curve fitting at all, but is instead a process of synthesis. Curve fitting is performed frequently for response functions with the physical and with the experiment–based approach.

Once a CE or a response function is established the constants of the CE and of the response function must be determined from experimental data of a given material. To fix terminology the author would call the determination of the constants of a Ramberg–Osgood or of a creep law curve fitting. Once the constants are fitted the Ramberg–Osgood or the creep law can only model the tests that were used to determine the constants. They have no predictive capability. Determining the constants for a combined isotropic/kinematic hardening incremental plasticity theory or for VBO would not constitute curve fitting. Once the constants are determined there is predictive capability. For example the behavior in cyclic loading could be calculated.

It is clear that a phenomenological approach cannot distinguish between competing micromechanisms. But it can identify important microstructure–based mechanisms that affect the specimen response. The following example illustrates the method. Annealed type 304 stainless steel is known to exhibit significant cyclic hardening and rate dependence at ambient temperature, see Krempl [10] for an initial paper on this subject.

The question arises as to how hardening influences the rate dependence, i.e. whether a hardened sample would show different rate dependence than an annealed specimen. One way to answer this question would be by the use of micromechanics. The other way is through the response of specimens subjected to a suitable input. A strain rate cycling test was chosen to get the response of the annealed material, see curve A in Fig. 2.1. The rate sensitivity can be estimated by the difference in the flow stresses at a certain strain once inelastic flow is established. Specimen B underwent cyclic loading in torsion until the changes in the hysteresis loop were very small. Unloading to the origin provided no residual strain. The so hardened specimen was then subjected to the same input as specimen A. The result is identified by B. To a first approximation the rate dependence remained unaltered and the hardening increased the flow stress. It is therefore termed plastic and rate independent.

The comparison of the outputs of two specimens subjected to the same input permits the conclusion that the hardening is predominantly rate independent or plastic. This result suggests that there are at least two mechanisms of hardening, one rate independent the other rate–dependent. The model should then have two repositories for modeling hardening. This is true for VBO, see [11, 12]. For a simulation of cyclic hardening under proportional and non– proportional loading including the hardening described in Fig. 2.1, see Krempl and Choi [13].

The information contained in Fig. 2.1 says clearly that the hardening is predominantly plastic or rate independent, which implies that the drag stress should not be the repository for hardening. It would change the rate dependence, see Krempl [11, 12]. Rather, the results suggest that the hardening affects the equilibrium stress which is the rate–independent contribution to the stress. Since the equilibrium stress is thought of as a repository for the changing defect structure VBO suggests that the hardening is predominantly coming from

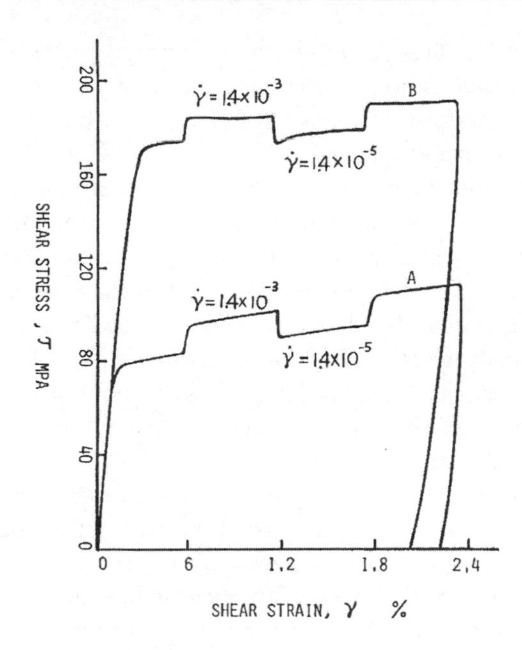

Figure 2.1. Strain rate cycling of type 304 stainless steel at room temperature. Curve A annealed, Curve B after prior torsional cycling. As a first approximation rate dependence is unchanged, hardening is plastic.

a change in dislocation density, from entanglement of dislocations and other mechanisms that increase the obstacle density for dislocation movement. Although we have identified a possible source of hardening, these mechanisms need not be known in detail in a continuum approach. Their total effect must and can be accounted for.

The combined experimental/theoretical approach proposed and used by the author takes advantage of the specimen or volume element to "integrate" the effects of all mechanisms. The real material behavior as it is influenced by the changing microstructure is accounted for and can directly be used in the continuum formulation. The writer has called this method "experiment based" and has used it in the development of VBO for metals and alloys and solid polymers.

3. INFLUENCE OF TEMPERATURE ON THE MECHANICAL BEHAVIOR OF METALS AND ALLOYS

3.1. Materials science point of view

Inelastic deformation of metals and alloys is considered a rate process and is caused by a change of the internal structure. Dislocation movements, generation and annihilation of dislocations and other changes of the defect structure alter the material and influence the response. The metal or alloy changes its internal structure and may become a material with different mechanical properties but with the same chemical composition. As deformation progresses an increased force is needed for plastic flow to develop; the material is hardening and exhibits an increasing yield strength.

At all temperatures time–dependent deformation occurs and is important (Nix, Gibeling et al. [14]). As the temperature increases thermal softening due to diffusion processes counteracts the hardening. One way to account for the influence of the diffusion is to introduce the homologous temperature defined as the ratio of operating temperature to melting temperature measured in °K. The homologous temperature will be designated by θ.

At $\theta < 0.4$ hardening by defect structure changes and dislocation interactions predominates and the effect of diffusion or recovery is small. This temperature range is the normal operating range of most structures.

As the temperature increases, say $0.6 > \theta > 0.4$, diffusion and low temperature mechanisms compete with one or the other prevailing. They can also be balanced to reach a steady state deformation such as found in creep with constant strain rate. The model for this behavior is the so–called Bailey–Orowan law, see LeMay [15], p. 187 and Gittus [16], p. 287. The operating temperature range for power generation and propulsion machinery is in this range. The deformation behavior is of considerable technological importance.

In the range $\theta > 0.6$, the diffusion–controlled recovery is dominant and very little hardening is observed. Forming processes are taking place in this temperature range. The transition from a solid to a fluid state is an interesting subject and is important in casting and welding simulations. These regions portray a continuous transition from low–temperature behavior to the melting of the alloy. No sharp demarcations of various regions are considered.

3.2. Engineering approach

In contrast, current engineering approaches have distinct demarcations. An example of this practice is to be found in design codes. In these codes the creep regime starts at a given temperature that is material specific. In the ASME Code for power plant materials, the creep range starts at 700 °F. Below this temperature creep is not considered to be important and the stress analysis uses elasticity and rate (time)–independent plasticity as material models. In the designated creep range creep rupture is limiting and classical creep theory is employed Bailey [1]; Odqvist and Hult [2]; Rabotnov [3]. A distinction between creep strain and plastic strain is made. In most of the cases plastic strain is negligible and creep theory alone is sufficient and no difficulties arise. But the appearance of plastic strain and the identification of creep and plastic strain cause difficulties with the concept and with the formulation, see Odqvist [17].

For other alloys the temperature at which creep analysis starts is different, but the method is the same. Rate (time)–independent plasticity and combined plasticity and creep are generally used. Plasticity and creep models were developed independently and were combined as soon as finite element programs were ready to accept these nonlinear models and as soon as technology demanded their use. This was the case in the 1970s when the high–temperature, fast breeder reactors were in the design stage and when jet engines needed the inelastic analysis for lifetime assessments.

After a discussion of the classical approaches using creep and plasticity modern state variable theories and their application to monotonic, cyclic loading with or without creep and relaxation are discussed. Both thermal and isothermal applications are included.

4. CLASSICAL PLASTICITY AND CREEP

4.1. The constitutive equations

The theory is developed for small, true strains $\boldsymbol{\epsilon}$ and true (Cauchy) stress $\boldsymbol{\sigma}$[1], bold face symbols denote tensors. It is assumed that the total true strain increment consists of

$$\mathrm{d}\boldsymbol{\epsilon} = \mathrm{d}\boldsymbol{\epsilon}^{\mathrm{el}} + \mathrm{d}\mathbf{e}^{\mathrm{pl}} + \mathrm{d}\mathbf{e}^{\mathrm{cr}} + \mathrm{d}\boldsymbol{\epsilon}^{\mathrm{th}} \tag{4.1}$$

where the superscripts denote the elastic, plastic, creep and thermal contributions to the overall true strain increment. Only isotropic material behavior is considered. The quantity e denotes a deviator obtained from $\mathbf{e} = \boldsymbol{\epsilon} - (1/3)\,\mathrm{tr}\,(\boldsymbol{\epsilon})\,\mathbf{I}$ where tr denotes the trace operation and $\mathbf{I}$ is the identity tensor. The use of deviators implies that the plastic and creep increments are volume preserving. The thermal strain increment is given by $\mathrm{d}\boldsymbol{\epsilon}^{\mathrm{th}} = \alpha\,\mathrm{d}T\,\mathbf{I}$ where α is the coefficient of thermal expansion and where T is the absolute temperature. The elastic strains are calculated from $\mathrm{d}\boldsymbol{\epsilon}^{\mathrm{el}} = \mathrm{d}\left[\frac{1+\nu}{E}\mathrm{s} + \frac{1}{3}\frac{1-2\nu}{E}\,(\mathrm{tr}\,\boldsymbol{\sigma})\,\mathbf{I}\right]$. The differential form insures that the elastic deformation is path–independent for temperature dependent elastic modulus E and Poisson's ratio ν, see Lee and Krempl [18].

A typical isotropic creep law is given by

$$\mathrm{de}^{\mathrm{cr}} = F\left[\bar{\sigma}, \bar{e}^{\mathrm{cr}}\right]\frac{\mathrm{s}}{\bar{\sigma}} \tag{4.2}$$

with F a positive, increasing function of the effective stress $\bar{\sigma} = \sqrt{\frac{3}{2}\,\mathrm{tr}\,(\mathrm{ss})}$ and the effective creep strain $\bar{e}^{\mathrm{cr}} = \sqrt{\frac{2}{3}\,\mathrm{tr}\,(\mathrm{e}^{\mathrm{cr}}\mathrm{e}^{\mathrm{cr}})}$. The stress deviator is s. We require that $F\left[0, \bar{e}^{\mathrm{cr}}\right] = 0$ as well as $F\left[\bar{\sigma}, 0\right] \neq 0$. The creep strain is zero at the start of any creep test. It is seen that the creep law is formulated independently of plasticity.

It has no influence on creep except through the initial conditions for the strain due to the incremental form of (4.1).

Plasticity in turn is independent of creep. A representative plasticity law is

$$\mathrm{de}^{\mathrm{pl}} = \mathrm{d}\lambda\frac{\partial f}{\partial \boldsymbol{\sigma}} \tag{4.3}$$

where f is the yield function. It depends on $\bar{\sigma}$ and can depend on accumulated inelastic strain or inelastic work and on the kinematic stress.

[1]It might be argued that a distinction between true and engineering quantities is not needed. However, a contemplated future extension to finite deformation suggests the introduction of true quantities. In the uniaxial case the theory can be applied for finite deformation without any problems and then a distinction between true and engineering quantities is essential.

4.2. Some qualitative predictions of creep and plasticity theories compared with experiments

There are numerous creep data that show the modeling capabilities of the above creep law for monotonic loading. Using a power law for $F[\,]$ and only a dependence on the effective stress secondary or Norton's creep can be reproduced. This Norton model has become the workhorse for creep analysis and is widely used.

However, when transient conditions and cyclic loadings are considered then some deficiencies of the creep equation become apparent. Examples are the following observations.

The creep equation (4.2) predicts that the creep increment is the same whenever the stress and the creep strain are the same. Consider two specimens subjected to the same stress. At the beginning of the creep test creep strain is zero so that the first creep strain increment is determined by the stress alone and is the same for the two specimens. In the second increment the stress and the creep strain are identical which results in an identical creep strain increment in both cases. Equation (4.2) predicts that whenever the creep stress is the same the creep curve is the same. Some experiments show different results, see Fig. 4.1. Four creep test results at a stress of 140 MPa at 650 °C with a stainless steel are shown. The first test designated by + shows a "normal" creep curve, i.e. the stress is reached on loading. For the other three the stress is reached after loading to a plastic strain and subsequent unloading. It is seen that the creep strain is reduced and that the reduction in creep strain increases with an increase of the prior plastic strain or prior maximum stress. These results are usually attributed to creep–plasticity interaction. The conventional explanation is that prior plasticity has hardened the alloy and as a consequence the creep rate is reduced when the creep test is performed during unloading. There are other indications of such behavior, see Krempl [11, 20]. These results show that the creep

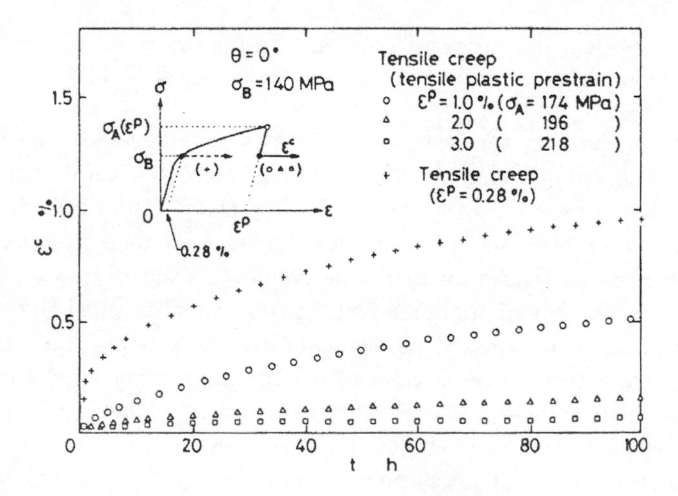

Figure 4.1. Creep rate at the same stress level is larger on loading than on unloading, from Ohashi, Kawai et al. [19].

deformation is path dependent, which is at variance with the prediction of (4.2).

A second example depicts the inadequacy of (4.2) in transient conditions is given in Fig. 4.2 from Kujawski, Kallianpur et al. [21]. Two room temperature experiments on Type 304 stainless steel are shown. The dashed curve represents a load controlled tensile test with an engineering stress rate of 0.7 MPa/s. The second specimen underwent multi–step 200 s creep tests. At the end of each creep test loading is resumed with the stress rate employed in the first specimen. The accumulated creep strain is a measure of the average creep rate. It is seen that the creep segments labeled b, d, and f do not start from a stress level on the dashed curve but creep segments a, c, e, and g do. Although the stress level increases in going from a to b, c to d and from e to f the average creep rate diminishes. The creep rate, therefore, need not increase with an increase in stress level. This "paradoxical" behavior indicates that (4.2) is not fully adequate for describing the transient behavior. The behavior is easily explained with an overstress dependence of the inelastic strain rate, see Krempl [20].

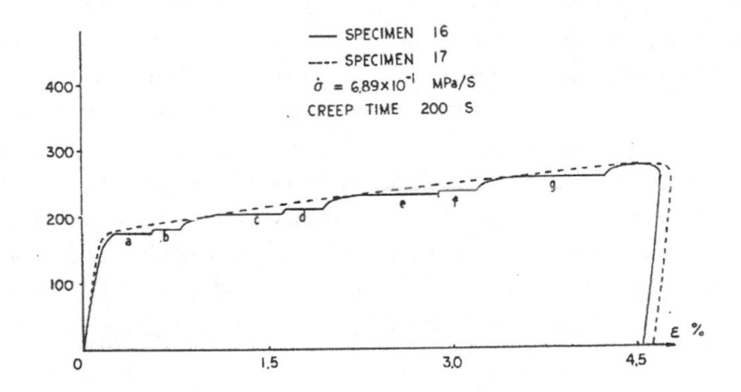

Figure 4.2. Creep rate does not necessarily increase with an increase in stress level. Type 304 Stainless Steel at room temperature, from Kujawski, Kallianpur et al. [21].

When the design tools for the fast breeder reactor were developed by Pugh, Liu et al. 1972 in the 1970s Oak Ridge National Laboratory performed several critical tests. In the first a specimen was exposed to a 1200 °F, 2500 h creep test in which it accumulated about 0.8% strain in creep. The specimen was cooled to ambient temperature and was subsequently subjected to tensile loading. The results are shown in Fig. 4.3. It is clearly seen that there is an influence of prior creep on the subsequent tensile behavior. The authors anticipated that the hardening could be accounted for by starting the tensile test at the accumulated creep strain. If the predictions of plasticity and creep theory would have been correct the tensile curve of the creep–tested specimen would have reached the stress level at A in Fig. 4.3. Instead the curve exceeds that stress level.

There is also an influence of prior room temperature cycling on the subsequent creep behavior see Fig. 4.4 from Pugh, Liu et al. [22]. The creep deformation at 12.5 ksi of the virgin *First Specimen* is much larger than that of the *Third Specimen*, which was cycled at room temperature. Clearly, prior plastic cycling has hardened the stainless steel so that the

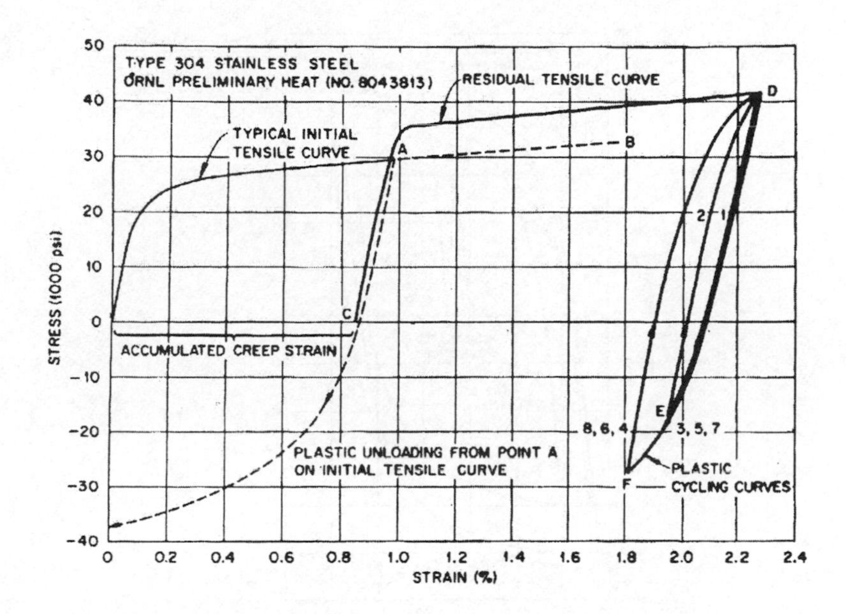

Figure 4.3. Prior creep deformation influences the stress behavior at room temperature, from Pugh, Liu et al. [22].

creep strain accumulation is reduced. Cyclic behavior at room temperature is also affected by creep. This is shown in Fig. 4.5 from Pugh, Liu et al. [22]. The stress amplitude for post–creep cyclic loading is much higher than for the virgin material.

These examples may suffice to demonstrate that there is plasticity–creep interaction at

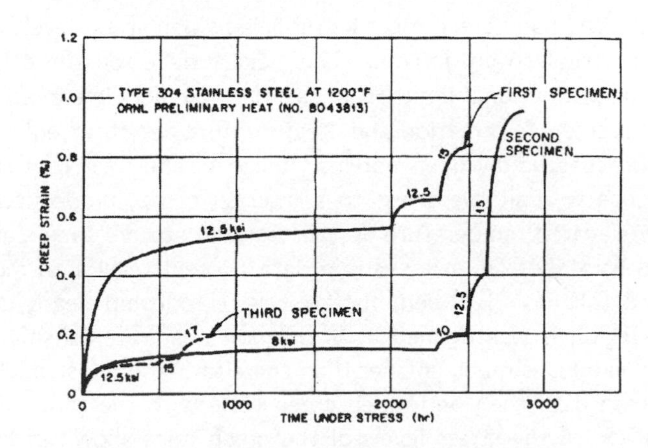

Figure 4.4. Prior inelastic deformation at room temperature influences subsequent creep behavior, from Pugh, Liu [22].

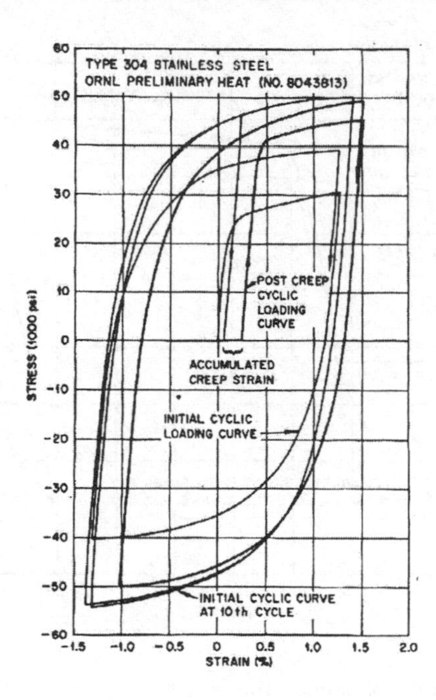

Figure 4.5. The influence of prior creep on the room temperature cyclic behavior from Pugh, Liu et al. [22].

small strains that are appropriate for design. The effects may be pronounced for the stainless steels but they can also be found in other steels, see Krempl [9]. Independent creep and plasticity formulations are not appropriate.

A further problem with separate creep and plasticity formulations is their contradictory requirements. On the one hand creep must be modeled when stress levels are in the quasi–linear portion of the stress–strain diagram. The yield surface, on the other hand, has to extend at least to the proportional limit. Plasticity then predicts linear, elastic behavior for stress levels at which creep deformation and creep rupture are observed!

This contradiction has hardly been noticed. It can be inferred from the fact that the normal design practice to limit the stress to a fraction of the proportional or yield limit does not suffice for elevated temperature design. Actual data are sparse and usually creep data are reported separately from stress–strain data. Recent results on modified 9Cr 1Mo steel by Yaguchi and Takahashi [23] demonstrate this relationship clearly, see Fig. 4.6. The initial quasi–linear region is well defined as is the nonlinear rate sensitivity in the region where the tangent modulus is much smaller than the elastic modulus which is almost zero. Yaguchi and Takahashi [23] then performed creep tests with the same alloy as shown in Fig. 4.7. A comparison of the stress levels of the creep tests shown in Fig. 4.7 with the stress–strain curves in Fig. 4.6 demonstrates that creep occurs in the quasi–linear region of the stress–strain diagrams. For example at a stress level of 240 MPa the creep rupture life is less than 500 h, see Fig. 4.7. This stress level is in the quasi–linear region as shown in Fig. 4.6. This relation is not an isolated occurrence, it is rather a common phenomenon.

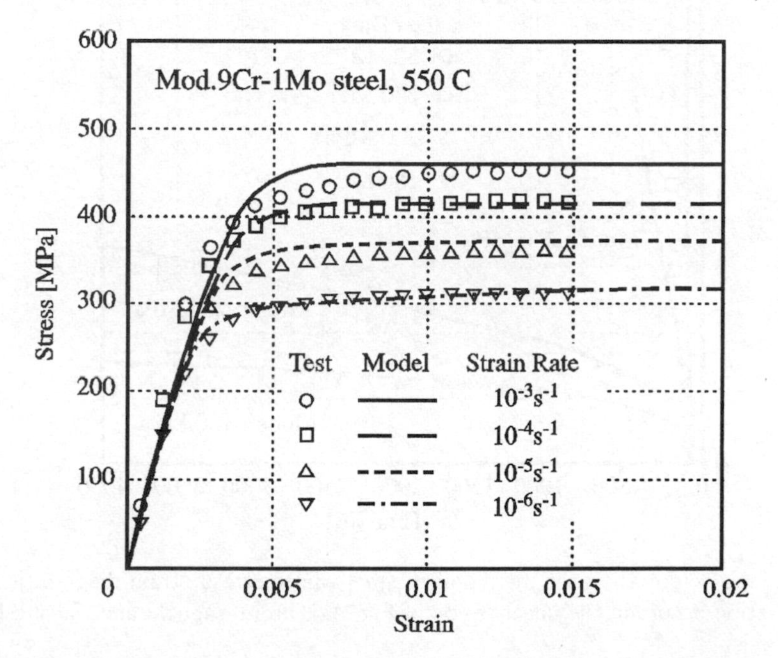

Figure 4.6. Stress strain behavior of a modified 9Cr–1Mo steel at 550 C showing rate dependence, from Yaguchi and Takahashi [23].

An old literature review and interpretive report states additional reasons for improved constitutive equations, see Krempl [9]. Appendix C shows an excerpt from this report. This report, the author's industrial experience and the general agreement for the need of improved constitutive equations for high–temperature environment have led to the "unified" VBO model to be introduced below.

5. THE "UNIFIED" STATE VARIABLE THEORIES

5.1. Need for new approaches

These and other observations led to the conclusion that new approaches were needed that would not concentrate on a single mode of deformation, say creep or monotonic or cyclic deformation. Rather the "total time–dependent deformation behavior" should be taken as a basis. The total time–dependent deformation picture consists of monotonic loading at different rates, cyclic loading involving hysteresis as well as creep and relaxation at any point during monotonic and cyclic loading. A good model should give, as a minimum, an acceptable qualitative response to these loadings.

The "unified" state variable models developed consider all inelastic deformation rate–dependent. This approach is in agreement with the notions of materials science where inelas-

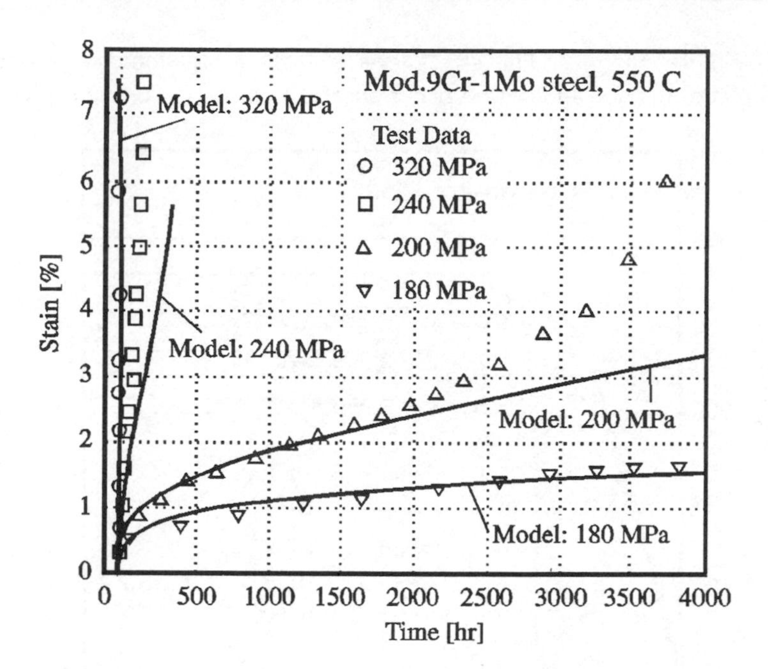

Figure 4.7. Creep deformation for the modified steel whose stress–strain diagram is Figure 4.6. Creep deformation occurs in the linear region of Fig. 4.6. From Yaguchi and Takahashi [23].

tic deformation is considered a rate–dependent process. Also the results shown in Fig. 4.2 demonstrate that creep and plastic strain are fully equivalent with regard to hardening. Identical results were obtained by Ikegami and Niitsu [24]. The correspondence between creep and plastic strains is not as good at high temperature than at room temperature. A possible explanation might be that some aging processes are taking place at elevated temperature.

The separation of creep and plastic deformation is more historic than the expression of a physical reality. Inelastic deformation is caused by changes in the defect structure such as dislocation movement and interaction. These mechanisms can act in every motion. No dislocation is specifically reserved to contribute to creep deformation.

We are going to use the term "inelastic strain" or "inelastic strain rate" to mean any strain or strain rate except the thermal strain, that is different from the linear elastic strain. Only the test boundary conditions for monotonic or cyclic loading at different rates (stress or strain rate is constant), creep (constant stress or zero stress rate) and relaxation (constant strain or zero strain rate) limit how the inelastic deformations can evolve. No distinct creep or relaxation strains are recognized.

The deformations are always inelastic and are not classified as creep or relaxation deformations. The zero strain rate boundary condition hides the fact that inelastic deformation develops during such a test. Unloading to zero stress after a relaxation test can demonstrate that an inelastic strain developed during relaxation.

The traditional formulation of creep laws only has information on the material behavior under constant stress. This may not be sufficient for applications. Even if the loading on the boundary of engineering structures is kept constant in time, the constant stress condition generally does not exist inside a structure, see Hult [25] for sample calculations.

5.2. General remarks

"Unified" models including VBO are basically continuum models and assume a representative volume element. At this level the contributions of the many possible microstructural mechanisms are only recognized through their aggregate effects. In experiments, the specimen representing a macroscopically homogeneous state of deformation (tensile bar or thin–walled tube) is the representative volume element. If it is of appropriate size such that it contains all the microstructural elements in sufficient numbers then it serves as an integrator of all mechanisms. The individual contributions are recognized in "smeared out" form.

In VBO and other state variable theories the so–called state variables and their growth laws accomplish the modeling of the changing microstructure. There is no one–to–one correspondence between the state variables and certain micromechanisms. However, the correspondence is rather diffuse. In VBO the state variables are motivated by experimental results and some qualitative considerations, Krempl [20]. Generally speaking, state variables for modeling work–hardening in monotonic loading, for cyclic hardening (softening) and for the Bauschinger effect are needed. These are all phenomena that develop as a consequence of inelastic deformation that causes a change in the internal structure.

Inelastic deformation in metals and alloys is primarily caused by dislocation motion and by other changes in the defect structure. In most of the cases the dislocation density increases with inelastic deformation and the increasing density impedes further movement of dislocations. As a consequence work hardening is observed macroscopically. In some cases, when cold–worked metals are subjected to cyclic inelastic loading for example, cyclic softening can occur, which indicative of an easing in the passage of dislocations and a decrease in dislocation density.

At low homologous temperatures diffusion is negligible and the defect structure acquired during inelastic deformation is stable in the absence of further inelastic mechanical loading. As the temperature increases diffusion processes become important. Defects can now change by "thermal action". Generally, diffusion tends to counteract the hardening effects of inelastic deformation. The model for such a behavior is the Bailey–Orowan model, see Gittus [16] Eqs. 6.8 and 8.120. Hardening due to inelastic deformation and softening due to diffusion can occur while external mechanical loads are applied. In the absence of external loads hardening essentially ceases and the effects of diffusion continue until equilibrium is attained[2]. The defect structure is observed to change with time, "static" recovery is said to occur. Subsequent loading shows that the hold periods with zero external loads can lead

[2]Although there are no external mechanical loads acting on the material, residual internal stresses with zero resultant exist and they influence the change in the microstructure. These internal stresses are self equilibrating and do not enter into a continuum formulation.

to a softening of the response[3].

Hardening due to inelastic deformation and softening due to diffusion (static recovery) act simultaneously. Depending on the loading and on the temperature, either hardening or softening may prevail. At low homologous temperatures hardening is dominant. Hardening and softening can also be in equilibrium as happens when the stress is constant and secondary creep can be observed. As the temperature increases diffusion effects become increasingly important and hardening ceases when the melting temperature is approached.

Diffusion may also cause chemical reactions such as carbide formation in the absence of any applied stress. This phenomenon is called "aging." The influence of "aging" can be ascertained by performing the same test with several specimens which were "aged" different times at the laboratory, for a formal definition, see Krempl [9, 26].

The viscoplasticity theory based on overstress (VBO) developed by the author and his students is a state variable theory without a yield surface and without loading and unloading conditions. Other unified approaches by Chaboche, Estrin and Miller are described in Krausz and Krausz [27] and in Gittus and Zarka [28]. For the model by Robinson, see [29]. A description of the Bodner model may be found in [30].

Only homogeneous states of stress will be considered from the view point of materials testing with servo–controlled testing machines and strain measurement on the gage section of a specimen (usually round bars or thin–walled tubes). Monotonic and cyclic loadings under displacement (strain) or load (stress) control are considered. Creep (relaxation) is said to occur when the stress (strain) rate is zero. When stress (strain) is imposed the strain (stress) is the "answer" of the material. It is believed that consistent and methodical selections of the imposed histories together with principles of continuum mechanics permit the development of a constitutive law for the inelastic behavior of metals and alloys.

The basic model of VBO was initially chosen because a model with stress, strain and their rates is the simplest model that can show creep and relaxation. In the linear case the standard linear solid is the simplest model that can show these properties. The properties of the standard linear solid and their relation to VBO are discussed in Appendix A.

6. SMALL STRAIN, ISOTROPIC VBO WITHOUT STATIC RECOVERY TERMS (LOW HOMOLOGOUS TEMPERATURE)

Here the three dimensional version of the viscoplasticity theory based on overstress (VBO) developed by the author and his students is introduced. First the low temperature version is presented which models essentially hardening[4]. Then the static recovery term will be introduced to model diffusion originated softening.

[3] It is assumed that the material behavior is "normal" and that effects like strain aging and age hardening do not occur.

[4] An exception is cyclic softening which can be found in cold worked materials at low homologous temperature and which can be modeled using the low temperature version of VBO.

6.1. The flow law

On the assumption of volume preserving inelastic deformation the deviatoric flow law can be written as

$$\dot{\mathbf{e}} = \dot{\mathbf{e}}^{\mathrm{el}} + \dot{\mathbf{e}}^{\mathrm{in}} = \frac{\mathrm{d}}{\mathrm{d}t}\left(\frac{1+\nu}{E}\mathbf{s}\right) + \frac{3}{2}F[\,]\frac{\mathbf{s}-\mathbf{g}}{\Gamma} \tag{6.1}$$

or

$$\dot{\mathbf{e}} = \dot{\mathbf{e}}^{\mathrm{el}} + \dot{\mathbf{e}}^{\mathrm{in}} = \frac{\mathrm{d}}{\mathrm{d}t}\left(\frac{1+\nu}{E}\mathbf{s}\right) + \frac{3}{2}\frac{\Gamma}{E k[\,]}\frac{\mathbf{s}-\mathbf{g}}{\Gamma} \tag{6.2}$$

where $\Gamma = \sqrt{\frac{3}{2}\operatorname{tr}\left[(\mathbf{s}-\mathbf{g})(\mathbf{s}-\mathbf{g})\right]}$ is the overstress invariant with $\mathbf{s}$ the true stress (Cauchy) deviator and $\mathbf{g}$ the equilibrium stress deviator. $\dot{\phi} = \sqrt{\frac{2}{3}\operatorname{tr}(\dot{\mathbf{e}}^{\mathrm{in}}\dot{\mathbf{e}}^{\mathrm{in}})} = F[\,] = \frac{\Gamma}{Ek[\,]}$ is the effective inelastic strain rate and $F[\,]$ is the positive, increasing flow function with the dimension of 1/time and $F[0] = 0$. A frequently used flow function is the power law. The positive, decreasing viscosity function, with $k[0] \neq 0$ and the dimension of time can be considered a variable relaxation time. Modified power laws and single and double exponential functions have been used here. The function $F[\,]$ or $k[\,]$ is the repository for modeling nonlinear rate sensitivity and nonlinear creep behavior. The theory does not impose further conditions on these functions, which must be determined from experimental results. The functions have constants, which can depend on temperature. The elastic, temperature dependent constants are E and ν. The [ ] following a symbol denotes "function of". The deviatoric true strain rate is $\dot{\mathbf{e}}$ and the superscripts ^{el} and ^{in} denote elastic and inelastic parts, respectively. The symbol t designates time.

The viscosity and the flow functions are usually functions of Γ/D where D stands for the positive drag stress with the dimension of stress. It can be considered a state variable with a growth law. Unless otherwise mentioned D will be a constant in this report.

The deviatoric formulation has to be augmented by the volumetric elastic relation

$$\operatorname{tr}\dot{\boldsymbol{\epsilon}} = \frac{\mathrm{d}}{\mathrm{d}t}\left(\frac{1-2\nu}{E}\operatorname{tr}\boldsymbol{\sigma}\right) + 3\alpha\dot{T} \tag{6.3}$$

where α is the coefficient of thermal expansion, T is the absolute temperature and a superposed dot designates differentiation with respect to time. $\boldsymbol{\epsilon}$ and $\boldsymbol{\sigma}$ are the true strain and true stress tensor, respectively. The time derivatives in (6.1) through (6.3) insure the path independence of the elastic deformation for variable temperature and temperature dependent material properties, see Lee and Krempl [18]. For constant temperature the expressions simplify to the stress rates.

It is seen from (6.1) and (6.2) that in the case of rest, when all time rates are zero, $\mathbf{s}-\mathbf{g} = \mathbf{0}$. This condition can be satisfied by setting $\mathbf{s} = \mathbf{g} = \mathbf{0}$. If the first condition holds, the stress $\mathbf{g}$ is the stress that can be sustained at rest. It is therefore called the equilibrium stress. For a discussion of the uses of the terms equilibrium stress and overstress, see Krempl [11]. When the applied stress is zero at rest the equilibrium stress must also be zero. On loading the difference of $\mathbf{s}-\mathbf{g}$ determines the sign of the inelastic strain rate. For loading in tension the sign must be positive. The equilibrium stress increases but has to be less than the applied stress. When not at rest, $\mathbf{g}$ represents the stress that must be overcome to generate inelastic deformation. It can be thought of a measure of the defect structure of a metal. The growth law of the equilibrium stress needs to be hysteretic so that realistic loading/unloading behaviors can be modeled. The evolution has to be path–dependent to represent plastic deformation properly.

The back stress and effective stress are used in the literature and their respective meanings are close to the meanings of the equilibrium stress and of the overstress, respectively. Subtle differences must be kept in mind. The back stress in plasticity is identified with kinematic hardening and is responsible for modeling the Bauschinger effect. The equilibrium stress, however, is not the repository for modeling of the Bauschinger effect. The kinematic stress of VBO has that function, see below.

The equilibrium stress cannot be the internal stress, which by equilibrium considerations must be equal to the applied stress. The designation internal stress for the equilibrium stress is incorrect and is unacceptable. There is no question that internal stresses exist in a specimen for which a macroscopically uniform stress distribution is assumed. The internal stress distribution is non–uniform. However, when integrated over the specimen cross section it must be equal to the applied stress.

There could be no overstress if the equilibrium stress was equal to the average internal stress. The equilibrium stress is a state variable, and like all true state variables it can neither be measured nor controlled. But the presence of such a state variable can be inferred from transient tests and a comparison of the observations with the predictions of a model, see [20, 12]. The usefulness of the equilibrium stress can be demonstrated by such tests. These tests have established the advantage of the equilibrium stress in modeling transient phenomena. Despite the usefulness of the equilibrium stress in explaining unusual phenomena, it cannot be said that the equilibrium stress is the only quantity that has such capabilities. But no other of equal utility has been found as of now.

6.2. Growth laws for the state variables

The growth law for the equilibrium stress consists of elastic and inelastic hardening terms, a dynamic recovery term that depends on the difference between the equilibrium stress and the kinematic stress and a term that involves the kinematic stress. It is needed so that the positive, negative or zero slope of the stress–strain diagram can be modeled at the maximum strain of interest. The growth law for the equilibrium stress is

$$\dot{\mathbf{g}} = \frac{\partial}{\partial T}\left(\frac{\psi[\Gamma]}{E}\right)\dot{T}\mathbf{s} + \frac{\psi[\Gamma]}{E}\left(\dot{\mathbf{s}} + \frac{\mathbf{s}-\mathbf{g}}{k[\Gamma]} - \frac{\Gamma}{k[\Gamma]}\frac{(\mathbf{g}-\mathbf{f})}{A}\right) + \left(1 - \frac{\psi[\Gamma]}{E}\right)\dot{\mathbf{f}} \tag{6.4}$$

The first term on the right hand side is only for variable temperature and insures the path independence of $\mathbf{g}$ during initial deformation. The positive, decreasing shape function ψ is bounded by $1 > \left(\frac{\psi[\Gamma]}{E}\right) > \frac{E_{\mathrm{t}}}{E}$. It controls the transition from initial quasi–elastic behavior to fully established inelastic flow. The first two terms in the parenthesis following $\frac{\psi[\Gamma]}{E}$ are the elastic and inelastic hardening terms, the term with the negative sign is the dynamic recovery term. This growth law can also be written in terms of the flow function F to yield

$$\dot{\mathbf{g}} = \frac{\partial}{\partial T}\left(\frac{\psi[\Gamma]}{E}\right)\dot{T}\mathbf{s} + \frac{\psi[\Gamma]}{E}\left[\left(\dot{\mathbf{s}} + \frac{\mathbf{s}-\mathbf{g}}{\Gamma} - \frac{(\mathbf{g}-\mathbf{f})}{A}\right)EF[\,]\right] + \left(1 - \frac{\psi[\Gamma]}{E}\right)\dot{\mathbf{f}} \quad (6.5)$$

The last term multiplied by the kinematic stress deviator $\dot{\mathbf{f}}$ is needed so that an appropriate long–term asymptotic solution can be developed, see below. The growth laws for the kinematic stress is given in the Prager hardening form by

$$\dot{\mathbf{f}} = \frac{2}{3}\hat{E}_{\mathrm{t}}\dot{\mathbf{e}}^{\mathrm{in}} = \frac{\hat{E}_{\mathrm{t}}}{E}\frac{\mathbf{s}-\mathbf{g}}{k[\Gamma]} \quad (6.6)$$

$\hat{E}_{\mathrm{t}}$ is the tangent modulus at the maximum inelastic strain of interest. E_{t} is the same quantity referred to the total strain. The two quantities are related by $\hat{E}_{\mathrm{t}} = E_{\mathrm{t}}/\left(1 - E_{\mathrm{t}}/E\right)$ and differ very little for most engineering materials. The isotropic stress A is given the simple growth law

$$\dot{A} = A_{\mathrm{c}}\left(A_{\mathrm{f}} - A\right)\dot{\phi} \quad (6.7)$$

where A_{c} and A_{f} are constants with no dimension and the dimension of stress, respectively. The first controls the speed with which the final value A_{f} is reached. A more complicated growth law for the isotropic stress is needed when complex cyclic behavior is to be modeled, see Ruggles and Krempl [31] and Choi and Krempl [32].

These equations constitute the VBO model for small strain and metals and alloys (the inelastic deformations are volume preserving). They are coupled, nonlinear stiff differential equations and for their solution initial conditions must be known.

The growth laws for the state variables are homogeneous of degree one in the rates and represent rate–independent behavior. This property is not obvious for (6.4). It is written in the stress form. Using the flow law (6.2), (6.4) can be rewritten as

$$\dot{\mathbf{g}} = \frac{\partial}{\partial T}\left(\frac{\psi(\Gamma)}{E}\right)\dot{T}\mathbf{s} + \frac{\psi(\Gamma)}{E}\left(\frac{\dot{\mathbf{e}}^{\mathrm{el}}}{1+\nu} + \frac{2}{3}\dot{\mathbf{e}}^{\mathrm{in}} - \dot{\phi}\frac{(\mathbf{g}-\mathbf{f})}{A}\right)E + \left(1 - \frac{\psi(\Gamma)}{E}\right)\frac{2}{3}\hat{E}_{\mathrm{t}}\dot{\mathbf{e}}^{\mathrm{in}} \quad (6.8)$$

where the homogeneity of degree one in the rates is apparent. It will be demonstrated below that the dependence of the shape function on the overstress invariant Γ introduces a rate dependence during the transition from the initial quasi elastic region to fully established plastic flow. The elastic and inelastic strain rate terms could also be combined to yield the deviatoric strain rate $\dot{\mathbf{e}}$.

6.3. Properties in very slow and very fast loading

The responses to very slow and very fast loading at constant temperature are of interest. To obtain the slow response the uniaxial version of (6.1) is transformed to an integral equation with $\frac{3}{2}s_{11} \equiv \sigma$ and $\frac{3}{2}g_{11} \equiv G$, see (8.5). For loading with a constant strain rate

ρ it is advantageous to change the integration variable from time to strain to yield

$$(\sigma - G) = (\sigma_0 - G_0)\exp\left(-\frac{1}{\rho}\int_0^{\varepsilon}\frac{1}{k[\,]}\mathrm{d}s\right) + \int_0^{\varepsilon}\left(E - \frac{\mathrm{d}G}{\mathrm{d}\varepsilon}\right)\exp\left(-\frac{1}{\rho}\int_{\zeta}^{\varepsilon}\frac{1}{k[\,]}\mathrm{d}s\right)\mathrm{d}\zeta \tag{6.9}$$

If the strain rate ρ is reduced to smaller and smaller values the right hand side tends to zero and we have $\lim_{\rho\to 0}(\sigma - G) = 0$ so that the stress is equal to the equilibrium stress for very slow loading. Since the growth law for g is essentially rate–independent a nonzero equilibrium stress exists and the growth law represents a solid. This property and the fact that $\sigma = G$ at rest suggests the name "equilibrium stress".

To obtain the response for infinitely fast loading the uniaxial version of the flow law is used again. After application of the chain rule we get

$$\left(1 - \frac{\mathrm{d}\sigma/\mathrm{d}\varepsilon}{E}\right) = \frac{1}{\rho}\left(\frac{\sigma - G}{Ek[\,]}\right) \tag{6.10}$$

Taking the limit for very fast strain rates we obtain $\lim_{\rho\to\infty}\left(1 - \frac{\mathrm{d}\sigma/\mathrm{d}\varepsilon}{E}\right) = 0$. The fast response is therefore elastic. The slow and the fast responses bracket the responses for all other strain rates. Fig. 6.1 gives a sketch of these results.

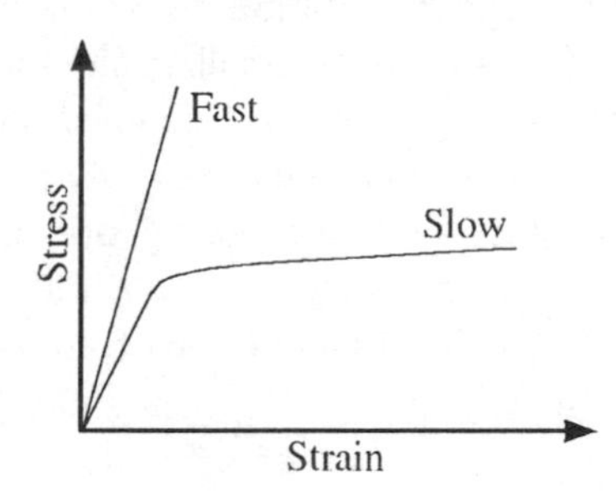

Figure 6.1: Very slow and very fast responses of VBO.

6.4. Fully established inelastic flow and asymptotic solutions of VBO

Reference is made to Hart and Solomon [33] who prefer to perform their relaxation tests when inelastic flow is fully established. This region is also identified as the region where the flow stress is reached, or where the tangent modulus is small compared to the elastic modulus. In demonstrating that VBO has a long–term solution that applies to the region of established flow stress isothermal conditions are considered. At all times the identity

$$\mathbf{s} = (\mathbf{s} - \mathbf{g}) + (\mathbf{g} - \mathbf{f}) + \mathbf{f} \tag{6.11}$$

holds. It shows that the stress is composed of the overstress $(\mathbf{s} - \mathbf{g})$, the difference between the equilibrium stress and the kinematic stress $(\mathbf{g} - \mathbf{f})$ and the contribution of the kinematic stress. Its slope can be positive (strain hardening), zero or negative (strain softening). Fig. 6.2 shows a sketch of (6.11) in the deviatoric stress space.

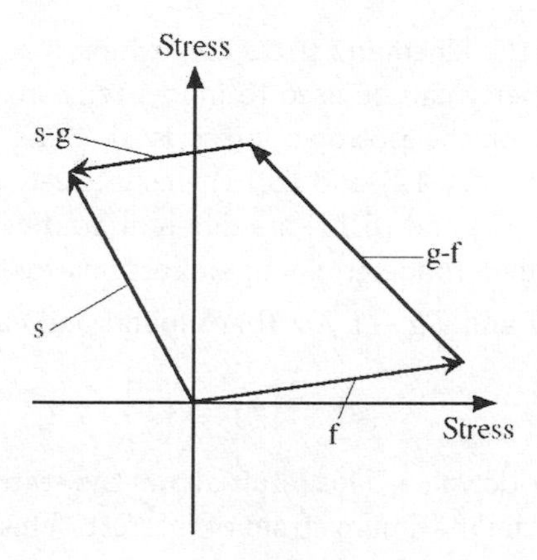

Figure 6.2: Normal condition. The stress differences can have different directions.

This method of presentation suggests investigating the stress differences. It will be shown that the two stress differences have a long–term, stationary solution which the stress need not have.

The flow law can, after some calculations be written as the integral equation

$$\begin{aligned}(\mathbf{s}-\mathbf{g}) = & \ (\mathbf{s}_0-\mathbf{g}_0)\exp\left(-\int_{t_0}^{t}\frac{3}{2(1+\nu)}\frac{1}{k[\,]}\mathrm{d}\zeta\right) \\ & +\int_{t_0}^{t}\left(\frac{E}{1+\nu}\dot{\mathbf{e}}-\dot{\mathbf{g}}\right)\exp\left(-\int_{\zeta}^{t}\frac{3}{2(1+\nu)}\frac{1}{k[\,]}\mathrm{d}s\right)\mathrm{d}\zeta\end{aligned} \tag{6.12}$$

Similar manipulations of the growth law of the equilibrium stress, (6.4), for isothermal conditions result in

$$\begin{aligned}(\mathbf{g}-\mathbf{f}) = & \ (\mathbf{g}_0-\mathbf{f}_0)\exp\left(-\int_{t_0}^{t}\psi\frac{\dot{\phi}}{A}\mathrm{d}\zeta\right) \\ & +\int_{t_0}^{t}\psi\left(\dot{\mathbf{s}}-\frac{\mathbf{s}-\mathbf{g}}{k[\,]}-\dot{\mathbf{f}}\right)\exp\left(-\int_{t_0}^{t}\psi\frac{\dot{\phi}}{A}\mathrm{d}s\right)\mathrm{d}\zeta\end{aligned} \tag{6.13}$$

The subscript zero denotes initial conditions. For the isotropic stress we have

$$A = A_{\mathrm{f}} + (A_0 - A_{\mathrm{f}})\exp\left(-\int_{t_0}^{t}A_{\mathrm{c}}\dot{\phi}\,\mathrm{d}\zeta\right) \tag{6.14}$$

The integral arguments of the exponential functions are positive. If time grows without bounds the initial conditions of the overstress and of $(\mathbf{g}-\mathbf{f})$ vanish since the exponential function reduces to zero. The final value A_{f} of the isotropic stress is also independent of the initial condition. Only the kinematic stress can exhibit a non–vanishing influence on the initial condition. This property can be used to model tension/compression asymmetry.

The asymptotic value of the isotropic stress is A_{f}. The long–term solutions for the overstress and for $(\mathbf{g}-\mathbf{f})$ in (6.12) and (6.13), respectively, are found following Ho and Krempl [34]. Equations (6.12) and (6.13) are differentiated with respect to time. For large times these equations result in indeterminate expressions for the rates which must be resolved. Then $\{\dot{\mathbf{s}}-\dot{\mathbf{g}}\}=\mathbf{0}$ and $\left\{\dot{\mathbf{g}}-\dot{\mathbf{f}}\right\}=\mathbf{0}$ are found and can be written as

$$\{\dot{\mathbf{s}}\}=\{\dot{\mathbf{g}}\}=\left\{\dot{\mathbf{f}}\right\} \tag{6.15}$$

where { } denotes asymptotic value. This result allows the statement that $\mathbf{f}$ sets the slope of the stress–strain curve at the maximum strain of interest. This result shows that ultimately the stress, equilibrium stress and kinematic stress will have the same rate and the distance between any two of the three curves will be constant. The results shown in (6.15) are used in evaluating the asymptotic solution of (6.12)

$$\frac{\{\mathbf{s}-\mathbf{g}\}}{k\left[\{\}\right]}=\frac{2}{3}\left[E\dot{\mathbf{e}}-\{\dot{\mathbf{g}}\}\left(1+\nu\right)\right]=\frac{2}{3}E\left\{\dot{\mathbf{e}}^{\mathrm{in}}\right\} \tag{6.16}$$

It is seen that the overstress is rate dependent and that this dependence is nonlinear and is determined by the viscosity function $k[\ ]$. This equation can be expressed in terms of the flow function

$$\left\{\frac{\mathbf{s}-\mathbf{g}}{\Gamma}\right\}F\left[\{\}\right]=\frac{2}{3}\left\{\dot{\mathbf{e}}^{\mathrm{in}}\right\} \tag{6.17}$$

Again the nonlinear rate sensitivity of the overstress is evident.

The arguments of the viscosity and the flow functions have not been explicitly written to allow for either pure overstress dependence, for a dependence on the drag stress or for a dependence on both. For the asymptotic solution to hold it is necessary for the drag stress to have reached the final value.

The constant stationary overstress is determined by the inelastic strain rate and the material properties expressed by the viscosity function $k[\ ]$ or the flow function $F[\ ]$. These functions determine the nonlinear rate sensitivity.

The long–term solution of (6.13) is

$$\{\mathbf{g}-\mathbf{f}\}=A_{\mathrm{f}}\frac{\{\mathbf{s}-\mathbf{g}\}}{\Gamma} \tag{6.18}$$

The rate–independent contribution to the stress is given by this expression since A_{f} and the expression are rate independent. The directions of the rate–dependent and of the rate–independent contributions to the stress are the same. Both point in the direction of the overstress. Their magnitudes can be easily determined to be $\Pi=A_{\mathrm{f}}$ with $\Pi=\sqrt{\frac{3}{2}\operatorname{tr}\left[\left(\{\mathbf{g}-\mathbf{f}\}\right)\left(\{\mathbf{g}-\mathbf{f}\}\right)\right]}$ from (6.18) and $\left\{\dot{\phi}\right\}=\mathbf{F}\left[\{\ \}\right]=\frac{\{\Gamma\}}{\mathbf{Ek}\left[\{\ \}\right]}$ from (6.17).

When the long–term solution holds the kinematic stress locates the center of a sphere in the deviatoric stress space whose radius consists of the isotropic stress and the overstress magnitudes. The direction of the inelastic strain rate is radial, perpendicular to the sphere. The situation is depicted in Fig. 6.3. In other than the asymptotic solution the directions of the overstress and of the isotropic stress are different, see Fig. 6.2.

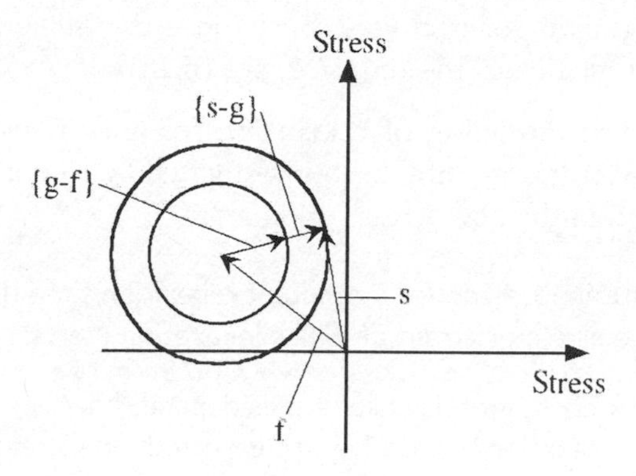

Figure 6.3: Asymptotic solution in deviatoric stress space.

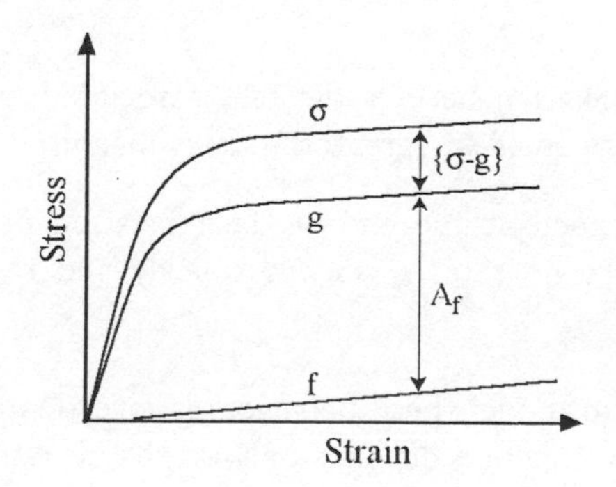

Figure 6.4: The asymptotic solution in one dimension.

The counterpart of Fig. 6.3 in one dimension is given in Fig. 6.4. It is seen that the long–term asymptotic solution of the stress consists of the work–hardening, rate–independent contribution $\mathbf{f}$, the rate–independent isotropic stress A_f and the rate–dependent overstress $\{\sigma - g\}$. VBO clearly separates these contributions when the long–term asymptotic solution holds. It should be emphasized that the stress may or may not reach an asymptotic solution when the stress differences $\{\mathbf{s} - \mathbf{g}\}$ and $\{\mathbf{g} - \mathbf{f}\}$ do. The stress can change according to the growth law for the kinematic stress, it can increase, decrease or stay constant. The stress differences can become stationary.

The asymptotic state of the VBO model is therefore a privileged state in the sense that the rate–dependent and the rate–independent contributions to the deviatoric stress have the same direction and are constant in magnitude.

For further tests it is assumed that the asymptotic solution is reached and that the deformation takes place in the flow stress or the transient free region. It is seen from (6.16) that the asymptotic solution applies strictly under circumstances when the rate of the equilibrium stress is constant. Only then can the right hand side of (6.16) be constant. However, small changes in the tangent modulus found in the experiment do not affect the applicability of the long–term solution very much since the influence of the equilibrium stress rate term is small relative to the term $E\dot{e}$, see (6.16).

6.5. Application to the modeling of relaxation for pure overstress dependence

The relaxation behavior has recently been investigated by Krempl and Nakamura [35]. The characteristics found in the flow stress region are:

1. For a given prior strain rate and for a constant relaxation time the stress drop is nearly independent of the stress or strain at which relaxation starts.

2. The stress drops for constant relaxation times depend, however, strongly on the strain rate preceding the relaxation tests. They increase with an increase of prior strain rates.

3. The increase with prior strain rate is nonlinear (Increasing the prior strain rate by a factor of ten increases the stress drop in a given relaxation time by far less than a factor of ten).

4. At the end of the relaxation periods the test associated with the highest (lowest) prior strain rate has the smallest (greatest) stress magnitude.

5. Once straining is resumed at the end of the relaxation periods, the "flow stress" characteristic of the strain rate is reached quickly and the relaxation periods are forgotten.

The capability of VBO to model these behaviors is investigated. We assume that the viscosity function or the flow function depends only on the overstress. The pure overstress behavior is of primary interest. Additional drag stress dependence can be postulated and its properties can be ascertained.

The relaxation tests are started after the transient free region has been reached. Then the overstress is theoretically constant but may in reality change very little as the stress–strain curve evolves to higher and higher strains. The specialized flow law for relaxation is

$$\mathbf{0} = \dot{\mathbf{e}}^{\mathrm{el}} + \dot{\mathbf{e}}^{\mathrm{in}} = \left(\frac{1+\nu}{E}\dot{\mathbf{s}}\right) + \frac{3}{2}F[\,]\frac{\mathbf{s}-\mathbf{g}}{\Gamma} \tag{6.19}$$

This equation can be rewritten as

$$\dot{\mathbf{s}} = -\frac{3E}{2\,(1+\nu)}F[\,]\frac{\mathbf{s}-\mathbf{g}}{\Gamma} = -\frac{3}{2\,(1+\nu)}\frac{\mathbf{s}-\mathbf{g}}{k[\,]} \tag{6.20}$$

It is seen that the relaxation rate depends only on the overstress. So all relaxation tests with identical prior strain rate and performed in the transient free region have initially the same overstress and the stress rate is initially the same. The growth laws for the other state variables depend only on the overstress $\{\mathbf{s}-\mathbf{g}\}$ and on $\{\mathbf{g}-\mathbf{f}\} = A_{\mathrm{f}}\dfrac{\{\mathbf{s}-\mathbf{g}\}}{\Gamma}$. These quantities do not depend on stress or strain directly. It follows that further values of the relaxation rate are also independent of the strain or stress at which the relaxation starts. The evolution of the relaxation curves is the same for any point in the transient free region as long as the prior strain rate is the same.

This is observed in the experiments, see item 1. above. In real experiments the tangent modulus is not strictly constant as required but changes somewhat. It is still assumed that the asymptotic solution is applicable.

Since the overstress increases nonlinearly with the strain rate the relaxation rate will therefore increase with an increase in prior strain rate. These prediction is found in experiments, see item 2. given previously.

Items 3. to 5. are also predicted by VBO as demonstrated in Krempl and Nakamura [35].

The modeling of the quick transition from the relaxation test to the flow stress characteristic of the strain rate mentioned under item 4. requires that the influence of initial conditions vanishes quickly. Once the flow stress is reached in a simulation any influence of the initial conditions will have vanished. The above analyses show that the influence of the initial conditions of the overstress and of $(\mathbf{g}-\mathbf{f})$ will vanish. The relaxation periods have no influence on the long–term behavior of $\mathbf{f}$. No permanent influence of the relaxation holds is predicted by VBO. Figs. 6–8 of Majors and Krempl [36] demonstrate that VBO is capable of modeling repeated relaxation tests with loading intervals between them.

Relaxation has also been observed in the quasi–elastic region of the stress–strain diagram at low homologous temperatures, see Krempl [5]; Kujawski, Kallianpur et al. [21]. It is small but noticeable. At these temperatures the metals and alloys can sustain a non–zero stress at rest at any strain level. If the homologous temperature increases the stress may relax to zero after a long time. The shape of the relaxation curve does not appear to be significantly affected by temperature.

6.6. "Cold" creep behavior for pure overstress dependence

The question arises whether the creep behavior can have the same regularity as the relaxation behavior. Again pure–overstress dependence is considered. As in the relaxation case the overstress in the transient free region is constant for a given prior strain rate. The flow law for creep is obtained from (6.1) or (6.2)

$$\dot{\mathbf{e}} = \dot{\mathbf{e}}^{\mathrm{in}} = \frac{3}{2}F[\,]\,\frac{\mathbf{s}_0-\mathbf{g}}{\Gamma_0} = \frac{3}{2}\frac{\mathbf{s}_0-\mathbf{g}}{Ek[\,]} \tag{6.21}$$

where the subscript 0 indicates that the stress is constant. It is seen that for all tests with the same initial overstress the initial creep rate is the same. Once started the creep strain rate will evolve identically if the evolution of the equilibrium stress is the same for all strain intervals of several creep tests. This is only true if the change in the equilibrium stress is linear. This would require that the stress–strain diagram have a constant slope. A constant slope of the stress–strain diagram is unlikely to be found over a large strain interval and therefore congruent creep curves can in general not be expected.

Note that there is a fundamental difference in the creep and the relaxation test. The evolution of the equilibrium stress with time in relaxation is at most between the present stress and zero stress and occurs at constant total strain. Failure of a specimen in relaxation is not to be expected. In a creep test creep can terminate, can change from primary to secondary and even to tertiary creep, considerable strain can develop and creep rupture can occur, even at room temperature, see Krempl and Lu [37]. The constant stress must be high enough, close to the ultimate stress to cause creep rupture at low homologous temperature. Also creep tests are usually performed at constant load which causes an increasing true stress when creep strain develops. The constant load creep test significantly enhances the creep process at high stresses. At stress levels around yield creep usually terminates and this phenomenon is usually termed "cold creep".

At low homologous temperatures either work hardening or elastic–perfectly plastic stress–strain curves $(E_t \geq 0)$ are observed in a tensile test. In these cases creep is either primary or secondary at stress levels below the ultimate strength. For a softening stress–strain diagram $(E_t < 0)$ VBO predicts tertiary creep only, Krempl [12]. Creep strains at stress levels that lie in the initial quasi–elastic region are negligibly small see Krempl [12]; Kujawski, Kallianpur et al. [21].

7. BEHAVIOR UNDER A JUMP IN STRESS (STRAIN) RATE AND UNLOADING BEHAVIOR

7.1. Behavior under a jump in rates

Before the advent of the personal computer and the advancement of electronics mechanical testing, machines with gearboxes and scales with weights for applying the load were used. Either crosshead motion or a mechanical clip–on extensometer with mirrors or similar devices enabled the recording of strain. There was no continuous read–out of a signal possible, rather discontinuous data taking by eye and recording these data on a sheet of paper was the standard.

The situation has changed completely, now digital data acquisition of load and displacement are routine. So is the use of feedback principles and computer control for the mechanical testing machine.

Servo–controlled testing machines, with computer control make new methods of testing possible. These types of tests can yield new information on material behavior. Instantaneous changes in the rates, some times called rate jumps, can be realized. It is possible to change the strain or stress rate by several magnitudes instantaneously and to measure the response. For example the strain rate can be cycled between two values to reveal the rate sensitivity of the material. To illustrate the point the response of a stainless steel at room temperature is shown in Fig. 7.1. The rate–sensitivity is seen to be nonlinear. A tenfold increase in the rate is accompanied by an increase in the stress level of much less than a factor of ten. It is seen that after a change in strain rate the flow stress characteristic of the strain rate is reached quickly and no strain rate history effect is apparent. The steel "forgets" the prior history and after a short transient behaves as if a constant strain rate would have been operating all the time.

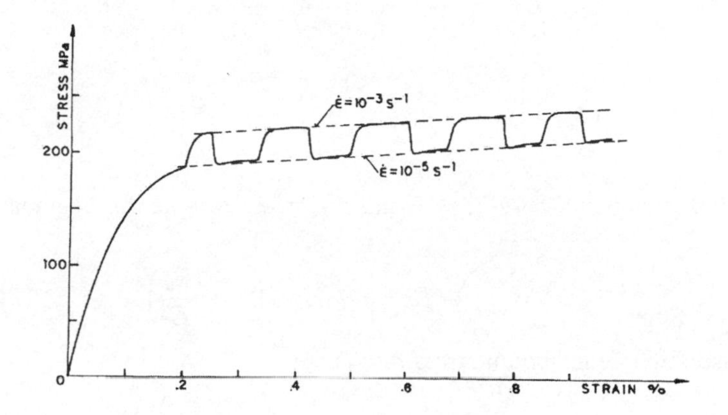

Figure 7.1: Response to strain rate cycling of a stainless steel at room temperature.

Jumps in rates are also experienced when the test condition changes from say constant strain (stress) rate to zero strain (stress) rate at the beginning of a relaxation (creep) test. Using the "mode switch" capabilities a relaxation (creep) test can be performed after prior stress (strain)–controlled test. In cyclic loading an instantaneous change in either the strain or stress rate takes place at each reversal if a saw–tooth waveform is used as a command signal. For a sine wave command signal there is no discontinuity.

The "mode switch" from say displacement to load control is also possible while under load. It corresponds to a switch of the independent variable from strain to stress for the model. A frequently used condition in evaluating integral representation is an instantaneous change in either stress or strain. Either condition is difficult to achieve in actual testing since it requires an infinite loading rate.

In a conventional creep test the loading rate to the creep stress level is not specified. Usual practice is to apply weights, one at a time. Therefore the loading rate will be very uneven and will depend on the physical condition of the operator at that time. The recorded information starts with the creep strain once the stress level of the creep test is reached. When the creep test is performed on a servo–controlled testing machine the loading rate, the strain accumulated during loading, and the creep strain are usually available and this information is needed for modeling.

All these new testing methods suggest the evaluation of the model response to instantaneous changes (jumps) in stress or strain rate. Let a superposed + and a superposed – designate the value just after and just before the change in rate, respectively. Under an instantaneous change in rate the stress and the state variables are continuous. Only the flow law and the growth law for the equilibrium stress can have instantaneous changes in the rates. Applying the uniaxial flow and the growth law for the equilibrium stress, see (8.5–8.7), before and after the jump yields

$$\dot{\varepsilon}^{+} - \dot{\varepsilon}^{-} = \frac{\dot{\sigma}^{+} - \dot{\sigma}^{-}}{E} \tag{7.1}$$

and

$$\dot{G}^{+} - \dot{G}^{-} = \frac{\psi}{E}\left(\dot{\sigma}^{+} - \dot{\sigma}^{-}\right) = \psi\left(\dot{\varepsilon}^{+} - \dot{\varepsilon}^{-}\right) \tag{7.2}$$

respectively.

For the relaxation test where $\dot{\varepsilon}^{+} = 0$ and the stress rate at the beginning is computed to be

$$\dot{\sigma}^{+} = \dot{\sigma}^{-} - E\dot{\varepsilon}^{-} = -E\dot{\varepsilon}^{-}\left(1 - \frac{E_{\mathrm{t}}}{E}\right) \tag{7.3}$$

and the corresponding equilibrium stress rate is

$$\dot{G}^{+} = \dot{G}^{-} - \psi\dot{\varepsilon}^{-} = \psi\dot{\varepsilon}^{-}\left(1 - \frac{E_{\mathrm{t}}}{\psi}\right) \tag{7.4}$$

In the second equation it was assumed that the long–term solution had been reached (The tangent modulus is $E_{\mathrm{t}} \ll E$) prior to the start of the relaxation test. It is seen that both the stress and the equilibrium stress rates are smaller in magnitude at the beginning of the relaxation test than immediately before.

The corresponding relations for the strain rate in a creep test are

$$\dot{\varepsilon}^{+} = \dot{\varepsilon}^{-} - \frac{\dot{\sigma}^{-}}{E} = \left(1 - \frac{E_{\mathrm{t}}}{E}\right)\dot{\varepsilon}^{-} \tag{7.5}$$

and for the equilibrium stress

$$\dot{G}^{+} = \dot{G}^{-} - \frac{\psi}{E\dot{\varepsilon}^{-}} = E_{\mathrm{t}}\dot{\varepsilon}^{-}\left(1 - \frac{\psi}{E}\right) \tag{7.6}$$

As in the case of relaxation, the creep rate and the equilibrium stress rate after the jump are reduced from those existing before the jump.

The results for instantaneous changes of the slope in a stress and a strain–controlled test are, respectively

$$\left(\frac{\mathrm{d}\sigma}{\mathrm{d}\varepsilon}\right)^{+} = E\left\{1 - \gamma\left[1 - \frac{E}{(\mathrm{d}\sigma/\mathrm{d}\varepsilon)^{-}}\right]\right\}^{-1} \tag{7.7}$$

and

$$\left(\frac{\mathrm{d}\sigma}{\mathrm{d}\varepsilon}\right)^{+} = E\left\{1 - \delta\left[1 - \frac{(\mathrm{d}\sigma/\mathrm{d}\varepsilon)^{-}}{E}\right]\right\} \tag{7.8}$$

The results are for any slope prior to the jump. If the asymptotic solution has been reached prior to the jump then $(\mathrm{d}\sigma/\mathrm{d}\varepsilon)^{-} = E_{\mathrm{t}}$ has to be substituted. The quantities $\delta = \dot{\varepsilon}^{-}/\dot{\varepsilon}^{+}$ and $\gamma = \dot{\sigma}^{-}/\dot{\sigma}^{+}$ are the strain and stress rate ratios, respectively. Under strain rate reversal $\delta = -1$ equation (7.8) predicts a slope of $\mathrm{d}\sigma/\mathrm{d}\varepsilon^{+} \approx 2E$ whereas for stress rate reversal from (7.7) $\mathrm{d}\sigma/\mathrm{d}\varepsilon^{+} \approx -E/100$. These predictions have been found to correspond to the experimentally determined values by Krempl and Kallianpur [38] for alloys under a variety of stress and strain rate ratios, see their figures 2 and 3.

It is important to note that the predictions do not involve the equilibrium stress. Equations (7.7) and (7.8) are valid for any model that has the structure $\dot{\varepsilon} = \dot{\sigma}/E + \dot{\varepsilon}^{\text{in}}$ where the inelastic strain rate depends only on the instantaneous values of the stress and the state variables, but not on any rate. The predictions are therefore applicable to the standard linear solid. Solid polymers were shown to have the same behavior as demonstrated by Bordonaro and Krempl [39] in their figures 3–6.

7.2. Unloading and cyclic behavior

The low and high homologous temperature behavior is reported together. First the real unloading behavior at room temperature and at high temperature is shown. Numerical experiments on a hypothetical VBO model without recovery and on two alloys at high homologous temperature are reported.

In Fig. 7.2 the cyclic hardening of type 304 Stainless Steel is shown in strain control. The cyclic hardening is very pronounced and unloading does not appear to be exactly straight. Small deviations from the straight line are noticeable. The transition from loading to unloading appears somewhat rounded so that the predicted slope of $2E$ at the first unloading step is possible. It is also important the keep in mind that the theoretically calculated slope is for the condition of leaving the reversal point. It does not apply at any other point.

The behavior of the same material in a stress–controlled test is depicted in Fig. 7.3. The negative slope upon leaving the maximum stress is very much pronounced and the evolution of the slope is quite different from the unloading in strain control. The negative slope turns

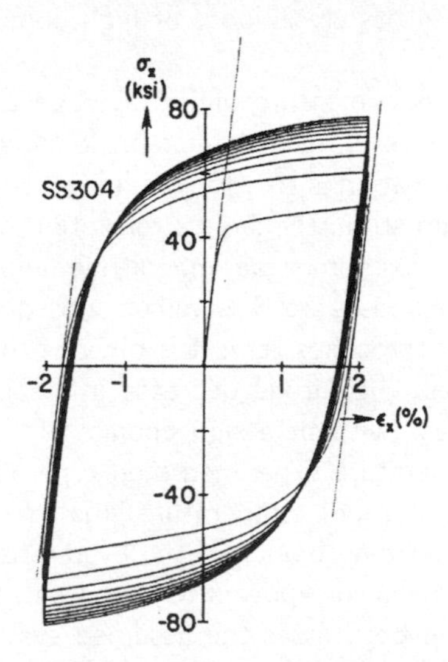

Figure 7.2. The unloading behavior of a stainless steel at room temperature in strain control, from Hassan and Kyriakides [40].

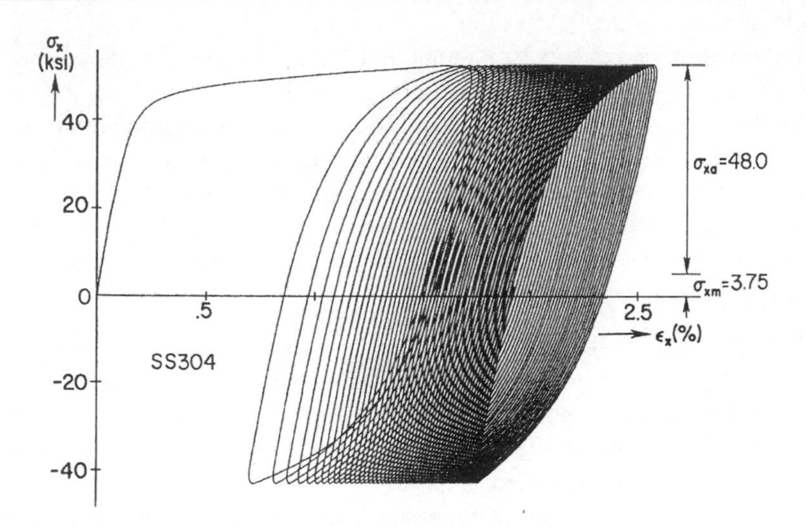

Figure 7.3: Load control for Type 304 stainless steel from Hassan and Kyriakides [40].

positive as the stress decreases and the unloading is less straight in this case than in the strain controlled unloading case. It is definitely nonlinear. The behavior predicted by the slope equations, see (7.7) and (7.8), especially the difference between strain (displacement) and stress (load) control is reproduced by the experiments. In load control the positive mean stress causes a ratcheting effect which is very pronounced.

The previous examples were at room temperature (low homologous temperature) for continuous cycling. In Fig. 7.4 hysteresis loops at high homologous temperature with 30 s hold–times are shown.

The stress drop in 30 s is a measure of the average relaxation rate. It is then seen that the average relaxation rate is much higher on loading (stress magnitude increases) than on unloading. This observation is true for the initial loops 1–2 and for the subsequent loops 20–22. At the maximum strain the largest relaxation drop is observed. At the end of the relaxation period at the maximum strain magnitude unloading continues and small but noticeable stress magnitude increases are observed on subsequent relaxation tests. The stress drops increase as the stress approaches zero. It is obvious that the relaxation rate changes sign upon unloading. A stress magnitude decrease immediately after unloading changes to a stress magnitude increase without a sign change of the stress. Similar observations were reported for room temperature experiments on type 304 stainless steel by Kujawski, Kallianpur et al. [21]. Even at room temperature small stress magnitude increases were found in the almost linear unloading branch of the hysteresis loop.

The observed relaxation behavior appears to be as paradoxical as the creep behaviors shown in Figs 4.1 and 4.2. In both cases the assumed symmetry of the behavior relative to the coordinate axes did not materialize. As in the case of the standard linear solid, see Appendix A, points at rest can be away from the zero stress axis.

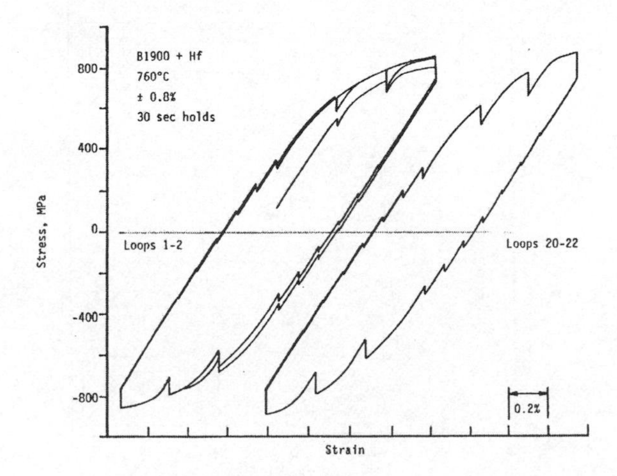

Figure 7.4. Hysteresis loops for alloy B1900 + Hf at 760 °C with intermittent hold times of 30 s, from Lindholm[41].

The unloading is now simulated using VBO. From (8.5) the slope can be computed as

$$\frac{\mathrm{d}\sigma}{\mathrm{d}\varepsilon} = E\left(1 - F\left[\Gamma\right]\frac{\sigma - g}{\Gamma\dot{\varepsilon}}\right) \tag{7.9}$$

An unloading event with strain rate reversal is considered, $\delta = -1$. According to (7.8) the slope at the beginning of unloading is equal to $2E$, the overstress is positive and the strain rate is negative. As unloading continues the overstress and the slope decrease. When the stress is equal to the equilibrium stress the overstress is zero, the slope is E and the overstress switches signs, see (7.9). This happens at some stress below the stress at reversal. The slope decreases further from this value as the stress decreases and the overstress becomes negative and its magnitude increases. Under strain rate reversal the slope continuously decreases from the initial value of $2E$ to a value less than but close to E at zero stress.

This behavior is demonstrated by a numerical experiment that simulates loading and unloading with a hypothetical metal close to 304 Stainless Steel with no static recovery terms. A hysteresis loop at a strain rate of 10^{-4} 1/s is shown in Fig. 7.5. The result of a stress–controlled experiment at 1 MPa/s with mean stress is shown in Fig. 7.6. The differences between strain and stress control are borne out in the numerical experiments. They also reproduce ratcheting.

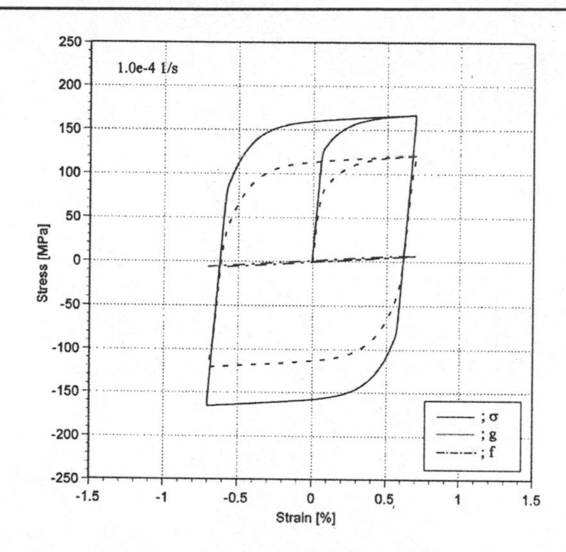

Figure 7.5. Simulation of a strain–controlled hysteresis loop using a hypothetical material similar to stainless steel at low homologous temperature. See Table 7.1 for constants used.

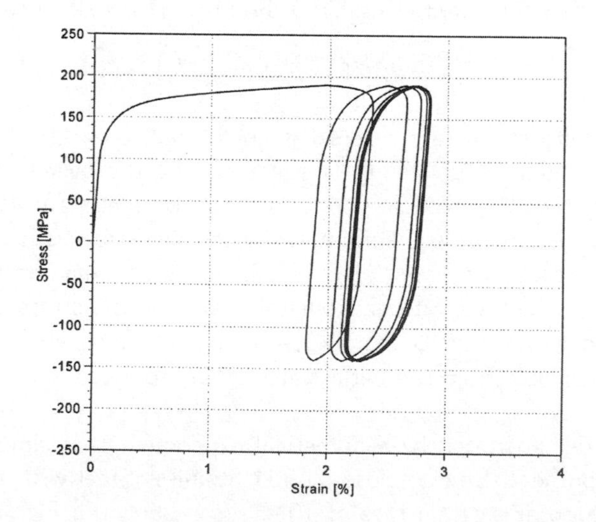

Figure 7.6. Simulation of the stress controlled cycling for the material of Fig. 7.5. Compare with Fig. 7.3. Stress rate = 1MPa/s.

The unloading slope appears to deviate less from a straight line in Fig. 7.5 than in Fig. 7.2. For the model it is comparatively easy to compute the slope of the stress–strain curve. The ratio $Alpha = (\mathrm{d}\sigma/\mathrm{d}\varepsilon)/E$ is plotted vs. strain in Fig. 7.7. Leaving the origin the ratio is one and then decreases rapidly and almost reaches zero after a very rapid change. From the material data of the simulation given in Table 7.1 it can be seen that $E_{\mathrm{t}} < E$

and $Alpha$ seems to be zero on this scale but it is on the on order of 0.005, see Table 7.1. At strain rate reversal, $Alpha$ almost reaches 2 and then passes through a region of unity. The equilibrium stress must be situated in this region at the time of stress traverse. Fig. 7.5 shows in addition to the stress the evolution of the equilibrium stress and of the kinematic stress. It is seen that there is an entire region where the stress equals the equilibrium stress. There is the region where the slope is equal to the elastic modulus, see Figs 7.5 and 7.7.

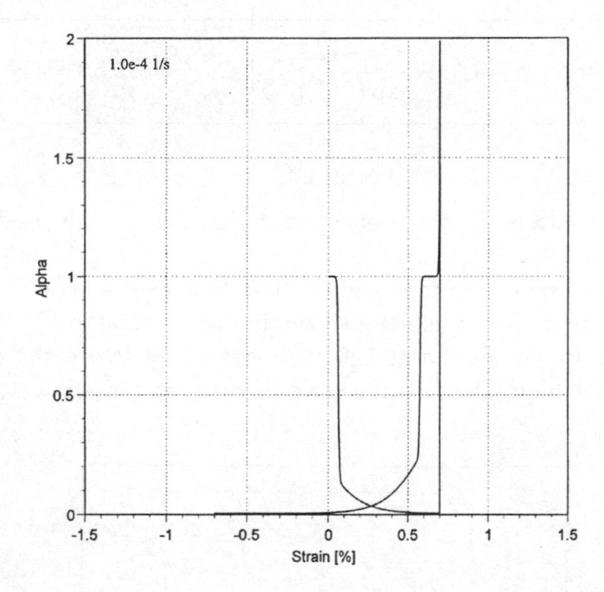

Figure 7.7. The change of the slope ratio for Fig. 7.5. Note the ratio greater than one immediately after reversal.

In the regions where the slope is equal to E the stress is equal to the equilibrium stress and it is then possible to find the equilibrium stress. However, it would not be possible to determine the equilibrium stress during loading, specifically at the reversal point, with this method.

The different unloading behavior for stress control as shown in Figs 7.3 and 7.6 is easily distinguished from the strain–controlled unloading. The slope immediately after reversal of the stress rate is small and negative. It turns positive as the stress magnitude decreases for the theoretical and the experimental curves. At some region after reversal the modulus is E and decreases further as in the case of strain control. The VBO modeling of the test in Fig. 7.6 shows the same qualitative appearance as Fig. 7.3.

The unloading behavior at high homologous temperatures is predicted next for two real alloys. Maciucescu et al. [43] and Tachibana and Krempl [42] have determined the constants for VBO with static recovery for a solder alloy and alloy 800 H at high homologous temperature, respectively. This offers an opportunity to check on the predictive capability of VBO for high homologous temperature where static recovery is important. No experimental

Table 7.1: Material constants used in the numerical experiments shown in Fig. 7.5 – 7.7.

Moduli	$E = 195$ GPa; $E_t = 1$ GPa
Isotropic stress, constant	$A_0 = 115$ MPa
Viscosity function $k = k_1 \left(1 + \frac{\Gamma}{k_2}\right)^{-k_3}$	$k_1 = 314,200$ s $k_2 = 60$ MPa $k_3 = 21$
Shape function $\psi = C_1 + \frac{C_2 - C_1}{\exp(C_3 \Gamma)}$	$C_1 = 50$ GPa $C_2 = 182.5$ GPa $C_3 = 0.05$ MPa^{-1}

relaxation data are available for these alloys, but Fig. 7.4 can be used as a qualitative comparison.

A complete cycle without hold time and another cycle with several hold–times of 300 s are numerically simulated. The results are shown in Fig. 7.8 and in Fig. 7.9 for the 800 H alloy at 1050 °C and for the Pb–Sn at 125 °C, respectively. When the relaxation drop is so small that it cannot be resolved on the scale of the graph insets are drawn and arrows

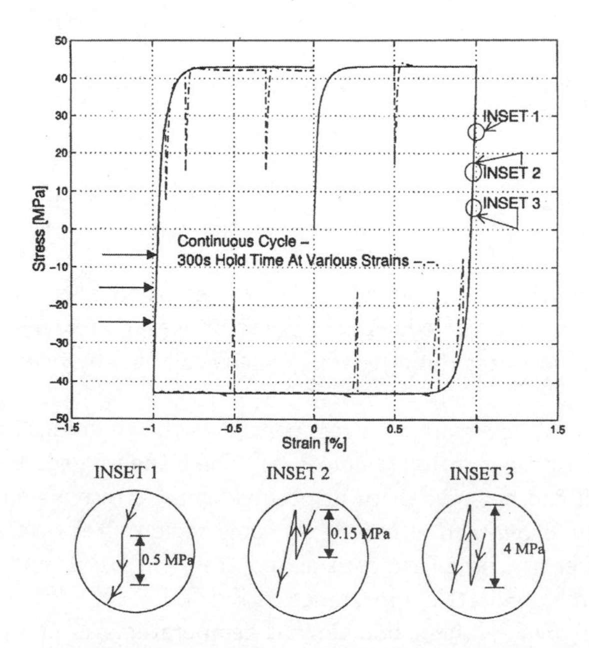

Figure 7.8. Hysteresis loop for Alloy 800 H at 1050 °C using the data of Tachibana and Krempl [42]. Hysteresis with 300 s relaxation periods (dash– dot) and with strain control (solid line). Strain rate magnitude during loading is 10–4 1/s in both cases. At inset 1 the stress decreases, but increases for insets 2, 3.

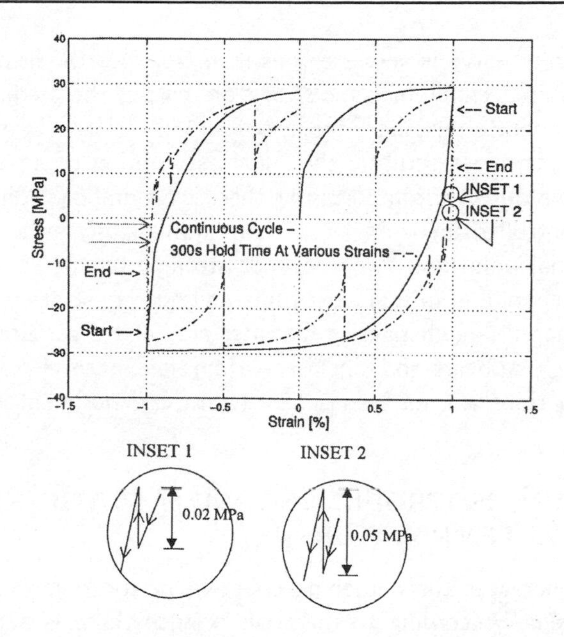

Figure 7.9. Same as Fig. 5.12. But the alloy is a Sn–Pb solder at 125 °C from Maciucescu et al. [43]. Stress magnitude increases slightly during unloading, see insets.

indicate the location of the hold time.

In Fig. 7.8 the hysteresis loop with and without hold time is almost identical. The characteristic constant strain rate curve is reached very quickly after straining is resumed at the end of the hold time. In some cases a slight overshoot is observed. This is not the case for the Sn–Pb alloy in Fig. 7.9. For the simulation using the solder alloy, see Fig. 7.9, the stress–strain curve is reached gradually in a manner exhibited by the 1900+Hf in Fig. 7.4.

The insets and arrows in Figs 7.8 and 7.9 indicate that a 300 s relaxation test has taken place there but the stress change is such that the stress drop cannot be seen using the scale of the figure. At inset # 1 in Fig. 7.8 the stress magnitude decreases whereas it increases at insets 2 and 3. A corresponding behavior is found for the solder alloy in Fig. 7.9. The first relaxation after stress reversal and marked by *Start* and *End* the stress magnitude decreases whereas Insets I and 2 show a stress magnitude increase which is very small.

It appears that the prediction of VBO compares favorably with the experimental results shown in Fig. 7.4.

In alloys the recovery is known to be very small at zero stress. The recovered strain is referred to as the inelastic strain, see Lubahn and Felgar [44], page 365. This inelastic strain can be computed using VBO. From (8.5) we obtain for isothermal creep at zero stress $\dot{\varepsilon} = F\,[|G|]\,\dfrac{-G}{|G|}$. Unloading from tension produces a positive equilibrium stress which in turn causes a negative strain rate. The equilibrium stress decreases rapidly as the strain is recovered. Recovery is complete when the equilibrium stress is zero at the strain axis. The equilibrium stress–strain curve is very steep as it approaches the strain axis, see Fig. 7.5, and the recovered strain is small. Inelastic strains are part of the predictions of VBO. Other models introduce a separate repository for it, see p. 158 of Hart, Li et al. [45].

It is clear that upon close scrutiny the usual assumption of an unloading curve with slope of E is not borne out. Most importantly, there is a significant difference between load and displacement controlled unloading, especially immediately after the rate reversal has taken place. The initial unloading slope depends strongly on the strain (stress) rate ratio before and after the change in rate. To experimentally demonstrate a significant change in the slopes order of magnitude changes in the rates are needed as demonstrated by Krempl and Kallianpur [38]. It appears that inelastic strains are naturally included in the VBO formulation. Its predictions are not inconsistent with the experimental results.

8. SMALL STRAIN, ISOTROPIC VBO WITH STATIC RECOVERY (HIGH HOMOLOGOUS TEMPERATURE)

As temperature increases the hardening effect of plastic deformation is counteracted by diffusion phenomena. According to materials science there is a competition between hardening due to deformation and softening due to diffusion which can give rise to steady–state type deformation as observed in secondary creep. The classical example for a model of that behavior is the Bailey–Orowan model, see Gittus [16] equations (6.8) and (8.120). The format is

$$\mathrm{d}\tau = h\mathrm{d}\gamma - r\mathrm{d}t \tag{8.1}$$

where τ is the athermal component of the shear stress, γ is the inelastic shear strain and h and r are the hardening and softening functions, respectively. They can depend on the state variables and stress, but do not depend on increments or rates.

This formalism is also used in the unified state variable models. The form of the flow law (6.2) is left unchanged but the material constants are allowed to change with temperature. This is also true for the constants of the state variable growth laws. The kinematic stress is the repository for setting the slope at the maximum strain of interest. It is already possible to model hardening, neutral behavior and softening by setting the slope to a positive, zero or negative number see Figs 1 and 4 in Krempl [12]. The addition of a recovery term would not alter the basic capabilities. Therefore the growth law for the kinematic stress is left unchanged.

Considering that the right hand side of the growth law for the equilibrium stress, see (6.4), represents hardening, the recovery term is subtracted to yield

$$\dot{\mathbf{g}} = \textit{low homologous temp. terms} \; - R[\bar{g}]\,\mathbf{g} \tag{8.2}$$

The positive, increasing function $R[\bar{g}]$ with the dimension of reciprocal time constitutes the static recovery term. The equilibrium stress invariant is $\bar{g} = \sqrt{\frac{3}{2}\operatorname{tr}(\mathbf{gg})}$ and $R[0] = 0$. The function R plays the role of r in the Bailey–Orowan formalism, the static recovery term reduces the growth of g and it is zero when the equilibrium stress is zero. It is immediately clear that with this modification relaxation ends at zero stress. The model with static recovery given in (8.2) cannot sustain a stress at rest. Majors and Krempl [36] have attempted to classify some of the softening phenomena that can be observed at high homologous temperature and are important for modeling. They include strain rate history dependence, cyclic softening and the modeling of the transition from primary, to secondary and tertiary creep. These can all be called dynamic softening effects. There is, in addition, the so–called static recovery effect where a specimen after deformation is kept at zero stress and where the time spent at zero stress has an influence on the subsequent response in reloading. Majors and Krempl [36] have shown that VBO with only a static recovery term in the growth law for the equilibrium stress is not capable of modeling the above listed phenomena. Only if a softening of the isotropic stress is introduced in addition to the recovery term in the growth law for the equilibrium stress can the above phenomena be modeled.

Necessary conditions for a satisfactory modeling of cyclic softening, strain rate history dependence and a permanent effect of a rest time at zero stress are the presence of a static recovery term in the growth law for the equilibrium stress and the softening of the isotropic stress.

Majors and Krempl [36] have demonstrated these capabilities. Different growth laws for the softening of the isotropic stress are used by Majors and Krempl [36], Tachibana and Krempl [8, 46, 42] and by Maciucescu et al. [43]. They show that VBO has the capabilities to model creep, relaxation, monotonic and cyclic loading at different loading rates.

When the static recovery term is present, the validity of long–term solution has to be checked. It is strictly speaking not applicable any more. However, it may still be useful as an estimate for modeling purposes, see Tachibana and Krempl [8].

8.1. Modeling of deformation behavior

The above equations represent the three dimensional version of VBO. To ascertain its modeling capabilities the material constants E_t, E, ν, and α and the functions $F[\Gamma]$, $\psi[\Gamma]$, and $R[\bar{g}]$[5] must be determined from appropriate tests. Included are monotonic and cyclic loading in displacement (strain) or load (stress) control, creep (stress is constant) and relaxation (displacement is constant). Such tests are usually performed in the uniaxial state of stress or, less frequently, in biaxial stress states. To start the determinations, the three–dimensional equations must be specialized for the appropriate stress state and the test conditions. Then values for the constants are assumed and the coupled, nonlinear, stiff differential equations are integrated numerically. The results are compared with the experiments. New constants are assumed and integration yields new results. This process is continued until a satisfactory match of simulation and experimental data is achieved.

This procedure, much criticized as the "curve fitting" portion of finding the model constants, is needed for every approach, see remarks in Section 2 on page 4. Once the constants have been determined the constitutive equations should be specialized for another

[5] They in turn contain other constants. It is possible to use constants instead of functions thus reducing the number of constants needed, see Maciucescu et al. [43].

set of tests which had not been used in the determination of the constants. To have confidence in the model it should render a good prediction for this new set. If the comparison with experiment is satisfactory, the model can be used with confidence in stress analyses.

8.2. Reduction to the uniaxial state of stress

It is advantageous to formulate the set of constitutive equations for the uniaxial state of stress. The model has been formulated that all constants can be determined from the uniaxial state of stress.

Using the definition of a deviator for the strain rate, equation (6.1–6.2) can be written as

$$\dot{\boldsymbol{\epsilon}} = \frac{\mathrm{d}}{\mathrm{d}t}\left(\frac{1+\nu}{E}\mathbf{s} + \frac{1-2\nu}{3E}\,\mathrm{tr}\,\boldsymbol{\sigma}\mathbf{I}\right) + \alpha\dot{T}\mathbf{I} + \frac{3}{2}F\,[\Gamma]\,\frac{\mathbf{s}-\mathbf{g}}{\Gamma} \tag{8.3}$$

where $\mathbf{I}$ is the identity matrix. For the uniaxial state of stress the stress deviator is given by

$$[\mathbf{s}] = \frac{\sigma}{3}\begin{bmatrix} 2 & 0 & 0 \\ 0 & -1 & 0 \\ 0 & 0 & -1 \end{bmatrix} \tag{8.4}$$

where σ is the uniaxial true stress so that $\sigma \equiv (3/2)\, s_{11}$. The same relation is assumed for the true equilibrium stress $G \equiv (3/2)\, g_{11}$ where$G$? is the counterpart of σ. The uniaxial component of the true strain rate $\dot{\varepsilon}$ is found to be

$$\dot{\varepsilon} = \frac{\mathrm{d}}{\mathrm{d}t}\left(\frac{\sigma}{E}\right) + \alpha\dot{T} + F\,[\Gamma]\,\frac{\sigma - G}{\Gamma} \tag{8.5}$$

It is easy to convert the growth laws for the state variables to the uniaxial state of stress. They are

$$\begin{aligned} \dot{G} = {} & \frac{\psi\,[\Gamma]}{E}\dot{\sigma} + \sigma\frac{\partial}{\partial T}\left(\frac{\psi\,[\Gamma]}{E}\right) \\ & + F\,[\Gamma]\left(\frac{\sigma - G}{\Gamma} - \frac{G - \bar{f}}{A}\right)\psi\,[\Gamma] + \left(1 - \frac{\psi\,[\Gamma]}{E}\right)\dot{\bar{f}} - R\,[|G|]\,G \end{aligned} \tag{8.6}$$

and

$$\dot{\bar{f}} = \hat{E}_{\mathrm{t}}F\,[\Gamma]\,\frac{\sigma - G}{\Gamma} \tag{8.7}$$

where $\bar{f}$ is the counterpart of σ and G and where $\Gamma = |\sigma - G|$. The growth law for A changes to allow for softening, see, Majors and Krempl [36].

Eqs. (8.5) through (8.7) are the uniaxial constitutive equations which must be specialized for the test conditions, say creep (stress rate equal to zero), relaxation (strain rate is zero) or for monotonic loading before they can be used to determine the material constants from tests.

The test conditions must be specified and there is a difference between the true and engineering quantities. With modern servo–controlled testing machines either engineering or true stress (or strain) can be imposed on the specimen. In most of the tests the rates are constant over certain time intervals but can change instantaneously. For example loading up to a creep stress level can be done with a certain stress or strain rate with a change to zero stress rate when the creep test starts.

It is useful to recall the relations between the true and the engineering quantities (designated by the same symbol but with a ^). With constant density assumed, these relations are $\varepsilon = \ln(1+\hat{\varepsilon})$ and $\sigma = \hat{\sigma}(1+\hat{\varepsilon})$. To simulate the described tests with VBO the conditions listed in the last column of the Table 8.1 must be substituted only into the flow law together with an appropriate value for the constant rates. The initial conditions of all variables must be known so that the formula can be applied.

Table 8.1: Expressions for strain rate and stress rate control

Type of control	Conditions	Expression to be substituted
Strain control	$\dot{\hat{\varepsilon}}=$ const	$\dot{\varepsilon} = \dot{\hat{\varepsilon}} / (1+\hat{\varepsilon}) = \dot{\hat{\varepsilon}} / \exp\varepsilon$
True strain control	$\dot{\varepsilon} =$ const	$\dot{\varepsilon}$
Stress control	$\dot{\hat{\sigma}}=$ const	$\dot{\sigma} = \dot{\hat{\sigma}}(1+\hat{\varepsilon}) + \hat{\sigma}\,\dot{\hat{\varepsilon}} = \dot{\hat{\sigma}}\exp\varepsilon + \sigma\dot{\varepsilon}$
True stress control	$\dot{\sigma} =$ const	$\dot{\sigma}$

Note that for relaxation both the engineering and the true strain rates are zero. There is a difference between the "constant load" creep test (engineering stress rate is zero) and the "constant stress" creep test (true stress rate is zero.). It is known that the experimentally observed constant load and constant true stress creep curves can be different. Therefore, a possibility exists to model these differences with VBO.

With these expressions the coupled nonlinear non–autonomous differential equations must be integrated numerically to simulate the specific tests. For a good model the curves obtained from the numerical test should be identical to the experimental one.

8.3. Isothermal creep

Loading to the creep stress level can be either in stress or strain control. In the transition from the loading to the creep test the rate quantities experience a jump. The jump is computed for a stress level when plastic flow is fully established and when the asymptotic solution holds (strictly speaking the recovery terms must be zero for the existence of the asymptotic solution). At short times the influence of the recovery term is usually negligible.

Loading up to the flow stress region is performed with a strain rate $\dot{\varepsilon}^-$ (although stress control can also be used). Then the true stress is kept constant. Applying the uniaxial equations immediately before and after the switch to constant stress enables the calculation of the initial creep rate and of the equilibrium stress rate. They are, see (7.5 and 7.6),

$$\dot{\varepsilon}^+ = \dot{\varepsilon}^- \left(1 - \frac{E_t}{E}\right) \tag{8.8}$$

and

$$\dot{G}^{+} = \dot{\varepsilon}^{-} E_{\mathrm{t}} \left(1 - \frac{\psi}{E}\right) \tag{8.9}$$

where $+$ designates the rate at the beginning of the creep test. The equilibrium stress rate before the start of the creep test is $\dot{G}^{-} = \dot{\varepsilon}^{-} E_{\mathrm{t}}$ where we have made use of the long–term asymptotic solutions of VBO. It is seen from (8.8) and (8.9) that the strain and equilibrium stress rates at the beginning of the creep test are reduced from the values just before the test. It is also seen that the equilibrium stress continues to increase in the creep test. The relations are different for stress levels less than that of the asymptotic solution.

8.4. Constant true stress and constant load creep tests

In applications, the majority of the creep tests are constant load tests. When strain is measured and is used in a feedback loop of a servo–controlled experiment a constant true stress test can be performed.

Using (8.5) for the isothermal condition and the definitions given in the above the expressions for the creep rate and its time derivative are found in Table 8.2 where a prime $'$ indicates derivative with regard to the argument of the function, the true stress in the constant load tests is given by $\sigma = \hat{\sigma}_0 \exp(\varepsilon - \varepsilon_0)$, Γ is the absolute value of the overstress and the subscript $_0$ indicates that the quantity is kept constant.

Table 8.2: Expressions for creep rate and creep acceleration

	Constant true stress	Constant load
$\dot{\varepsilon}$	$F[\Gamma]\dfrac{\sigma_0 - G}{\Gamma}$	$\dfrac{F[\Gamma]}{1-\sigma/E}\dfrac{\sigma - G}{\Gamma}$
$\ddot{\varepsilon}$	$F'[\Gamma]\left(-\dot{G}\right)$	$\dfrac{\dfrac{\sigma\dot{\varepsilon}^2}{E} + F'[\Gamma]\left(\sigma\dot{\varepsilon} - \dot{G}\right)}{1 - \dfrac{\sigma}{E}}$

Usually creep curves are classified as primary (creep rate decreases in magnitude and creep acceleration magnitude decreases), secondary (creep rate magnitude is constant and the acceleration is zero) and tertiary (creep rate magnitude increases and creep acceleration increases in magnitude)[6].

For the positive and constant true stress test the sign of the creep acceleration is determined by the sign of the equilibrium stress rate. For primary creep $\dot{G} > 0$, for secondary creep $\dot{G} = 0$ and $\dot{G} < 0$ is required for tertiary creep. For a constant load test the relations are similar but not as easy to see.

At low homologous temperature and when the long–term solution holds the sign of the equilibrium stress rate is determined by the sign of the tangent modulus, see Krempl [12].

[6] We refer to magnitude to make the description valid for tension and compression.

If $E_t > 0$ only primary creep can be modeled. A negative tangent modulus signifies tertiary creep and if the tangent modulus is zero then secondary creep is modeled. This classification is valid for creep stress levels that are in the flow stress region. In the quasi–linear region creep is always primary.

This observation is at odds with experiments, see for example Fig. 4.7 where secondary and tertiary creep are experienced even if the tangent modulus is positive. These results pertain to stress levels in the quasi–linear region of the stress–strain diagram and to high homologous temperature where the effects of softening and recovery are important. In this temperature region the sign changes in the equilibrium stress rate are affected by the static recovery term, the last term in (8.6) and by the softening of the isotropic stress. For tensile loading both quantities enter with a negative sign in the growth law for the equilibrium stress. When the isotropic stress is decreasing then the dynamic recovery term in (8.6) increases. This fact and the negative static recovery term can render the growth of the equilibrium stress zero and negative. Substituting the inelastic strain rate in the growth law for the equilibrium stress, (8.6) reveals that all the terms in the growth law for the equilibrium stress, except the static recovery term, are homogeneous of degree one in the rates. Rate and elapsed time have an influence only through the contribution of the static recovery term. It is therefore possible to model secondary and tertiary creep in the quasi–elastic regions even if the tangent modulus is positive. The capabilities of VBO in this respect are demonstrated by Tachibana and Krempl [8, 46, 42] on Alloy 800 H.

The situation is more complicated for the constant load creep test. The true stress is given by $\sigma = \hat{\sigma}_0 \exp(\varepsilon - \varepsilon_0)$ and is increasing with creep strain. In addition there are extra terms that increase the creep rate. Even if the equilibrium stress rate is zero, tertiary creep can be modeled due to the stress increase. In some instances when the stresses are high the recovery term and the softening of the isotropic stress are not needed to induce tertiary creep. At engineering stress levels above yielding Krempl and Lu [37] report tertiary creep for ductile steels even at ambient temperature. There is no doubt that this can also occur at high homologous temperatures.

At every temperature the constant load creep test causes an acceleration of the creep process relative to that of the constant true stress test. For a constant load creep test different behaviors in tension and compression are predicted due to the presence of the term $(1 - \sigma/E)$ which decreases in tension and increases in compression. For a constant true stress the equations in Table 8.2 predict tension–compression symmetry.

8.5. Relaxation

In relaxation the engineering and the true strain rate are zero and the true stress rate can be calculated easily. There are no separate conditions for the true or for the engineering strain rate control. The constitutive equations can be numerically integrated to yield the true stress vs. time from which the engineering stress can be easily calculated.

It has been shown by Krempl and Nakamura [35] that the low homologous temperature formulation leads to a nonzero stress at rest. The introduction of the static recovery term leads to zero equilibrium stress at rest.

The equivalent expressions to (8.8) and (8.9) at the beginning of the relaxation test are:

Table 8.3: Initial stress rates in relaxation

Stress rates for $\psi \neq 0$

E_t	$\dot{\sigma}^+$	$\dot{G}^+$	$\dot{\sigma}^+/\dot{G}^+$
0	$-E\dot{\varepsilon}^-$	$-\psi\dot{\varepsilon}^-$	E/ψ
$\neq 0$	$-E\dot{\varepsilon}^-\left(1-\frac{E_t}{E}\right)$	$-\psi\dot{\varepsilon}^-\left(1-\frac{E_t}{\psi}\right)$	$\frac{E-E_t}{\psi-E_t}$

It is seen that at the beginning of relaxation the stress as well as the equilibrium stress decrease. The equilibrium stress is not stationary during relaxation. This property gives rise to some interesting modeling capabilities, see Krempl and Nakamura [35].

8.6. Possible simplifications

In looking at (8.5) through (8.7) we see that aside from the elastic constants, the tangent modulus E_t and the two constants in the growth law for A, three functions, namely, the flow function $F[\Gamma]$, the shape function $\psi[\Gamma]$ and the static recovery function $R[\bar{g}]$ must be determined. Due to the nonlinear rate sensitivity of metals and alloys it is not possible to linearize the flow function$F[\Gamma]$. Maciucescu et al. [43] have replaced the shape function $\psi[\Gamma]$ by a constant and have linearized the recovery function by setting $R[\bar{g}] = R_1\bar{g}$ where R_1 is a constant. With this simplified version Maciucescu et al. [43] were able to model the behavior of a solder alloy at various high homologous temperatures. With similar success the same method was employed for Alloy 800 H which had been previously modeled using all three functions. The simplified version had essentially the same modeling capabilities but needed only half the constants, see Tachibana and Krempl [8, 46, 42]. In the future it is recommended that high homologous temperature modeling be started with the simplified version. Only if the simplified version is incapable of reproducing the data should the full theory be applied.

8.7. Variable temperature

A. Experimental results

Alloys designed for high homologous temperature service can grow new constituents while exposed to the high–temperature environment. Carbides can precipitate to name just one of many possibilities. It is of course natural to assume that these internal chemical reactions will alter the deformation behavior compared to the original material. If these changes are only influenced by the time spent in the high temperature environment and not by the deformation then we speak of aging. This property is modeled by an explicit dependence on age, i.e. time spent in high temperature environment. If the chemical reactions are influenced by deformation as well, experimental results are needed before modeling can be attempted see Marquis [47] for a study on the influence of aging and the deformation behavior of age–hardening Al alloys. In the following discussion it is assumed that aging is not present unless it is specifically mentioned.

An example of an ideal behavior is given in Fig. 8.1. Two isothermal stress–strain curves shown for 760 °C and 982 °C are the benchmark. In two other tests the temperature was changed from 760 °C (982 °C) to 982 °C (760 °C) during straining. It is seen that the variable temperature tests join their isothermal counterparts after a transition period. The change in temperature is forgotten. From this evidence it is assumed that no aging or any other micromechanism has acted since the material behavior is apparently unaffected by the history. The tests were performed on the superalloy B1900+Hf.

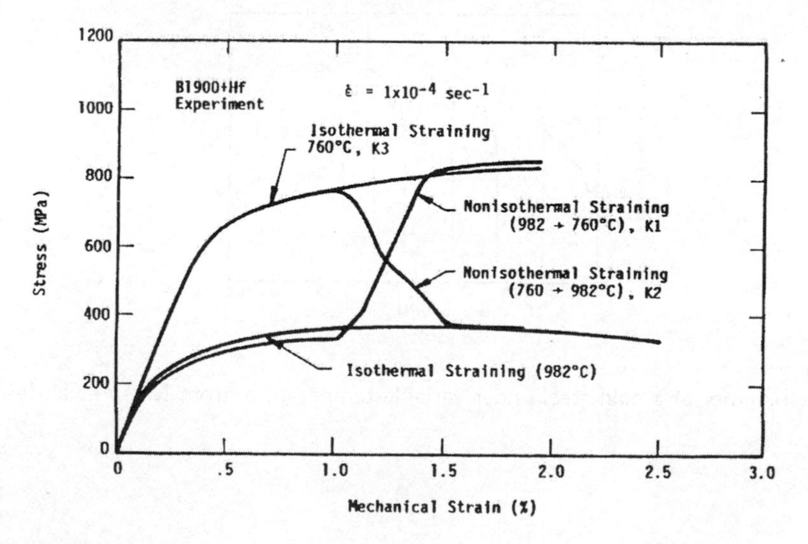

Figure 8.1. Influence of temperature change on the stress–strain behavior of B1900+Hf, from Chan and Lindholm [48].

Another example given in Fig. 8.2 is for mild steel and exhibits quite different results. The test starts at 400 °C and continues to 200 °C, to room temperature, to 200 °C to 400 °C and finally to 200 °C. The isothermal curves at ambient, 200 °C, and 400 °C are also depicted. It is seen that the respective isothermal curves are not always reached, especially at the last change to 200 °C. The behavior of the mild steel is quite different from that of B1900+Hf.

The results in Fig. 8.2b are for the mild steel and the temperature is cycled between 200 °C and 400 °C. The variable temperature stress–strain curve approximately coincides with the 400 °C isothermal curve but never reaches the isothermal 200 °C curve. At this temperature the stress strain curve has a "jerky" appearance indicative of the presence of dynamic strain aging.

The unusual behavior is most likely attributable to dynamic strain aging at 200 °C. The stress–strain curve at 200 °C in Fig. 8.3 shows the jerky appearance and a stress level that exceeds that of the room temperature curve. These are all signs of dynamic strain aging

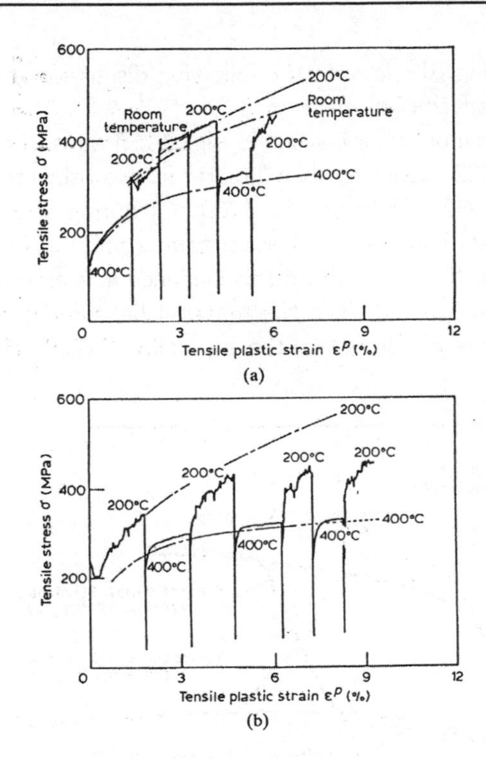

Figure 8.2. The behavior of a mild steel under variable temperature, from Niitsu et al. [49].

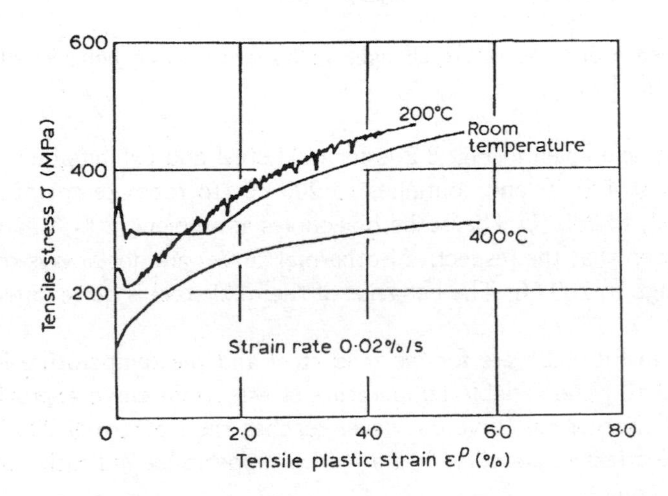

Figure 8.3. Stress strain diagrams showing the anomalous behavior at 200 °C. from Niitsu et al. [49].

which can be found in almost every engineering alloy in a certain temperature range. Mild steel is known to be susceptible to dynamic strain aging at the indicated temperature. The writer does not know whether the increase in stress level at 200 °C is a consequence of dynamic strain aging or whether another mechanism was operating.

Dynamic strain aging is generally thought to be caused by attachment and breaking loose of interstitial or solute atoms and their interaction with dislocations. As the stress increases the dislocations break loose, travel and get fixed again. This process then gives rise to the jerky motions. The behavior shown in Fig. 8.1 is designated as normal. The behavior shown in Figs 8.2 and 8.3 deserves the adjective pathological.

B. Modeling

The normal behavior can be easily reproduced by VBO, see Lee and Krempl [18]. In normal behavior the influence of prior loading is forgotten and the response is indistinguishable from the isothermal stress–strain diagram. It has been shown that for a constant loading rate and isothermal conditions the influence of the initial conditions vanishes, see Section 6. The temperature rate term is zero for isothermal conditions. VBO therefore predicts that the stress–strain curve with variable temperature ultimately coincides with the isothermal counter part.

The modeling of dynamic strain aging is in its infancy. Recently VBO has been extended to model rate independence and negative loading rate sensitivity. These properties are found when dynamic strain aging is present. The modeling of these phenomena is discussed Appendix B.

During the transient temperature region and with temperature dependent elastic properties the influence of the temperature rate terms in the flow law and in the growth law for the equilibrium stress become important. In their absence the isothermal curve will be reached with a considerable delay, see Figs 3a and 3b of Lee and Krempl [18]. In thermo–mechanical loading the temperature rate terms avoid a drift of the hysteresis loop as shown in Figure 9 of Lee and Krempl [18] and Figure 6 of Chaboche [50].

9. SYNOPSIS

Inelastic analyses are needed to ensure safety, reliability and predictability of performance of one–of–a–kind processes or structures. The availability of inexpensive computing power make inelastic analyses in the design office possible. The modeling of the inelastic behavior of engineering metals and alloys, however, has not kept up with the materials development and with the advancement of the computer. It is generally acknowledged that the material models or constitutive equations are the weakest link in the inelastic stress analysis package.

The comparison of the qualitative prediction of separately developed creep and plasticity models with experimental results reveals a fundamental inability to model creep–plasticity and plasticity–creep interactions.

New developments are the formulation of state variable models that do not have separate repositories for creep and plasticity. All inelastic deformation is rate dependent and creep (relaxation) is predicted by inserting the condition of creep (relaxation), stress rate is zero (total strain rate is zero), into the constitutive model. The subsequent integration yields the desired creep or relaxation equations.

The unified state variable theory called "viscoplasticity theory based on overstress (VBO)" is introduced and its properties are discussed. It is formulated without a yield surface and without loading and unloading conditions. Rate dependence is, following recent experimental results, introduced as fundamental. The equilibrium stress and the kinematic stress are two tensor valued state variables of VBO. It also has the scalar valued isotropic stress that has modeling functions close to the isotropic stress of classical plasticity.

At low homologous temperature there is hardening due to inelastic deformation. For steady loading a long–term asymptotic solution is possible and is thought to correspond to fully established inelastic flow. The flow stress (the stress in the region where the asymptotic solution applies) consists of rate–dependent and rate–independent contributions. "Cold" creep and relaxation are modeled as well as nonlinear rate sensitivity.

The same structure of the VBO prevails at high homologous temperature. Introducing a static recovery term according to the Bailey–Orowan format recognizes the softening effects of diffusion. The isotropic stress is made to decrease. With these modifications VBO is shown to model the high homologous temperature behavior including tertiary creep of a solder alloy and of Alloy 800 H.

The VBO structure allows a seamless transition from low homologous temperature to high homologous temperature. One concept applies. The form of the equations does not change. Several numerical experiments demonstrate the capability. The VBO could be used in inelastic analyses of structures and processes.

Acknowledgement The author acknowledges the support of the US Department of Energy, Grant DE– FG02–96ER14603. D. P. Alexander, K. Ho, H. Irizarry–Quiońes and M. Makaraci helped in the preparation of the figures. F. Khan provided editorial and word–processing assistance.

A THE STANDARD LINEAR SOLID AS A BASIS FOR VBO

The simplest linear viscoelastic solid that shows creep, relaxation and loading rate sensitivity is the "standard linear solid," or SLS, see Fig. A.1 and Fluegge [51]. Accordingly this model is used as starting point by writing its constitutive equation in the "overstress" format.

$$\dot{\varepsilon} = \frac{\dot{\sigma}}{E_1} + \frac{\sigma - aE_2\varepsilon}{a\eta} \tag{A.1}$$

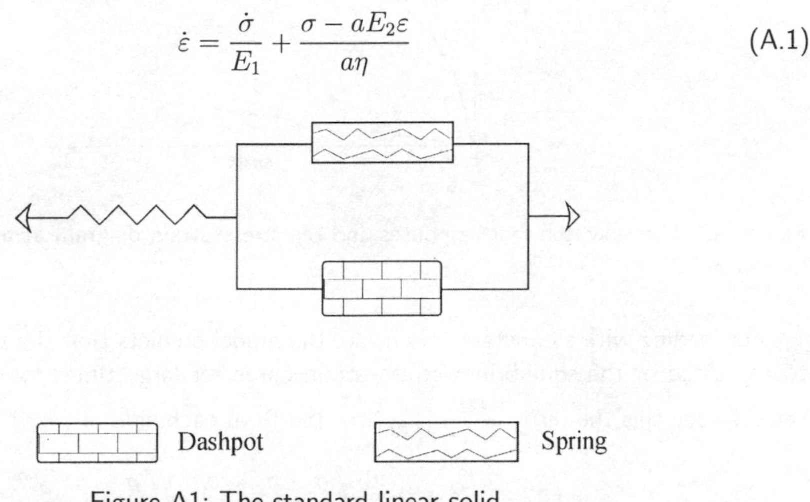

Figure A1: The standard linear solid

A superposed dot denotes material time derivative. σ and ε are true (Cauchy) stress and true strain, respectively; E_1 and E_2 is the modulus of the spring in front and of the spring in the Kelvin element, respectively. The quantity η is the viscosity coefficient of the dashpot with the dimension of stress$\times$time; and $a = E_1/(E_1 + E_2)$. The effective modulus of the two springs in series is equal to aE_2. The difference $\sigma - aE_2\varepsilon$ is known as the overstress. (A.1) can also be written as $\dot{\varepsilon} = \dot{\varepsilon}^{\mathrm{el}} + \dot{\varepsilon}^{\mathrm{in}}$ where the inelastic strain rate is a linear function of the overstress.

The stress response for infinitely slow loading is given by the two springs in series since the dashpot does not support any stress under these conditions. A mathematical derivation of this result is given in the main part of this report for VBO. For SLS the procedure is similar. It is seen from (A.1) that $\sigma = aE_2\varepsilon$ is the response in the limit as the strain and the stress rates approach zero. The straight line $aE_2\varepsilon$ is called the equilibrium stress. It will be shown below that the creep and relaxation tests come to rest at this line after infinite time.

The response of the SLS for infinitely fast strain rate is given by the response of the spring in front since the dashpot is rigid for infinitely fast loading. This can be demonstrated by writing

$$1 - \frac{\mathrm{d}\sigma/\mathrm{d}\varepsilon}{E_1} = \frac{\sigma - aE_2\varepsilon}{a\eta\dot{\varepsilon}} \tag{A.2}$$

and letting the strain rate $\dot{\varepsilon}$ go to infinity. For any other strain rate the response is between the infinitely fast and the infinitely slow response as shown in Fig. A.2.

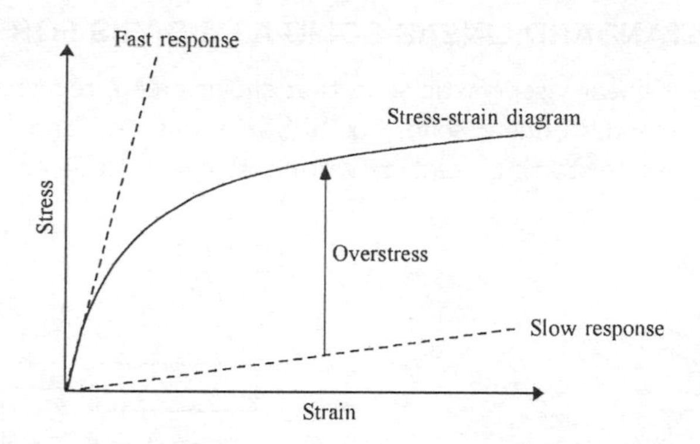

Figure A2. The slow, the fast responses and the stress–strain diagram at an intermediate strain rate of SLS.

For loading with a constant strain rate the model predicts that the slope becomes equal to the slope of the equilibrium stress–strain curve for large times for any constant strain rate. To see this the term $\frac{aE_2\dot{\varepsilon}}{E_1}$ is subtracted from each side of (A.1) so that

$$\dot{\varepsilon}\left(1-\frac{aE_2}{E_1}\right)=\frac{\mathrm{d}}{\mathrm{d}t}\left(\frac{\sigma-aE_2\varepsilon}{E_1}\right)+\left(\frac{\sigma-aE_2\varepsilon}{a\eta}\right) \tag{A.3}$$

For a constant strain rate this equation permits a solution for constant overstress. Then the first term on the right hand side of (A.3) vanishes and

$$\frac{\{\sigma-aE_2\varepsilon\}}{a\eta}=\dot{\varepsilon}\left(1-\frac{aE_2}{E_1}\right) \tag{A.4}$$

where $\{\}$ indicates the asymptotic solution. In this case the overstress is constant and linearly related to the strain rate. The SLS therefore exhibits linear rate sensitivity. By transforming the differential equation to an integral equation it can be shown that the constant overstress solution is the long–term solution that applies when time grows without bounds. The integral equation approach is demonstrated with the VBO model in the main part of the report.

When the expression of constant overstress is substituted into (A.2) the slope for constant overstress, long–term solution can be calculated as

$$\left\{\frac{\mathrm{d}\sigma}{\mathrm{d}\varepsilon}\right\}=aE_2 \tag{A.5}$$

For the SLS the slopes of all stress–strain diagrams of constant strain rate tests are ultimately independent of the strain rate and equal to that of the equilibrium stress. In the region prior to the long–term solution nonlinear rate sensitivity and a rate dependent slope are modeled.

For creep the stress rate is zero and (A.1) yields

$$\dot{\varepsilon} = \frac{\sigma_0 - aE_2\varepsilon}{a\eta} \tag{A.6}$$

For relaxation the strain rate is zero and we have

$$\dot{\sigma} = -E_1\frac{\sigma - aE_2\varepsilon_0}{a\eta} \tag{A.7}$$

In the above the subscript 0 indicates that this quantity is constant. Examination of (A.6) and (A.7) shows that the creep and relaxation rates depend on the difference between the stress and the equilibrium stress, the so–called overstress. It is positive in tensile loading as can be seen from (A.4) and from Fig. A2. It follows that for prior tensile loading the creep rate is positive and the relaxation rate is negative. Both, creep and relaxation motions end when the stress reaches the equilibrium stress. It will take infinite time to reach the equilibrium stress. This will be demonstrated for the creep case. To this end (A.6) is transformed to the integral

$$\varepsilon\left[t > t_0\right] = \varepsilon\left[t_0\right]\exp\left(-\frac{E_2}{\eta}\left(t - t_0\right)\right) + \exp\left(-\frac{E_2}{\eta}\left(t - t_0\right)\right)\int_0^t \frac{\sigma_0}{a\eta}\exp\left(\frac{E_2}{\eta}\left(\tau - \tau_0\right)\right)\mathrm{d}\tau \tag{A.8}$$

For large values of time the influence of the initial condition vanishes and the second term on the right hand side becomes indeterminate. Resolving this indeterminacy by applying the l'Hospital rule yields

$$\varepsilon\left[t \to \infty\right] = \frac{\sigma_0}{aE_2} \tag{A.9}$$

This shows that it takes infinite time for the creep motion to reach the equilibrium stress. Similarly it can be shown that a relaxation test terminates on the equilibrium stress after infinite time has passed. The equilibrium straight line is the response for infinitely slow loading and the termination point for creep and relaxation tests. It is clear that only primary creep can be modeled with the SLS.

Suppose, several creep (relaxation) tests are started at different stresses when the long–time solution in monotonic loading holds and the overstress is constant. Although the tests are started at different stresses and strains, the creep (relaxation) behavior is identical since the overstress is the same for each test. This unique behavior does not appear to have been discussed. In the regions where the long–time asymptotic solution holds, the curves of creep strain (defined as current strain minus strain at the start of the creep test) vs. time and the curves of the stress–drop (initial stress minus current stress) vs. time are congruent, respectively. This appears to be a surprising and little known result. VBO shares this property with the SLS.

From (A.2) we see that the slope of the stress–strain curve is equal to the elastic modulus of the spring in front of the model E_1 when the overstress is zero. This can happen at the origin and when the stress–strain curve crosses the equilibrium straight line. Stress–strain curves cross the equilibrium line with slope E_1.

A special test is the creep test at zero stress, sometimes called the recovery test. Setting the stress equal to zero in (A.6) shows a negative (positive) strain rate for positive (negative) strains. The creep rate becomes zero when the strain is zero, i. e. when the strain is zero and completely recovered.

A relaxation test started from the strain axis (stress is equal to zero) results in a positive(negative) stress rate for positive (negative) strains. In this case the stress magnitude increases as can be seen from (A.7). As usual the relaxation stops at the equilibrium stress. If the initial stress of the relaxation test is negative (positive) for positive (negative) strain the stress relaxes through zero before it stops at the equilibrium stress.

The SLS model exhibits a linear dependence of the overstress on the strain rate and a unique and linear equilibrium stress–strain behavior. A long–term asymptotic solution in a test with constant strain rate exists and at that time the slopes are equal to the slope of the equilibrium stress–strain line. In a creep and in a relaxation test it takes infinite time until the equilibrium line is reached at which the tests terminate. The model exhibits complete recovery and the tangent modulus is equal to modulus of the spring in front when the equilibrium line is crossed and at the origin. In the SLS the springs and the dashpot are linear. In VBO the spring in front of the Kelvin element is still linear, but the properties of the Kelvin element are radically changed. The dashpot is made nonlinear and depends on the overstress. This change is responsible for nonlinear rate dependence, which is controlled by the viscosity or the flow function. In VBO the spring of the Kelvin element is nonlinear and can model hysteresis. SLS and VBO exhibit a long–term, asymptotic solution in monotonic loading. In regions where the long–term solution holds and the overstress is constant relaxation curves started at different strains will be independent of the strain. Whereas SLS can only model primary creep with linear stress dependence VBO can model primary, secondary and tertiary creep and a nonlinear stress dependence. In relaxation tests and in creep tests that terminate, it takes infinite time to reach the equilibrium stress–strain curve in either model. If one were to test either model in a thought experiment it would not be possible to determine the current equilibrium stress. However, it can be inferred and approximately determined using one or more of the properties discussed. This has been done for real materials by Krempl [20].

The drawing depicting the model in Fig. A1 has different fills to indicate that the properties of the Kelvin element change in going from the SLS to the VBO model.

In the main part of the text a distinction is made between a response function and a constitutive equation. (A.1) is the constitutive equation. It cannot be solved in this form. Once the condition of a test is inserted as has been done for (A.6) and (A.7) the remaining part of the CE can be integrated to yield the response function for creep and relaxation, respectively. The response function, see (A.8), is quite different from the CE which is (A.1).

In creep theory and with most physically based approaches the manipulations are almost exclusively with the response functions. The CE is almost never established.

The SLS is a simple, linear CE that is not a realistic model for real materials. VBO is more complicated than the SLS and has modeled some materials very well. It is easy to demonstrate the difference between a CE and a response function using this model. It is also clear that the creep and relaxation motions are intertwined as the viscosity function or the flow function is involved for creep and relaxation.

B MODELING OF RATE–INDEPENDENCE AND OF NEGATIVE RATE SENSITIVITY

B1. General

Most engineering alloys show a decrease in strength and an increase in rate dependence and ductility with an increase in temperature. In many, if not most engineering alloys a region of temperature exists where the strength increases and ductility decreases with an increase in temperature. At the same time rate independence or negative rate sensitivity together with serrated yielding or jerky flow is found. The jerky flow is also known as Portervin–le Chatelier effect. These phenomena together are referred to as dynamic strain aging. Before and after this temperature region dynamic strain aging disappears and the material behavior is normal, i.e. positive rate sensitivity, decrease of strength and an increase of ductility with temperature. The dynamic strain aging effects are significant and can cause brittle fracture. It is therefore important that strain aging be modeled to avoid the often destructive and costly consequences of failure.

In his Ph.D. thesis K. Ho has found a way to model rate–independence, negative rate sensitivity with a slight modification to VBO, see Ho [52]. An increase in the flow stress with temperature can easily be modeled using the isotropic stress A, see Ho and Krempl [34]. Here we give a short synopsis of the modeling of zero rate sensitivity and of negative rate dependence.

B2. Modified version of the growth law for the equilibrium stress

The growth law for the equilibrium stress can be considered an interpolation between the initial quasi–elastic behavior and fully established inelastic flow when $\hat{E}_t$ (the tangent modulus when plastic flow is fully established) $\ll E$ (the elastic modulus). In this region the flow stress has been reached in an experiment and the long–term solution can be applicable for VBO. The terms that affect the asymptotic solution can be modified and this modification fully affects this region. To model changes in rate sensitivity the isotropic stress A is replaced by $A + (1 + \beta\Gamma)$ in 6.4 and 6.5. The dimensionless quantity β is called the strain rate sensitivity parameter. The long–term solution is, see Ho and Krempl [34] and 6.12– 6.14

$$\{\mathbf{s} - \mathbf{f}\} = \{A + (1 + \beta)\,\Gamma\} \left(\left\{ \frac{\mathbf{s} - \mathbf{g}}{\Gamma} \right\} \right) \tag{B.1}$$

When $\beta = 0$ the previous VBO formulation is regained and in this case the difference $\{\mathbf{s} - \mathbf{f}\}$ is the sum of the rate–independent (plastic) contribution A_f and the rate–dependent or viscous part Γ, see Fig. 1 of Krempl [12] and Figs 6.3 and 6.4. It is necessary that the term $\{A + (1 + \beta)\,\Gamma\} > 0$ at all times to obtain stable and physically realistic solutions, see Nakamura [53].

For normal VBO the rate sensitivity parameter $\beta = 0$ and the asymptotic solution contains viscous (the overstress) and rate–independent or plastic contributions through the isotropic stress A Krempl [12] Fig. 1 and Fig. 6.4 of this paper. The modified version of VBO allows the modeling of regular rate dependence with plastic (rate–independent) and viscous contributions. It also permits negative rate sensitivity with a rate–independent contribution through the isotropic stress.

Table B.1 shows that it is also possible to set the isotropic state variable A equal to zero so that the entire difference $\{\mathrm{s} - \mathbf{f}\}$ is made up by only viscous contributions. Then it is no longer possible to model rate–independent or plastic hardening such as extra hardening in out–of–phase loading, see Choi and Krempl [32]; Krempl and Choi [13]. See also Fig. 2.1 for experimental results that need plastic (rate–independent) hardening for their modeling. If $A = 0$ is selected then hardening can only be modeled by changing the viscous contributions, for example by a growth law for the drag stress as it is done for the Peirce, Shih et al. [54] model that is used frequently in polycrystal plasticity finite element applications. The modified VBO allows the modeling of viscoplastic or of viscous behavior depending on the values of the isotropic stress A and the rate sensitivity parameter β. A summary of the capabilities is as follows:

Table B1: Properties of modified VBO

$A > 0$	$\beta = 0$	Normal VBO
$A > 0$	$\beta > -1$	Positive rate sensitivity
$A > 0$	$\beta = -1$	Rate insensitivity
$A > 0$	$\beta < -1$	Negative rate sensitivity
$A = 0$	$\beta > -1$	No plasticity effects

The addition of the β–term has enriched the modeling capabilities of VBO considerably.

Before results of numerical experiments are given to demonstrate the increased modeling capabilities it should be remarked that the long–time, asymptotic limit of the overstress is not affected by the β–term. To demonstrate this property the term $\frac{1+\nu}{E}\dot{\mathbf{g}}$ is subtracted from (6.1) to yield

$$\left(\mathbf{I} - \frac{1+\nu}{E}\frac{\mathrm{d}\mathbf{g}}{\mathrm{d}\mathbf{e}}\right)\dot{\mathbf{e}} = \frac{1+\nu}{E}\left(\dot{\mathbf{s}} - \dot{\mathbf{g}}\right) + \frac{3}{2}F\left[\Gamma\right]\frac{\mathbf{s} - \mathbf{g}}{\Gamma} \tag{B.2}$$

where $\mathbf{I}$ is the fourth order identity tensor. It can be seen that an asymptotic solution for a constant overstress, indicated by $\{\}$, is possible for a constant strain rate if the slope of the equilibrium stress curve becomes constant. The asymptotic value of the overstress $\{\mathbf{s} - \mathbf{g}\}$ is obtained as the solution of the nonlinear equation

$$\left(\mathbf{I} - \frac{1+\nu}{E}\left\{\frac{\mathrm{d}\mathbf{g}}{\mathrm{d}\mathbf{e}}\right\}\right)\dot{\mathbf{e}} = \frac{3}{2}F\left[\{\Gamma\}\right]\left\{\frac{\mathbf{s} - \mathbf{g}}{\Gamma}\right\} \tag{B.3}$$

where we have assumed an ultimately constant value of the slope of the equilibrium curve. It is seen that the asymptotic solution of the overstress models the desired nonlinear dependence of the overstress on strain rate but does not depend on the quantities β and A.

The conditions for creep (stress rate is zero) or relaxation (strain rate is zero) require a specialization of (B.2). VBO predicts that creep and relaxation are independent of the isotropic stress A and of the β–term. **Even when the modified VBO is made to model rate independence or negative strain rate sensitivity by setting β and A equal to the appropriate value this theory predicts that "normal" creep and relaxation will continue to be observed.**

B3. Experimental evidence of creep or relaxation in the presence of anomalous rate sensitivity

The notion carried over from plasticity theory is that creep and relaxation are absent for rate independence. With this background the prediction of the modified VBO of rate independence with regular creep and relaxation seems paradoxical. However, if a new theory predicts new and paradoxical results it has to be compared with experiments. The competence of the new theory can only be confirmed or falsified by relevant experimental results.

An extensive search of the materials science literature yielded many papers on dynamic strain aging or the Portevin Le Chatelier effect but no reference was found that treated *stress–strain behavior and creep/relaxation on the same material.* In the engineering literature only three references were found and they do not contradict the modified VBO prediction.

Fig. 2.1.2 of Stouffer and Dame [55] shows the results of two tensile tests run at two different strain rates (0.8 and 0.001 1/s). The age–hardening Al alloy showed rate independence. However, at the end of the tests the strain was kept constant and surprisingly a stress drop was observed for both tests indicative of relaxation. It has to be remarked that the Al alloy was age hardening. It is not known whether it exhibited dynamic strain aging as well.

Cheng and Krempl [56] tested an Al–Mg alloy at room temperature and made similar observations. The stress–strain curve showed serrated yielding and nearly rate–independent behavior. Relaxation was observed again, see their Figs 2, 3 and 5.

An extensive investigation of a modified 9Cr–1Mo steel by Yaguchi and Takahashi [23] shows the most convincing results yet, see Fig. B.1.

The influence of strain rate on the stress–strain and on the relaxation behavior of modified 9Cr–1Mo steel as a function of temperature is shown. At temperatures above 400 °C positive rate sensitivity is observed together with "normal" relaxation behavior. This is also true at 200 °C. (It should be noted that for "normal" relaxation behavior there is an influence on prior strain rate on the relaxation behavior. At the end of the relaxation tests of equal duration the stress associated with the highest prior strain rate is lowest. For a discussion of these "normal" effects, see Krempl and Nakamura [35]. At temperatures below 400 °C dynamic strain aging is observed as evidenced by the rate independence of the

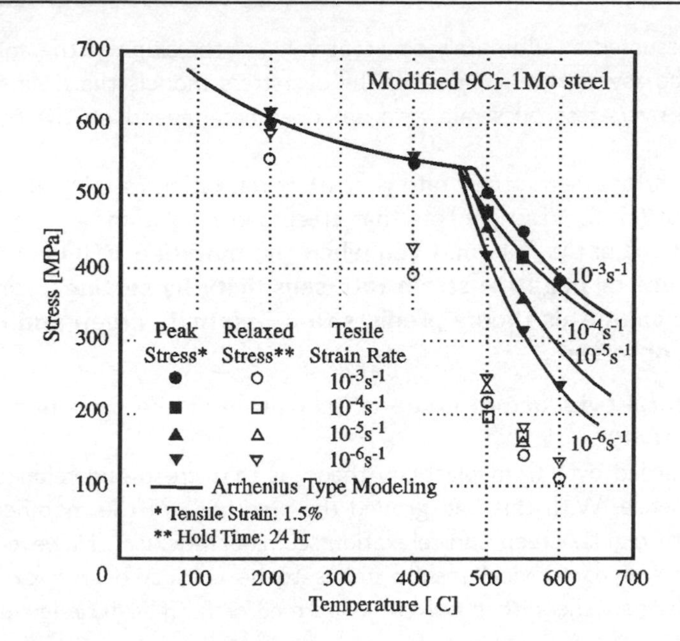

Figure B1. Rate dependence and relaxation behavior of modified 9Cr 1Mo steel as a function of temperature, from Yaguchi and Takahashi [23].

flow stress. Despite this rate independence of the stress–strain behavior "normal" relaxation is observed. This behavior is in full agreement with the predictions of the modified VBO. For this model creep and relaxation are to be found even if there is a zero or negative rate sensitivity.

Tests on an Al alloy are underway at the Mechanics of Materials Laboratory of RPI to get a consistent picture of the interrelation between creep, relaxation and the stress–strain behavior when dynamic strain aging is present.

B4. Some predictions of the modified VBO

In the following the capabilities of the modified VBO in modeling the influence of dynamic strain aging is demonstrated for a hypothetical material whose constants are given in Table B.2. The purpose is to show the qualitative capabilities of the modified VBO model rather than modeling a specific material.

As mentioned previously one or all of the following phenomena are considered indicative of the occurrence of dynamic strain aging or possibly age hardening, see above.

- Serrated yielding, Portevin le Chatelier effect,
- Zero rate sensitivity or negative rate sensitivity,
- Increase of strength with an increase in temperature,
- Decrease in ductility.

Table B2. Material properties of the hypothetical material used in the numerical experiments.

Moduli	$E = 195$ GPa; $E_{\text{t}} = 200$ MPa
Isotropic stress, constant	$A_0 = 115$ MPa
Viscosity function $k = k_1\left(1 + \dfrac{\Gamma}{k_2}\right)^{-k_3}$	$k_1 = 314,200$ s $k_2 = 60$ MPa $k_3 = 21.98$
Shape function $\psi = C_1 + \dfrac{C_2 - C_1}{\exp(C_3\Gamma)}$	$C_1 = 30$ GPa $C_2 = 182.5$ GPa $C_3 = 0.11$ (MPa)$^{-1}$

VBO is not capable of producing serrated stress–strain curves and modeling of ductility but is competent at representing the other two properties. These effects will be demonstrated by numerical experiments.

B5. Rate sensitivity parameter depends on effective inelastic strain

Experiments on the occurrence of serrated yielding show that it can depend on strain. Nortmann and Schwink [57, 58] show this very clearly. For the modeling of such an event a strain rate sensitivity parameter is made a function of the accumulated effective inelastic strain $p[t] = \int_0^t |\dot{\varepsilon}^{\text{in}}|\, d\tau$

$$\beta = \beta_1 \exp(-\beta_2 p) + \beta_3 \tag{B.4}$$

It is seen that $\beta[0] = \beta_1 + \beta_3$ and $\beta[\infty] = \beta_3$. By selecting the constants β_1 and β_3 appropriately a transition from negative rate sensitivity to positive rate sensitivity can be modeled whereby the constant β_2 controls the speed of transition. Fig. B.2 shows the results of strain rate cycling experiments where the strain rate changes abruptly between 10^{-3} and 10^{-6} 1/s. In Fig. B.2 the material exhibits initially negative rate sensitivity, which changes to rate independence followed by regular rate sensitivity as inelastic strain accumulates. For Fig. B.3 regular strain rate sensitivity changes to inverse rate sensitivity as straining progresses.

Each figure exhibits two curves to show the influence of a constant in the shape function $\psi[\Gamma]$ listed in Table B.2. When the constant $C_3 = 0.11$ an overshoot occurs upon the change in strain rate. This overshoot is diminished for $C_3 = 0.03$. Note that this constant affects only the transient region. Both curves merge as the strain increases. The overshoots can be almost eliminated by choosing a different constant in the shape function. Such overshoots have been reported, see the experiment with a 61 ST age hardening Al alloy by Lubahn and Felgar [44], p. 188, Figs 6–53.

The experiment and the simulations show an overshoot upon the sudden change in strain rate that quickly dies out in both cases.

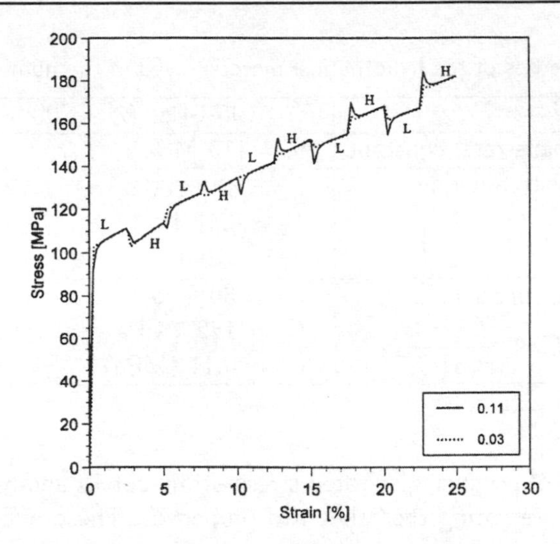

Figure B2. Negative rate sensitivity changing to positive rate sensitivity. H and L designate high and low strain rate respectively.

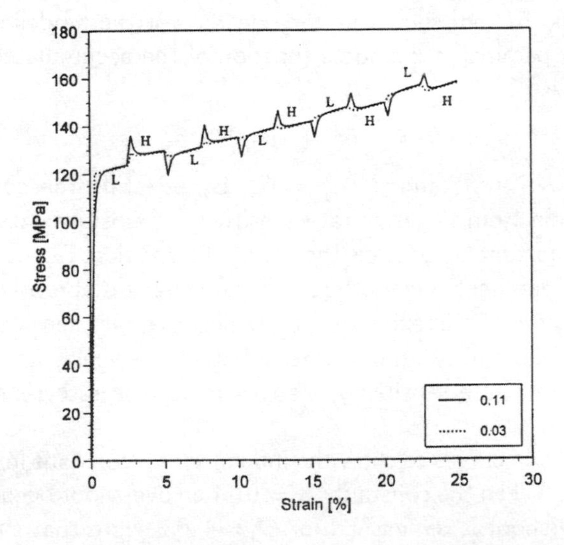

Figure B3: Positive rate sensitivity changing to negative rate sensitivity.

B6. Discussion

The numerical experiments on modeling rate independence and negative rate sensitivity have shown that the modified VBO is competent at modeling these phenomena. The rate sensitivity coefficient controls the modeling of anomalous rate–dependent behavior and has a negligible influence on the initial, quasi–elastic portion of the stress–strain diagram, see the numerical experiments by Ho and Krempl [34]. Its effect on the long–time, asymptotic behavior is completely known.

The model predicts that even when there is rate independence or inverse strain rate sensitivity there is "normal" creep and relaxation behavior. When the creep periods are introduced during tension or compression loading, the creep strain increases in magnitude. On the other hand, the stress magnitude decreases in relaxation for prior monotonic loading. The few available experiments confirm the VBO prediction. Experiments are in progress to further verify on this behavior.

It was very surprising that the search of the materials literature did not produce a single paper that dealt with rate sensitivity, creep and relaxation together. The author's point of view is that a creep and/or relaxation investigation should always complement the study of the influence of the loading rate. After all rate sensitivity, creep and relaxation are just different manifestations of rate dependent behavior of materials.

Miller and Sherby [59] modeled solute strengthening effects in the framework of the MATMOD equations. The inverse strain rate sensitivity was modeled well for solute strengthening alone. Their Fig. 14 exhibits a very "square" transition, no overshoots are modeled as shown in Fig. B.2. In their simulations the strengthening due to deformation was neglected. It appears to be difficult to clearly separate the hardening contributions due to deformation and solute strengthening in engineering applications. In this case a material is selected and its behavior has to be determined by suitable macroscopic tests. The test results contain the combined influence of deformation and of solute strengthening. It is difficult to separate the two contributions as required by the model. It is also very interesting to note that the MATMOD equations predict regular creep and relaxation even when inverse rate sensitivity or rate independence is modeled.

Details and other applications can be found in Ho [52]. The demonstration of rate insensitivity is taken from a report by Ho and Krempl [34] that explores other modeling possibilities.

In some alloys cycling reveals an initial almost rate independent behavior that gradually changes to normal rate sensitivity, see Nakamura [53] who has modeled this behavior using VBO.

C EXCERPTS FROM WRC BULLETIN 195

C1. Introduction

For historical reasons an excerpt from an early report of the author is, see Krempl [9] given below. The points to be made are

I. Both time–hardening and strain–hardening theories are inadequate as models for variable loading. They contain only information on the deformation behavior under constant stress.

II. The manipulations of the strain or time–hardening theories are modifications of a response function. For different loading conditions, say cyclic, the constitutive equation

(CE) of the material should be exercised rather than the response function. See Appendix A for the difference between a CE and a response function.

III. The manipulations are subject to inconsistencies and therefore the result is to be viewed with suspicion.

Here is an excerpt of the beginning of a chapter illustrating the argument. This report was the starting point of the writer's involvement with constitutive equations. It is for the most part outdated but contains arguments and conclusions that are still valid.

C2. Excerpt

Time–hardening vs. strain–hardening theory.

"In attempts to apply creep equations, such as specific forms of (28), to conditions of varying stress, one "attributes universal validity to such a relationship" (Rabotnov [3], p. 200). In doing so one takes a response function to be a constitutive law. This causes immediate difficulties since this constitutive law now contains time explicitly, indicative of "aging" (Rabotnov [3], p.200). Further, for a step change, the prediction of time–hardening theory frequently leads to unrealistic results on the basis of the initial conditions. They are usually explained in terms of a schematic diagram such as Fig. 46 on p.202 of Rabotnov [3]. These are the usual two arguments to refute time–hardening theory and to favor strain–hardening theory, which will now be discussed.

"Following Finnie [60] and Odqvist [61] , the strain–hardening theory is obtained in the following way. Suppose (28) has the following specific form

$$p = A\sigma^n t^m \tag{C.1}$$

where A, n, m are constants. Differentiating (C.1) yields

$$\dot{p} = \sigma^n m t^{m-1} \tag{C.2}$$

and elimination of time between (C.1) and (C.2) yields the strain–hardening theory expression

$$\dot{p} = m\left(A\sigma^n p^{m-1}\right)^{1/m} \tag{C.3}$$

Expression (C.3) is now considered to be valid for varying stress, whereas in Eqs. (C.1) and (C.2) stress was considered to be a parameter, i.e., a constant with respect to differentiation with respect to time. Strain hardening is then usually favored on the basis of a diagram such as Fig. 2 in Finnie [60] or Fig. 49 in Rabotnov [3] and the comparison of the prediction of both theories with experimental results.

"It should be noted that an equivalent to the derivation given above could not be found in Rabotnov [3]. There the treatment starts from the hypothesis of the equation of state, p. 206 of Rabotnov [3], of which Eq. 83 is a special form. Thus, mathematically, the difficulty of treating σ parameter or a variable depending on how it fits into the present scope is avoided. However, even the equation of state has to be fitted to test data, which in all these cases come from constant stress creep curves. (We are interested only in strains up to 2%, so that there is no difference between constant load and constant stress creep curves.)

In the approach of Rabotnov [3], the difficulty arises at a point where it is not quite as obvious as it is above that constant stress data are used for variable stress application.

"A prime application of time and strain–hardening theory is the prediction of relaxation from creep data. Again conceptual difficulties arise, and an example from Rabotnov [3], p.203 should suffice, ..."

REFERENCES

1. Bailey R.W. : Design aspect of creep, Trans. ASME, Journal of Applied Mechanics, (1936), pp. A1–A6.
2. Odqvist F.K.G. and Hult J. : Kriechfestigkeit Metallischer Werkstoffe, Springer Verlag, Berlin, 1962.
3. Rabotnov Y.N. : Creep Problems in Structural Members, North Holland, Amsterdam, London, 1969.
4. Sehitoglu H. : Thermal and Thermomechanical Fatigue of Structural Alloys, in: ASM Handbook. Volume 19. Fatigue and Fracture (Edited by G. Dieter), ASM International, Metals Park, Ohio, 1996 pp. 527–556.
5. Krempl E. : Design for Fatigue Resistance, in: ASM Handbook. Volume 20. Materials Selection and Design (Edited by G. Dieter), ASM International, Metals Park, Ohio, 1997 pp. 516–532.
6. Woodford D.A. : Design for High-Temperature Applications, in: ASM Handbook. Volume 20. Materials Selection and Design (Edited by G. Dieter), ASM International, Metals Park, Ohio, 1997 pp. 571–588.
7. Frost H.J. and Ashby M.F. : Deformation–Mechanism Maps, Pergamon Press, Oxford, 1982.
8. Tachibana Y. and Krempl E. : Modeling of high homologous temperature deformation behavior using the viscoplasticity theory based on overstress (VBO): Part I — Creep and tensile behavior, Journal of Engineering Materials and Technology, 117(1995), pp. 456–461.
9. Krempl E. : Cyclic Creep — An Interpretive Literature Survey, in: Welding Research Council Bulletin 195, New York, N. Y., 1974 pp. 63–123.
10. Krempl E. : An experimental study of room–temperature rate sensitivity creep and relaxation of AISI Type 304 Stainless Steel, Journal of the Mechanics and Physics of Solids, 27(1979), pp. 363–385.
11. Krempl E. : Models of viscoplasticity. Some comments on equilibrium (back) stress and drag stress, Acta Mechanica, 69(1987), pp. 25–42.
12. Krempl E. : A small strain viscoplasticity theory based on overstress, in: Unified Constitutive Laws of Plastic Deformation (Edited by A. Krausz and K. Krausz), Academic Press, San Diego, 1996 pp. 281–318.
13. Krempl E. and Choi S.H. : Viscoplasticity theory based on overstress: The modeling of ratcheting and cyclic hardening of AISI Type 304 stainless steel, Nuclear Engineering and Design, 133(1992), pp. 401–410.

14. Nix W.D. , Gibeling J.C. and Hughes D.A. : Time-dependent behavior of metals, Metallurgical Transactions A, 16A(1985), pp. 2215–2226.
15. LeMay I. : Principles of Mechanical Metallurgy, Elsevier North Holland, Inc, New York, NY, 1981.
16. Gittus J. : Creep, Viscoelasticty and Creep–Rupture in Solids, Halsted Press a Division of John Wiley and Sons, Inc, New York, 1975.
17. Odqvist F.K.G. : Mathematical Theory of Creep and Creep Rupture, Clarendon Press, Oxford, 1966.
18. Lee K.D. and Krempl E. : Uniaxial thermomechanical loading. Numerical experiments using the thermal viscoplasticity theory based on overstress, European Journal of Mechanics, A/Solids, 10(1991), pp. 173–192.
19. Ohashi Y. and Kawai M. : Effects of prior plasticity on subsequent creep of Type 316 stainless steel at elevated temperature, Journal of Engineering Materials and Technology, 108(1986), pp. 68–74.
20. Krempl E. : The overstress dependence of the inelastic rate of deformation inferred from transient tests, Materials Sience Research International, 1(1995), pp. 3–10.
21. Kujawski D. , Kallianpur V.V. and Krempl E. : An experimental study of uniaxial creep, cyclic creep and relaxation of AISI Type 304 stainless steel at room temperature, Journal of the Mechanics and Physics of Solids, 28(1980), pp. 129–148.
22. Pugh C.E. and Liu K. : Currently recommended constitutive equations for inelastic design analysis of FFTF components, Technical Report ORNL 3602, Oak Ridge National Laboratory, Oak Ridge, TN, 1972.
23. Yaguchi M. and Takahashi Y. : Unified inelastic constitutive model for modified 9Cr-1Mo steel incorporating dynamic strain aging effect, JSME International Jouranl, Series A, 42(1999), pp. 1–10.
24. Ikegami K. and Niitsu Y. : Effect of creep prestrain on subsequent plastic deformation, International Journal of Plasticity, 1(1985), pp. 331–345.
25. Hult J. : Creep in Engineering Structures, Blaisdell Publishing Company, Waltham, Toronto, London, 1966.
26. Krempl E. : Plasticity and variable heredity, Archives of Mechanics, 33(1981), pp. 289–306.
27. Krausz A. and Krausz K. : Unified Constitutive Laws of Plastic Deformation, Academic Press, San Diego, 1996.
28. Gittus J. and Zarka J. , (eds.): Modeling Small Deformations of Polycrystals, Barking and New York, 1996, Elsevier Applied Science Publisher.
29. Robinson D.N. and Swindeman R.W. : Unified creep plasticity constitutive equations for 2 1/4 Cr-1 Mo steel at elevated temperature, Technical Report ORNL 8444, Oak Ridge National Laboratory, Oak Ridge, TN, 1982.
30. Bodner S.R. : Review of a unified elastic–viscoplastic theory, in: Unified Constitutive Equations for Plastic Deformationand Creep of Engineering Alloys (Edited by A. Miller), Elsevier Science Publisher, 1987 pp. 273–301.
31. Ruggles M.B. and Krempl E. : The interaction of cyclic hardening and ratchetting for

AISI Type 304 stainless steel at room temperature — I Experiments and II Modeling with the viscoplasticity theory based on overstress, Journal of the Mechanics and Physics of Solids, 38(1990), pp. 575–597.
32. Choi S.H. and Krempl E. : Viscoplasticity theory based on overstress: The modeling of biaxial hardening using irreversible plastic strain, in: Advances in Multiaxial Fatigue, American Society for Testing and Materials, STP 1191, San Francisco, 1993 .
33. Hart E.W. and Solomon H.D. : Load relaxation studies of polycrystalline high purity Aluminium, Acta Metallurgica, 21(1972), pp. 295–307.
34. Ho K. and Krempl E. : Modeling of rate independence and of negative rate sensitivity using the viscoplasticity theory based on overstress (VBO), Technical Report MML 98–3, Rensselaer Polytechnic Institute and Mechanics of Materials Laboratory, Troy, N.Y., 1998.
35. Krempl E. and Nakamura T. : The influence of the equilibrium stress growth law formulation on the modeling of recently observed relaxation behaviors, JSME International Journal, Series A, 41(1998), pp. 103–111.
36. Majors P.S. and Krempl E. : The isotropic viscoplasticity theory based on overstress applied to the modeling of modified 9Cr-1Mo steel at 538C, Materials Science and Engineering, A186(1994), pp. 23–34.
37. Krempl E. and Lu H. : The rate (time) dependence of ductile fracture at room temperature, Engineering Fracture Mechanics, 20(1984), pp. 629–632.
38. Krempl E. and Kallianpur V.V. : The uniaxial unloading behavior of two engineering alloys at room temperature, Journal of Applied Mechanics, 52(1985), pp. 654–658.
39. Bordonaro C.M. and Krempl E. : The rate–dependent mechanical behavior of plastics: A comparison between 6/6 Nylon, Polyetherimid and Polyetheretherketone, in: Use of Plastics and Plastic Composites: Materials and Mechanics Issues, ASME, 1993 .
40. Hassan T. and Kyriakides S. : Ratcheting in cyclic plasticity, Part I: Uniaxial behavior, International Journal of Plasticity, 8(1992), pp. 91–116.
41. Lindholm U.S. : Experimental basis for temperature-dependent viscoplastic constitutive equations, Applied Mechanics Reviews, 43(1990), pp. S338–S344.
42. Tachibana Y. and Krempl E. : Modeling of high homologous temperature deformation behavior using the viscoplasticity theory based on overstress (VBO): Part III — A simplified model, Journal of Engineering Materials and Technology, 120(1998), pp. 193–196.
43. Maciucescu L. : Modeling the deformation behavior of a Pn-Pb solder alloy using the simplified viscoplasticty theory based on overstress (VBO), Journal of Electronic Packaging, (1999), in press.
44. Lubahn J.D. and Felgar R.P. : Plasticity and Creep of Metals, John Wiley and Sons, New York, London, 1961.
45. Hart E.W. and C. Y. Li et al : Phenomenological Theory: A Guide to Constitutive Relations and Fundamental Deformation Properties, in: Constitutive Equations in Plasticity (Edited by A. Argon), MIT Press, Cambridge, MA, 1975 pp. 149–196.
46. Tachibana Y. and Krempl E. : Modeling of high homologous temperature deforma-

tion behavior using the viscoplasticity theory based on overstress (VBO): Part II — Characteristics of the VBO Model, Journal of Engineering Materials and Technology, 119(1997), pp. 1–6.
47. Marquis D. : Phenomenology and Thermodynamics: Coupling between Thermoelasticity, Plasticity, Ageing and Damage, Technical Report 307, NASA, Washington D. C., 1990.
48. Chan K.S. and Lindholm U.S. : Inelastic deformation under nonisothermal loading, Journal of Engineering Materials and Technology, 112(1990), pp. 15–25.
49. Niitsu Y. , Horigushi A. and Ikegami K. : Effect of temperature variation on plastic behavior of S25C mild steel, International Journal of Pressure Vessel & Piping, 53(1993), pp. 511–523.
50. Chaboche J.-L. : Viscoplastic constitutive equations, Part I: A thermodynamically consistent formulation, Journal of Applied Mechanics, 60(1993), pp. 813–821.
51. Fluegge W. : Viscoelasticity, Blaisdell Publishing Company, Waltham, MA, Toronto, London, 1967.
52. Ho K. : Application of the viscoplasticity theory based on overstress theory to the modeling of dynamic strain again of metals and to solid polymers, specially Nylon 66, Ph.D. thesis, Rensselaer Polytechnic Institute, Troy, N.Y., 1998.
53. Nakamura T. : Application of viscoplasticity theory based on overstress (VBO) to high temperature cyclic deformation of 316 FR steel, JSME International Journal, Series A, 41(1998), pp. 539–546.
54. Peirce D. and Shih C. F. : A tangent modulus method for rate-dependent solids, Computers and Structures, 18(1984), pp. 875–887.
55. Stouffer D.S. and Dame L.T. : Inelastic Deformation of Metals, John Wiley and Sons, New York, N. Y., 1996.
56. Cheng S. and Krempl E. : Experimental determination of strain–induced anisotropy during nonproportional straining of an Al/Mg alloy at room temperature, International Journal of Plasticity, 7(1991), pp. 805–810.
57. Nortmann A. and Schwink C. : Characteristics of dynamic strain aging in binary f.c.c. copper alloys II. Comparison and analyses of experiments on CuAl and CuMn, Acta Materialia, 45(1997), pp. 2051–2058.
58. Nortmann A. and Schwink C. : Characteristics of dynamic strain aging in binary f.c.c. copper alloys-I. results on solid solution sof CuAl, Acta Materialia, 45(1997), pp. 2043–2050.
59. Miller A.K. and Sherby O.D. : A simplified phenomenological model for non–elastic deformation: Predictions of pure Aluminum behavior and incorporation of solute strengthening effects, Acta Metallurgica, 26(1978), pp. 289–304.
60. Finnie I. : Stress analysis in the presence of creep, Applied Mechanics Reviews, 13(1960), pp. 705–712.
61. Odqvist F.K.G. : Nonlinear Solid Mechanics, Past, Present and Future, in: Proceedings 12th Int. Congress of Applied Mechanics, Springer Verlag, 1969 .